Topics in Current Chemistry Collections

Aims and Scope

The series *Topics in Current Chemistry Collections* presents critical reviews from the journal *Topics in Current Chemistry* organized in topical volumes. The scope of coverage is all areas of chemical science including the interfaces with related disciplines such as biology, medicine and materials science.

The goal of each thematic volume is to give the non-specialist reader, whether in academia or industry, a comprehensive insight into an area where new research is emerging which is of interest to a larger scientific audience.

Each review within the volume critically surveys one aspect of that topic and places it within the context of the volume as a whole. The most significant developments of the last 5 to 10 years are presented using selected examples to illustrate the principles discussed. The coverage is not intended to be an exhaustive summary of the field or include large quantities of data, but should rather be conceptual, concentrating on the methodological thinking that will allow the non-specialist reader to understand the information presented.

Contributions also offer an outlook on potential future developments in the field.

More information about this series at http://www.springer.com/series/14181

Youhong Tang • Ben Zhong Tang
Editors

Aggregation-Induced Emission

With contributions from

Alessandro Aliprandi • Luisa De Cola • Jun Dai • Lei Dong
Suzanne Fery-Forgues • Amin Jamshidi Ghahfarokhi • Ting Han
Jing-Jing Hu • Vishal Kachwal • Inamur Rahaman Laskar
Dangyuan Lei • Siqi Li • Xiaoding Lou • Fan Luo • Dongge Ma
Guillermo Moreno-Alcántar • Li-Ya Niu • Hui-Qing Peng
Harini A. Perera • Ana Clara B. Rodrigues • J. Sérgio Seixas de Melo
Ben Zhong Tang • Youhong Tang • Javad Tavakoli
Corinne Vanucci-Bacqué • Dong Wang • Xinnan Wang
Zhaoyu Wang • Jiaming Wei • Fan Xia • Mingdi Yan
Juliang Yang • Qing-Zheng Yang • Ruquan Ye • Haoke Zhang

Springer

Editors
Youhong Tang
College of Science and Engineering
Flinders University
Adelaide, SA, Australia

Ben Zhong Tang
School of Science and Engineering
The Chinese University of Hong Kong
Shenzhen, China

Partly previously published in Topics in Current Chemistry Volume 378 (2020); Topics in Current Chemistry Volume 379 (2021).

ISSN 2367-4067
Topics in Current Chemistry Collections
ISBN 978-3-030-89932-5

This Springer imprint is published by the registered company Springer Nature Switzerland AG
The registered company address is: Gewerbestrasse 11, 6330 Cham, Switzerland

Contents

Preface

Aggregation-induced emission (AIE) is a new branch of luminescent materials research that aims to explore luminescence at the aggregate level. AIE luminogens (AIEgens) often show no emission in solutions at the molecular level but become brightly emissive when their molecular species are aggregated. Aggregation processes generally involve the interplay of competitive or cooperative effects, and new structures, properties, and functions usually appear and diversify in the aggregates. Over the course of the last 20 years, AIE research has made great strides in material development, mechanistic study, and high-tech applications. Meanwhile, research on AIE is an ideal entry point for scientists to explore and unravel the mysteries in the meso world, since AIE is a typical phenomenon showing a nonlinear property change from molecules to aggregates. The past two decades has witnessed booming progress in AIE research. Some examples of the remarkable advances in the field of AIE study in recent years are presented in this topical collection.

Niu *et al.* summarize recent progress in the design strategy of AIE-based energy transfer systems for light-harvesting, fluorescent probes and theranostic systems, with an emphasis on design strategies to achieve desirable properties. The limitations, challenges, and future opportunities of AIE-excitation energy transfer systems are briefly outlined. The discovery of aggregation-induced electrochemiluminescence (AIECL) in 2017 has opened new research avenues in the quest for novel, more efficient emitters and offered platforms for biological and environmental sensing applications. The great abundance of fluorophores presenting AIE in aqueous media renders AIECL a potentially powerful tool for future diagnostics. De Cola *et al.* give their reflections about the future directions to take for the development of sensing devices based on AIECL. Ye *et al.* highlight recent developments in clusteroluminescence, including mechanistic studies, the disclosure of room-temperature phosphorescence, and the extension of light emission to the longer-wavelength region and also demonstrate a few applications in various fields. Kachwal and Laskar review AIE-active organic mechanofluorochromic materials, i.e., AIEgens with changes in their emissions under anisotropic and isotropic pressures. The relationship between the chemical structures of AIEgens and the change in the emission behaviours under pressures is discussed.

Fery-Forgues and Vanucci-Bacqué encompass the various ways to use benzazole units in solid-state luminescence enhancement systems. They underline the significant progress recently made in the understanding of the photophysical mechanisms involved. Benzazoles are robust building blocks, easily incorporated into a variety of structures which demonstrates advantages taken from these small heterocycles for the design of increasingly efficient AIEgens. Ma reviews the status and envisions the prospects of AIEgens in organic light-emitting diodes, optically pumping organic lasers, organic solar cells and organic photodetectors. It demonstrates that AIEgens are important organic optoelectronic materials.

The enhancement of photoluminescence through formation of molecular aggregates in organic oligomers and conjugated organic polymers is reviewed by Rodrigues and de Melo. Future perspectives for the rational design of AIEgen structures are discussed. Fluorescent gels have numerous properties that are intrinsic to the gel structures, with additional light-emitting properties making them attractive for different applications. Y Tang *et al.* summarize current studies associated with the development of fluorescent gels using AIEgens and propose new directions for future research, especially the biomedical applications in drug delivery, biosensors, bioimaging, and tissue engineering. B Z Tang *et al.* review the recent advances in the design, preparation, performance, and applications of functional synthetic polymer systems with AIE attribute and stimuli responses. Various AIE-based polymer systems responding to single or multiple physical or chemical stimuli are illustrated with specific examples. The current challenges and perspectives on the future development of this research area are also discussed.

Enzyme-responsive peptide-based AIE bioprobes used for biomedical applications are summarized by Lou *et al.* How each aggregation strategy detects enzyme activity and treats the diseases under imaging guidance are discussed and the current problems and future prospects of enzyme-responsive peptide-based AIE bioprobes are explored in this review. Perera and Yan highlight examples where AIEgens are employed as molecular probes in the imaging, discrimination and killing of bacteria, viruses and fungi.

As guest editors, we are sincerely thankful for the great contributions from all the authors and the professional assistance from the editorial office of *Topics in Current Chemistry*. We hope this topical collection will attract broad attention among scientists and technologists and inspire scientific exploration at the aggregate level and high-tech innovation utilizing AIEgens.

Youhong Tang
College of Science and Engineering
Flinders University, Australia

Ben Zhong Tang
School of Science and Engineering
The Chinese University of Hong Kong, Shenzhen, China

REVIEW

Modulation of Aggregation-Induced Emission by Excitation Energy Transfer: Design and Application

Lei Dong[1] · Hui-Qing Peng[2] · Li-Ya Niu[1] · Qing-Zheng Yang[1]

Received: 10 January 2021 / Accepted: 11 March 2021 / Published online: 7 April 2021

Abstract

Excitation energy transfer (EET) as a fundamental photophysical process is well-explored for developing functional materials with tunable photophysical properties. Compared to traditional fluorophores, aggregation-induced emission luminogens (AIEgens) exhibit unique advantages for building EET systems, especially serving as energy donors, due to their outstanding photophysical properties such as bright fluorescence in aggregation state, broad absorption and emission spectra, large Stokes shift, and high photobleaching resistance. In addition, the photophysical properties of AIEgens can be modulated by energy transfer for improved luminescence performance. Therefore, a variety of EET systems based on AIEgens have been constructed and their applications in different areas have been explored. In this review, we summarize recent progress in the design strategy of AIE-based energy transfer systems for light-harvesting, fluorescent probes and theranostic systems, with an emphasis on design strategies to achieve desirable properties. The limitations, challenges and future opportunities of AIE–EET systems are briefly outlined.

Chapter 1 was originally published as Dong, L., Peng, H-Q., Niu, L-Y. & Yang, Q-Z. Topics in Current Chemistry (2021) 379: 18. https://doi.org/10.1007/s41061-021-00330-0.

✉ Hui-Qing Peng
hqpeng@mail.buct.edu.cn

✉ Li-Ya Niu
niuly@bnu.edu.cn

1 Key Laboratory of Radiopharmaceuticals, Ministry of Education, College of Chemistry, Beijing Normal University, Beijing 100875, China

2 Beijing Advanced Innovation Center for Soft Matter Science and Engineering, State Key Laboratory of Chemical Resource Engineering, Beijing University of Chemical Technology, Beijing 100029, China

Graphic Abstract
Design strategies and applications (light-harvesting, fluorescent probe and theranostics) of AIEgen-based excitation energy systems are discussed in this review.

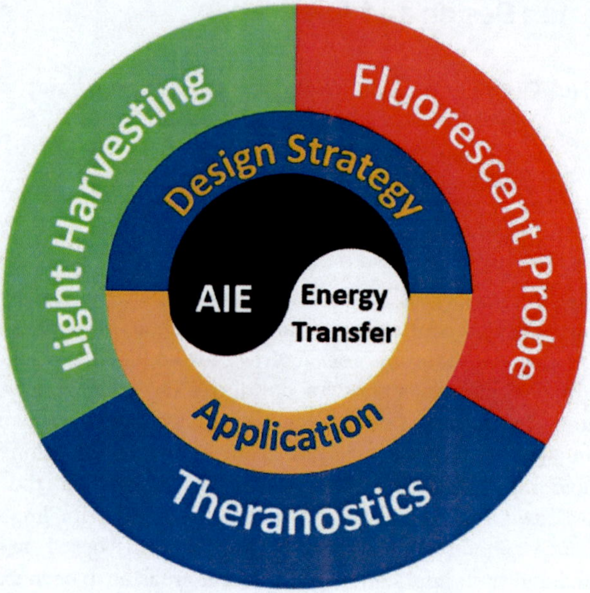

Keywords Aggregation-induced emission · Energy transfer · Light-harvesting · Fluorescent probe · Theranostics

1 Introduction

Excitation energy transfer (EET) is an essential process, in which a chromophore (donor, D) absorbs light to generate an excited state, subsequently transferring the excitation energy to another molecule (acceptor, A) [1–3]. In the energy transfer process, the donor and acceptor can exchange energy by Förster resonance energy transfer (FRET) through dipole–dipole interactions or electron exchange (Dexter energy transfer) [4–6]. EET is well-suited for modulating the photophysical properties of fluorophores, and thus has been widely employed in fluorescent imaging and sensing [7, 8], organic light emitting diodes (OLEDs) [9, 10], and other emerging areas [11]. Meanwhile, energy transfer plays a crucial role in photosynthesis, where the energy of photons absorbed by antenna chromophores of the light-harvesting system (LHS) flows to the reaction center by highly efficient EET [9, 10, 12]. To make efficient use of light with satisfactory energy collection efficiency, one of the most noteworthy features of LHSs is their densely packed antenna chromophores. However, the tight and compact arrangement of donors is usually subject to intermolecular aggregation-caused fluorescence quenching (ACQ) due to π–π interactions [13]. On the

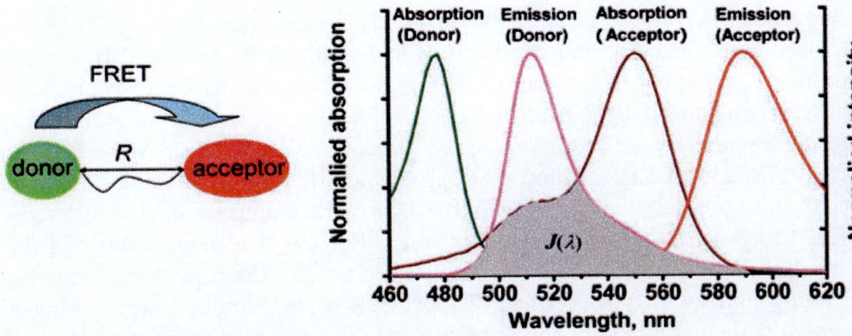

Fig. 1 Förster resonance energy transfer (FRET) and relative spectra overlap between donor emission and acceptor absorption. Reproduced with permission from Ref. [4] © 2013 American Chemical Society

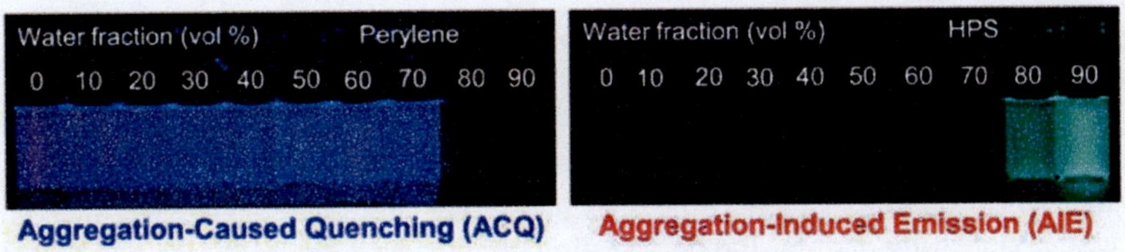

Fig. 2 Fluorescence photographs of perylene (*left*) and hexaphenylsilole (HPS, *right*) in tetrahydrofuran (THF)/water mixtures with different fractions of water. Reproduced with permission from Ref. [24] © 2015 American Chemical Society

other hand, the efficiency of EET is extremely sensitive to the distance and spectral overlap between donor and acceptor, which offers a unique opportunity to construct fluorescent probes for detecting biologically important analytes by measuring fluorescence changes, helping to uncover the physiological and pathological functions of these analytes [14]. As common chromophores naturally aggregate in aqueous media, the phenomenon of ACQ is a problem in biological applications (Fig. 1).

Luminogens with aggregation-induced emission (AIEgens) have emerged as promising candidates for building energy transfer systems to address the ACQ problem [13, 15–18]. Compared with ACQ dyes, AIEgens show weak or inappreciable emission in solution but emit intense fluorescence in aggregate or solid state (Fig. 2) [19–21]. The free rotation of intramolecular aromatic rings (e.g., phenyl rings) of AIEgens in dilute solution releases excitation energy in a nonradiative way, thereby quenching the fluorescence [18, 22, 23]. Upon aggregate formation, the emission is rejuvenated by blocking nonradiative energy dissipation via restriction of intramolecular motion (RIM). Meanwhile, the twist molecular conformation of AIEgens can preclude their intermolecular π–π stacking interaction to avoid excitation energy losses [24]. Moreover, to construct an EET system, the absorption spectrum of the acceptor should overlap with the emission spectrum of the donor. AIEgens usually have a wide absorption band and broad full width at half maximum of the luminescence spectra; thus, they have been considered as a superior candidates for energy donors [4, 25–29]. For these reasons, a variety of aggregation-induced emission (AIE)–EET systems have been prepared, and their applications in different areas such as LHSs, high-resolution bioimaging, white-emitting luminescent materials, fluorescent probes and theranostics, have been explored.

In this review, we will summarize the latest developments in AIE–EET systems in the field of LHSs, fluorescent probes and theranostic systems. We will discuss the construction of AIE-based LHSs with emphasis on the strategies required to achieve high energy collection efficiency. The fabrication of EET-based probes by introducing recognized sites on the donor/acceptor moieties or linking units will be described. This review also includes a description of image-guided chemotherapy and photodynamic therapy by combining AIE–EET systems with therapeutic modalities. Finally, we will briefly shed light on the current limitations, challenges, and future opportunities in this area.

2 Artificial LHSs

In nature, LHSs of green plants and photosynthetic bacteria have successfully evolved photosynthetic processes to harvest sunlight to generate chemical energy. Generally, natural light-harvesting complexes contain a large number of closely packed chlorophyll molecules as antenna chromophores [30, 31]. When donor chromophores absorb light, the excitation energy is unidirectionally and efficiently transferred to the reaction center to induce electron transfer reactions. The mimicking of natural LHSs has attracted increasing interest on account of its essential role in photosynthesis as well as its potential application in various fields [32].

An ideal artificial LHS should meet two requirements simultaneously: (1) high energy transfer efficiency and (2) high donor to acceptor ratio. To achieve a high energy transfer efficiency, the system requires a tight and compact arrangement of antenna chromophores. The introduction of AIEgens as donor chromophores can avoid fluorescence self-quenching in the tightly packed system and minimize the excitation energy loss, ensuring a high energy collection efficiency when incorporated with proper acceptors [33]. Therefore, over the past few years, AIE-based artificial LHSs have been prepared in many fields. Artificial LHSs with covalent bonds contain very few donor chromophores per acceptor because of synthetic difficulties. For instance, Atilgan et al. introduced only three tetraphenylethene (TPE) units on one borondipyrromethene (BODIPY) by CuAAC cycloaddition due to the limited number of reaction sites [34]. Therefore, a wide variety of artificial LHSs have been fabricated by non-covalent interactions. In this review, such artificial LHSs with aggregation-induced emission are generally classified by the driving force used to construct the self-assemblies, including hydrophobic interaction, host–guest interaction, hydrogen bonds and metal coordination (Fig. 3). Their further applications in biological imaging, white light emission materials and photocatalysis will be discussed.

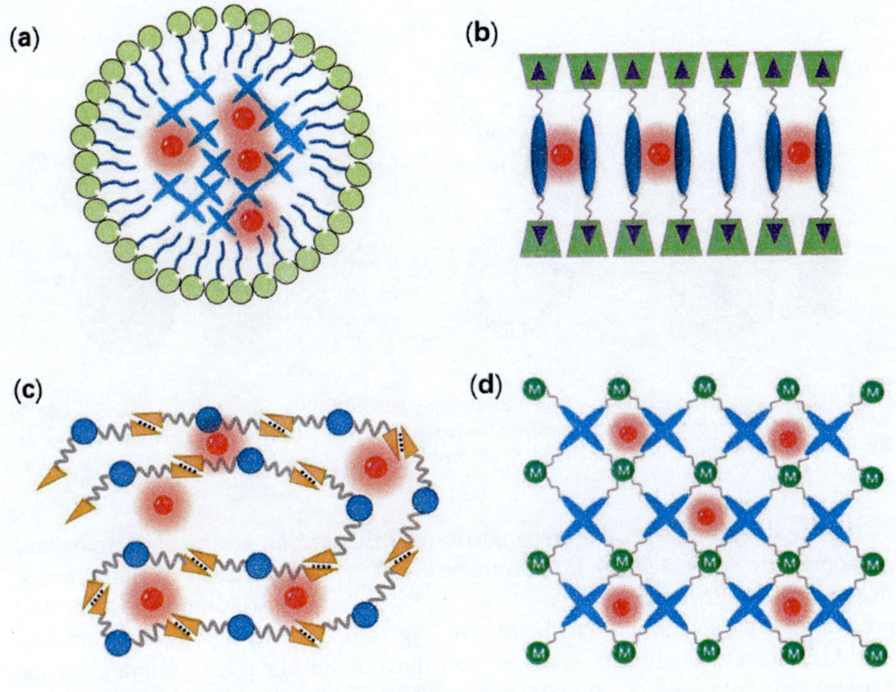

Fig. 3 Illustration of construction of luminogens with aggregation-induced emission (AIEgen)-based light-harvesting systems (LHSs) by **a** hydrophobic interaction, **b** host–guest interaction, **c** hydrogen bonds, and **d** metal coordination interaction

2.1 Hydrophobic Interaction

Self-assembly of donors and acceptors by hydrophobic effects to form nanoparticles is a convenient strategy for the fabrication of EET systems. Commonly, the amphiphilic micelle formed from surfactants can encapsulate low water-soluble donor and acceptor molecules in its hydrophobic core, and expose the hydrophilic surface in water, thereby allowing dispersal in aqueous solution [35]. Through hydrophobic interactions, the energy donors and acceptors were brought together in very close intermolecular distances for energy transfer [36]. Besides, this strategy efficiently simplifies the synthetic procedures for fluorophores pre-functionalization, which can be easily employed to obtain targeting nanoparticles for light-harvesting.

Zhao et al. [37] developed a TPE-based LHS through the controlled aggregation of donor and acceptor in a confined nanosphere. TPE-based energy donor (**1**) and acceptor (**2**) with similar structures tended to aggregate in aqueous solution (Fig. 4a). They were protected by self-assemblies of amphiphilic surfactant in water, which could be further stabilized by the participation of *N*-hydrosuccinimide as the cross-linker. This artificial LHS reached 95% energy transfer efficiency at a donor/acceptor ratio of 100:1. The high white-light quantum yield ($\Phi_F = 38\%$), outstanding water-dispersibility and structural stability of the nanoparticles make them very promising candidates for illumination and bioimaging applications.

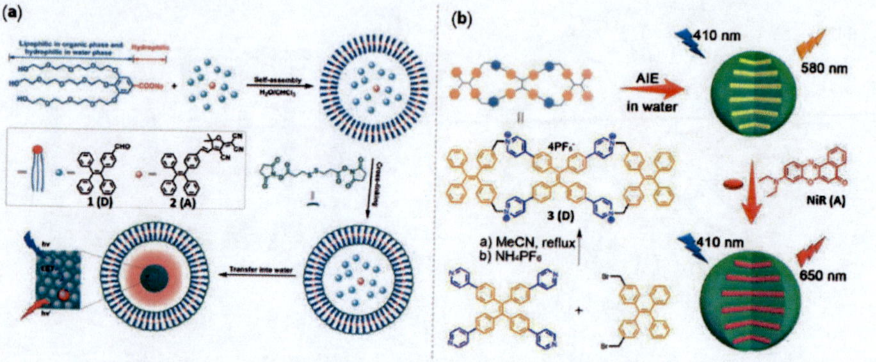

Fig. 4 **a** Surfactant-assisted LHS based on AIEgens. Reproduced with permission from Ref. [37] © 2019 Wiley. **b** Light-harvesting nanospheres self-assembled from a tetraphenylethene (TPE)-based framework. Reproduced with permission from Ref. [38] © 2019 American Chemical Society

Cao et al. reported a LHS prepared from TPE-based tetracationic dicyclophane monomers by self-assembly [38]. Tetrapyridyl TPE derivative, as a central cross-linker, was employed to react with two equivalents of bis(bromomethyl) TPE to produce the tetracationic TPE donor (**3**) (Fig. 4b). Due to the specific structure and AIE effect, **3** could self-assemble into three-dimensional (3D) supramolecular nanospheres with a strong emission (Φ_F=97.7%) in aqueous solution. The energy acceptor Nile red (**NiR**) could enter the hydrophobic center of the nanospheres to construct LHSs with energy transfer efficiency up to 77.5% and an antenna effect of 14.3.

The hydrophobic interactions driving self-assembly strategies are usually used for fabricating light-harvesting fluorescent materials for high-performance bioimaging. Liu et al. [39] reported a specific nanoparticle with amplified NIR fluorescence for bioimaging. The AIEgen (**4**, donor) and **NIR775** were encapsulated in the hydrophobic core of 1,2-distearoyl-sn-glycero-3-phosphoethanolamine-N-amino(polyethylene glycol)-2000 (DSPE-PEG$_{2000}$) micelles to obtain nanoparticles (Fig. 5a). In comparison with direct excitation of **NIR775** at 760 nm, a 47-fold emission enhancement could be found upon excitation of **4** at 510 nm. Narrow band depth (20 nm), large Stokes Shift (275 nm), good photostability and low cytotoxicity endowed the nanoparticles with high NIR fluorescence contrast in living animals. He et al. [40] further improved this self-assembly system. They utilized F127 (surfactant) to encapsulate the AIE-donor (**5**) and **NIR775** together (Fig. 5b). The resulting nanoparticles have a large three-photon absorption cross-section, allowing their NIR emission to appear under light excitation at wavelength of 1550 nm. After injection, the nanoparticles realized a vivid 3D reconstruction of mouse brain vasculature with small blood vessel imaging, which would be useful for the design of other NIR systems for deep-tissue imaging. Wang et al. reported NIR nanoparticles constituting amino tetraphenylene and triphenylamino-based fluorophores [41]. Instead of surfactant, the bovine hemoglobin (BHb) molecule was used to encapsulate fluorophores for good biocompatibility, low cytotoxicity and high brightness of nanoparticles. In combination

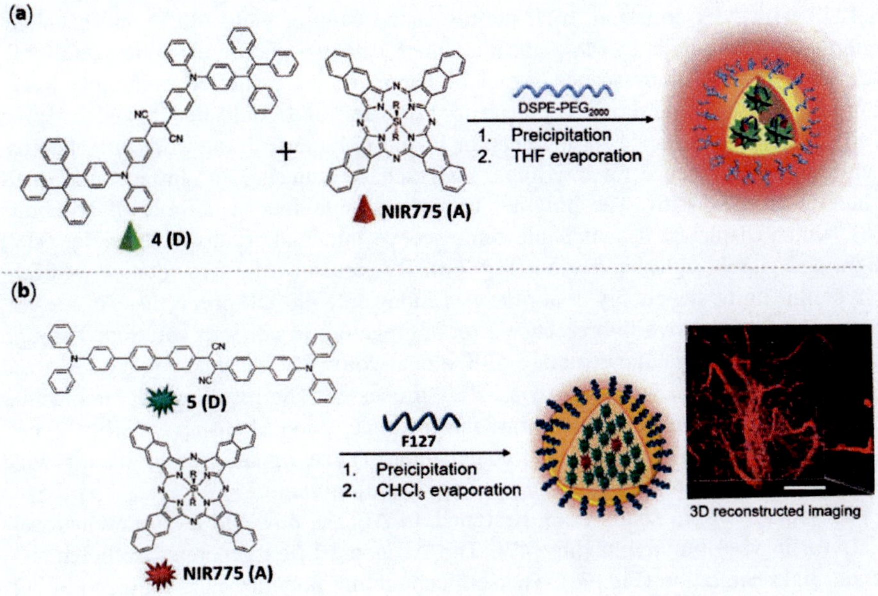

Fig. 5 **a** AIEgen-based nanoparticles with near infrared (NIR) fluorescence signal for bioimaging. Reproduced with permission from Ref. [39] © 2014 The Royal Society of Chemistry. **b** AIEgen-based nanoparticles for three-dimensional (3D) imaging of blood vessels in a mouse brine. Reproduced with permission from Ref. [40] © 2018 The Royal Society of Chemistry

with the large Stoke shift, the nanoparticles could avoid self-absorption by biological tissues, thus making them well-suited to cell imaging.

Polymeric networks are a splendid scaffold to provide multifunctional platforms for efficient energy transfer. In comparison with simple encapsulation by surfactants, polymeric networks can accommodate multiple donors or acceptors by stable covalent bonds. The stronger connection enables stabilization of energy transfer systems whether in vitro or in vivo. Normally, there are two popular strategies for the design of such LHSs: (a) decorating donor and acceptor on the same polymer to realize energy transfer [42]; (b) multiple donor units linked on the prepared polymeric network through chemical bonds and acceptor self-assembly into the hydrophobic region of the network [43, 44]. Later, these polymeric frameworks containing donor and acceptor are employed as scaffold or surfactant to construct nanoparticles by hydrophobic interactions. When the hydrophobic polymer acts as linker, the additional surfactant should participate in the nanoparticles to provide sufficient water solubility. For this strategy, a high donor–acceptor ratio can be afforded from the polymeric scaffold with multiple functional positions. The main challenge is the synthesis of different kinds of polymeric frameworks.

Tian et al. developed polymeric hybrid nanoparticles enabling two-photo excitation and near infrared (NIR) emission [45]. The AIE monomers were conjugated together to obtain a polymeric donor (**6**) for two-photon excitation at 800 nm (Fig. 6a). The energy acceptor (**7**) could be encapsulated by **6** and F127 to establish a LHS with NIR emission, high fluorescence quantum yield (45%) and excellent energy transfer efficiency ($\Phi_{ET}=95\%$). Under NIR two-photon excitation (800 nm), the efficient NIR fluorescence signal of nanoparticles exhibited preferable imaging ability with penetration depth up to 1200 μm, and could be utilized in HeLa cells. Zhang et al. reported an acceptor-linked polymeric framework with photoswitchable fluorescence for rewritable fluorescence pattering and intracellular dual-channel imaging [46]. The polymer framework contained a number of acceptors (**8**), which displayed a switchable fluorescence under alternative ultraviolet (UV) light and visible light irradiation (Fig. 6b). By means of the free rotation of triphenylamine units, the energy donor (**9**) was endowed with AIE properties. As a result, it could emit intensive fluorescence after aggregation in aqueous solution. Through controlling the molecular structure of **8**, a dual-color fluorescence could be realized in self-assembled nanoparticles via FRET processes. The resulting supramolecular nanoparticles exhibited high-contrast fluorescence, good photo-reversibility, excellent stability and a distinct dual-channel fluorescence signal, thereby having wide applications in rewritable materials and intracellular imaging.

In another work, Zhang et al. designed an AIEgen-powered afterglow nanoparticle for in vivo tumor imaging [47]. The AIEgen **10** first generated sufficient $^1O_2$ under light excitation (Fig. 6c). The semiconducting polymer **11** is oxidized by $^1O_2$ to generate amounts of high-energy 1,2-dioxetane intermediate, resulting in chemiluminescence during its slow decomposition to excite **10**. Thereafter, red afterglow

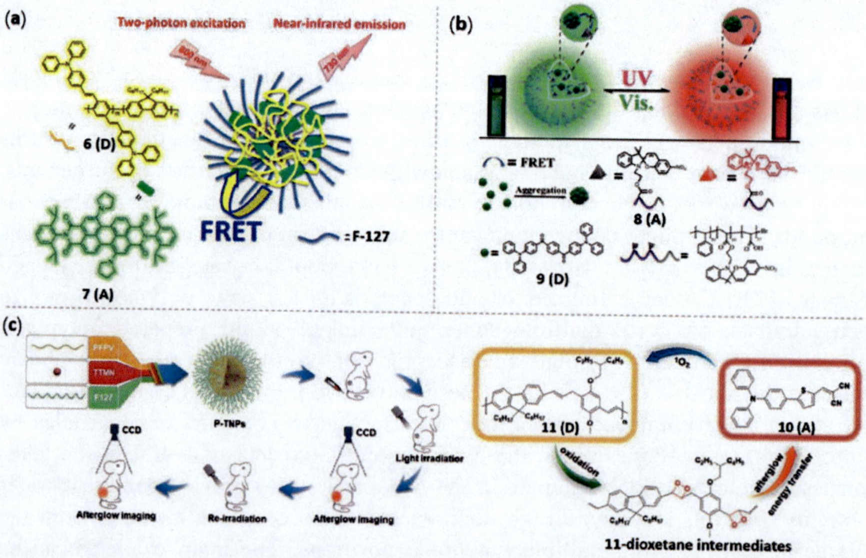

Fig. 6 **a** Nanoparticles self-assembled from AIEgen-based polymers and NIR acceptors for bioimaging. Reproduced with permission from Ref. [45] © 2015 American Chemical Society. **b** Photoswitchable fluorescent nanoparticles for rewritable patterning and intracellular dual-color imaging. Reproduced with permission from Ref. [46]. © 2015 American Chemical Society. **c** An AIEgen-powered afterglow nanoparticle for tumor imaging. Reproduced with permission from Ref. [47] © 2020 The Royal Society of Chemistry

emission and $^1O_2$ generation of **10** was accomplished simultaneously through an energy transfer process. This closed-loop circulation could provide afterglow emission in tumor imaging with low-autofluorescence background and high signal-to-noise ratio. This study presented a novel AIE-facilitated afterglow system and was expected to develop more innovative designs for cancer theranostics.

Li et al. developed a NIR-emissive LHS for Golgi apparatus sensing [48]. A naphthyl-1,8-diphenyl pyridinium derivative (**12**) was capable of conjugating with amphiphilic dodecyl-modified sulfonato-calix[4]arene to afford a nanosphere, which manifested fluorescence enhancement and wavelength blueshift to 550 nm (Fig. 7). It is worth noting that the main driving force for nanoparticle formation is hydrophobic interaction between the phenyl naphthyl group in **12** and the long dodecyl groups in sulfonato-calix[4]arene, instigating host–guest interaction. Subsequently, Nile blue (**NiB**) was encapsulated in the hydrophobic interspace of the nanospheres, which exhibited intensive NIR emission due to energy transfer from **12** to **NiB**. This LHS displayed excellent antenna effect (33.1) at a high donor/acceptor ratio (250:1), thereby amplifying fluorescence in the NIR region. Such light-harvesting nanoparticles have been utilized in the imaging of Golgi apparatus. Outstanding sensing ability of the nanospheres for golgi apparatus has been demonstrated by co-staining PC-3 cells with a Golgi apparatus reagent. This interesting study provided a useful strategy towards imaging and diagnosis with high-performance.

White-light emissive materials play essential roles in many applications, including fluorescent sensors, information display, and optical-recoding systems [49, 50]. According to the Commission International de L'Eclairage (CIE), combination of three primary RGB (red, green and blue) colors or two complementary colors (blue and yellow) can afford white-light emission [51, 52]. The self-assembly strategy can simplify the construction of light-harvesting platforms and modulate emission color by regulating the ratio of each fluorophore [53–58].

Wan et al. [59] developed a hyperbranched polymeric fluorophore for facile fabrication of OLED and light-harvesting films. Through Barbier hyperbranching polymerization, a series of AIE polytriphenylmethanols and polydiphenylmethanols

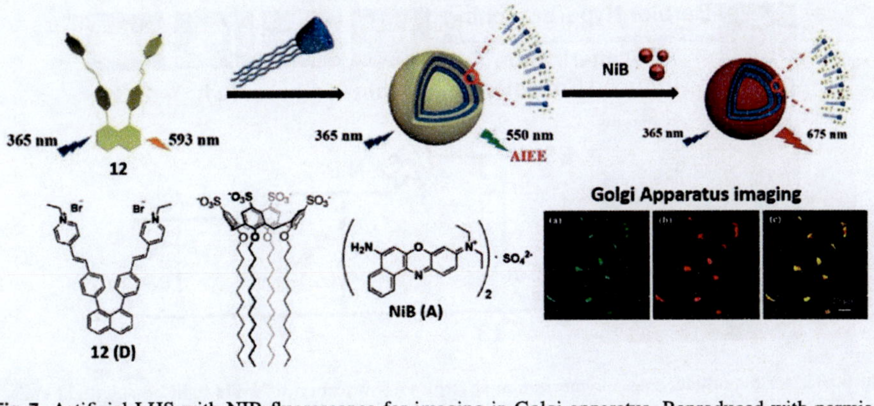

Fig. 7 Artificial LHS with NIR fluorescence for imaging in Golgi apparatus. Reproduced with permission from Ref. [48] © 2020 Wiley

was synthesized. Among them, polymers (**13**) with favourable AIE properties and tunable fluorescence colors from blue to yellow were obtained through regulation of polymerization time. By mising a suitable ratio of blue/yellow polymers, white light-emitting materials for OLED could be obtained (Fig. 8). Meanwhile, light-harvesting films were afforded with energy transfer from blue emissive polymer to **NiR** and then to **IR-780**. The films constituted from **13** and **IR-780** emitted fluorescence at 780 nm with an antenna effect of 14.3 and Φ_{ET} of 29.8%. Insufficient overlap between absorption of **IR-780** and emission of **13** led to low Φ_{ET}. Therefore, as an energy bridge, **NiR** improved the energy transfer efficiency to 88.6% (Φ_{1stET}) and 90.1% (Φ_{2ndET}) at the mixed **13/NiR/IR-780** ratio of 1000:1:2. These polymerization-controlled emission strategies could inspire scientists to design more luminescent materials.

2.2 Host–Guest Interaction

Through controlling host–guest interactions, self-assembled nano-platforms can display stimuli-responsive optical characteristics with tunable chemical and biological properties [16, 60, 61]. Crown ether, cyclodextrins, pillararenes and other macrocyclic structures are favorable candidates to build host skeletons [62]. Prefabricated fluorophores with host–guest interaction units are necessary. For instance, He et al.

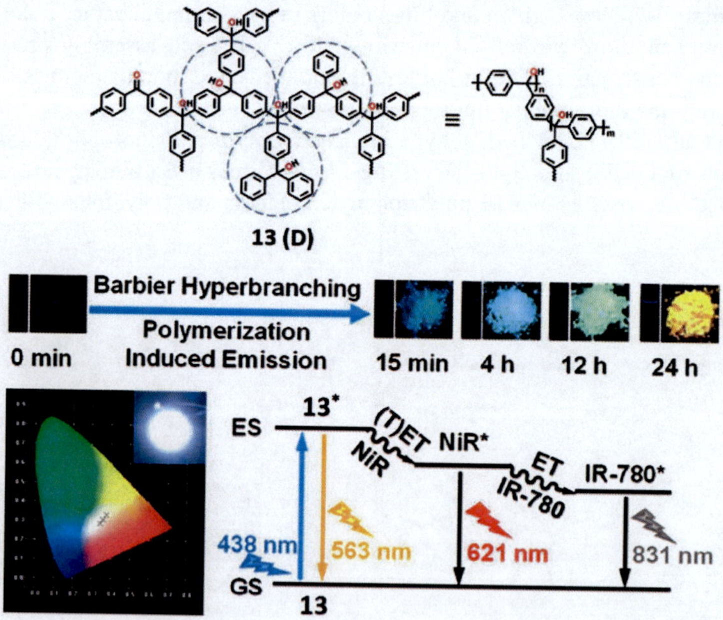

Fig. 8 Barbier hyperbranching polymerization strategy for fabrication of white-light organic light emitting diodes (OLEDs) and light-harvesting films. Reproduced with permission from Ref. [59] © 2019 American Chemical Society

designed a highly efficient LHS based on the host–guest interaction of crown ether and BODIPY [63]. Crown ether decorated TPE as the host was allowed to associate with a BODIPY-functionalized guest to produce the supramolecular system with excellent AIE properties, high energy transfer efficiency (~93%) and outstanding tolerance to pH variation (Fig. 9a). Instead of decorating host/guest groups on fluorophores directly, Liu et al. developed an artificial LHS based on the host–guest interaction of sulfato-*β*-cyclodextrin (**SCD**) and oligo(phenylenevinylene) derivative [64]. The oligo(phenylenevinylene) derivative (**14**), as the guest with AIE properties, could enter the **SCD** cavity to form a supra-amphiphile as the energy donor, which further self-assembled into nanoparticles (Fig. 9b). The hydrophobic dye (**NiR**) was loaded in the hydrophobic layer of the nanoparticles as the energy acceptor. This light-harvesting scaffold with a tunable ratio of donor and acceptor showed a high energy transfer efficiency and satisfying antenna effect (with trace amount of acceptor). Tang et al. developed a leaf-like lamellar LHS. After host–guest interaction between energy donor and *β*-cyclodextrin, different morphologies of self-assembled scaffolds, including leaf-like lamellar structure, nanoribbons, vesicles, nanofibers, helical nanofibers and toroids, were obtained by manipulating solvent proportion and temperature in the self-assembly process. A *β*-cyclodextrin-decorated rhodamine was mixed in the resultant scaffold as energy acceptor, which provided 94.2% energy transfer efficiency through the FRET process [65].

Wang et al. described an artificial LHS constructed in aqueous solution [66]. As the AIEgen-based donor, a salicylaldehyde azine derivative (**15**) with long alkyl chains and quaternary ammonium terminals formed host–guest complexes with pillar[6]arene (Fig. 10). The resulting supra-amphiphilic host–guest complexes further self-assembled to form spherical nanoparticles with a sandwich structure. The remarkable fluorescence of the nanoparticles at 513 nm was ascribed to the AIE

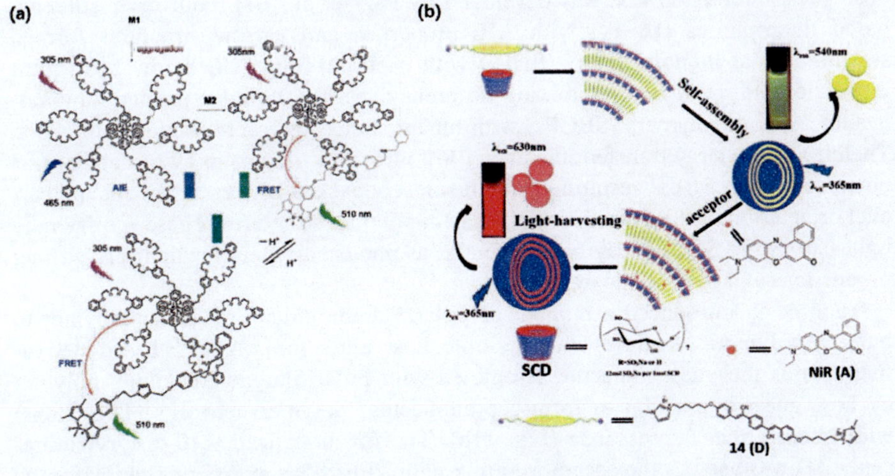

Fig. 9 **a** Highly efficient FRET from AIEgens to borondipyrromethene (BODIPY) based on host–guest interaction for mimicking the LHS. Reproduced with permission from Ref. [63] © 2017 The Royal Society of Chemistry. **b** Energy donor incorporated by macrocycles and then self-assembled with acceptor to obtain AIE and FRET properties. Reproduced with permission from Ref. [64] © 2017 Wiley

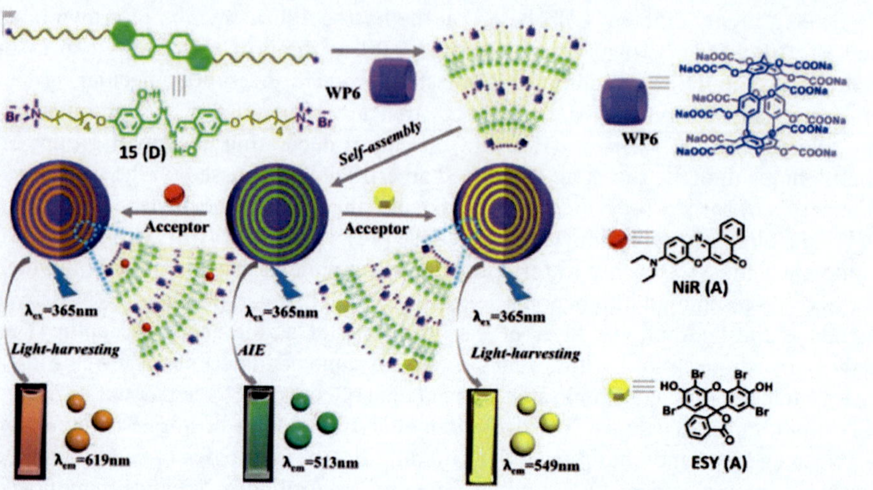

Fig. 10 Supramolecular LHSs based on the AIE effect and host–guest interaction. Reproduced with permission from Ref [66]. © 2018 Wiley

and excited-state intramolecular proton transfer (ESIPT) properties of **15** in aqueous solution. The fluorescent dyes of **NiR** and eosin Y (**ESY**), as ideal acceptors, were encapsulated within the hydrophobic interiors of the nanoparticle, respectively, to obtain two LHSs. The energy transfer efficiency of 55% for **NiR** and 42% for **ESY**, and antenna effect of 25.4 for **NiR** and 28.0 for **ESY** were achieved in the two self-assemblies with high donor/acceptor ratio. These water-dispersible artificial LHSs provided a versatile platform for further applications.

Utilizing cucurbiturils to fabricate nanobundles is another smart strategy for host–guest interaction that was developed by Park et al. [67]. Four cyanostilbene-based fluorophores (**16–19**) with AIE properties and narrow structures formed supramolecular homopolymers (SHPs) with cucurbit[8]uril (CB[8]) by host–guest interaction (Fig. 11a). After mixing different dyes, the light-harvesting supramolecular block copolymers (SBCPs) with tunable emissions were produced in water. Therein, high energy transfer efficiency (96%) from **16** (donor) to **19** (acceptor) was obtained in their SBCP, resulting from the closely packed D/A arrangement and parallel orientation of the transition dipoles in copolymer structures. These eco-friendly light-harvesting SBCPs may have potential as photosensitizers for light-driven fuel productions and therapeutic treatments.

Yang et al. introduced a number of pillar[5]arene units on a linear polymer to build a polymeric scaffold with multiple host units [68]. A TPE-based derivative (**20**) as the guest molecule assembled with pillar[5]arene-containing polymer by host–guest interaction to form supramolecular polymer networks (TPE-SPNs) with intense blue fluorescence (Fig. 11b). The functionalized 9,10-distyrylanthracene (**21**) was used as the acceptor to mix with TPE-SPNs at different ratios for the construction of spherical LHSs. The AIE characteristics of **20** and **21** resulted in a high fluorescence quantum yield ($\Phi_F = 98.22\%$) of the nanoparticles. This strategy

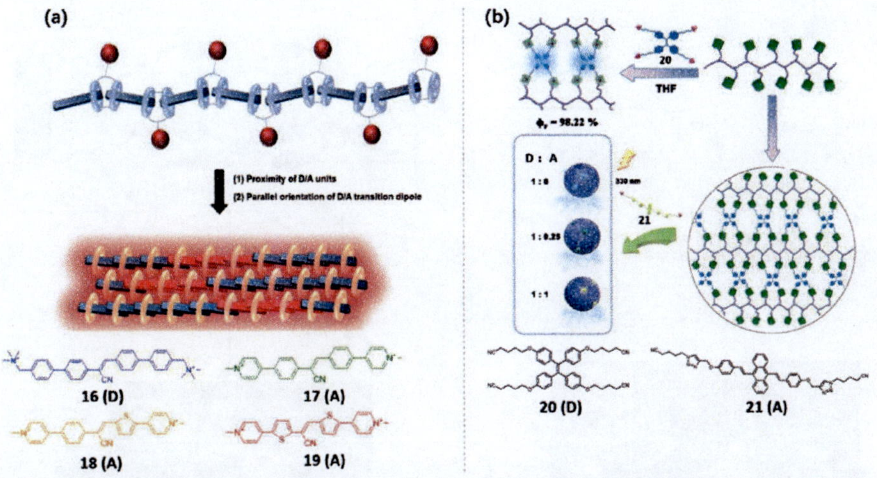

Fig. 11 **a** Supramolecular block copolymers constituted by cucurbit[8]urils and cyanostilbene derivatives with AIE properties for light-harvesting. Reproduced with permission from Ref. [67] © 2017 Wiley. **b** An efficient LHS with pillar[5]arene-based polymers and TPE derivatives. Reproduced with permission from Ref. [68] © 2019 Wiley

suggested another representative approach for the design of high-performance light-harvesting nanoparticles.

In the design of white-light emission materials, host–guest interactions can not only easily change the emission color by controlling the donor/acceptor ratio, but can also further improve the stability of the composite materials. A binary light-harvesting platform-based host–guest interactions was reported by Tang et al. [69]. The amphiphilic organic cage was synthesized by conjugation of tetraphenylpyrazine (**22**) with triglycol monomentyl ether-substituted triazine (Fig. 12). Strong blue fluorescence was emitted from the resulting cage both in solution and aggregate state, due to the rotation restriction of phenyl groups on tetraphenylporphyrin (TPP). The hydrophobic core had the ability to encapsulate the energy acceptor, diketopyrrolopyrrole (**23**), to obtain light-harvesting nanoparticles. By preventing π–π interactions in the cage, compound **23** could still release yellow-light emission in aggregate or solid state. Through controlling the ratio of D/A, the **23@22**-based cage was enabled to provide stable white-light emission both in aggregate and film. High photostability and bright white-light emission was still maintained 30 days after film preparation.

2.3 Hydrogen Bonds

A quadruple hydrogen bonding unit, 2-ureido-4[1*H*]-pyrimidinone (UPy), is one of the building blocks used most widely for the development of supramolecular assembly due to its synthetic accessibility and high association constant ($K_{dimer} > 10^7$ in $CHCl_3$) [70–72, 73, 74]. The intermolecular hydrogen bonds ensure dependable

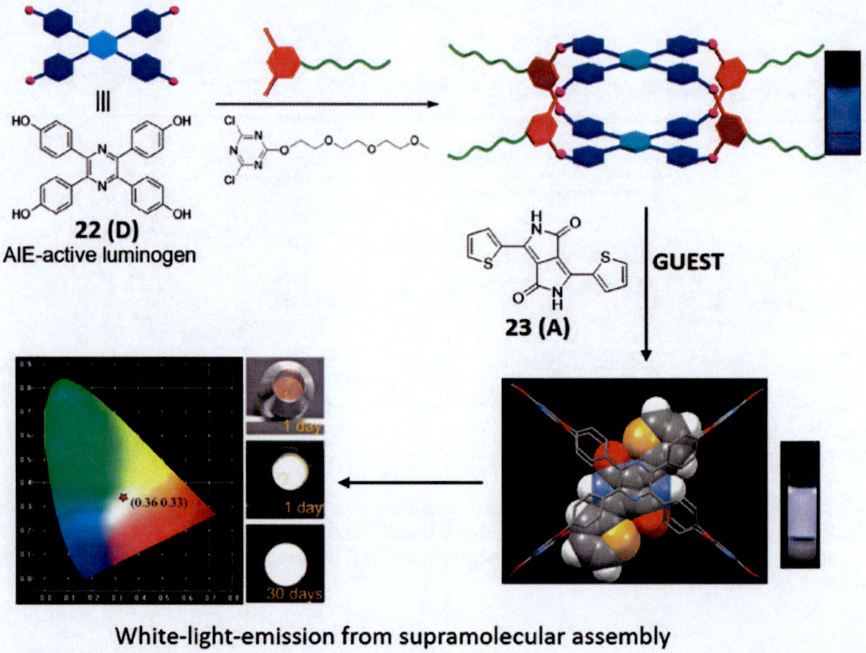

Fig. 12 Amphiphilic organic cage as a light-harvesting platform for white-light emission. Reproduced with permission from Ref [69]. © 2017 American Chemical Society

incorporation between energy donor and acceptor, and thus UPy is well-suited to fabricate supramolecular polymers for LHSs. In 2014, our group reported the first quadruply hydrogen-bonded nanospheres by using a mini-emulsion method in water, which was used to mimic natural LHSs [75]. Subsequently, we explored a convenient strategy to functionalize bis-UPy on alkyne-containing groups by a simple thiol-ene reaction [76]. For further applications, our group developed a series of fluorescent nanoparticles through multicomponent self-assembly based on quadruple hydrogen bonds for cell imaging and ratiometric oxygen sensing [77, 78].

To overcome the limitation of AIEgens displaying broad emission spectrum, we fabricated a variety of light-harvesting nanoparticles with narrower emission band and brighter fluorescence than conventional AIEgens [79]. The light-harvesting nanoparticles have been constructed from supramolecular copolymers comprised of quadruply hydrogen-bonded monomers of AIEgens as energy donor (**24–26**) and BODIPY with sharp emission peaks as acceptor (**27–30**), respectively (Fig. 13a). By changing donor/acceptor or modulating their ratio, the resulting light-harvesting nanoparticles could exhibit tunable fluorescence colors of blue, green, yellow, deep red, near-infrared and pure white in aqueous solution. The sufficient energy transfer from TPE to BODIPY in the supramolecular copolymer resulted in up to sixfold enhancement of fluorescence intensity. In comparison to the broad emission band of the TPE monomer, the emission spectra of copolymers have a narrowed full width at half-maxima (FWHM), decreased from 148 to 32 nm after excitation energy

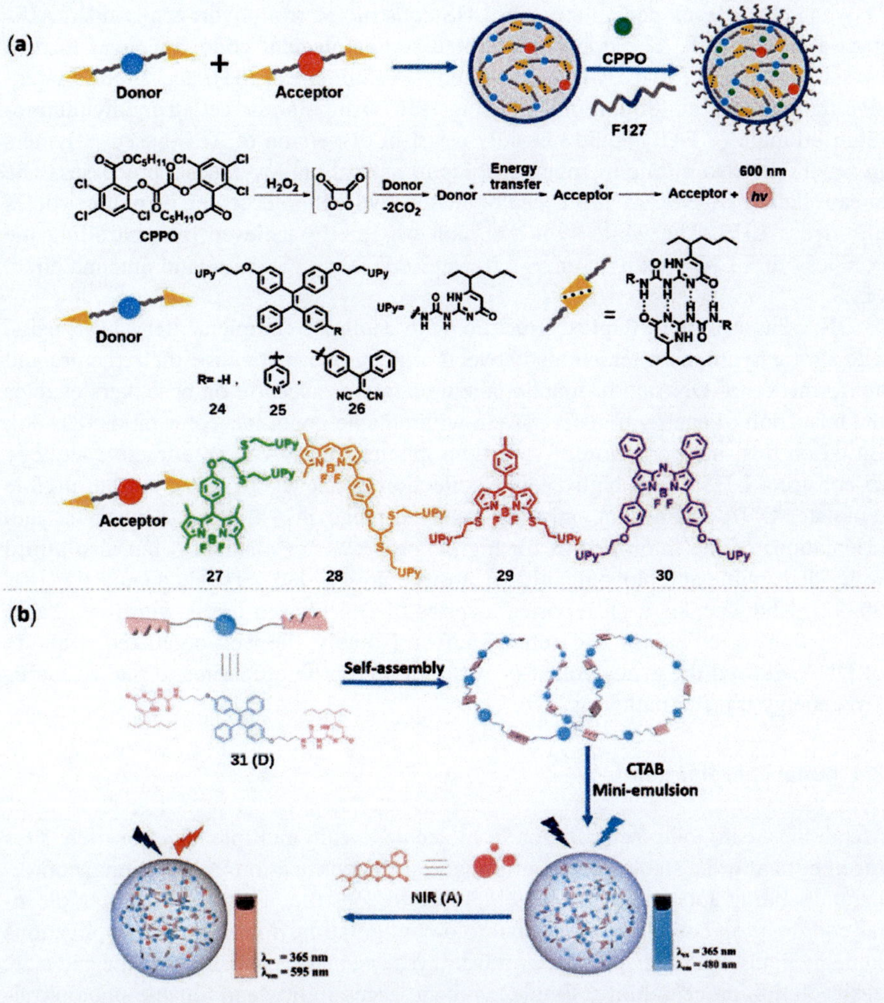

Fig. 13 **a** Light-harvesting nanoparticles formed from supramolecular copolymers comprised of 2-ureido-4[1*H*]-pyrimidinone (UPy)-functionalized TPE and BODIPY for fluorescence bioimaging through chemiluminescence-resonance-energy transfer (CRET), AIE and FRET processes. Reproduced with permission from Ref [79]. © 2019 American Chemical Society. **b** Light-harvesting nanoparticles prepared from UPy-functionalized TPE and Nile red (**NiR**). Reproduced with permission from Ref. [80] © 2020 The Royal Society of Chemistry

transferring to BODIPY. Moreover, these resulting nanoparticles displayed ~ 5-fold higher brightness than other commercial quantum dots in water, and have been used successfully for fluorescence imaging in live animals. Interestingly, these LHSs could be excited by the chemiluminescence of bis(2,4,5-trichloro-6-carbopentoxyphenyl)oxalate (CPPO). This strategy realized H_2O_2-responsive chemiluminescence imaging in mice with low autofluorescence interference and high signal-to-noise ratio.

Wang et al. developed an efficient LHS constructed from hydrogen-bonded AIEgens and **NiR** [80]. The donor-constituted supramolecular copolymer was formed by self-assembly of di-UPy-functionalized TPE monomers (**31**), and emitted intensive fluorescence at 480 nm in water (Fig. 13b). Amphiphilic cetyltrimethyl ammonium bromide (CTAB) could not only assist in dispersion of TPE-based polymers in water, but also mimic membrane lipids in natural photosynthesis processes. The encapsulation of **NiR** as the acceptor in the hydrophobic center of nanoparticles afforded a LHS. The white-light emission was easily achieved by controlling the composition of NPs with an energy transfer efficiency of 53.4% and antenna effect of 28.

DNA has a unique duplex structure with multiple complementary base pairs. The strong hydrogen interactions between base pairs can stabilize the structure and conformation of DNAs. Pre-functionalizing donor or acceptor on base pairs enables the formation of energy transfer system with tunable donor–acceptor ratios [81–86]. Based on this, incorporation of AIE fluorophores in DNA is an attractive strategy to construct LHSs with high energy collection efficiency [87–89]. As an eligible template, DNA chains not only accurately arrange intermolecular distances and orientations of the fluorophores for high energy transfer efficiency, but also afford water-dispersible and biocompatible nanostructures with diverse shape and size [81, 90–92]. Klymchenko et al. reported a series of DNA-based LHSs with remarkable energy transfer efficiency and antenna effect. Notably, the well-organized structure of DNA allowed the arrangement of several different fluorophores to realize multilevel energy transfer pathways [86].

2.4 Metal Coordination

Metal ions can joint several donor molecules with multiple electron-rich sites together to afford a stable skeleton through metal coordination. A directional coordination bonding approach has been employed to construct a variety of supramolecular coordination complexes with manifold configurations for numerous applications in many fields. These display many advantages, such as well-defined shape and size, easy solution processability, simple synthetic accessibility, and tunable photophysical properties, and are thereby welcomed for development of LHSs [93–95].

Stang et al. constructed artificial LHSs by self-assembly of fluorescent Pt(II) metallacycles with **ESY** [96]. The energy donor (**32**) was constituted from a triphenylamine core and cyanstilbene moiety, which displayed interesting aggregation-induced emission enhancement (AIEE) optical properties (Fig. 14a). The two Pt(II)-based linkers chelated with donor **32** in mixed solvent of CH_2Cl_2–MeOH at room temperature to obtain the hexagonal [3 + 3] metallacycle and hexagonal [6 + 6] metallacycle, respectively. The resulting two metallacycles have similar absorption bands at around 300–400 nm, and a similar emission maximum at 521 nm upon excitation at 410 nm in DMSO. They exhibited no obvious solvatochromic characteristics and significant fluorescence enhancements at aggregate in 80% H_2O–DMSO or solid state. Scanning electron microscopy (SEM) and transmission electron microscopy (TEM) analysis showed agglomerate spherical morphology of

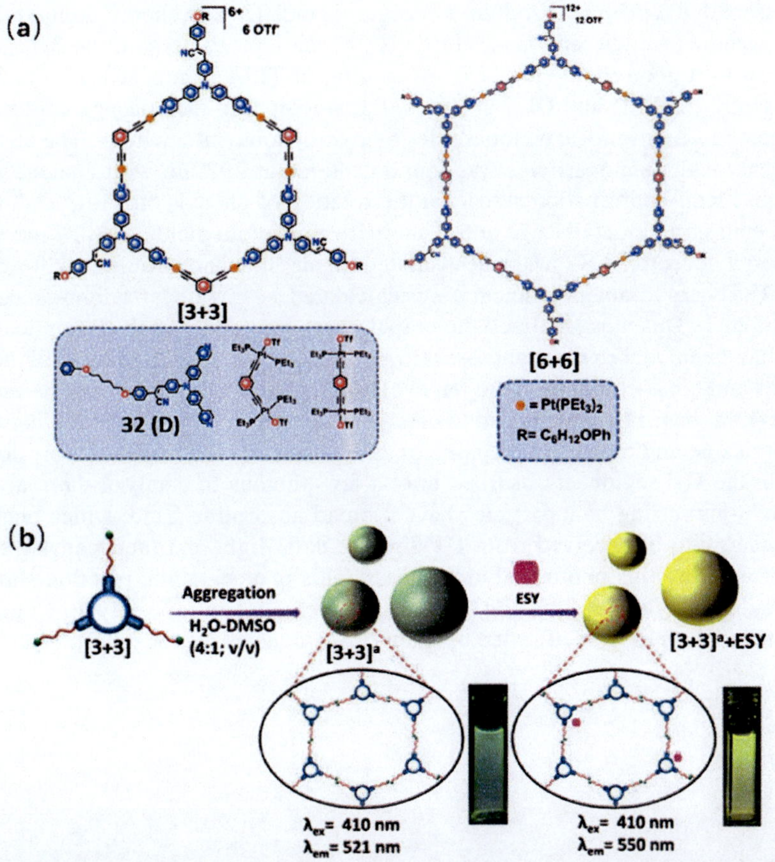

Fig. 14 **a** The structures of [3+3] and [6+6] metallacycles, and related building blocks. **b** LHS self-assembled from metallacycles and eosin Y (**ESY**). Reproduced with permission from Ref [96] © 2019 American Chemical Society

two metallacycles. According to dynamic light scattering (DLS) analysis, the average hydrodynamic diameters of [3+3] and [6+6] metallacycle in aggregate state were around 250 nm and 174 nm, respectively. Then, **ESY** was selected as energy acceptor for co-assembly with [3+3] and [6+6] metallacycle to construct LHSs. In comparison with direct excitation of **ESY**, the fluorescence emission intensity was substantially increased in the metallacycle–ESY complexes. The energy transfer efficiencies for [3+3]+**ESY** and [6+6]+ **ESY** at a 10:1 donor/acceptor ratio were calculated as 65% and 52%, respectively. The fluorescence lifetime of the metallacycle with **ESY** presence was shorter than the lifetime of metallacycle alone. Such differences indicated the existence of the intermolecular energy transfer processes (Fig. 14b).

Zhang et al. further developed the metal coordination-based LHS and used it in photocatalytic cross-coupling hydrogen evolution reactions [97]. They self-assembled TPE-based sodium benzoate ligands (**33**) as energy donors, dipyridyl ligand as studdle and *cis*-Pt$(PEt_3)_2(OTf)_2$ as joint to prepare the 3D metallacage with a good yield (Fig. 15). According to TEM, confocal laser scanning microscopy (CLSM) and DLS analyses, the amphiphilic metallacage could self-assemble in water to form nanoparticles by hydrophobic interactions. The absorption peak of the nanoparticles was found centered at 292 nm with a shoulder at 356 nm. Metal-coordination restricted the rotation of phenyls on TPE, leading to bright emission of metallacage at 511 nm with a quantum yield of 23.8% in water. As energy acceptor, **ESY** was encapsulated inside the nanoparticles to afford the LHS. The energy transfer efficiency was calculated as 45% under a donor:acceptor ratio of 20:1. This work realized the output energy for photocatalysis application. **ESY** has been reported as photosensitizer to catalyze C–H arylation of heteroarenes and cross-coupling hydrogen evolution reactions. In this work, the metallacage-**ESY**-based LHS was used to catalyze the cross-coupling reaction between benzothiazole and diphenylphosphine oxide. Because of the narrow absorption of **ESY** in the UV region, it absorbed only a few photons to catalyze the reaction. The light-harvesting nanoparticles have a broad absorption band, which enabled more energy to be received from UV light or solar light and then transferred to **ESY**, triggering this photocatalytic process. This photocatalytic reaction showed conversion and yield of 88% and 65%, respectively. In contrast, only 40% conversion and 33% yield were afforded by using **ESY** alone as photocatalyst.

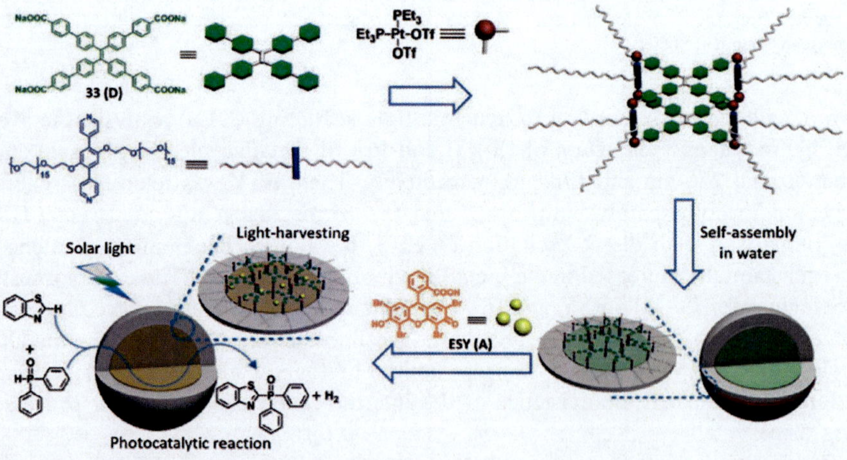

Fig. 15 Water-soluble LHS self-assembled from a TPE-based metallacage and **ESY** for photocatalysis. Reproduced with permission from Ref. [97] © 2019 Wiley

3 Fluorescent Probes

Fluorescent probes are employed widely in the detection of analytes and direct visualization of biomolecules through emission signal changes with high spatial and temporal resolution. However, in a highly heterogeneous biological milieu, quantitative detection of fluorescent probes that show only a single emission suffer is challenging because emission intensities are also affected by other extraneous factors such as pH, temperature and polarity probe concentrations, and cellular context. Ratiometric fluorescent sensors relying on the intensity of two or more emission bands would provide an effective internal reference that would greatly increase sensitivity and improve quantification. Energy transfer processes provide the most prominent approach to construct ratiometric probes. After responding to analytes, the optical properties or distances of the donor/acceptor pairs are regulated to realize ratiometric sensing. The general strategy of designing EET-based fluorescent probes is to anchor the donor–acceptor in close proximity in a single system or molecule, which transfers energy through FRET or through-bond energy transfer (TBET). Commonly, the responsive sites are introduced either on the donor or acceptor chromophore (Fig. 16a), or on the linking units (Fig. 16b). When the target analytes interact with the responsive sites on donor or acceptor chromophores, the chemical structure of the acceptor alters, which affects the overlap between the absorption band of the acceptor and the emission band of the donor, thus the EET process is initiated or prevented. When the target analytes cleave the linking units of the EET system, the distance between donors and acceptors increase significantly, blocking the energy transfer process. As a result, a change in fluorescence signal can be observed, allowing for the detection of the presence of target analytes. Here, modulation of fluorescent probe fluorescence signal by utilizing FRET or TBET processes will be introduced.

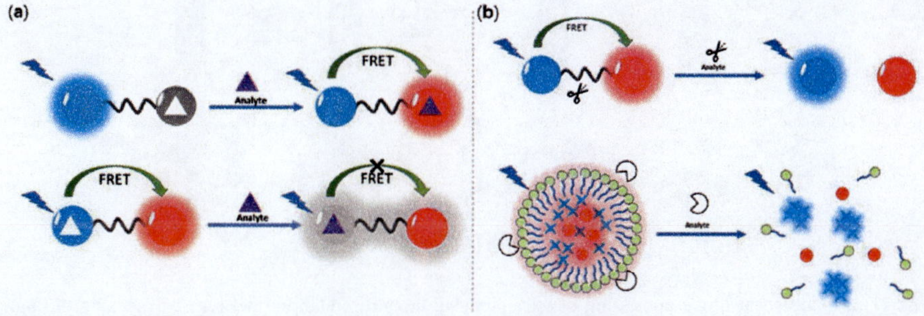

Fig. 16 **a,b** Illustration of the two common strategies for constructing FRET-based fluorescent probes. **a** Fluorescence-switch by alteration of the chemical structures of donor or acceptor. **b** Fluorescence-switch by alteration of the distances between donor and acceptor

3.1 FRET-Based Fluorescent Probes

FRET is a fundamental photophysical process [6, 98]. The excited fluorescent donor transfers its excitation energy to the acceptor through dipole–dipole interactions, accompanied by a ratiometric variation in fluorescence signals. The center-to-center distance between donor and acceptor is the determining factor for energy transfer. Thus, incorporation of the two advantages of AIE and FRET effect is advisable when designing fluorescent probes and improving sensing sensitivity.

Gao et al. [99] designed AIEgen-modified rhodamine derivatives for Fe^{3+} detection with ratiometric fluorescence variation. They combined tetraphenylsilole and rhodamine 6G through a flexible chain to obtain probe **34** (Fig. 17a) [99]. In the absence of Fe^{3+}, the probe emitted negligible fluorescence from the energy donor in good solvent and blue fluorescence in aggregate. After coordination of Fe^{3+}, the absorption peak at 526 nm of rhodamine 6G ascended followed by increased fluorescence at 550 nm. The excitation energy was transferred from tetraphenylsilole to rhodamine 6G through dark resonance energy transfer (DRET) in solution and

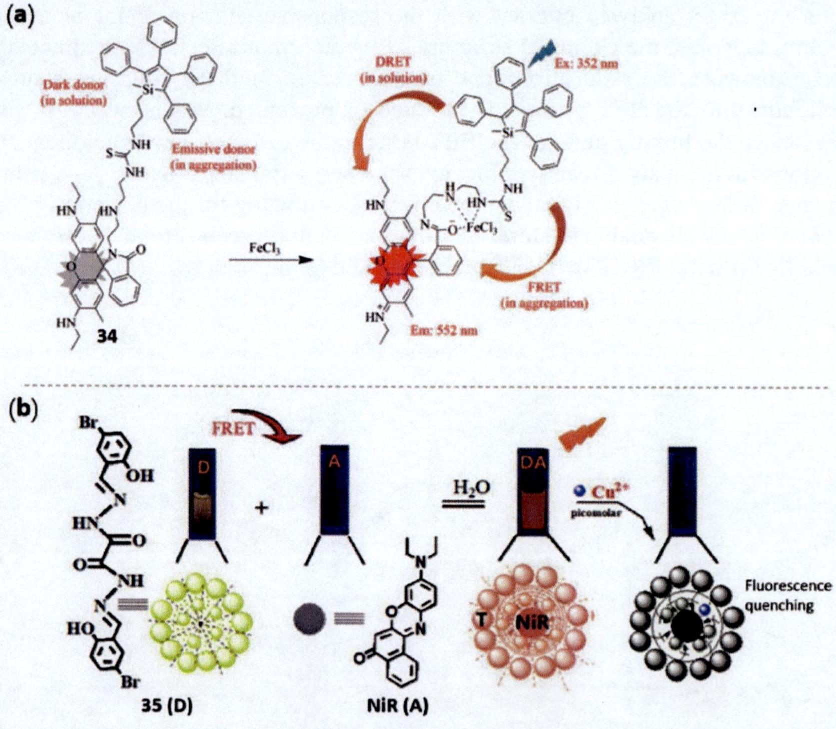

Fig. 17 **a** Fluorescent probe consisting of tetraphenylsilole and rhodamine 6G for Fe^{3+} sensing through the combination of dark resonance energy transfer (DRET) and FRET processes. Reproduced with permission from Ref. [99] © 2020 Elsevier B.V. **b** FRET-based supramolecular nanoparticles with AIE molecules for the sensitive detection of Cu^{2+}. Reproduced with permission from Ref. [100] © 2019 Elsevier B.V.

FRET in aggregate, resulting in **34** possessing excellent selectivity and high sensitivity for Fe^{3+} detection. A ratiometric fluorescence signal (blue to red) observed in HeLa cells indicated that the probe could be used in detection of Fe^{3+} in living cells.

Liu et al. fabricated a nanoprobe by supramolecular assembly for the Cu^{2+} detection [100]. The AIE-fluorophore **35** and **NiR** self-assembled into nanoparticles through hydrophobic interactions (Fig. 17b). The nanoparticles showed fluorescence at 630 nm upon excitation at 380 nm with high energy transfer efficiency ($\Phi_{ET}=82.52\%$) and antenna effect (24.9) at a donor/acceptor ratio of 250:1. In the presence of Cu^{2+}, both donor and acceptor fluorescence were quenched and the energy transfer process disappeared, since Cu^{2+} was able to insert between donor and acceptor to form a stable polymer system and block energy transfer processes. However, such a fluorescence-off response model was inadequate for further biological applications.

For biomolecule detection, Vidal et al. reported water-soluble glyco-dots self-assembled from TPE-based glycoclusters and dicyanomethylene-4*H*-pyran based fluorescent probes (DM) for exogenous and endogenous $ONOO^-$ response [101]. Due to ACQ interference, the fluorescence of DM was quenched in aqueous solution even though it responded to $ONOO^-$. TPE-glycoclusters were augmented with several glycosyls by CuAAC cycloaddition to obtain outstanding water solubility. The amphiphilic structures of TPE-glycoclysters aggregated easily and constructed nanoglyco-dots, which had a hydrophobic core to capture DM probes inside. The close distances from TPE to DM allowed efficient EET. In the presence of $ONOO^-$, intensive fluorescence of DM was observed, with high energy transfer efficiency ($\Phi_{ET}=89\%$) upon excitation of TPE-glycoclusters at 320 nm.

Liu et al. developed probe **36**, with dual-fluorescence amplification and high signal-to-noise ratio (Fig. 18) [102]. The probe is composed of a coumarin as energy donor and a red-emissive AIEgen as energy acceptor. The DEVD substrate was a water-soluble peptide between the donor and acceptor, which could be cleaved by caspase 3. Initially, the free motion of AIEgen dissipated the energy from coumarin, leading to fluorescence disappearance of the probe in aqueous solution. Upon caspase-3 hydrolysis of the peptide linker, the probe was divided into two parts, which displayed strong green and red fluorescence emission, respectively, due to disruption

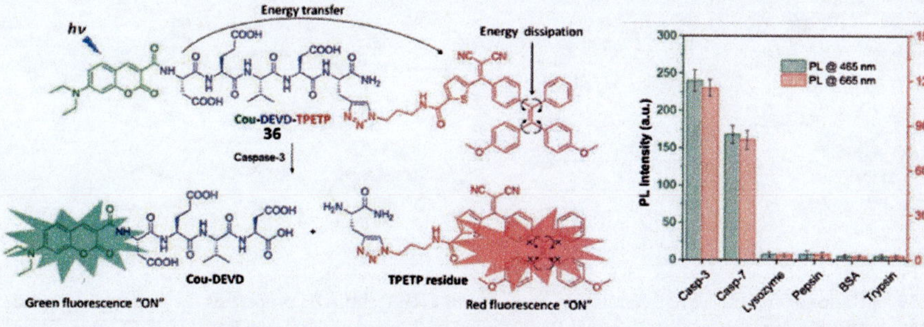

Fig. 18 Linker-responsive FRET probe for intracellular caspase-3 detection. Reproduced with permission from Ref. [102] © 2016 The Royal Society of Chemistry

of FRET systems and aggregation of AIEgen. The dual turn-on fluorescence signal allowed real-time enzyme detection, which avoided interference from the biological background. Moreover, this probe sensitively monitored the intercellular caspase-3 level generated from cell apoptosis, thereby potentially offering new opportunities to evaluate the efficiency of novel anticancer drugs.

Tian et al. developed AIE-based glyconanoparticles for the ratiometric detection of *β*-amyloid [103]. The nanoparticles were formed from aggregation of silole-diyne (**37**) with two dicyanomethylene-4*H*-pyran (DCM)-based glycosidic probes (**38** and **39**) in aqueous solution (Fig. 19a). They exhibited red fluorescence via an energy transfer process from **37** to DCM probes. The *β*-amyloid peptides could insert in the middle of the donors and acceptors to block their FRET process, leading to blue-shifted emissions at 491 nm (from **37**) and 525 nm (from **38** interacting with *β*-amyloid). The addition of lectins removed DCM probes from the donor, resulting in a single emission remaining at 491 nm. This design provides a novel multifunctional strategy with ratiometric detection, thus allowed extension to other AIEgens and materials.

Xia et al. reported DNA-based probes for intracellular telomerase sensing. AIEgen with a positive charge could bind to a negatively charged DNA with phosphate anions backbone by electrostatic attraction. The DNA chain was molecularly labelled with a quencher group Dabcyl (4-(4-(dimethylamino)phenylazo)benzoic acid). Before the telomerase extension reaction, the emission of tetraphenylsilole (**40**) was quenched efficiently due to energy transfer from **40** to the quencher. After telomerase binding to more extension repeat units, the donors were relatively far away from the quencher, thus the fluorescence could be turned on with a significant signal increase. This probe enabled sensitive detection of telomerase activity in living cells [104]. These authors then utilized the TPE-Py fluorophore with long-wavelength emission and high fluorescence quantum yield to develop a facile, fast-responsive and photostable probe for telomerase sensing [105]. This probe not only achieved monitoring of intracellular telomerase activities, but also evaluated the pesticide effect of telomerase-targeted drugs in vivo.

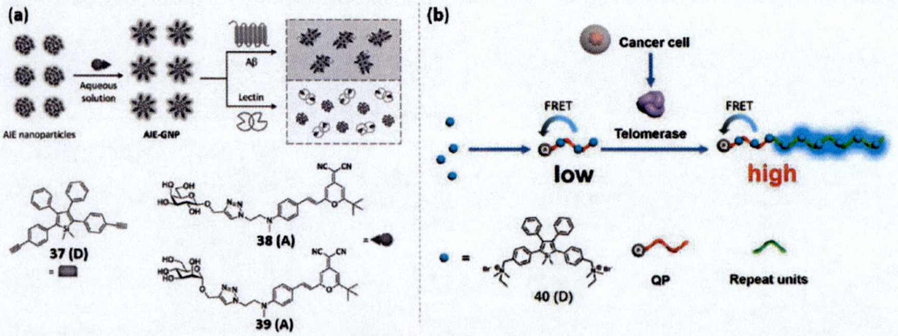

Fig. 19 **a** Ratiometric self-assembled nanoparticles with FRET and AIE properties for *β*-amyloid peptide detection and discrimination from lectins. Reproduced with permission from Ref. [103] © 2016 Wiley. **b** Facile, fast-responsive and photostable FRET system for telomerase activity sensing in living cells. Reproduced with permission from Ref. [104]. © 2016 American Chemical Society

3.2 TBET-Based Fluorescent Probes

TBET is another useful choice to provide ratiometric fluorescence signals. This design strategy requires that the energy gap between the highest occupied molecular orbital (HOMO) and lowest unoccupied molecular orbital (LUMO) of the donor should be higher than that of the acceptor. Compared to FRET, it is not necessary to pursue spectral overlap of donor emission and acceptor absorption, or expend much effort to explore an ideal intermolecular distance. The TBET system can have a larger pseudo-Stokes shift and higher energy transfer efficiency (~100-fold) than conventional FRET probes [106]. However, the structural design of TBET systems is limited in utilizing π-bonds to conjugate energy donor and acceptor, and the distance between donor and acceptor cannot be regulated to tune energy transfer efficiency, making it less applied compared to FRET.

Introducing AIEgens on the phenyl of rhodamine is a convenient strategy to afford TBET-controlled fluorescent probes. Tang et al. developed AIE-based ratiometric sensors for Hg^{2+} detection through dark through-bond energy transfer (DTBET) [107]. High energy transfer efficiency ($\Phi_{ET}=99\%$) and large pseudo-Stokes shifts (> 280 nm) were observed. In the presence of Hg^{2+}, the sensors exhibited well-resolved emission signals and > 6000-fold fluorescence enhancements at 595 nm, which contributed to the low detection limit of Hg^{2+} at 0.03 ppb. The DTBET strategy solved the ACQ problem and could be explored for various ratiometric probes for analyte sensing. Based on the DTBET process, Ho et al. developed a TPE-rhodamine probe for Hg^{2+} detection and imaging [108]. Probe **41** displayed an ultra-sensitive sensing ability for Hg^{2+} with a detection limit as low as 43 pM (Fig. 20). Distinct fluorescence changes, wide pH

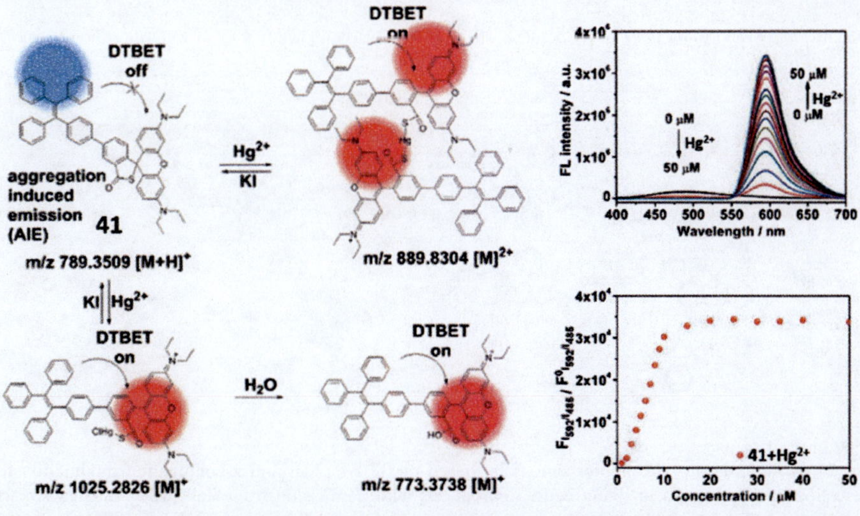

Fig. 20 TPE-rhodamine-based ratiometric fluorescence sensor for Hg^{2+} detection and bioimaging applications. Reproduced with permission from Ref. [108] © 2019 The Royal Society of Chemistry

application range, excellent selectivity and fast response-time (< 10 s) endowed **41** with great potential for biological applications. In addition to TPE, hexaphenylbenzene was also introduced on rhodamine to afford an AIEE-active fluorescent probe for sensing Hg^{2+} in living cells [109].

Liu et al. developed a NIR probe (**42**) based on TBET process between TPE and pH-responsive hemicyanine moieties for monitoring microenvironmental pH variation (Fig. 21a) [109]. As the pH values changed from 7.0 to 3.0, the broken-lactam-induced hemicyanine enabled excitation energy to be accepted from the donor. The fluorescence of TPE decreased and fluorescence of hemicyanine increased simultaneously with a large pseudo-Stokes shift of 361 nm and an intensive fluorescence enhancement of 238-fold at 737 nm. This ratiometric platform could be employed to monitor the pH variation of Hela cells with outstanding photostability and reversibility. Then, based on a similar strategy, the authors further designed another pH probe by taking advantage of TBET between TPE and a cyanine-based pH probe that was responsive to pH changes from 5.0 to 10.7 [110].

Qu et al. [111] developed a nanoprobe for the detection of hypochlorite (ClO^-). Amphiphilic TPE (donor) and phenyl-diketopyrrolopyrrole (acceptor) co-assemble to generate micelles with fluorescence emission at 523 nm. After ClO^- incubation, the structure of the acceptor decomposed, leading to inhibition of the energy transfer process and leaving blue fluorescence from TPE at 478 nm. Sun et al. designed an ultra-sensitive ratiometric fluorescent probe (**43**) for HClO response (Fig. 21b) [112]. After responding to HClO, energy transfer occurred from TPE to the rhodamine moiety through a TBET process followed by colormetric changes in solution. Based on the ultrasensitive limit of detection (LOD) (1.29 nM), this probe was applied successfully to sense exogenous/endogenous HClO in different cell lines.

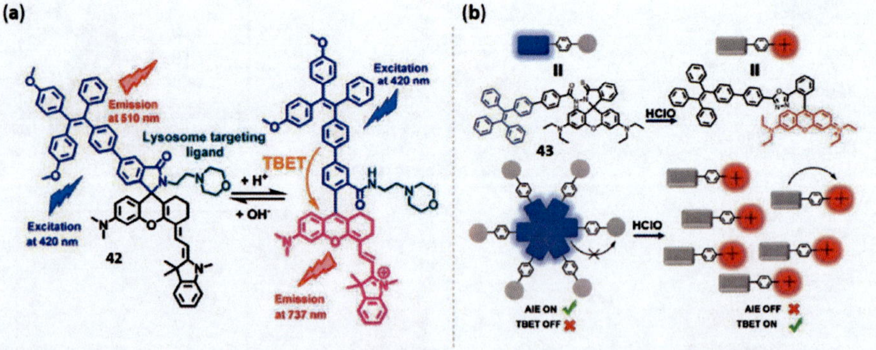

Fig. 21 **a** Ratiometric NIR fluorescent probe based on TBET, AIE and π-conjugation modulation for detection of pH changes in living cells. Reproduced with permission from Ref. [109] © 2018 American Chemical Society. **b** Ultra-sensitive fluorescent probe based on TBET and AIE effect for exogenous/endogenous HOCl detection. Reproduced with permission from Ref. [112] © 2019 The Royal Society of Chemistry

4 Theranostic Systems

Theranostics are dual function systems that offer both therapeutic promise and potential for concurrent diagnosis. Fluorescence-based theranostics enable expectant therapeutic schedules to be accomplished, and simultaneously monitor feedback of therapeutic process at nidus in real time. Here, we summarize two therapeutic strategies with energy transfer process from AIEgens: (a) AIEgens guiding a drug delivery system for chemotherapy; (b) AIEgens as photosensitizers providing functions in fluorescence imaging and photodynamic therapy.

4.1 Imaging-Guided Drug Delivery

Development of fluorophore-marked drug delivery systems for precise imaging and monitoring of drug release is significant for the investigation of pharmacodynamic mechanisms and metabolism processes. Some anticancer drugs, such as Doxorubicin (DOX), can emit fluorescence. They are usually employed as energy acceptors with suitable energy donors to form FRET-based drug delivery systems. In general, such systems can be activated by triggers in pathological states, the drug is released and the FRET process is disturbed with fluorescence variation, allowing real-time monitoring of drug delivery (Fig. 22) [23].

Gao et al. [113] developed a self-assembled nanoparticle for observing DOX delivery. Aggregates of a TPE derivative (donor) had a hydrophobic core to capture DOX (acceptor) to obtain nanoparticles with fluorescence at 590 nm by energy transfer. The resulting nanoparticles were subsequently encased in an amphiphilic and pH responsive cationic amino poly(glyco methacrylate) (PGMA) derivative to regulate drug release. Several folic acids were introduced on the amino groups of PGMA to improve tumor targeting of the nanoparticles. After incubating with tumor tissues, the acid microenvironment or excessive endogenous glutathione (GSH) decomposed the PGMA and damaged the nanoparticle structure, consequently resulting in drug release and fluorescence appearance of TPE at 460 nm since FRET from TPE to DOX was hampered. This study allowed the real-time monitoring of a pH-responsive drug release process.

He et al. [114] fabricated a polymeric micelle-based drug delivery system. The hydrophobic AIE fluorophore (**44**) was esterified with a hydrophilic segment to

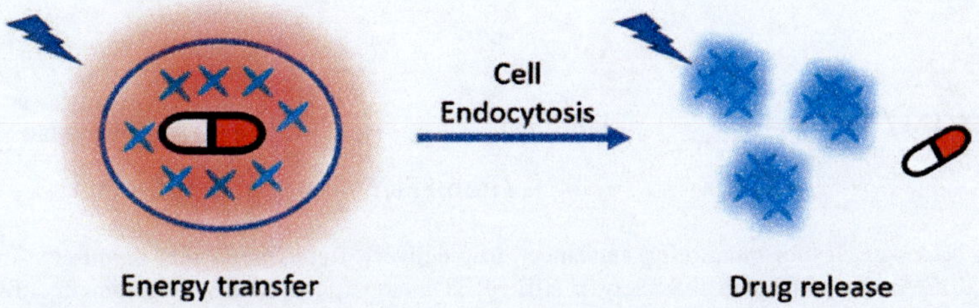

Fig. 22 Illustration of the AIE-FRET system for real-time monitoring of drug delivery

afford an amphiphilic structure, which further self-assembled to form micelles in water (Fig. 23a). DOX could enter the hydrophobic core of the micelles and act as the energy acceptor to trap excitation energy from **44**. The drug release speed of the micelles in pH 5.5 was higher than that in pH 7.4, accompanied with a blue-shifted fluorescence of **44**. These nanoparticles presented good physiological stability, high drug-loading capacity (10.4%) and outstanding encapsulation efficiency (86%).

Gao et al. [115] reported nanoparticles for cancer therapy and real-time monitoring of drug delivery. TPE-CHO and D-peptide (FFKY) were modified on hydrophilic PEG chains with disulfide bonds as the linker to afford the amphiphilic **45**

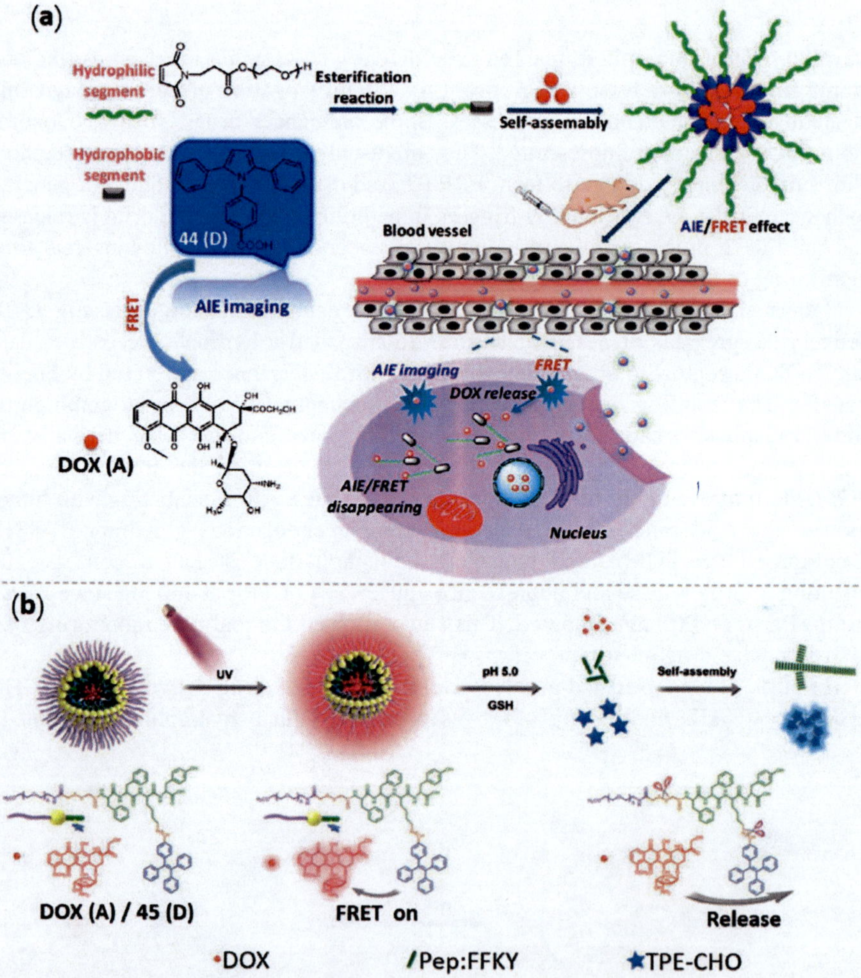

Fig. 23 **a** Nanoparticles for monitoring anticancer drug delivery. Reproduced with permission from Ref. [114] © 2017 American Chemical Society. **b** AIE-FRET nanoparticles with synergistic effect from peptide and DOX for cancer therapy and monitoring drug release. Reproduced with permission from Ref. [115] © 2020 The Royal Society of Chemistry

(Fig. 23b). Self-assemblies of **45** could load DOX in the hydrophobic core. The resultant nanoparticles showed red fluorescence due to the FRET process from TPE-CHO to DOX. When incubated with cancer cells, disulfide bonds were cleaved by the endogenous GSH, thereby releasing DOX accompanied by the appearance of blue fluorescence from TPE-CHO. This platform first detected and quantified changes of FRET signals in A549 cells to monitor the cellular uptake process and intracellular drug release intuitively. Meanwhile, the accumulating peptide-TPE residue might aggregate and interact with a cytoskeleton-associated protein to inhibit cancer cells, thus efficiently promoting the synergistic anticancer efficacy of the nanoparticles.

Wu et al. [116] reported fluorescent polymeric micelles for the encapsulation of DOX. A pentaphenylsilane derivative was polymerized with hydrophilic linkers to produce amphiphilic polymers, which could form micelles to load DOX and emit fluorescence at 559 nm. After cellular uptake, the FRET process is suspended due to DOX release, leading to the appearance of blue fluorescence of the donor at $\lambda_{max} = 503$ nm. Zhao et al. [42] explored redox-responsive polymeric micelles for tracing intracellular curcumin release. TPE was selected as an energy donor, which was chemically decorated on an amphiphilic polymer framework via the amide bond. Subsequently, curcumin as the acceptor was linked on the polymer by connection of a disulfide bond. The polymer self-assembled to micelles with a hydrodynamic size of 125 nm, and the fluorescence of Cur by donor excitation was quenched due to its ACQ effect in the micelles. Disulfide bond cleavage by glutathione (GSH) or tris(2-carboxyethyl)phosphine (TCEP) resulted in drug release and fluorescence enhancement of TPE and curcumin (Fig. 24a).

Cheng et al. [117] developed an AIE-active polymer as a drug carrier for monitoring drug release and cancer therapy. The energy donor and two energy acceptors were conjugated together to construct the fluorescent polymer. Under excitation of 370 nm, a two-stage FRET process occurred whereby excitation energy was first transferred from the fluorene to TPE and subsequently from TPE to the benzothiazole derivative with red fluorescence emission. These drug-loaded nanoparticles

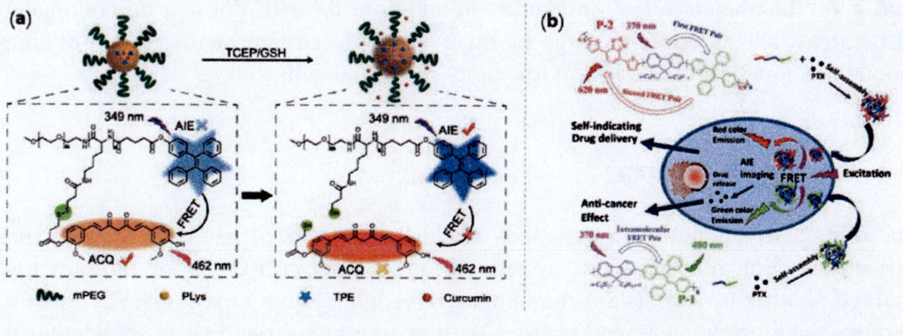

Fig. 24 **a** Redox-responsive polymeric micelles for monitoring curcumin delivery. Reproduced with permission from Ref [42]. © 2018 Wiley. **b** AIE-active polymeric nanoparticles with a two-stage FRET process as drug carriers for cancer therapy. Reproduced with permission from Ref. [117] © 2018 The Royal Society of Chemistry

exhibited effective cytotoxicity and red fluorescence in HeLa cells and A549 cells. (Fig. 24b).

Utilization of host–guest interactions to construct fluorescent scaffolds to enfold drug molecules in the hydrophobic core will form supramolecular nanoparticles that can be transported efficiently in cells. Huang et al. [118] reported a pillararene-based amphiphilic drug-delivery system. Biotin ligands with PEG-linkers were introduced on pillar[5]arene to afford a "host" molecule. TPE as the fluorescence reporter and methyl viologen as the guest molecule, were decorated on polymeric linkers. Host–guest interaction promoted the construction of amphiphilic nanoparticles, whose hydrophobic core could encapsulate DOX as the acceptor and result in fluorescence quenching of TPE by FRET and ACQ effects. DOX was released from the nanoparticles after being triggered by viologen dissociation in low pH or in the presence of reductase. This strategy displayed higher cytotoxicity towards to cancer cells than to normal cells because of the contribution from targeting moieties. Chen et al. [119] further utilized such host–guest systems for tracing drug release processes. They obtained a naphthalene derivative by modifying PEG chains on naphthalene (Fig. 25a). In the self-assembly process, the naphthol derivative interacted with the methyl viologen on polymer **46**, and then co-entered in the cavity of cucurbit[8]uril (CB[8]) to afford nanoparticles encapsulating DOX in aqueous solution with "silent" fluorescence signal due to FRET from TPE to DOX. The drug was released due to instability of the viologen moiety in the presence of reducing agents or in low pH environments. The nanoparticles not only accumulated highly in tumor tissues through the enhanced permeability and retention effect, but also displayed a long blood circulation time.

Wang et al. developed a GSH-responsive nanoparticle for the visualization of drug release [120]. The organic AIE dot (**47**) introduced disulfide bonds by amidation (Fig. 25b). The diverging branches crossed several α-cyclodextrin (α-CD) moieties to obtain nanoparticles with outstanding water-dispersibility and biocompatibility, which subsequently captured DOX for red fluorescence through EET from the AIE dot. Abundant GSH and acidic environments could damage the host–guest framework, leading to DOX release and blue-shifted fluorescence due to the interruption of FRET processes. This drug delivery system exhibited rapid drug release and good therapeutic effect on cancer. In addition, the AIE-dot had no conjugated structures such as benzene rings or other aromatic structures, thereby providing good biocompatibility and low cytotoxicity in normal cells.

4.2 Photodynamic Therapy

In addition to chemotherapy, photodynamic therapy (PDT) has been a promising strategy in clinical applications to cure various tumors. In PDT, upon illumination, excited photosensitizers (PSs) may undergo an intersystem crossing (ISC) to form a more stable triplet state and transfer their energy to surrounding O_2 molecules to generate an excited singlet state oxygen ($^1O_2$), which causes oxidative damage to cells. Traditional PSs tend to interact with adjacent molecules through strong π–π stacking, resulting in their ability to generate reactive oxygen species (ROS) being

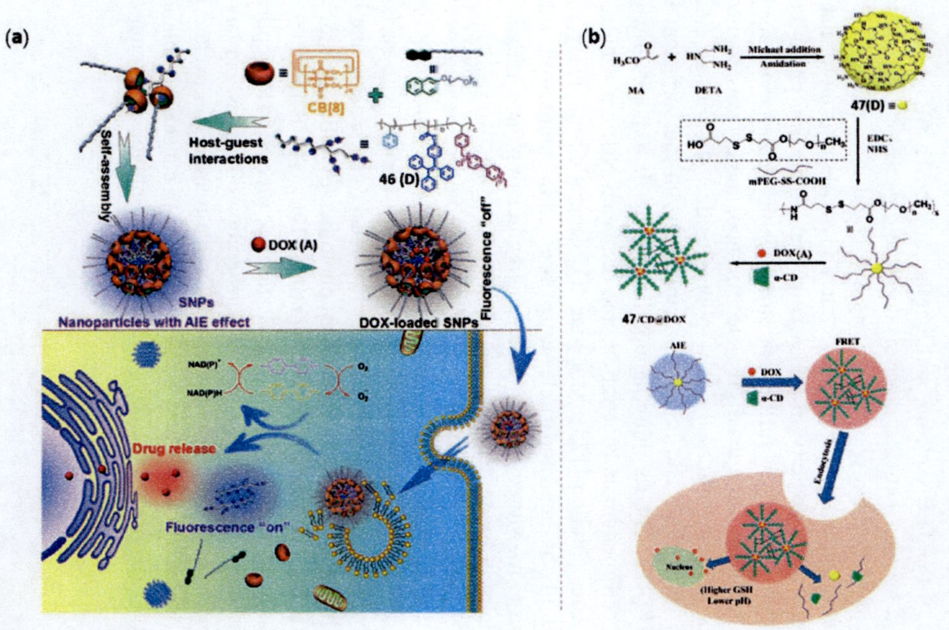

Fig. 25 **a** Cucurbit[8]uril-based amphiphilic nanomedicine self-assembled by host–guest interaction for cancer therapy. Reproduced with permission from Ref. [119] © 2017 American Chemical Society. **b** Stimuli-responsive nanoparticles self-assembled from AIE-dots and α-cyclodextrin for the visualization of drug release processes. Reproduced with permission from Ref [120]. © 2019 American Chemical Society

markedly reduced [121]. AIEgens with a twisted configuration represent a new class of PSs due to the suppression of strong intermolecular interactions. Moreover, they also have the ability to fluoresce, allowing the identification of cancerous lesions and monitoring of therapeutic effects during treatment (Fig. 26) [27, 122].

To endow AIEgens with the capability of PSs, reducing the energy gap between the lowest excited singlet state (S_1) to T_1 (ΔE_{S1-T1}) is required to facilitate the ISC process. To reduce the energy gap on the basis of the TPE skeleton, commonly used strategies include incorporating electron donating and withdrawing moieties on different sides of the TPE skeleton, and lengthening the HOMO–LUMO space by incorporating a π-linker between electron donor and acceptor [123]. Liu et al. reported a series of TPE-based photosensitizers that could form nanocrystals with improved brightness and $^1O_2$ generation efficiency [124]. Furthermore, nanocrystals exhibited higher phototoxicity in HeLa cells than their amorphous counterparts. Subsequently, these authors further optimized the molecular structures of PSs for their NIR emission. In more detail, the electron-poor moiety was connected to the phenyl ring of TPE and afforded a new photosensitizer **48** with an electronic pull–push system (Fig. 27a). The PS was precipitated with the surfactant DSPE-PEG-maleimide (MAL). The resulting nanoparticles were then modified with specific peptides for cell membrane penetration, and displayed NIR emission centered at 820 nm in aqueous solution [125]. Moderate $^1O_2$ generation efficiency and outstanding photostability of the nanoparticles were observed in comparison with other commercial PSs in vitro. Moreover, PS **48** presented efficient photocytotoxicity and low dark cytotoxicity in living cells and mouse models, while also realizing synergetic fluorescence imaging. This study revealed the significance of fluorescent PSs in imaging-guided PDT applications. Besides, liposomes were introduced on TPE-based PSs to yield AIEgen-lipid conjugates. After self-assembly with surfactants, the amphiphilic AIEsome could be used for in vitro cancer cell sensing and in vivo tumor targeting. It performed with satisfactory phototherapy efficiency for cancer cell ablation and antitumor therapy in the meantime [126]. The latter authors also reported a polymeric PS with two-photon excitation for precise PDT applications. The PS **49** possessed good brightness, high $^1O_2$ generation efficiency and large two-photon-absorption (2PA) cross-sections for an effective two-photon excited photodynamic therapy, which exhibited splendid performances in cancer cell ablation and zebrafish liver tumor treatment (Fig. 27b) [127]. The two-photon excitation at long

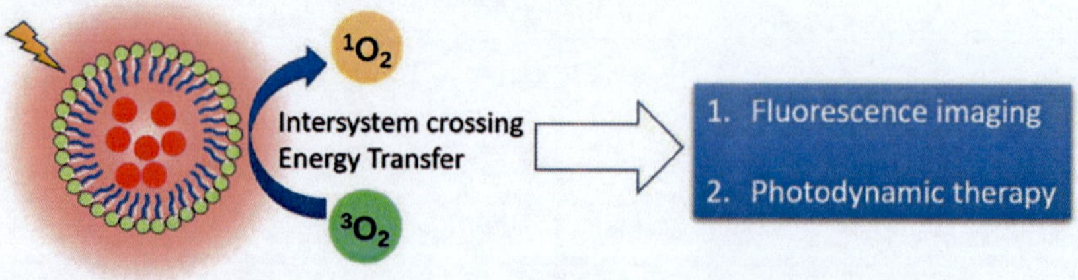

Fig. 26 Illustration of AIEgen-based nanoparticles for fluorescence imaging and photodynamic therapy (PDT)

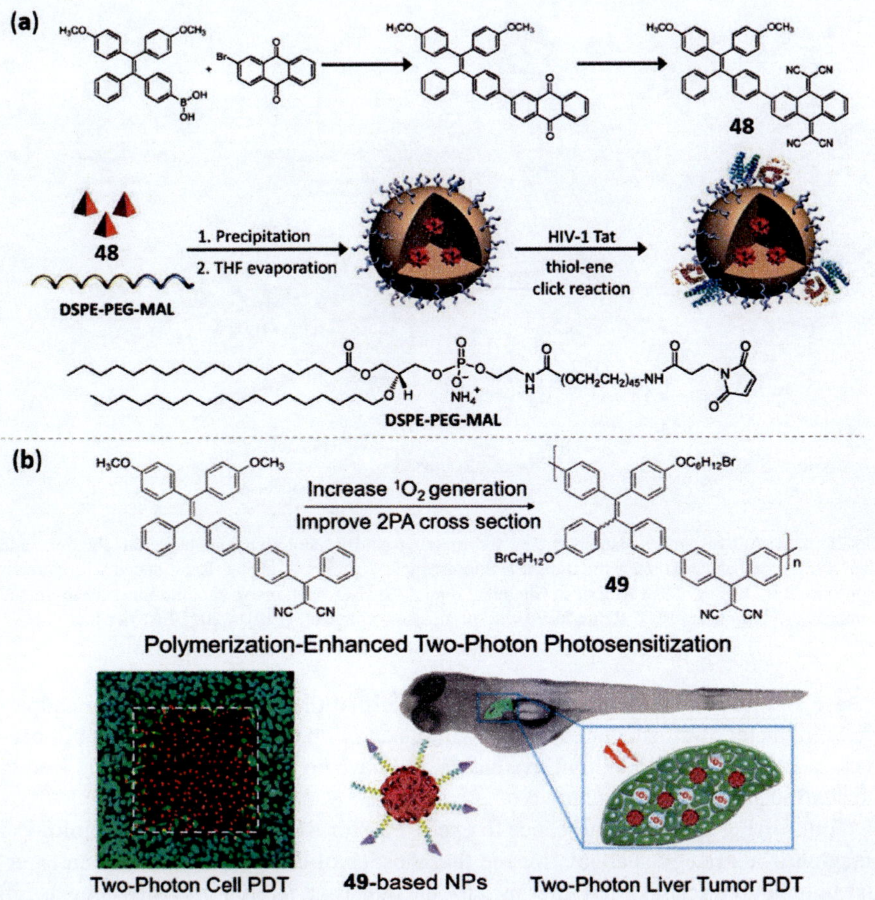

Fig. 27 **a** NIR AIE photosensitizer (PS) for imaging-guided photodynamic anticancer therapy. Reproduced with permission from Ref. [124] © 2016 Wiley. **b** Two-photon excited polymeric PSs for PDT in cancer cells and in zebrafish. Reproduced with permission from Ref. [127] © 2019 American Chemical Society

wavelengths led to high tissue penetrability, decreased tissue damage and negligible biological interference for **49**, thereby being of significance for developing PDT applications.

Liu et al. developed stimuli-responsive nanoparticles for specific imaging-guided photodynamic bacterial ablation in vivo [128]. The amphiphilic TPE-derivative (**50**) with AIE characteristics self-assembled to form nanoparticles that could capture the NIR fluorophore (**51**) into the hydrophobic core (Fig. 28a). Through FRET process, **51** emitted NIR fluorescence by accepting energy transfer from **50**, leading to interruption of the ISC process of **50**. In the presence of $ONOO^-$ or ClO^-, decomposition of **51** hampered the FRET process and resulted in the recovery of fluorescence and ROS generation of **50**. A fluorescence signal at 650 nm from **50** was observed in

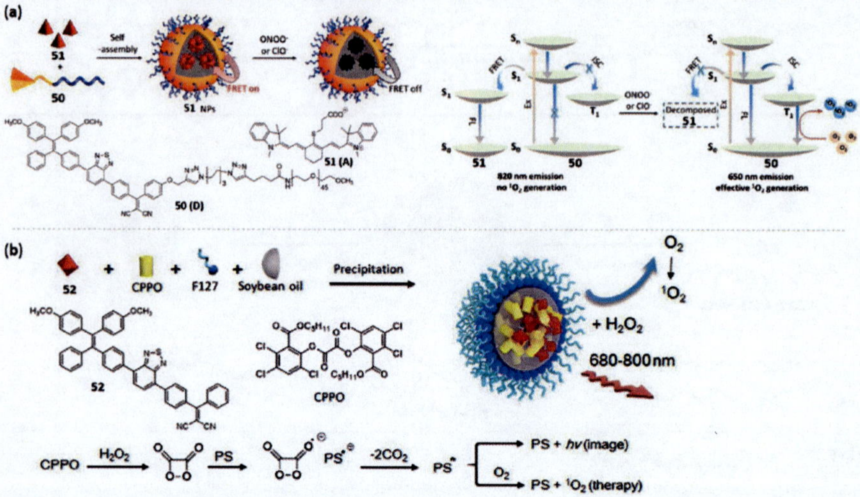

Fig. 28 **a** Stimuli-responsive nanoparticles for imaging-guided photodynamic bacterial ablation, and their working mechanism. *Ex* Excitation, *FL* fluorescence, *S* singlet, *T* triplet. Reproduced with permission from Ref [128]. © 2018 American Chemical Society. **b** H_2O_2-responsive chemiluminescence-guided nanoparticles for cancer PDT. Reproduced with permission from Ref. [129] © 2017 Elsevier Inc.

mouse and its corresponding organs, which verified the responsiveness of nanoparticles in bacterial-affected areas. Meanwhile, nanoparticles were enabled to kill bacteria in infected skin upon light irradiation, but have no dark-cytotoxicity and lead to no significant tissue damage in vivo.

Transferring chemiluminescence to excite PSs for ROS generation and biological imaging is a strategy available for the theranostics of cancer cells. Use of an external light source for PS excitation usually encounters a photon intensity decrease in deep tissue due to the biological self-absorption and diffuse scattering. Chemiluminescence enables AIE-PS to be excited in situ to generate fluorescence signals and ROS, which avoids auto-fluorescence interference in imaging and tissue injury from prolonged light-irradiation. Furthermore, chemiluminescence is generated from the substrate reacting with H_2O_2, which is overexpressed in cancer cells. Hence, H_2O_2-responsive PSs are worthy of wide exploration for PDT applications. Liu et al. [129] explored chemiluminescence-guided nanoparticles for cancer therapy. With the assistance of F127 surfactant, bis[2,4,5-tricholoro-6-(pentyloxycarbonyl)phenyl]oxalate (CPPO) was encapsulated with the AIE-PS **52** (Fig. 28b). The resulting nanoparticles served as a sensitive H_2O_2 probe to precisely monitor tumors in vivo. Meanwhile, sufficient $^1O_2$ was generated from **52** when it was excited by the intermediate after CPPO reacted with H_2O_2. The $^1O_2$ caused tumor cell apoptosis and tumor growth inhibition.

Gao et al. [130] reported a nano-assembly of upconversion nanoparticles (UCNPs) and AIE PS for PDT applications. The UCNPs ($NaYF_4$:Yb/Tm) served as the energy donor performing the fluorescence upconversion process, which could absorb excitation light at 980 nm and emit fluorescence at $\lambda_{max} = 500$ nm. An AIE

PS was chosen as the energy acceptor. After AIEgen and UCNPs were co-assembled with Ginsenoside Rf saponin, the nanoparticles displayed red fluorescence emission and generated ROS from the AIE acceptor. When incubated with HeLa cells, the nanoparticles showed appreciable biocompatibility, low dark cytotoxicity, and high photocytotoxicity with a half maximal inhibitory concentration (IC_{50}) of 4.82 μg/ml. This PDT nanoparticle has potential in biomedical and diagnostic applications.

5 Conclusion

In this review, we have summarized the approaches used to design AIE-based energy transfer systems in the fields of light-harvesting, fluorescent probes and theranostic systems. Firstly, we discussed the construction of artificial LHSs with AIE through non-covalent interactions categorized according to the driving forces of molecular self-assembly, including hydrophobic interactions, host–guest interactions, hydrogen-bonding and metal coordination. Then, we introduced the structural design of EET-based fluorescent probes with AIE and how their fluorescence signals were modulated to detect target analytes. Finally, we described the design of multifunctional theranostic systems by introducing anticancer drugs or AIE photosensitizers to achieve imaging-guided drug delivery and photodynamic therapy. Despite the rapid development and impressive performance of AIE-based energy transfer systems, there are still challenges and perhaps future opportunities to further advance AIE-EET systems for real applications:

1. Most of the AIEgens as donors are based on the TPE moiety, whose absorption spectra are located mainly in the short wavelength region. The short wavelength excitation is unfavorable for bioapplications because of their high tissue autofluorescence, low photon penetration in tissues and photo-damage to biological samples. Although two-photon excitation can address these issues, the requirement for an expensive pulsed laser source still hampers its wider exploitation. The development of AIEgens with absorption at long wavelength is promising for achieving deep tissue imaging and theranostics in vivo.
2. Fluorescence imaging in the second NIR (NIR-II, 1000–1700 nm) window is an emerging research field [131, 132]. AIEgen-based NIR-II bioimaging exhibits deep tissue imaging with high spatial resolution and enhanced precision, which shows great potential in advanced imaging-guided surgery [133]. However, the maximum emission wavelength of the existing NIR-II AIEgens are below 1100 nm, with quantum yields below 1%. AIE-based EET systems might provide an opportunity for developing advanced NIR-II imaging to obtain emission at 1100–1500 nm with enhanced fluorescence signal.
3. In comparison with single-molecular systems, self-assembled energy transfer systems are composed of multiple components. The nanostructure might decompose in complex biological environments and during metabolism processes. To achieve energy transfer systems with high stability, a balance of a covalent synthesis strategy and a non-covalent self-assembly protocol is an alternative approach.

In summary, considering the widespread potential applications, we can expect that the combination of AIE and EET will provide new opportunities to facilitate their future application in biological analysis, biomedicine, and luminescent materials. We hope that, with this review, we have provided a knowledge base of the area and inspired the reader to contribute to the further development of this field.

Acknowledgments This work was supported financially by National Natural Science Foundation of China (21971023 and 21525206) the start-up funding from Beijing Advanced Innovation Center for Soft Matter Science and Engineering, Beijing University of Chemical Technology.

Declarations

Conflict of interest The authors declare that they have no conflict of interest.

References

1. Nadort A, Zhao J, Goldys EM (2016) Lanthanide upconversion luminescence at the nanoscale: fundamentals and optical properties. Nanoscale 8(27):13099–13130. https://doi.org/10.1039/C5NR08477F
2. Mirkovic T, Ostroumov EE, Anna JM, van Grondelle R, Govindjee SGD (2017) Light absorption and energy transfer in the antenna complexes of photosynthetic organisms. Chem Rev 117(2):249–293. https://doi.org/10.1021/acs.chemrev.6b00002
3. Jiang Y, McNeill J (2017) Light-harvesting and amplified energy transfer in conjugated polymer nanoparticles. Chem Rev 117(2):838–859. https://doi.org/10.1021/acs.chemrev.6b00419
4. Yuan L, Lin W, Zheng K, Zhu S (2013) FRET-based small-molecule fluorescent probes: rational design and bioimaging applications. Acc Chem Res 46(7):1462–1473. https://doi.org/10.1021/ar300273v
5. Wu L, Huang C, Emery BP, Sedgwick AC, Bull SD, He XP, Tian H, Yoon J, Sessler JL, James TD (2020) Forster resonance energy transfer (FRET)-based small-molecule sensors and imaging agents. Chem Soc Rev 49(15):5110–5139. https://doi.org/10.1039/c9cs00318e
6. Peng HQ, Niu LY, Chen YZ, Wu LZ, Tung CH, Yang QZ (2015) Biological applications of supramolecular assemblies designed for excitation energy transfer. Chem Rev 115(15):7502–7542. https://doi.org/10.1021/cr5007057
7. Jia X, Chen Q, Yang Y, Tang Y, Wang R, Xu Y, Zhu W, Qian X (2016) FRET-based mito-specific fluorescent probe for ratiometric detection and imaging of endogenous peroxynitrite: dyad of Cy3 and Cy5. J Am Chem Soc 138(34):10778–10781. https://doi.org/10.1021/jacs.6b06398
8. Aron AT, Loehr MO, Bogena J, Chang CJ (2016) An endoperoxide reactivity-based FRET probe for ratiometric fluorescence imaging of labile iron pools in living cells. J Am Chem Soc 138(43):14338–14346. https://doi.org/10.1021/jacs.6b08016
9. Kim HS, Lee SH, Lee JY, Yoo S, Suh MC (2019) Mitigating the trade-off between triplet harvesting and roll-off by opening a dexter-type channel in OLEDs. J Phys Chem C 123(30):18283–18293. https://doi.org/10.1021/acs.jpcc.9b05007
10. Lai Y-S, Pan F, Su Y-H (2018) Firefly-like water splitting cells based on FRET phenomena with ultrahigh performance over 12%. ACS Appl Mater Interfaces 10(5):5007–5013. https://doi.org/10.1021/acsami.7b18003
11. Wu D, Sedgwick AC, Gunnlaugsson T, Akkaya EU, Yoon J, James TD (2017) Fluorescent chemosensors: the past, present and future. Chem Soc Rev 46(23):7105–7123. https://doi.org/10.1039/c7cs00240h
12. Ma L, Fang W-H, Shen L, Chen X (2019) Regulatory mechanism and kinetic assessment of energy transfer catalysis mediated by visible light. ACS Catal 9(4):3672–3684. https://doi.org/10.1021/acscatal.9b00146

13. Wang X, Gao Z, Zhu J, Gao Z, Wang F (2016) Aggregation induced emission of a cyanostilbene amphiphile as a novel platform for FRET-based ratiometric sensing of mercury ions in water. Polym Chem 7(33):5217–5220. https://doi.org/10.1039/c6py01109h
14. Teunissen AJP, Pérez-Medina C, Meijerink A, Mulder WJM (2018) Investigating supramolecular systems using Förster resonance energy transfer. Chem Soc Rev 47(18):7027–7044. https://doi.org/10.1039/C8CS00278A
15. Feng H-T, Yuan Y-X, Xiong J-B, Zheng Y-S, Tang BZ (2018) Macrocycles and cages based on tetraphenylethylene with aggregation-induced emission effect. Chem Soc Rev 47(19):7452–7476. https://doi.org/10.1039/C8CS00444G
16. Li J, Wang J, Li H, Song N, Wang D, Tang BZ (2020) Supramolecular materials based on AIE luminogens (AIEgens): construction and applications. Chem Soc Rev 49(4):1144–1172. https://doi.org/10.1039/C9CS00495E
17. Peng H-Q, Liu B, Wei P, Zhang P, Zhang H, Zhang J, Li K, Li Y, Cheng Y, Lam JWY, Zhang W, Lee C-S, Tang BZ (2019) Visualizing the initial step of self-assembly and the phase transition by stereogenic amphiphiles with aggregation-induced emission. ACS Nano 13(1):839–846. https://doi.org/10.1021/acsnano.8b08358
18. Gao M, Tang BZ (2017) Fluorescent sensors based on aggregation-induced emission: recent advances and perspectives. ACS Sens 2(10):1382–1399. https://doi.org/10.1021/acssensors.7b00551
19. Luo M, Wang S, Li C, Miao W, Ma X (2019) Aggregation-induced emission organogel formed by both sonication and thermal processing based on tetraphenylethylene and cholesterol derivative. Dyes Pigments 165:436–443. https://doi.org/10.1016/j.dyepig.2019.02.041
20. Wang B, Wu Z, Fang B, Yin M (2020) Blue-shifted mechanochromism of a dimethoxynaphthalene-based crystal with aggregation-induced emission. Dyes Pigments 182:108618. https://doi.org/10.1016/j.dyepig.2020.108618
21. Xu L, Yu Y, Shi J, Cui W, Lv X, Cang M, Sun Q, Xue S, Yang W (2020) Highly efficient nondoped blue organic light-emitting diodes based on a star-group tetraphenylethylene-substituted aggregation-induced-emission-active organic fluorescent small molecules. Dyes Pigments 175:108082. https://doi.org/10.1016/j.dyepig.2019.108082
22. Hu R, Leung NL, Tang BZ (2014) AIE macromolecules: syntheses, structures and functionalities. Chem Soc Rev 43(13):4494–4562. https://doi.org/10.1039/c4cs00044g
23. Yuan Y, Liu B (2017) Visualization of drug delivery processes using AIEgens. Chem Sci 8(4):2537–2546. https://doi.org/10.1039/c6sc05421h
24. Mei J, Leung NL, Kwok RT, Lam JW, Tang BZ (2015) Aggregation-induced emission: together we shine, united we soar! Chem Rev 115(21):11718–11940. https://doi.org/10.1021/acs.chemrev.5b00263
25. Gu K, Qiu W, Guo Z, Yan C, Zhu S, Yao D, Shi P, Tian H, Zhu WH (2019) An enzyme-activatable probe liberating AIEgens: on-site sensing and long-term tracking of beta-galactosidase in ovarian cancer cells. Chem Sci 10(2):398–405. https://doi.org/10.1039/c8sc04266g
26. Zhang W, Huang Y, Chen Y, Zhao E, Hong Y, Chen S, Lam JWY, Chen Y, Hou J, Tang BZ (2019) Amphiphilic tetraphenylethene-based pyridinium salt for selective cell-membrane imaging and room-light-induced special reactive oxygen species generation. ACS Appl Mater Interfaces 11(11):10567–10577. https://doi.org/10.1021/acsami.9b00643
27. Chen H, Li S, Wu M, Kenry HZ, Lee CS, Liu B (2020) Membrane-anchoring photosensitizer with aggregation-induced emission characteristics for combating multidrug-resistant bacteria. Angew Chem Int Ed Engl 59(2):632–636. https://doi.org/10.1002/anie.201907343
28. Chen Y, Lam JWY, Kwok RTK, Liu B, Tang BZ (2019) Aggregation-induced emission: fundamental understanding and future developments. Mater Horizons 6(3):428–433. https://doi.org/10.1039/c8mh01331d
29. He Z, Ke C, Tang BZ (2018) Journey of aggregation-induced emission research. ACS Omega 3(3):3267–3277. https://doi.org/10.1021/acsomega.8b00062
30. Michel H, Epp O, Deisenhofer J (1986) Pigment–protein interactions in the photosynthetic reaction centre from *Rhodopseudomonas viridis*. EMBO J 5(10):2445–2451
31. Wasielewski MR (1992) Photoinduced electron transfer in supramolecular systems for artificial photosynthesis. Chem Rev 92(3):435–461. https://doi.org/10.1021/cr00011a005
32. Miller RA, Presley AD, Francis MB (2007) Self-assembling light-harvesting systems from synthetically modified tobacco mosaic virus coat proteins. J Am Chem Soc 129(11):3104–3109. https://doi.org/10.1021/ja063887t

33. Kühlbrandt W (1995) Many wheels make light work. Nature 374(6522):497–498. https://doi.org/10.1038/374497a0
34. Sen E, Meral K, Atilgan S (2016) From dark to light to fluorescence resonance energy transfer (FRET): polarity-sensitive aggregation-induced emission (AIE)-active tetraphenylethene-fused BODIPY dyes with a very large pseudo-stokes shift. Chem 22(2):736–745. https://doi.org/10.1002/chem.201503457
35. Yao J, Yang M, Duan Y (2014) Chemistry, biology, and medicine of fluorescent nanomaterials and related systems: new insights into biosensing, bioimaging, genomics, diagnostics, and therapy. Chem Rev 114(12):6130–6178. https://doi.org/10.1021/cr200359p
36. Dong R, Pang Y, Su Y, Zhu X (2015) Supramolecular hydrogels: synthesis, properties and their biomedical applications. Biomater Sci 3(7):937–954. https://doi.org/10.1039/c4bm00448e
37. Li C, Zhang J, Zhang S, Zhao Y (2019) Efficient light-harvesting systems with tunable emission through controlled precipitation in confined nanospace. Angew Chem Int Ed Engl 58(6):1643–1647. https://doi.org/10.1002/anie.201812146
38. Li Y, Dong Y, Cheng L, Qin C, Nian H, Zhang H, Yu Y, Cao L (2019) Aggregation-induced emission and light-harvesting function of tetraphenylethene-based tetracationic dicyclophane. J Am Chem Soc 141(21):8412–8415. https://doi.org/10.1021/jacs.9b02617
39. Geng J, Zhu Z, Qin W, Ma L, Hu Y, Gurzadyan GG, Tang BZ, Liu B (2014) Near-infrared fluorescence amplified organic nanoparticles with aggregation-induced emission characteristics for in vivo imaging. Nanoscale 6(2):939–945. https://doi.org/10.1039/c3nr04243j
40. Liu W, Wang Y, Han X, Lu P, Zhu L, Sun C, Qian J, He S (2018) Fluorescence resonance energy transfer (FRET) based nanoparticles composed of AIE luminogens and NIR dyes with enhanced three-photon near-infrared emission for in vivo brain angiography. Nanoscale 10(21):10025–10032. https://doi.org/10.1039/c8nr00066b
41. Tang P, Wang Y, Wang K (2020) Preparation of high-efficiency near-infrared aggregation-induced emission nanoparticles based on FRET and their use in bio-imaging. Methods Appl Fluoresc 8(1):015007. https://doi.org/10.1088/2050-6120/ab6704
42. Wang X, Li J, Yan Q, Chen Y, Fan A, Wang Z, Zhao Y (2018) In situ probing intracellular drug release from redox-responsive micelles by united FRET and AIE. Macromol Biosci 18(3):170039. https://doi.org/10.1002/mabi.201700339
43. Xu L, Wang Z, Wang R, Wang L, He X, Jiang H, Tang H, Cao D, Tang BZ (2019) A Conjugated polymeric supramolecular network with aggregation-induced emission enhancement: an efficient light-harvesting system with an ultrahigh antenna effect. Angew Chem Int Ed Engl 59:9908–9913. https://doi.org/10.1002/anie.201907678
44. Wang Y, Xu JF, Chen YZ, Niu LY, Wu LZ, Tung CH, Yang QZ (2014) Photoresponsive supramolecular self-assembly of monofunctionalized pillar[5]arene based on stiff stilbene. Chem Commun 50(53):7001–7003. https://doi.org/10.1039/c4cc02760d
45. Lv Y, Liu P, Ding H, Wu Y, Yan Y, Liu H, Wang X, Huang F, Zhao Y, Tian Z (2015) Conjugated polymer-based hybrid nanoparticles with two-photon excitation and near-infrared emission features for fluorescence bioimaging within the biological window. ACS Appl Mater Interfaces 7(37):20640–20648. https://doi.org/10.1021/acsami.5b05150
46. Zhong W, Zeng X, Chen J, Hong Y, Xiao L, Zhang P (2017) Photoswitchable fluorescent polymeric nanoparticles for rewritable fluorescence patterning and intracellular dual-color imaging with AIE-based fluorogens as FRET donors. Polym Chem 8(33):4849–4855. https://doi.org/10.1039/c7py00834a
47. Xu Y, Yang W, Yao D, Bian K, Zeng W, Liu K, Wang D, Zhang B (2020) An aggregation-induced emission dye-powered afterglow luminogen for tumor imaging. Chem Sci 11(2):419–428. https://doi.org/10.1039/c9sc04901k
48. Chen XM, Cao Q, Bisoyi HK, Wang M, Yang H, Li Q (2020) An efficient near-infrared emissive artificial supramolecular light-harvesting system for imaging in the Golgi apparatus. Angew Chem Int Ed Engl 59(26):10493–10497. https://doi.org/10.1002/anie.202003427
49. Kamtekar KT, Monkman AP, Bryce MR (2010) Recent advances in white organic light-emitting materials and devices (WOLEDs). Adv Mater 22(5):572–582. https://doi.org/10.1002/adma.200902148
50. Farinola GM, Ragni R (2011) Electroluminescent materials for white organic light emitting diodes. Chem Soc Rev 40(7):3467–3482. https://doi.org/10.1039/c0cs00204f

51. Geng W-C, Liu Y-C, Wang Y-Y, Xu Z, Zheng Z, Yang C-B, Guo D-S (2017) A self-assembled white-light-emitting system in aqueous medium based on a macrocyclic amphiphile. Chem Commun 53(2):392–395. https://doi.org/10.1039/C6CC09079F
52. Gong Q, Hu Z, Deibert BJ, Emge TJ, Teat SJ, Banerjee D, Mussman B, Rudd ND, Li J (2014) Solution processable MOF yellow phosphor with exceptionally high quantum efficiency. J Am Chem Soc 136(48):16724–16727. https://doi.org/10.1021/ja509446h
53. Yang Q-Y, Lehn J-M (2014) Bright white-light emission from a single organic compound in the solid state. Angew Chem Int Ed Engl 53(18):4572–4577. https://doi.org/10.1002/anie.201400155
54. Zhang Z, Chen Y-A, Hung W-Y, Tang W-F, Hsu Y-H, Chen C-L, Meng F-Y, Chou P-T (2016) Control of the reversibility of excited-state intramolecular proton transfer (ESIPT) reaction: host-polarity tuning white organic light emitting diode on a new thiazolo[5,4-d]thiazole ESIPT system. Chem Mater 28(23):8815–8824. https://doi.org/10.1021/acs.chemmater.6b04707
55. Samanta S, Manna U, Das G (2017) White-light emission from simple AIE–ESIPT-excimer tripled single molecular system. New J Chem 41(3):1064–1072. https://doi.org/10.1039/C6NJ03070J
56. Vijayakumar C, Sugiyasu K, Takeuchi M (2011) Oligofluorene-based electrophoretic nanoparticles in aqueous medium as a donor scaffold for fluorescence resonance energy transfer and white-light emission. Chem Sci 2(2):291–294. https://doi.org/10.1039/C0SC00343C
57. Giansante C, Raffy G, Schäfer C, Rahma H, Kao M-T, Olive AGL, Del Guerzo A (2011) White-light-emitting self-assembled nanofibers and their evidence by microspectroscopy of individual objects. J Am Chem Soc 133(2):316–325. https://doi.org/10.1021/ja106807u
58. Ni X-L, Chen S, Yang Y, Tao Z (2016) Facile cucurbit[8]uril-based supramolecular approach to fabricate tunable luminescent materials in aqueous solution. J Am Chem Soc 138(19):6177–6183. https://doi.org/10.1021/jacs.6b01223
59. Jing YN, Li SS, Su M, Bao H, Wan WM (2019) Barbier hyperbranching polymerization-induced emission toward facile fabrication of white light-emitting diode and light-harvesting film. J Am Chem Soc 141(42):16839–16848. https://doi.org/10.1021/jacs.9b08065
60. Bottari G, Trukhina O, Ince M, Torres T (2012) Towards artificial photosynthesis: supramolecular, donor–acceptor, porphyrin- and phthalocyanine/carbon nanostructure ensembles. Coordin Chem Rev 256(21):2453–2477. https://doi.org/10.1016/j.ccr.2012.03.011
61. Lou X-Y, Song N, Yang Y-W (2017) Fluorescence resonance energy transfer systems in supramolecular macrocyclic chemistry. Molecules 22(10):1640
62. Zhang M, Yin X, Tian T, Liang Y, Li W, Lan Y, Li J, Zhou M, Ju Y, Li G (2015) AIE-induced fluorescent vesicles containing amphiphilic binding pockets and the FRET triggered by host–guest chemistry. Chem Commun 51(50):10210–10213. https://doi.org/10.1039/c5cc02377g
63. Wang S, Ye J-H, Han Z, Fan Z, Wang C, Mu C, Zhang W, He W (2017) Highly efficient FRET from aggregation-induced emission to BODIPY emission based on host–guest interaction for mimicking the light harvesting system. RSC Adv 7(57):36021–36025. https://doi.org/10.1039/c7ra05925f
64. Li JJ, Chen Y, Yu J, Cheng N, Liu Y (2017) A supramolecular artificial light-harvesting system with an ultrahigh antenna effect. Adv Mater 29:30. https://doi.org/10.1002/adma.201701905
65. Fu S, Su X, Li M, Song S, Wang L, Wang D, Tang BZ (2020) Controllable and diversiform topological morphologies of self-assembling supra-amphiphiles with aggregation-induced emission characteristics for mimicking light-harvesting antenna. Adv Sci 7(20):2001909. https://doi.org/10.1002/advs.202001909
66. Guo S, Song Y, He Y, Hu XY, Wang L (2018) Highly efficient artificial light-harvesting systems constructed in aqueous solution based on supramolecular self-assembly. Angew Chem Int Ed Engl 57(12):3163–3167. https://doi.org/10.1002/anie.201800175
67. Kim H-J, Nandajan PC, Gierschner J, Park SY (2018) Light-harvesting fluorescent supramolecular block copolymers based on cyanostilbene derivatives and cucurbit[8]urils in aqueous solution. Adv Func Mater 28(4):1705141. https://doi.org/10.1002/adfm.201705141
68. Wang XH, Song N, Hou W, Wang CY, Wang Y, Tang J, Yang YW (2019) Efficient aggregation-induced emission manipulated by polymer host materials. Adv Mater 31(37):e1903962. https://doi.org/10.1002/adma.201903962
69. Feng H-T, Zheng X, Gu X, Chen M, Lam JWY, Huang X, Tang BZ (2018) White-light emission of a binary light-harvesting platform based on an amphiphilic organic cage. Chem Mater 30(4):1285–1290. https://doi.org/10.1021/acs.chemmater.7b04703
70. Peng H-Q, Zheng X, Han T, Kwok RTK, Lam JWY, Huang X, Tang BZ (2017) Dramatic differences in aggregation-induced emission and supramolecular polymerizability of

tetraphenylethene-based stereoisomers. J Am Chem Soc 139(29):10150–10156. https://doi.org/10.1021/jacs.7b05792

71. Peng HQ, Liu B, Liu J, Wei P, Zhang H, Han T, Qi J, Lam JWY, Zhang W, Tang BZ (2019) "Seeing" and controlling photoisomerization by (Z)-/(E)-isomers with aggregation-induced emission characteristics. ACS Nano 13(10):12120–12126. https://doi.org/10.1021/acsnano.9b06578
72. Sijbesma RP, Beijer FH, Brunsveld L, Folmer BJB, Hirschberg JHKK, Lange RFM, Lowe JKL, Meijer EW (1997) Reversible polymers formed from self-complementary monomers using quadruple hydrogen bonding. Science 278(5343):1601–1604. https://doi.org/10.1126/science.278.5343.1601
73. Söntjens SHM, Sijbesma RP, van Genderen MHP, Meijer EW (2000) Stability and lifetime of quadruply hydrogen bonded 2-ureido-4[1H]-pyrimidinone dimers. J Am Chem Soc 122(31):7487–7493. https://doi.org/10.1021/ja000435m
74. Sijbesma RP, Meijer EW (2003) Quadruple hydrogen bonded systems. Chem Commun 1:5–16. https://doi.org/10.1039/B205873C
75. Peng HQ, Xu JF, Chen YZ, Wu LZ, Tung CH, Yang QZ (2014) Water-dispersible nanospheres of hydrogen-bonded supramolecular polymers and their application for mimicking light-harvesting systems. Chem Commun 50(11):1334–1337. https://doi.org/10.1039/c3cc48618d
76. Peng HQ, Sun CL, Xu JF, Niu LY, Chen YZ, Wu LZ, Tung CH, Yang QZ (2014) Convenient synthesis of functionalized bis-ureidopyrimidinones based on thiol-yne reaction. Chem 20(37):11699–11702. https://doi.org/10.1002/chem.201402955
77. Wang R-F, Peng H-Q, Chen P-Z, Niu L-Y, Gao J-F, Wu L-Z, Tung C-H, Chen Y-Z, Yang Q-Z (2016) A hydrogen-bonded-supramolecular-polymer-based nanoprobe for ratiometric oxygen sensing in living cells. Adv Func Mater 26(30):5419–5425. https://doi.org/10.1002/adfm.201601831
78. Peng H-Q, Sun C-L, Niu L-Y, Chen Y-Z, Wu L-Z, Tung C-H, Yang Q-Z (2016) Supramolecular polymeric fluorescent nanoparticles based on quadruple hydrogen bonds. Adv Func Mater 26(30):5483–5489. https://doi.org/10.1002/adfm.201600593
79. Zhu X, Wang J-X, Niu L-Y, Yang Q-Z (2019) Aggregation-induced emission materials with narrowed emission band by light-harvesting strategy: fluorescence and chemiluminescence imaging. Chem Mater 31(9):3573–3581. https://doi.org/10.1021/acs.chemmater.9b01338
80. Xiao T, Wu H, Sun G, Diao K, Wei X, Li Z-Y, Sun X-Q, Wang L (2020) An efficient artificial light-harvesting system with tunable emission in water constructed from a H-bonded AIE supramolecular polymer and Nile Red. Chem Commun 56:12021–12024. https://doi.org/10.1039/d0cc05077f
81. Heilemann M, Tinnefeld P, Sanchez Mosteiro G, Garcia Parajo M, Van Hulst NF, Sauer M (2004) Multistep energy transfer in single molecular photonic wires. J Am Chem Soc 126(21):6514–6515. https://doi.org/10.1021/ja049351u
82. Wagner RW, Lindsey JS (1994) A molecular photonic wire. J Am Chem Soc 116(21):9759–9760. https://doi.org/10.1021/ja00100a055
83. Dutta PK, Varghese R, Nangreave J, Lin S, Yan H, Liu Y (2011) DNA-directed artificial light-harvesting antenna. J Am Chem Soc 133(31):11985–11993. https://doi.org/10.1021/ja1115138
84. Klein WP, Díaz SA, Buckhout-White S, Melinger JS, Cunningham PD, Goldman ER, Ancona MG, Kuang W, Medintz IL (2018) Utilizing HomoFRET to extend DNA-scaffolded photonic networks and increase light-harvesting capability. Adv Opt Mater 6(1):1700679. https://doi.org/10.1002/adom.201700679
85. Boulais É, Sawaya NPD, Veneziano R, Andreoni A, Banal JL, Kondo T, Mandal S, Lin S, Schlau-Cohen GS, Woodbury NW, Yan H, Aspuru-Guzik A, Bathe M (2018) Programmed coherent coupling in a synthetic DNA-based excitonic circuit. Nat Mater 17(2):159–166. https://doi.org/10.1038/nmat5033
86. Trofymchuk K, Reisch A, Didier P, Fras F, Gilliot P, Mely Y, Klymchenko AS (2017) Giant light-harvesting nanoantenna for single-molecule detection in ambient light. Nat Photon 11(10):657–663. https://doi.org/10.1038/s41566-017-0001-7
87. Han X, Chen Q, Lu H, Ma J, Gao H (2015) Probe intracellular trafficking of a polymeric DNA delivery vehicle by functionalization with an aggregation-induced emissive tetraphenylethene derivative. ACS Appl Mater Interfaces 7(51):28494–28501. https://doi.org/10.1021/acsami.5b09639
88. Li H, Wang C, Hou T, Li F (2017) Amphiphile-mediated ultrasmall aggregation induced emission dots for ultrasensitive fluorescence biosensing. Anal Chem 89(17):9100–9107. https://doi.org/10.1021/acs.analchem.7b01797

89. Zhao Z, Yang H, Deng S, Dong Y, Yan B, Zhang K, Deng R, He Q (2019) Intrinsic conformation response-leveraged aptamer probe based on aggregation-induced emission dyes for aflatoxin B1 detection. Dyes Pigments 171:107767. https://doi.org/10.1016/j.dyepig.2019.107767
90. Malinovskii VL, Wenger D, Häner R (2010) Nucleic acid-guided assembly of aromatic chromophores. Chem Soc Rev 39(2):410–422. https://doi.org/10.1039/B910030J
91. Dutta PK, Levenberg S, Loskutov A, Jun D, Saer R, Beatty JT, Lin S, Liu Y, Woodbury NW, Yan H (2014) A DNA-directed light-harvesting/reaction center system. J Am Chem Soc 136(47):16618–16625. https://doi.org/10.1021/ja509018g
92. Garo F, Häner R (2012) A DNA-based light-harvesting antenna. Angew Chem Int Ed Engl 51(4):916–919. https://doi.org/10.1002/anie.201103295
93. Chen L-J, Yang H-B (2018) Construction of stimuli-responsive functional materials via hierarchical self-assembly involving coordination interactions. Acc Chem Res 51(11):2699–2710. https://doi.org/10.1021/acs.accounts.8b00317
94. Wei P, Yan X, Huang F (2015) Supramolecular polymers constructed by orthogonal self-assembly based on host–guest and metal–ligand interactions. Chem Soc Rev 44(3):815–832. https://doi.org/10.1039/C4CS00327F
95. Cook TR, Stang PJ (2015) Recent developments in the preparation and chemistry of metallacycles and metallacages via coordination. Chem Rev 115(15):7001–7045. https://doi.org/10.1021/cr5005666
96. Acharyya K, Bhattacharyya S, Sepehrpour H, Chakraborty S, Lu S, Shi B, Li X, Mukherjee PS, Stang PJ (2019) Self-assembled fluorescent Pt(II) metallacycles as artificial light-harvesting systems. J Am Chem Soc 141(37):14565–14569. https://doi.org/10.1021/jacs.9b08403
97. Zhang Z, Zhao Z, Hou Y, Wang H, Li X, He G, Zhang M (2019) Aqueous platinum(II)-cage-based light-harvesting system for photocatalytic cross-coupling hydrogen evolution reaction. Angew Chem Int Ed Engl 58(26):8862–8866. https://doi.org/10.1002/anie.201904407
98. Stryer L, Haugland RP (1967) Energy transfer: a spectroscopic ruler. Proc Natl Acad Sci USA 58(2):719–726. https://doi.org/10.1073/pnas.58.2.719
99. Zhang L, Gao J, Qi A, Gao Y (2020) A novel DRET and FRET combined fluorescent molecule and its applications in sensing and bioimaging. Sens Actuat B-Chem 320:128457. https://doi.org/10.1016/j.snb.2020.128457
100. Guan P, Yang B, Liu B (2020) Fabricating a fluorescence resonance energy transfer system with AIE molecular for sensitive detection of Cu(II) ions. Spectrochim Acta A Mol Biomol Spectrosc 225:117604. https://doi.org/10.1016/j.saa.2019.117604
101. Dong L, Fu M, Liu L, Han H-H, Zang Y, Chen G-R, Li J, He X-P, Vidal S (2020) Supramolecular assembly of TPE-Based glycoclusters with dicyanomethylene-4H-pyran (DM) fluorescent probes improve their properties for peroxynitrite sensing and cell imaging. Chem-Eur J 26(63):14445–14452. https://doi.org/10.1002/chem.202002772
102. Yuan Y, Zhang R, Cheng X, Xu S, Liu B (2016) A FRET probe with AIEgen as the energy quencher: dual signal turn-on for self-validated caspase detection. Chem Sci 7(7):4245–4250. https://doi.org/10.1039/c6sc00055j
103. Zhang JD, Mei J, Hu XL, He XP, Tian H (2016) Ratiometric detection of beta-amyloid and discrimination from lectins by a supramolecular AIE glyconanoparticle. Small 12(47):6562–6567. https://doi.org/10.1002/smll.201601470
104. Zhuang Y, Zhang M, Chen B, Duan R, Min X, Zhang Z, Zheng F, Liang H, Zhao Z, Lou X, Xia F (2015) Quencher group induced high specificity detection of telomerase in clear and bloody urines by AIEgens. Anal Chem 87(18):9487–9493. https://doi.org/10.1021/acs.analchem.5b02699
105. Zhuang Y, Huang F, Xu Q, Zhang M, Lou X, Xia F (2016) Facile, fast-responsive, and photostable imaging of telomerase activity in living cells with a fluorescence turn-on manner. Anal Chem 88(6):3289–3294. https://doi.org/10.1021/acs.analchem.5b04756
106. Qiu J, Jiang S, Guo H, Yang F (2018) An AIE and FRET-based BODIPY sensor with large Stoke shift: novel pH probe exhibiting application in CO_3^{2-} detection and living cell imaging. Dyes Pigments 157:351–358. https://doi.org/10.1016/j.dyepig.2018.05.013
107. Chen Y, Zhang W, Cai Y, Kwok RTK, Hu Y, Lam JWY, Gu X, He Z, Zhao Z, Zheng X, Chen B, Gui C, Tang BZ (2017) AIEgens for dark through-bond energy transfer: design, synthesis, theoretical study and application in ratiometric Hg(2+) sensing. Chem Sci 8(3):2047–2055. https://doi.org/10.1039/c6sc04206f

108. Jiang Y, Duan Q, Zheng G, Yang L, Zhang J, Wang Y, Zhang H, He J, Sun H, Ho D (2019) An ultra-sensitive and ratiometric fluorescent probe based on the DTBET process for Hg(2+) detection and imaging applications. Analyst 144(4):1353–1360. https://doi.org/10.1039/c8an02126k
109. Wang J, Xia S, Bi J, Fang M, Mazi W, Zhang Y, Conner N, Luo FT, Lu HP, Liu H (2018) Ratiometric near-infrared fluorescent probes based on through-bond energy transfer and pi-conjugation modulation between tetraphenylethene and hemicyanine moieties for sensitive detection of pH changes in live cells. Bioconjug Chem 29(4):1406–1418. https://doi.org/10.1021/acs.bioconjchem.8b00111
110. Fang M, Xia S, Bi J, Wigstrom TP, Valenzano L, Wang J, Mazi W, Tanasova M, Luo FT, Liu H (2018) A cyanine-based fluorescent cassette with aggregation-induced emission for sensitive detection of pH changes in live cells. Chem Commun 54(9):1133–1136. https://doi.org/10.1039/c7cc08986d
111. Nie K, Yuan Y, Peng X, Song J, Qu J (2020) A diketopyrrolopyrrole-based hybrid organic nanoprobe for ratiometric imaging of endogenous hypochlorite in live cells. Sens Actuat B-Chem 307:127632. https://doi.org/10.1016/j.snb.2019.127632
112. Duan Q, Zheng G, Li Z, Cheng K, Zhang J, Yang L, Jiang Y, Zhang H, He J, Sun H (2019) An ultra-sensitive ratiometric fluorescent probe for hypochlorous acid detection by the synergistic effect of AIE and TBET and its application of detecting exogenous/endogenous HOCl in living cells. J Mater Chem B 7(33):5125–5131. https://doi.org/10.1039/c9tb01279f
113. Han X, Liu DE, Wang T, Lu H, Ma J, Chen Q, Gao H (2015) Aggregation-induced-emissive molecule incorporated into polymeric nanoparticulate as FRET donor for observing doxorubicin delivery. ACS Appl Mater Interfaces 7(42):23760–23766. https://doi.org/10.1021/acsami.5b08202
114. Hao N, Sun C, Wu Z, Xu L, Gao W, Cao J, Li L, He B (2017) Fabrication of polymeric micelles with aggregation-induced emission and forster resonance energy transfer for anticancer drug delivery. Bioconjug Chem 28(7):1944–1954. https://doi.org/10.1021/acs.bioconjchem.7b00274
115. Wang TT, Wei QC, Zhang ZT, Lin MT, Chen JJ, Zhou Y, Guo NN, Zhong XC, Xu WH, Liu ZX, Han M, Gao JQ (2020) AIE/FRET-based versatile PEG-Pep-TPE/DOX nanoparticles for cancer therapy and real-time drug release monitoring. Biomater Sci 8(1):118–124. https://doi.org/10.1039/c9bm01546a
116. Chen JI, Wu WC (2013) Fluorescent polymeric micelles with aggregation-induced emission properties for monitoring the encapsulation of doxorubicin. Macromol Biosci 13(5):623–632. https://doi.org/10.1002/mabi.201200396
117. Wang Z, Wang C, Fang Y, Yuan H, Quan Y, Cheng Y (2018) Color-tunable AIE-active conjugated polymer nanoparticles as drug carriers for self-indicating cancer therapy via intramolecular FRET mechanism. Polym Chem 9(23):3205–3214. https://doi.org/10.1039/c8py00329g
118. Yu G, Zhao R, Wu D, Zhang F, Shao L, Zhou J, Yang J, Tang G, Chen X, Huang F (2016) Pillar[5]arene-based amphiphilic supramolecular brush copolymer: fabrication, controllable self-assembly and application in self-imaging targeted drug delivery. Polym Chem 7(40):6178–6188. https://doi.org/10.1039/C6PY01402J
119. Wu D, Li Y, Yang J, Shen J, Zhou J, Hu Q, Yu G, Tang G, Chen X (2017) Supramolecular nanomedicine constructed from cucurbit[8]uril-based amphiphilic brush copolymer for cancer therapy. ACS Appl Mater Interfaces 9(51):44392–44401. https://doi.org/10.1021/acsami.7b16734
120. Dong Z, Bi Y, Cui H, Wang Y, Wang C, Li Y, Jin H, Wang C (2019) AIE supramolecular assembly with FRET effect for visualizing drug delivery. ACS Appl Mater Interfaces 11(27):23840–23847. https://doi.org/10.1021/acsami.9b04938
121. Teng K-X, Niu L-Y, Kang Y-F, Yang Q-Z (2020) Rational design of a "dual lock-and-key" supramolecular photosensitizer based on aromatic nucleophilic substitution for specific and enhanced photodynamic therapy. Chem Sci 11(35):9703–9711. https://doi.org/10.1039/D0SC01122C
122. Liu Z, Jiang Z, Yan M, Wang X (2019) Recent progress of BODIPY dyes with aggregation-induced emission. Front Chem 7:712. https://doi.org/10.3389/fchem.2019.00712
123. Hu F, Xu S, Liu B (2018) Photosensitizers with aggregation-induced emission: materials and biomedical applications. Adv Mater 30(45):1801350. https://doi.org/10.1002/adma.201801350
124. Fateminia SMA, Kacenauskaite L, Zhang CJ, Ma S, Kenry MPN, Chen J, Xu S, Hu F, Xu B, Laursen BW, Liu B (2018) Simultaneous increase in brightness and singlet oxygen generation of an organic photosensitizer by nanocrystallization. Small 14(52):e1803325. https://doi.org/10.1002/smll.201803325
125. Wu W, Mao D, Hu F, Xu S, Chen C, Zhang CJ, Cheng X, Yuan Y, Ding D, Kong D, Liu B (2017) A Highly efficient and photostable photosensitizer with near-infrared aggregation-induced

emission for image-guided photodynamic anticancer therapy. Adv Mater 29:33. https://doi.org/10.1002/adma.201700548
126. Cai X, Mao D, Wang C, Kong D, Cheng X, Liu B (2018) Multifunctional liposome: a bright AIEgen-lipid conjugate with strong photosensitization. Angew Chem Int Ed Engl 57(50):16396–16400. https://doi.org/10.1002/anie.201809641
127. Wang S, Wu W, Manghnani P, Xu S, Wang Y, Goh CC, Ng LG, Liu B (2019) Polymerization-enhanced two-photon photosensitization for precise photodynamic therapy. ACS Nano 13(3):3095–3105. https://doi.org/10.1021/acsnano.8b08398
128. Wu W, Mao D, Cai X, Duan Y, Hu F, Kong D, Liu B (2018) ONOO– and ClO– responsive organic nanoparticles for specific in vivo image-guided photodynamic bacterial ablation. Chem Mater 30(11):3867–3873. https://doi.org/10.1021/acs.chemmater.8b01320
129. Mao D, Wu W, Ji S, Chen C, Hu F, Kong D, Ding D, Liu B (2017) Chemiluminescence-guided cancer therapy using a chemiexcited photosensitizer. Chem 3(6):991–1007. https://doi.org/10.1016/j.chempr.2017.10.002
130. Fu H, Huang Y, Lu H, An J, Liu DE, Zhang Y, Chen Q, Gao H (2019) A theranostic saponin nano-assembly based on FRET of an aggregation-induced emission photosensitizer and photon up-conversion nanoparticles. J Mater Chem B 7(35):5286–5290. https://doi.org/10.1039/c9tb01248f
131. Ding F, Zhan Y, Lu X, Sun Y (2018) Recent advances in near-infrared II fluorophores for multi-functional biomedical imaging. Chem Sci 9(19):4370–4380. https://doi.org/10.1039/c8sc01153b
132. Li C, Chen G, Zhang Y, Wu F, Wang Q (2020) Advanced fluorescence imaging technology in the near-infrared-II window for biomedical applications. J Am Chem Soc 142(35):14789–14804. https://doi.org/10.1021/jacs.0c07022
133. Xu W, Wang D, Tang BZ (2020) NIR-II AIEgens: a Win-Win integration towards bioapplications. Angew Chem Int Ed Engl 59:2–14. https://doi.org/10.1002/anie.202005899

REVIEW

Recent Advances in Clusteroluminescence

Zhaoyu Wang[1] · Haoke Zhang[2] · Siqi Li[3] · Dangyuan Lei[3] · Ben Zhong Tang[1,4,5] · Ruquan Ye[6]

Received: 10 December 2020 / Accepted: 8 February 2021 / Published online: 3 March 2021

Abstract

Clusteroluminescence is a phenomenon whereby the aggregation or clustering of non-conjugated electron-rich units leads to the emission of light at long wavelengths. This phenomenon was first discovered in poly(amido amine) (PAMAM) dendrimers. In recent years, clusteroluminescence has attracted growing research interest and its photophysical properties and mechanism have been thoroughly studied. In this review, we first briefly introduce the development of different types of clusteroluminogens. Then we highlight recent developments in clusteroluminescence, including mechanistic studies, the disclosure of room-temperature phosphorescence, and the extension of emission to the longer-wavelength region. Lastly, we demonstrate a few applications in various fields. With advantages such as being earth-abundant, biocompatible and biodegradable, clusteroluminogens are envisioned to be commonplace in the future.

Chapter 2 was originally published as Wang, Z., Zhang, H., Li, S., Lei, D., Tang, B. Z. & Ye, R. Topics in Current Chemistry (2021) 379: 14. https://doi.org/10.1007/s41061-021-00326-w.

✉ Ruquan Ye
ruquanye@cityu.edu.hk

1 Department of Chemistry, Hong Kong Branch of Chinese National Engineering Research Center for Tissue Restoration and Reconstruction, The Hong Kong University of Science and Technology, Clear Water Bay, Kowloon, Hong Kong, China

2 MOE Key Laboratory of Macromolecular Synthesis and Functionalization, Department of Polymer Science and Engineering, Zhejiang University, Hangzhou 310027, China

3 Department of Materials Science and Engineering, City University of Hong Kong, Hong Kong, China

4 Center for Aggregation-Induced Emission, State Key Laboratory of Luminescent Materials and Devices, SCUT-HKUST Joint Research Institute, South China University of Technology, Guangzhou 510640, China

5 AIE Institute, Guangzhou Development District, Guangzhou 510530, China

6 Department of Chemistry, City University of Hong Kong, Hong Kong, China

1 Introduction

Luminescence is a process that involves the excitation of an electron from the ground state to excited states, followed by the relaxation of energy to the ground state with a radiative pathway. The luminescent properties of materials not only depend on their structure, but can also be greatly affected by the local environment. In 1853, George Stokes observed that platinocyanides emitted light only in the solid state, but their solutions looked like mere water [1]. Some similar phenomena were also later observed in organic molecules. In 2001, Tang et al. [2] discovered that 1-methyl-1,2,3,4,5-pentaphenylsilole was non-emissive in good solvents, but that the emission became very strong upon the addition of bad solvents. This was coined the aggregation-induced emission (AIE) effect, which describes the phenomenon whereby chromophores at aggregated states show strong luminescence, whereas their solution states are weakly or not emissive [3]. The AIE effect is attributed to the restricted intramolecular motions that suppress the non-radiative decay pathway and benefit luminescence [4]. This phenomenon has aroused great research interest. Advances in aggregate science, both in fundamental photophysics and applied science, have been delightful to behold.

AIE luminogens (AIEgens) commonly refer to luminophores with extended conjugation. Their aggregates, either from the molecule itself or from intermolecular interactions, lead to strong luminescence. However, some materials without the traditional aromatic or extended π-conjugation structures were also found to emit light in the aggregated state [5, 6]. These materials feature some electron-rich units, such as amine, amide, carboxylate, or other heteroatoms [7, 8]. Upon aggregation, these units interact with each other through space, thereby turning on emission in the visible light region, resembling the AIE phenomenon. A major difference from AIEgens is that the emission of these materials, from an electronic structure of single-molecule aspect, is, in theory, not within the visible light region. Instead, the luminescence comes from the entity of electron-rich units. A new term, non-traditional intrinsic luminescence (NTIL), was proposed by Tomalia et al. [5] to describe this unusual phenomenon. NTIL refers to luminescence that occurs in the absence of traditional luminophores and involves the aggregation/clustering and/or physico-chemical confinement of normally non-emissive, electron rich, hetero-atomic, functionalized moieties. The definition mainly covers the non-traditional luminescence from small or macromolecules. In a broader context, clusterization-triggered emission (CTE), as denoted by Tang et al. [6], further expanded the scope to aromatic small molecules that emit at long wavelengths as clusters. The phenomenon was called clusteroluminescence and the luminogens were termed as clusteroluminogens. In this minireview, we will first discuss the mechanism of clusteroluminescence, followed by showcasing a few examples of CTE. The clusteroluminogens are cataloged as either nonaromatic or aromatic clusteroluminogens. Lastly, their applications in bioimaging, detection, and single-molecule wires will be highlighted.

2 General Mechanism of Clusteroluminescence

The phenomenon of NTIL could date back to the 1970s, when Halpern observed the emission from saturated amine [9]. However, this phenomenon was not capitalized upon until the 2000s [10–13], when a large variety of CTE materials were disclosed. Concerns were raised in the early studies that the clusteroluminescence phenomenon could be due to the presence of impurities, yet with increasing evidence from experimental observations and theoretic calculations, CTE is now more widely accepted.

A prerequisite for clusteroluminescence is the ability for electrons to communicate through space. This has been observed across different experiments. For example, in 2018, Tang and coworkers measured the conductance of single molecules [14]. Two types of molecules were used: one had a hexaphenylbenzene unit (HPB(OM)-SM) and the other a linear polyphenyl unit (PP-SM). Crystal structures show that the torsion angle of HPB(OM)–SM is higher than that of PP–SM, which should be unfavorable for the conductance of electrons. Yet due to the presence of the hexaphenylbenzene unit, a through space channel is created, leading to a higher conductance of HPB(OM)–SM. A more intriguing example is the conductance of poly(4-glycidyloxy-2,2,6,6-tetramethylpiperidine-1-oxyl). The polymer is nonconjugated, but its conductivity is comparable to that of commercial conducting polymers [15]. The conductivity is attributed to the through-space conductance of electrons of the radicals. These experiments suggest that the through-space communication of electrons is possible and could even be comparable to those of conjugated systems.

A second prerequisite is that the interaction should decrease the energy gap accounting for emission at longer wavelengths. Historically, electron interactions in molecules have been well documented by molecular orbital (MO) theory, in which the overlapping of orbitals leads to a reduction in energy gap between the highest occupied molecular orbital (HOMO) and lowest unoccupied molecular orbital (LUMO). The MO theory holds for chemical bonds, which are usually on the scale of Ångstroms. In more recent discoveries, a similar phenomenon, coined "plasmon hybridization", was also observed and theoretically rationalized on plasmonic nanoparticles [16, 17]. The interaction could be up to a few nanometers, or tens of a chemical bond. These two theories suggest that the interactions of electrons through space can reduce the energy gap; it depends on both the electronic structure of a single moiety (i.e., an atom of a covalent bond or nanoparticle) and their distance. Similar to these two theories, the through-space interaction of electrons also shows a reduced energy gap. This was observed as early as 1963, when Lumry et al. observed the emission of isotactic polystyrene at longer wavelengths than atactic polystyrene [18]. Even for non-conjugated polymers, interactions of amide and carboxylates enabled emission in the visible light region. For example, Maser et al. reported that polymerization of citric acid and ethylenediamine produced a bright blue polymer dot. Theoretic calculations suggested that the clustering of polymers and their through space interaction chains would reduce the energy gap [19]. This interaction is affected by the polymer chain length, clustering sizes, and distance.

To sum up, as illustrated in Fig. 1, clusteroluminescence comes from the interaction of electron-rich units through space, which reduces the energy gap and

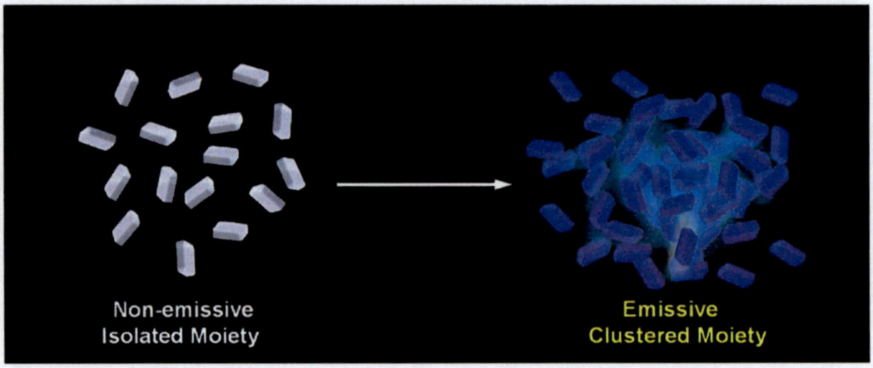

Fig. 1 A general illustration of clusterization-triggered emission (CTE) mechanism

turns on emission in the visible light region. The induction of clustering can be from covalent bonding or non-covalent bonding, which is discussed case-by-case in the sections on Nonaromatic Clusteroluminogens and Aromatic clusteroluminogens without extended π-conjugation. Due to this mechanism, clusteroluminogens usually possess the following characteristics: (1) the excitation spectrum is different from the absorption spectrum; (2) the emission is excitation-dependent due to the existence of different clustering sizes and conformations; and (3) the emission is affected by the environment. The dissolution of clusteroluminogen or the interference of intramolecular interaction will weaken or dismiss the luminescence at a longer wavelength.

3 Nonaromatic Clusteroluminogens

3.1 Development of Nonaromatic Clusteroluminogens

The very first type of clusteroluminogens is the hyperbranched dendrimers. As mentioned above, these date back to the 1970s, and were based on a saturated amine structure [9]. Since then, a large variety of structural derivatives has been disclosed. In 1985, Tomalia et al. [5] revisited the topic and found weak emission by poly(amido amine) (PAMAM) dendrimers (Fig. 2a). Tucker et al. [13] confirmed the intrinsic fluorescence (440 nm) of carboxylated-terminated PAMAM dendrimers in 2001 and attributed it to the n–π* transition from the amido groups. Dickson et al. [20] used OH-terminated PAMAM to encapsulate and stabilize blue-emissive Au_8 nanodots, which achieved more than 100-fold enhancement in quantum yield. The authors explained the phenomenon as being due to the protection role of the dendrimer cage and the size of the Au nanoclusters. However, afterward, Bard et al. [12] found that strong blue-emission could also be obtained from OH-terminated PAMAM via simple oxidization in the absence of Au nanoparticles. They also applied the same treatment to amine-terminated PAMAM dendrimers, which exhibited relatively weak emission. Therefore, the authors proposed that the terminal

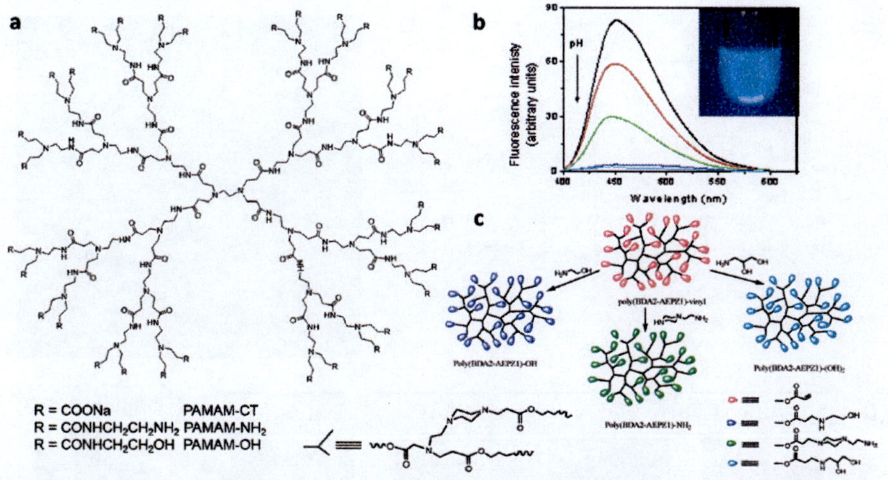

Fig. 2 **a** Structure of the poly(amido amine) (PAMAM) dendrimer with various terminal groups (adapted with permission from [24]). **b** Photoluminescence spectra of PAMAM-NH_2 dendrimer at different pH values: pH 2, 4, 5, 9, and 11 (from top to bottom, excitation: 390 nm). *Inset* Photograph of aqueous 0.7 mM solution at pH 2 under UV lamp (adapted with permission from [11]). Copyright 2004 American Chemical Society. **c** Reaction scheme for preparing hyperbranched poly(amino ester)s with different terminal groups (adapted with permission from [23]). Copyright 2005 American Chemical Society

–OH groups, rather than backbone, played a key role in blue-emissive species. In the same year, Imae and colleagues [22] discovered that NH_2-termimated PAMAM dendrimer also could exhibit strong fluorescence under acidic conditions (Fig. 2b). Furthermore, they also found that PAMAM dendrimers with other terminal groups (OH and carboxylate) and poly(propyleneimine) (PPI) dendrimers produced similar fluorescent emission and intensities at the same concentration at pH 6 [11]. Therefore, the authors confirmed the critical role of the dendrimer backbone in the novel fluorescence. In addition, Yang et al. also found the same phenomenon of blue emission under acidic conditions for hyperbranched PAMAM [21]. Imae et al. [22] further investigated the emission moiety on PAMAM and PPI dendrimers and claimed that it was related with protonated tertiary amine groups. A series of hyperbranched poly(amino ester)s (Fig. 2c) were synthesized by Liu and coworkers [23], and they expressed a different opinion from the above authors, suggesting that the coexistence of tertiary amines and carbonyl groups in the core and compacted dendritic architecture with abundant suitable terminal groups were an indispensable prerequisite for blue fluorescence.

Another important catalogue of clusteroluminogens is the biogenic or synthetic oligo- and polypeptides. In 2001, Swaminathan et al. [25] discovered an unusual blue fluorescence from a type of amino acid, L-lysine, at a high concentration in aqueous medium. Subsequently, Florence et al. [26] synthesized lysine-base dendrimer, Gly-$Lys_{63}(NH_2)_{64}$, with a glycine in the core, 63 lysine in the branching units, and 64 free amino groups in the surface (Fig. 3a). The polylysine dendrimer was demonstrated to be blue-emissive and used as nanoprobes for cell transport [26].

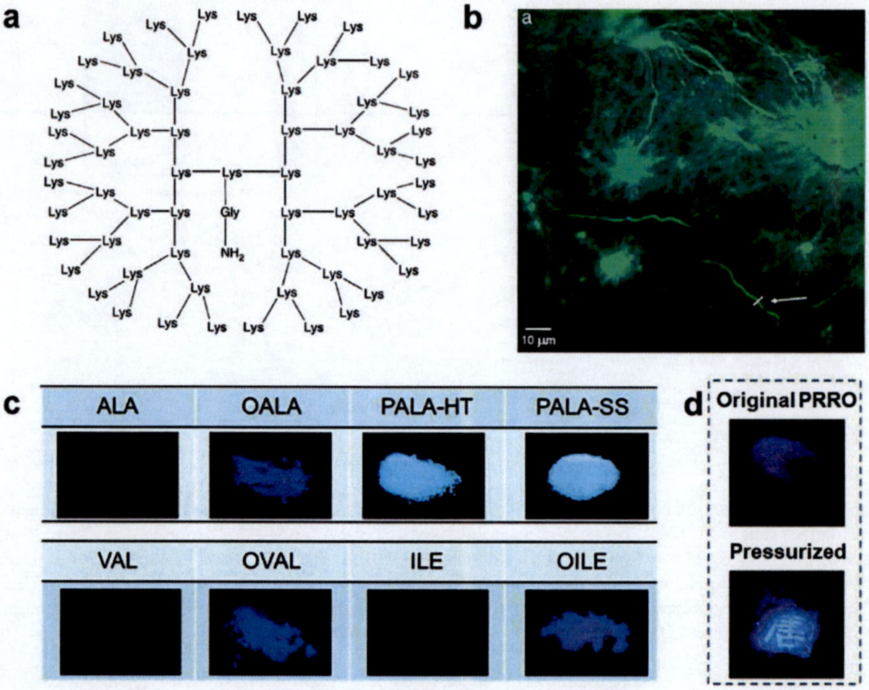

Fig. 3 **a** The chemical structure of Gly–$Lys_{63}(NH_2)_{64}$ (adapted with permission from [26]). Copyright 2006 Taylor & Francis **b** Confocal microscopy image of the poly(ValGlyGlyLeuGly) amyloidlike fibrils. Excitation: 405 nm. (adapted with permission from [27]) Copyright 2007, National Academy of Sciences. **c** Photos of amino acids, oligopeptides, and polypeptides under a 365 nm UV lamp. **d** Photos of fluffy and pressurized poly-L-proline (PRRO) using a Chinese stamp (**c** and **d** are adapted with permission from [28])

Apart from small molecules and dendrimers, amyloid-like fibrils were also reported to possess intrinsic blue-fluorescence, which were made from self-assembly of polypeptides, poly(ValGlyGlyLeuGly) (Fig. 3b) [27]. Interestingly, the emissive fibrils also exhibit conductive property at ambient conditions, in comparison with common conducting polymers with highly conjugated backbones. The authors also believed that water molecules play a critical role in the charge-transfer mechanism in the fibrillar structures. Since the fluorescent polypeptide-based dendrimers and fibrils usually have complicated structures, it remained unclear whether the polypeptides per se could be fluorescent under irradiation. To go further, Tang and colleagues synthesized a series of oligopeptides of L-alanine (OALA), L-valine (OVAL), and L-isoleucine (OILE), and polypeptides of L-alanine (PALA). The oligopeptides in the solid state were blue emissive (~400 nm) under UV irradiation and the polypeptides displayed more red-shifted cyanic emission at ~445 nm (Fig. 3c). Notably, poly-L-proline (PRRO) (Fig. 3d) is fluffy and weakly emissive originally, but emission could be greatly enhanced after pressing, which implicated the proximity of amide groups in turning on fluorescence.

3.2 Hyperbranched and Linear Polymer

The third type of clusteroluminogens is hyperbranched or linear polymers. Both hyperbranched polymer and dendrimers have dendritic structures. However, hyperbranched polymers are different from dendrimers in topological structures and synthetic approaches [29]. Namely, hyperbranched polymers have three types of structural units: dendritic units, linear units, and terminal units, while dendrimers possess only two types of structural units. Also, hyperbranched polymers can be obtained through one polymerization step, but dendrimers are usually synthesized step-by-step in an iterative fashion.

In the early studies, it was commonly believed that a dendritic structure is a prerequisite for fluorescent species. Apart from PAMAM dendrimers, it was demonstrated that a wide range of hyperbranched polymers and dendrimers, such as polyethylenimine (PEI), polypropyleneimine (PPI) [22], poly(amine ester) [30], polyurea (PU) [31], and siloxane-PAMAM dendrimers [32], displayed intrinsic fluorescence. Moreover, they shared the same property that the emission intensity could be affected by pH value as well as molecular weight. Usually, acidic conditions and high molecular weight are desirable for fluorescence [31]. Nevertheless, in 2007, Stiriba and Chen [33] compared the photophysical properties of hyperbranched and linear PEI (Fig. 4a), and surprisingly observed better emission performance from linear than hyperbranched PEI, which indicates the irrelevance of dendritic structure to create fluorescent species. A similar phenomenon was also reported in linear and hyperbranched PAMAM dendrimers [34]. In addition, other linear nonconjugated and nonaromatic polymers were also found to be fluorescent in the aggregated and solid state. For example, polyacrylonitrile (PAN) is free from hydrocarbon or heterocyclic aromatic building blocks, with a simple polyethylene backbone and cyano pendants (Fig. 4b), but becomes highly emissive with a quantum yield of 16.9% and present a distinct AIE characteristic [35]. The phenomenon is not limited to amine or carbonyl-containing systems, but can be observed in merely oxygen moieties (i.e., ether, hydroxyl). Yuan et al. [36] chose the poly(ethylene glycol) (PEG) and PEO–PPO–PEO triblock copolymer (F127) as the pure oxygen-containing models to reveal the CTE effect (Fig. 4c).

3.3 Mechanism and Photophysical Property

Although a series of mechanisms of unusual luminescence from nonconjugated and nonaromatic systems have been proposed to explain the observed phenomena in different materials such as oxidation [12], the formation of unsaturated hydroxylamine [37], and the role of tertiary amine functionality [22], a more generalized principle to account for the CTE phenomena is needed. It was proposed that intra/intermolecular interaction clusters generate new emissive species with smaller energy gaps to emit visible light upon UV irradiation. As shown in Fig. 5a, the theoretical calculation of a fragment in PAMAM with two repeating units was carried out using Gaussian 09 based on density functional theory (DFT) [34]. The delocalization of

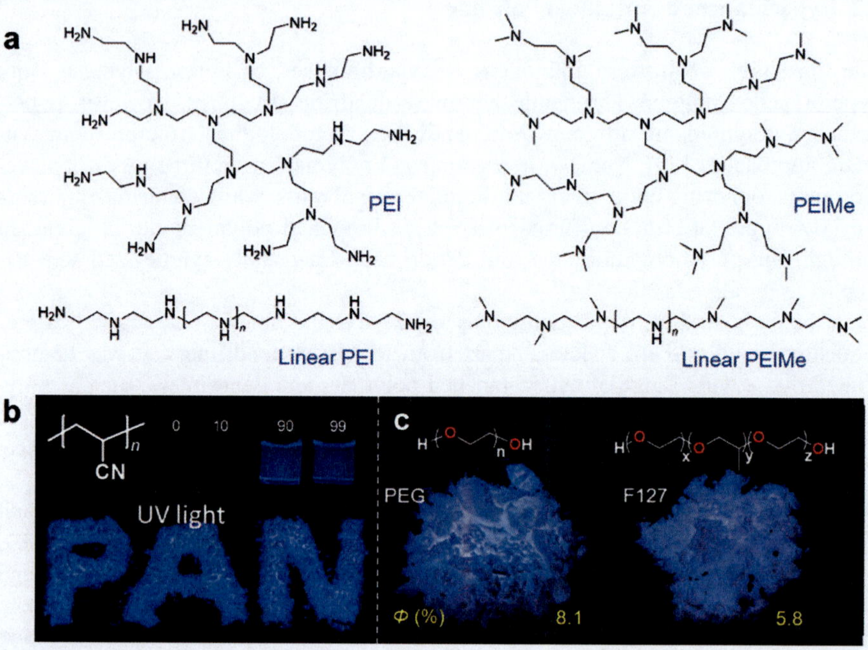

Fig. 4 **a** Chemical structures of hyperbranched and linear polyethylenimines (adapted with permission from [33]). Copyright 2007, WILEY-VCH Verlag GmbH & Co. KGaA, Weinheim. **b** Chemical structure of polyacrylonitrile (PAN) and the photos of PAN in the solid state (adapted with permission from [35]). Copyright 2016, WILEY-VCH Verlag GmbH & Co. KGaA, Weinheim. **c** Chemical structure of poly(ethylene glycol) (PEG) and PEO–PPO–PEO triblock copolymer (F127), and their photos under an ultraviolet (UV) lamp (adapted with permission from [36]). Copyright 2018, WILEY-VCH Verlag GmbH & Co. KGaA, Weinheim

lone-pair n electrons and π electrons leads to n–π and π–π interactions in the aggregated and solid states, resulting in through-space conjugation (TSC). This principle was also proved in Zhang's work, which elucidated the unusual emission phenomenon of PAN [35]. The unique emission of PAN was ascribed to the aggregation of cyano terminals, namely the formation of cyano clusters. The polymer chains are well extended and the cyano groups are almost isolated in dilute solution. In contrast, the aggregated polymer backbones enable cyano groups to be clustered in close proximity, resulting in electronic TSC, including electron overlap between lone pairs and π electrons, dipole–dipole interactions, and n–π interactions (Fig. 5b). The principle could also be extended to other nonconjugated but heteroatom-containing clusteroluminogens, such as nitrogen, oxygen, and phosphorus atoms (Fig. 5c) [6]. Aside from the role of n electrons on heteroatoms, Wang and coworkers proposed an interesting opinion that electron density delocalization can also arise from hydrogen bond (H-bond) interactions [38], which were traditionally believed to be simple electrostatic interactions. The simplest but the most typical example is water clusters (Fig. 5d), in which researchers studied theoretically the cyclic water rings from

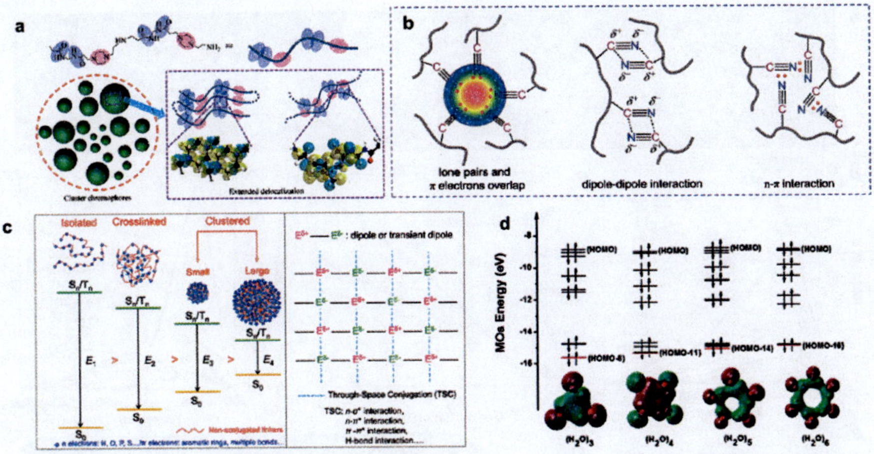

Fig. 5 **a** Schematic illustration of the emission mechanism of PAMAM dendrimer based on using Gaussian 09 based on density functional theory (DFT) (adapted with permission from [34]). Copyright 2015, Springer Nature. **b** Possible intra- and intermolecular interactions within cyano clusters: electron overlap between lone pairs and π electrons, dipole–dipole interactions, and n–π interactions (adapted with permission from [35]). Copyright 2016, WILEY-VCH Verlag GmbH & Co. KGaA, Weinheim. **c** Schematic illustration of the process of clusters formation accompanied with the clusterization-triggered emission properties and proposed mechanistic picture for clusteroluminescence (adapted with permission from [6]). Copyright 2019 Elsevier Inc. **d** Energy level diagrams of the various water rings. *Red lines* Energy levels of the characteristic occupied molecular orbitals (MO) presented in the bottom diagram (adapted with permission from [39]). Copyright 2015, Royal Society of Chemistry

trimers to hexamers using first-principles DFT methods. Interestingly, after examining their MOs, the electron density delocalization effect could be found readily in the water clusters and an increased number of water units in a ring leads to more localized MOs [39]. This work reflects the delocalized electronic distribution characteristics related to H-bonding. Markland and coworkers also found the quantum delocalization of protons in the H-bond network via studying the active site of an enzyme, ketosteroid isomerase, based on molecular dynamics simulations [40]. Afterward, Hassanali et al. [41, 42] attributed the intrinsic fluorescence of protein fibrils to charge delocalization in the H-bond network via experimental and theoretical results. In a nutshell, the mechanism of TSC and H-bond interactions has been proposed to account for the phenomenon of some types of unconventional chromophores.

Since the clusteroluminescence of nonaromatic materials comes from the interaction of multiple units, they share some common photophysical characteristics. The first one is analogous to the AIE effect, although their structures are completely different from classical AIEgens, with aromatic rings, such as tetraphenylethene (TPE) and its derivatives, as rotors [43]. This property is commonly characterized via concentration-enhanced emission. Namely, the chromophores are emissive in the aggregated or solid state, yet weakly emissive or non-emissive in dilute solution. The example of bovine serum albumin (BSA) is shown in Fig. 6a, where it can be clearly observed that the emission intensity of BSA solution boosts as the concentration increases (Fig. 6a–c)

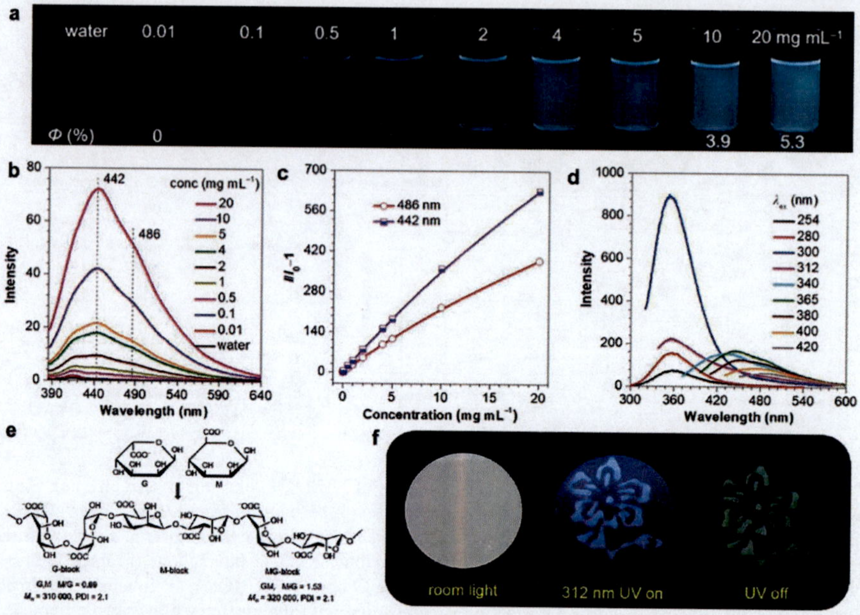

Fig. 6 **a** Photographs taken under 365 nm UV light, **b** emission spectra (excitation: 365 nm), and **c** net increase of the peak intensity (I_0/I) at 442 and 486 nm for bovine serum albumin (BSA) solutions at various concentration. **d** Emission spectra of 20 mg/ml BSA solution with different excitation wavelengths (**a–d** are adapted with permission from [44]). Copyright 2019, WILEY-VCH Verlag GmbH & Co. KGaA, Weinheim. **e** Chemical structure of two sodium alginates. **f** Photographs of a flower drawn on a polytetrafluoroethylene plate using 8 wt% sodium alginate taken at ambient conditions under room light, 312 nm UV light, and after switching off UV irradiation. (**e**, **f** are adapted with permission from [45]) Copyright 2018 American Chemical Society

[44]. Second, excitation-dependent emission is another common feature of clusteroluminogens. Traditional chromophores usually exhibit fixed emission peaks under excitation at various wavelengths due to the well-defined structure of luminescent species. When it comes to clusteroluminogens, the emission from concentrated solution stems from diverse cluster species. As depicted in Fig. 6d, the emission peak of BSA solution in high concentration varies from ~ 350 to ~ 480 nm with increasing excitation wavelength. Third, the excitation spectrum of clusteroluminogens often does not align with their absorption spectrum. Namely, a clusteroluminogen shows negligible absorption in the near-ultraviolet and visible region (300–900 nm) due to lack of extended π-conjugation, but can be excited at 365 nm to emit blue light, such as with BSA and PAN. Last but not least, some clusteroluminogens possess the fascinating property of persistent room-temperature phosphorescence (RTP) in the solid state, e.g., sodium alginate (Fig. 6e,f) [45].

3.4 Recent Progress in Structural Diversity and Emission Colors

With thorough studies of the mechanism and properties of clusteroluminogens across different groups, scientists can now systematically design new nonconjugated luminogens. A diversity of clusteroluminogens comprising different functional groups with electron-rich heteroatoms have been reported, including hydroxyl (OH), amino (NH_2) [46], carbonyl (C=O), cyano (CN), carboxyl (COOH), amide (CONH) [47], anhydride, and sulfinyl (S=O) groups (Fig. 7a) [48]. Recently, an interesting CTE phenomenon from succinimide was reported by Yang et al. [49]. Despite the fact that succinimide is known as a fluorescence quencher due to electron transfer [49], modification of dithiol or amino substituents can turn succinimides into the fluorescent molecules 2-aminosuccinimide (AS), and 2,3-dithiosuccinimide (DTS) as shown in Fig. 7b [50]. Theoretical calculation shows that emission from AS and DTS is associated with separated spatial overlaps of MO between the HOMO and the LUMO. Afterward, Yang et al. [49] further demonstrated enhanced emission from succinimide copolymers (Fig. 7c).

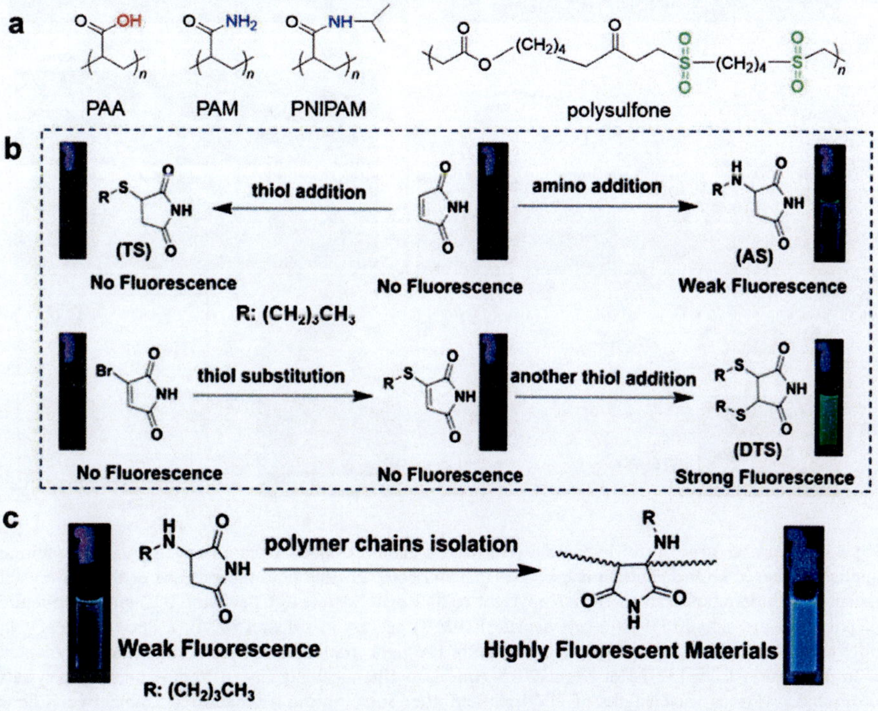

Fig. 7 **a** Examples of clusteroluminogens with typical and common functional groups (adapted with permission from [46, 47]). **b** Succinimides become fluorescent through functionalization with dithiol/amino substituents. **c** Fluorescence enhancement of amino-succinimides by polymer chain isolation. (**b**, **c** are adapted with permission from [50]) Copyright 2015, Royal Society of Chemistry. *PAA* Polyacrylic acid, *PAM* poly(acrylamide), *PNIPAM* poly(N-isopropyl acrylamide)

In addition to the structural diversity, the emission color of clusteroluminogens has also been expanded. Most reported clusteroluminogens emit light at short wavelength (i.e., blue and cyan), because the bandgap cannot be decreased easily without extended π-conjugation. Here, three strategies have been reported to break the limitations and achieve longer wavelength emission. The first is to apply the RTP property of clusteroluminogens. For example, Yuan and coworkers [51] reported that hydantoin (HA), which resembles thymine in DNA, would emit phosphorescence at a longer wavelength (yellowish-green) in comparison with that of fluorescence (sky-blue) (Fig. 8a). The second strategy is to make use of the solvatochromic effect for tunable emission and absorption color in solution. In Tang's work, a nonconjugated polymer, poly-[(maleic anhydride)-alt-(vinyl acetate)] (PMV) (Fig. 8b), showed blue emission in aromatic and oxygenic solvents [toluene and tetrahydrofuran (THF)],

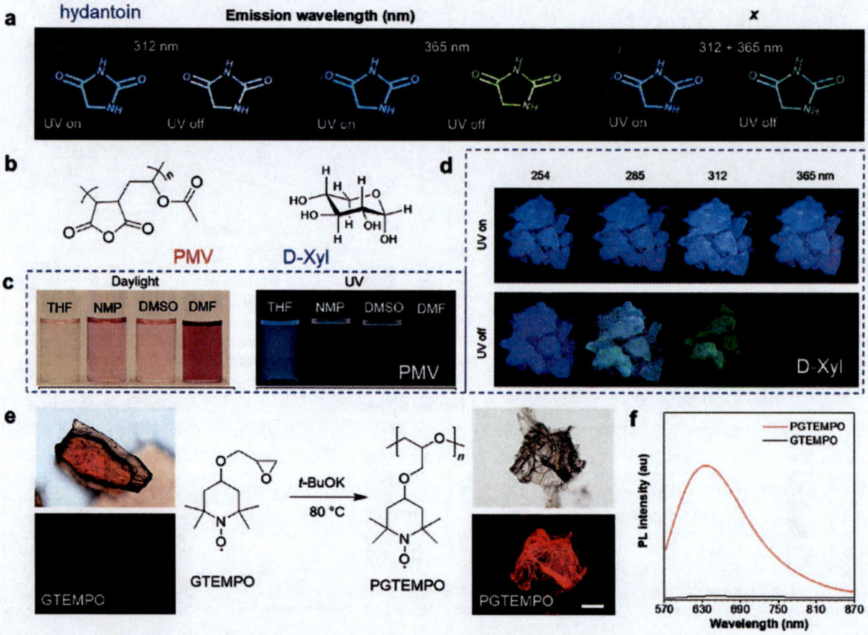

Fig. 8 **a** Chemical structure of hydantoin (HA) and a demonstration of the advanced anticounterfeiting application based on the photoluminescence (PL) and color-tunable phosphorescence of the HA crystal (adapted with permission from [51]). Copyright 2020, Royal Society of Chemistry. **b** Chemical structure of poly-[(maleic anhydride)-alt-(vinyl acetate)] (PMV) and D-(+)-xylose (D-Xyl). **c** Photos of PMV in different solvents taken under daylight and 365 nm UV light irradiation. Concentration: 5 mM. (adapted with permission from [52]) Copyright 2015 American Chemical Society. **d** Photos of D-Xyl crystals taken under various wavelengths of UV light and after stopping the irradiation at ambient conditions (adapted with permission from [54]). Copyright 2020, Royal Society of Chemistry. **e** Synthetic route of the poly(4-glycidyloxy-2,2,6,6-tetramethylpiperidine-1-oxyl) (PGTEMPO) radical polymer and the photos of 4-glycidyloxy-2,2,6,6-tetramethylpiperidine-1-oxyl (GTEMPO) and PGTEMPO in the solid state under (*upper panel*) daylight and (*lower panel*) 510–560 nm excitation. *Bar* 200 μm. **f** PL spectra of PGTEMPO and GTEMPO solid excited at 532 nm. (**e**, **f** are adapted with permission from [56])

but red-shifted emission in solvents with electron-rich atoms (NMP, DMSO, and DMF) (Fig. 8c) [52]. Qiao et al. [53] conducted a follow-up work via modifying PMV with NaOH into PMV/Na, which was emissive in magenta color under excitation. In addition, the aforementioned excitation-dependent could also be applied as another approach in combination with phosphorescence. An example of a small molecule, D-(+)-xylose (D-Xyl) (Fig. 8b), is presented in Fig. 8d [54]. Distinct emission colors of blue, cyan, and green were obtained after turning off the UV light of 254, 285, and 312 nm, respectively. The last strategy is an unusual way to realize a relatively long wavelength (red color) emission from a novel class of clusteroluminogens. Organic stable radicals commonly exhibit strong non-radiative energy decay to result in a weakly or non-luminescent property [55]. Nevertheless, Tang et al. [56] reported that a typical chromophore quencher, 2,2,6,6-tetramethylpiperidin-1-yl (TEMPO), could be turned into red-luminescent material via grafting polymer backbone (Fig. 8e). The red emission of nonconjugated polymer, poly(4-glycidyloxy-2,2,6,6-tetramethylpiperidine-1-oxyl) (PGTEMPO), proved to be associated with the interactions of nitroxide radicals. The presence of an unpaired electron creates a new MO, the singly occupied molecular orbital (SOMO), which lies exactly between HOMO and LUMO. According to theoretical calculation, the emission was generated from the SOMO–HOMO transition, instead of the traditional LUMO–HOMO pathway, from which the narrower bandgap enables long-wavelength emission. As shown in Fig. 8e,f, PGTEMPO is non-conjugated and nonaromatic, but readily emits red luminescence at 630 nm. Thus, clusteroluminogens are not limited to short-wavelength emission, and longer wavelength emission on nonconjugated materials has been realized with different strategies.

4 Aromatic Clusteroluminogens Without Extended π-Conjugation

4.1 Development of Aromatic Clusteroluminogens

Aromatic clusteroluminogens emit light from the inter- or intramolecular interactions of two or more aromatic units. Emission from a single aromatic unit or clustered units can coexist, whereas clusteroluminescence often results in a longer emission wavelength than that of the single aromatic unit. This is different from the photophysical properties of nonaromatic clusteroluminogens as introduced in the section Development of nonaromatic clusteroluminogens, where it was stated that nonaromatic clusteroluminogens are often non-luminescent in the solution state. In comparison with the first discovery of nonaromatic clusteroluminogens from dendrimers [13], the observation of unusual fluorescence from nonconjugated polymers with aromatic rings was much earlier. Bovey and Lumry reported in 1963 that isotactic polystyrene (PS) (Fig. 9a) exhibited a more red-shifted emission (324 nm) than that of atactic PS (292 nm), but they shared almost the same absorption [18]. Later, Hirayama [57] studied the fluorescence spectra of a series of diphenyl and triphenyl alkanes, and proposed that intramolecular excimer formation accounted for emission at a longer wavelength. These phenomena indicate the early attention to the role of TSC on

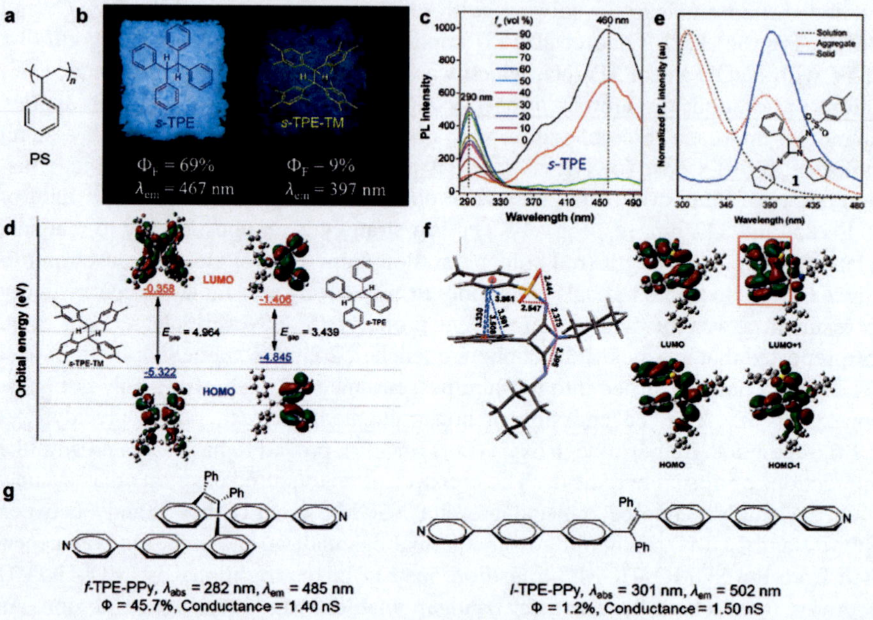

Fig. 9 **a** Chemical structure of isotactic polystyrene (PS). **b** Chemical structure of 1,1,2,2-tetraphenylethane (*s*-TPE) and 1,1,2,2-tetrakis-(2,4,5-trimethylphenyl)ethane (*s*-TPE-TM), and photos taken under 365 nm UV light irradiation. **c** PL spectra of *s*-TPE in tetrahydrofuran (THF)/water mixtures with different water fractions (f_w). Solution concentration: 100 μM; excitation: 340 nm. **d** Frontier MOs and energy levels of *s*-TPE and *s*-TPE-TM in the excited state calculated by time-dependent density-functional theory (TD-DFT) B3LYP/6-31G(d), Gaussian 09 program. (**b–d** are adapted with permission from [59]) Copyright 2017 American Chemical Society. **e** Normalized PL spectra of the THF solution, aggregates, and solid powders of compound **1**. The aggregates were obtained in THF/water mixture with water fraction of 80%. Solution concentration: 10 μM; excitation: 260 nm. **f** Crystal structure of compound **1** with indicated distances (Å) between π–π stacked aromatic rings and distances between atoms with lone pairs; MOs of compound 1 in the ground state calculated by B3LYP/6-31G(d). *Red rectangles* Through-space π–π interaction. (e and f are adapted with permission from [60]) Copyright 2018 American Chemical Society. **g** Chemical structure of *s*-DPE and *s*-DPE-TM (adapted with permission from [61]). Copyright 2015, WILEY-VCH Verlag GmbH & Co. KGaA, Weinheim. *HOMO* highest occupied molecular orbital, *LUMO* lowest unoccupied molecular orbital

photophysical behavior. The concept of TSC can be dated back to 1949 [58], when studies showed that electrons can delocalize on [2.2]paracyclophane between two close parallel aromatic rings with a distance of 3.1 Å. This type of through-space interaction serves as the basis of the study of aromatic clusteroluminogens. Aromatic clusteroluminogens can be from both polymeric materials and small molecules; nonetheless, here we focus on the recently developed small molecules system, as their well-defined structures can help to elucidate clusteroluminescent mechanisms.

4.2 Aromatic Clusteroluminogens from Intramolecular Interaction

We classify the first type of aromatic clusteroluminogens from the intramolecular interaction of aromatic units. Recently, Tang and coworkers [59] synthesized non-conjugated small molecules, 1,1,2,2-tetraphenylethane (*s*-TPE) and 1,1,2,2-tetrakis-(2,4,5-trimethylphenyl)ethane (*s*-TPE-TM), which possess similar chemical structures but exhibit different photophysical properties (Fig. 9b). Under the excitation of 240 nm, *s*-TPE and *s*-TPE-TM in a THF/water mixture with low water fraction (10^{-4} M) exhibited a peak only at short wavelength (290 and 300 nm, respectively) assigned to benzene emission. However, once aggregates formed at $f_w > 60\%$, a marvellous longer-wavelength emission at 460 nm was observed from *s*-TPE. Both *s*-TPE and *s*-TPE-TM presented a typical AIE property in THF/water mixture at longer excitation wavelength (340 nm) (Fig. 9c), but *s*-TPE was more emissive and more red-shifted than *s*-TPE-TM. To understand the different photophysical properties of *s*-TPE and *s*-TPE-TM, time-dependent density-functional theory (TD-DFT) was carried out and the frontier MOs of *s*-TPE and *s*-TPE-TM are illustrated in Fig. 9d. The calculation indicated the phenyl rings in *s*-TPE and *s*-TPE-TM interacted intramolecularly in a through-space manner with overlapping between phenyl ring electrons, resulting in a narrower bandgap and the formation of a new excimer. The TSC effect for *s*-TPE is stronger than for *s*-TPE-TM due to the absence of steric hindrance from methyl groups. Therefore, as shown in Fig. 9c, the emission of *s*-TPE in the aggregated and solid state generated a new emission peak attributed to the intramolecular TSC interaction. This also explains why *s*-TPE bears higher quantum yield (Φ_F) as well as a more red-shifted emission wavelength than does *s*-TPE-TM.

Another example of aromatic clusteroluminogens reported recently is shown in Fig. 9e [60]. Similar to *s*-TPE, the small molecule, compound **1**, showed an emission peak at ~ 300 nm in solution from the single benzene structure, whereas at aggregated or solid state, a new peak at ~ 400 nm merged. The crystal structure of compound **1** provided solid evidence of the conformation of two phenyl rings stacked in a nearly face-to-face fashion with a small angle and short distance at 3.33 Å. TD-DFT calculation (Fig. 9f) also indicated the overlapping of orbitals between the two parallel benzene rings, which accounted for emission at the longer wavelength. These types of intramolecular TSC as shown in the examples of *s*-TPE and compound **1**, explained the origin of long-wavelength emission with aggregated and solid state aromatic clusteroluminogens.

In addition to theoretical calculations, the TSC effect can also be inferred from the conductance of small molecules. Figure 9g shows the structure of a folded molecule, *f*-TPE-2-phenylpyridine (PPy), and its linear counterparts, *l*-TPE-PPy [61]. The folded conformation of *f*-TPE-PPy was confirmed by crystallographic data, in which two biphenyls adopted a face-to-face tilted structure with a close distance of 3.1 Å. The linear molecule, *l*-TPE-PPy, showed excellent through-bond π-conjugation, whose absorption (301 nm) and emission band are similar to those of TPE—a common AIEgen [62]. In contrast, the conformation of twisted biphenyls on *f*-TPE-PPy broke the efficient π-orbital overlap between biphenyl and central vinyl groups, leading to a shorter absorption wavelength at 282 nm. When excited by 282 nm, *f*-TPE-PPy displays a PL peak at 485 nm in THF solution, which was blue-shifted by only 17 nm compared to the emission of *l*-TPE-PPy. The large Stokes shifts of 203 nm support the hypothesis of weak through-bond conjugation, but efficient TSC [63]. In addition, due to the rigidity caused by the twisted structure, the quantum yield of *f*-TPE-PPy was greatly enhanced in solution. Theoretically, the reduction of π-conjugation would reduce the conductivity of organic molecules. However, the conductance of *f*-TPE-PPy was comparable to that of better-conjugated *l*-TPE-PPy, which was because of the intramolecular TSC effect [61]. The spectroscopic data and molecular conductance evidenced the existence of TSC and its effect in shifting the emission to a longer wavelength.

4.3 Aromatic Clusteroluminogens from Intermolecular Interaction

The second type of aromatic clusteroluminogen is classified as the result of intermolecular interaction. In the early study of diphenyl alkanes in 1965, Hirayama [57] concluded that 1,2-diphenylethane (*s*-DPE) had emission only from the single benzene unit and no excimer emission was observed. However, recently, the development of clusteroluminescence inspired the revisit of *s*-DPE and its derivative, 1,2-bis(2,4,5-trimethylphenyl)ethane (*s*-DPE-TM) (Fig. 10a), which disclosed its long-wavelength emission in aggregated and solid state. Figure 10b–d showed the emission of *s*-DPE and *s*-DPE-TM in a water/THF mixture with various fractions and in crystalline state. The clusteroluminescence of *s*-DPE and *s*-DPE-TM was

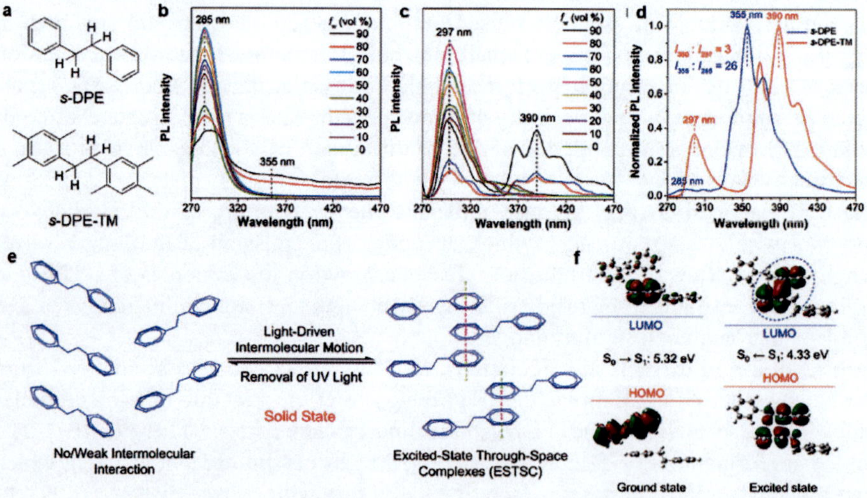

Fig. 10 **a** Chemical structure of 1,2-diphenylethane (*s*-DPE) and 1,2-bis(2,4,5-trimethylphenyl)ethane (*s*-DPE-TM). **b, c** PL spectra of (**b**) *s*-DPE and (**c**) *s*-DPE-TM in THF/water mixtures with different water fractions (f_w). Concentration: 10^{-4} M; excitation: 260 nm for s-DPE and 280 nm for s-DPE-TM. **d** PL spectra of their crystalline powder. Excitation: 260 nm. **e** Schematic illustration of the light-driven intermolecular motion of *s*-DPE in the solid state. **f** Frontier MOs of the *s*-DPE dimer in the ground and excited states calculated by the B3LYP/6-311G** method. (**a–f** are adapted with permission from [64]) Copyright 2019 American Chemical Society

proposed to originate from the through-space complexes generated at the excited state rather than the ground state [64]. Crystal structure and absorption data suggested that the neighboring molecules are far away from each other, resulting in weak intermolecular interaction. However, upon UV irradiation, their intermolecular motions construct the excited-state through-space complexes, whose radiative decay emitted fluorescence (Fig. 10e). The calculated orbitals illustrated in Fig. 10f show that the packing of neighboring phenyl rings transformed from perpendicular to parallel position driven by light. This transformation formed the TSC, which led to a smaller energy gap and a longer wavelength emission. This work proposed a new mechanism of clusteroluminescence and added to the family of clusteroluminogens.

5 Applications of Clusteroluminogens

The luminescent characteristics, as outlined in the section of mechanism of clusteroluminescence, have inspired the development of various applications, such as bacteria and cell imaging, ion detection, anticounterfeiting, and single-molecule conductors. One significant advantage of clusteroluminogens is that their constituent units can come from naturally abundant and biocompatible resources, such as peptides and cellulose. Therefore, non-conjugated luminogens show great potential in biological detection and imaging. For example, hyperbranched dendrimers are featured with water-soluble and biodegradable properties. By combining the intrinsic fluorescence characteristics, Pan and coworkers [65] applied hyperbranched mannose-attached poly(amido amine) (M-HPAMAM) as a probe for bacteria detection and imaging (Fig. 11a). M-HPAMAM could bind to bacteria due to the strong interaction between mannose and *Escherichia coli*. Using this property, Pan et al. [66] demonstrated the application of HPAMAM nanoparticles for cell imaging. Apart from dendrimers, sodium alginate—a natural anionic polysaccharide—was used to stain HeLa cells. Bright blue emission from the stained Hela cells was observed on confocal microscopy under excitation at 405 nm (Fig. 11b) [45]. In addition, clusteroluminogens also possess the potential for the detection of metal ions. For instance, Yuan et al. reported the detection of Fe^{2+} ions using PEG [36]. The detection was based on the fluorescence quenching mechanism that the clusteroluminescence intensity would decrease linearly in response to the increase in Fe^{2+} concentration (Fig. 11c) [67].

The second example of applications is for information encryption. For clusteroluminogens that exhibit RTP property, they could be utilized for the application of anticounterfeiting. Yang et al. developed a class of highly emissive polymer carbon dots (PCDs), which were synthesized from polyacrylic acid (PAA) and ethylenediamine (EDA) by a one-step hydrothermal method [68]. PCDs show excellent RTP performance after irradiation with a common UV lamp. As shown in Fig. 11d, a colorful butterfly and a sentence of "*Just Love U*" were visible under 365 nm UV irradiation. Once the lamp was switched off, the color combination of the butterfly was different and the words of "*Just Love U*" changed to "*JLU*", resulting from phosphorescence.

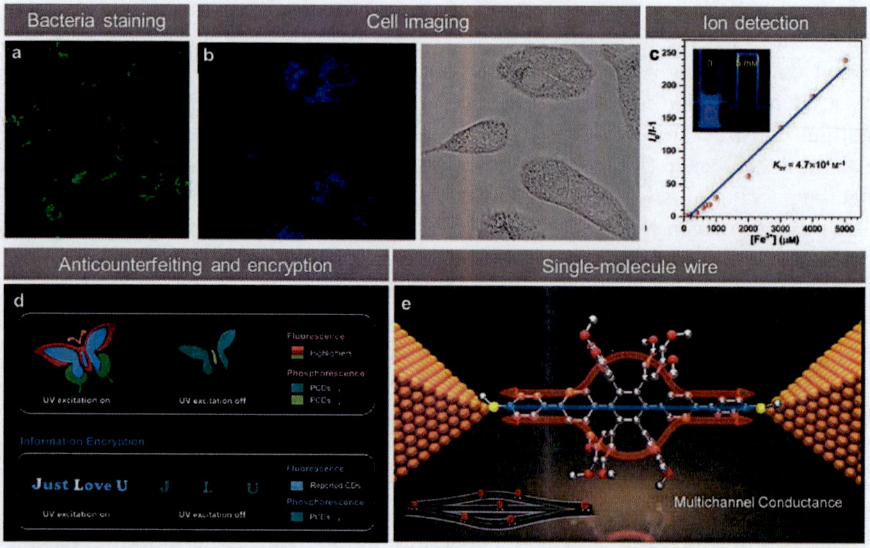

Fig. 11 **a** Laser scanning confocal microscopy (LSCM) image of *Escherichia coli* (OD 1.0)/hyperbranched mannose-attached poly(amido amine) (M-HPAMAS) aggregates in PBS solution, excitation: 488 nm (adapted with permission from [65]) Copyright 2010 American Chemical Society. **b** LSCM images of Hela cells after incubation with 0.5 wt% sodium alginate in Dulbecco's modified Eagle's medium (DMEM) for 1.5 h under 405 excitation and bright field (adapted with permission from [45]). 2018 American Chemical Society. **c** Stern–Volmer plots of the net increase of peak intensity ($I_0/I - 1$) at 396 nm against [Fe^{2+}] in aqueous polyethylene glycol (PEG) solutions. *Inset* Solutions with [Fe^{2+}] of 0 and 5 mM taken under 312 nm UV light (adapted with permission from [36]). Copyright 2018, WILEY-VCH Verlag GmbH & Co. KGaA, Weinheim. **d** Photographs of graphic security and information encryption made from polymer carbon dots (PCD) (adapted with permission from [68]). Copyright 2018, WILEY-VCH Verlag GmbH & Co. KGaA, Weinheim. **e** Schematic representation of circuits of hexaphenylbenzene (HPB) derivative anchored onto gold electrodes as single-molecule wire (adapted with permission from [14]). Copyright 2018 American Chemical Society

Another potential application of clusteroluminogens is molecular conductors. Single-molecule wires are the fundamental elements of molecular electronics. Most single-molecule wires are through-bond π-conjugated organic and organometallic molecules with delocalized π-electrons. In contrast, aromatic nonconjugated luminogens possess unusual conductive properties owing to TSC. Recently, Tang and Zhao demonstrated an examples of novel single-molecule wires built on through-space conjugated hexaphenylbenzene (HPB) (Fig. 11e) [69]. The TSC from six face-to-face aligned phenyl rings in HPB compensated exceedingly well for the conductance loss caused by weak through-bond conjugation to obtain Au–molecule–Au junctions. Similar molecules with TSC effect are also great candidates for optoelectronic materials and devices [70].

6 Conclusion

Clusteroluminescence is an unusual photophysical phenomenon from conventional materials without extended π-conjugations. With years of development, the materials have expanded from PAMAM dendrimers at the first discovery, to a large variety of synthetic and biogenic materials such as peptides, polysaccharides, and even small molecules. Both fluorescence and phosphorescence have been reported and the emission wavelength has extended from blue to red. Various mechanisms have been proposed to account for the clusteroluminescence, such as intra/intermolecular interactions, hydrogen bonding, and TSC. A series of applications from clusteroluminogens have been developed. The clusteroluminogens could be constructed from biocompatible and biodegradable materials, thereby showing advantages over some conventional synthetic dyes.

Looking ahead, the development of clusteroluminescence in the following aspects is suggested. First, due to its structural complexity, a simplified model is usually used for macromolecules. Although the mechanisms on macromolecule clusteroluminogens have been studied widely, it remains worth diving into the detailed working mechanism of macromolecules via more state-of-the-art characterization and calculation techniques. Second, the relatively low quantum efficiency of clusteroluminogens limits their biological applications. Thus, developing highly emissive clusteroluminogens remains challenging, especially in the long-wavelength region. Third, in terms of unique luminescence from natural materials, rational design of clusteroluminogens for wide applications in different fields is envisioned. We hope this review will inspire more researchers with diverse expertise to become involved in this intriguing research area and promote clusteroluminogens as powerful materials.

Supplementary Information The online version contains supplementary material available at https://doi.org/10.1007/s41061-021-00326-w.

References

1. Stokes GG (1854) On the change of refrangibility of light. Abstr Pap Commun R Soc Lond 6:195–200
2. Luo J, Xie Z, Lam JWY, Cheng L, Tang BZ, Chen H, Qiu C, Kwok HS, Zhan X, Liu Y, Zhu D (2001) Aggregation-induced emission of 1-methyl-1,2,3,4,5-pentaphenylsilole. Chem Commun 18:1740–1741
3. Zhang H, Zhao Z, Turley AT, Wang L, McGonigal PR, Tu Y, Li Y, Wang Z, Kwok RTK, Lam JWY, Tang BZ (2020) Aggregate science: from structures to properties. Adv Mater 32(36):2001457
4. Mei J, Leung NLC, Kwok RTK, Lam JWY, Tang BZ (2015) Aggregation-induced emission: together we shine, united we soar! Chem Rev 115(21):11718–11940
5. Tomalia DA, Klajnert-Maculewicz B, Johnson KAM, Brinkman HF, Janaszewska A, Hedstrand DM (2019) Non-traditional intrinsic luminescence: inexplicable blue fluorescence observed for dendrimers, macromolecules and small molecular structures lacking traditional/conventional luminophores. Prog Polym Sci 90:35–117

6. Zhang H, Zhao Z, McGonigal PR, Ye R, Liu S, Lam JWY, Kwok RTK, Yuan WZ, Xie J, Rogach AL, Tang BZ (2020) Clusterization-triggered emission: uncommon luminescence from common materials. Mater Today 32:275–292
7. He B, Zhang J, Zhang J, Zhang H (2020) Clusteroluminescence from cluster excitons in small heterocyclics free of aromatic rings. ChemRxiv. https://doi.org/10.26434/chemrxiv.13200491.v1
8. Wang Y, Zhao Z, Yuan WZ (2020) Intrinsic luminescence from nonaromatic biomolecules. ChemPlusChem 85(5):1065–1080
9. Halpern AM (1970) The vapor state emission from a saturated amine. Chem Phys Lett 6(4):296–298
10. Fan Y, Fan Y, Wang Y, Ma J (2007) Unexpected fluorescence emission of poly(α, β-L-malic acid) in aqueous medium. J Appl Polym Sci 106(3):1640–1647
11. Wang D, Imae T (2004) Fluorescence emission from dendrimers and its PH dependence. J Am Chem Soc 126(41):13204–13205
12. Lee WI, Bae Y, Bard AJ (2004) Strong blue photoluminescence and ECL from OH-terminated PAMAM dendrimers in the absence of gold nanoparticles. J Am Chem Soc 126(27):8358–8359
13. Larson CL, Tucker SA (2001) Intrinsic fluorescence of carboxylate-terminated polyamido amine dendrimers. Appl Spectrosc 55(6):679–683
14. Zhen S, Mao J-C, Chen L, Ding S, Luo W, Zhou X-S, Qin A, Zhao Z, Tang BZ (2018) Remarkable multichannel conductance of novel single-molecule wires built on through-space conjugated hexaphenylbenzene. Nano Lett 18(7):4200–4205
15. Joo Y, Agarkar V, Sung SH, Savoie BM, Boudouris BW, Zhang H, Zheng X, Kwok RTK, Wang J, Leung NLC, Shi L, Sun JZ, Tang Z, Lam JWY, Qin A, Tang BZ (2018) A nonconjugated radical polymer glass with high electrical conductivity. Science 359:1391–1395
16. Nordlander P, Oubre C, Prodan E, Li K, Stockman MI (2004) Plasmon hybridization in nanoparticle dimers. Nano Lett 4(5):899–903
17. Prodan E, Radloff C, Halas NJ, Nordlander P (2003) A hybridization model for the plasmon response of complex nanostructures. Science 302:419–422
18. Yanari SS, Bovey FA, Lumry R (1963) Fluorescence of styrene homopolymers and copolymers. Nature 200(4903):242–244
19. Vallan L, Urriolabeitia EP, Ruipérez F, Matxain JM, Canton-Vitoria R, Tagmatarchis N, Benito AM, Maser WK (2018) Supramolecular-enhanced charge transfer within entangled polyamide chains as the origin of the universal blue fluorescence of polymer carbon dots. J Am Chem Soc 140(40):12862–12869
20. Zheng J, Petty JT, Dickson RM (2003) High quantum yield blue emission from water-soluble Au8 nanodots. J Am Chem Soc 125(26):7780–7781
21. Cao L, Yang W, Wang C, Fu S (2007) Synthesis and striking fluorescence properties of hyperbranched poly(amido amine). J Macromol Sci Part A Pure Appl Chem 44(4):417–424
22. Wang D, Imae T, Miki M (2007) Fluorescence emission from PAMAM and PPI dendrimers. J Colloid Interface Sci 306(2):222–227
23. Wu D, Liu Y, He C, Goh SH (2005) Blue photoluminescence from hyperbranched poly(amino ester)s. Macromolecules 38(24):9906–9909
24. Fernández L, Sigal E, Otero L, Silber JJ, Santo M (2011) Solubility improvement of an anthelmintic benzimidazole carbamate by association with dendrimers. Braz J Chem Eng 28:679–689
25. Homchaudhuri L, Swaminathan R (2001) Novel absorption and fluorescence characteristics of L-lysine. Chem Lett 8:844–845
26. Al-Jamal KT, Ruenraroengsak P, Hartell N, Florence AT (2006) An intrinsically fluorescent dendrimer as a nanoprobe of cell transport. J Drug Target 14(6):405–412
27. Ellis B, Brignola P, Brashear RL, Thomas R, Dickerson SH, Dickson HD, Kelly H, Gaul M, Griffin RJ, Hassell AM, Keith B, Mullin R, Petrov KG, Reno MJ, Rusnak DW, Tadepalli SM, Ulrich JC, Craig D, Vanderwall DE, Waterson AG, Williams JD, White WL, Uehling DE, Pompa PP, Maruccio G, Della A, Sabella S, Tamburro AM, Laureana L, Pompa PP, Maruccio G, Torre X, Della A, Sabella S, Tamburro AM, Cingolani R, Rinaldi R (2007) Charge transport and intrinsic fluorescence in amyloid-like fibrils. Proc Natl Acad Sci USA 104(46):18019–18024
28. Ye R, Liu Y, Zhang H, Su H, Zhang Y, Xu L, Hu R, Kwok RTK, Wong KS, Lam JWY, Goddard WA, Tang BZ (2017) Non-conventional fluorescent biogenic and synthetic polymers without aromatic rings. Polym Chem 8(10):1722–1727
29. Caminade AM, Yan D, Smith DK (2015) Dendrimers and hyperbranched polymers. Chem Soc Rev 44:3870–3873

30. Sun M, Hong CY, Pan CY (2012) A unique aliphatic tertiary amine chromophore: fluorescence, polymer structure, and application in cell imaging. J Am Chem Soc 134(51):20581–20584
31. Restani RB, Morgado PI, Ribeiro MP, Correia IJ, Aguiar-Ricardo A, Bonifácio VDB (2012) Biocompatible polyurea dendrimers with PH-dependent fluorescence. Angew Chem 124(21):5252–5255
32. Lu H, Feng L, Li S, Zhang J, Lu H, Feng S (2015) Unexpected strong blue photoluminescence produced from the aggregation of unconventional chromophores in novel siloxane-poly(amidoamine) dendrimers. Macromolecules 48(3):476–482
33. Pastor-Pérez L, Chen Y, Shen Z, Lahoz A, Stiriba SE (2007) Unprecedented blue intrinsic photoluminescence from hyperbranched and linear polyethylenimines: polymer architectures and PH-effects. Macromol Rapid Commun 28(13):1404–1409
34. Wang R, Yuan W, Zhu X (2015) Aggregation-induced emission of non-conjugated poly(amido amine)s: discovering, luminescent mechanism understanding and bioapplication. Chin J Polym Sci (English Ed) 33(5):680–687
35. Zhou Q, Cao B, Zhu C, Xu S, Gong Y, Yuan WZ, Zhang Y (2016) Clustering-triggered emission of nonconjugated polyacrylonitrile. Small 12(47):6586–6592
36. Wang Y, Bin X, Chen X, Zheng S, Zhang Y, Yuan WZ (2018) Emission and emissive mechanism of nonaromatic oxygen clusters. Macromol Rapid Commun 39(21):1–6
37. Lin SY, Wu TH, Jao YC, Liu CP, Lin HY, Lo LW, Yang CS (2011) Unraveling the photoluminescence puzzle of PAMAM dendrimers. Chem A Eur J 17(26):7158–7161. https://doi.org/10.1002/chem.201100620
38. Zhang Z, Li D, Jiang W, Wang Z (2018) The electron density delocalization of hydrogen bond systems. Adv Phys X 3(1):298–315
39. Wang B, Xin M, Dai X, Song R, Meng Y, Han J, Jiang W, Wang Z, Zhang R (2015) Electronic delocalization in small water rings. Phys Chem Chem Phys 17(5):2987–2990
40. Wang L, Fried SD, Boxer SG, Markland TE (2014) Quantum delocalization of protons in the hydrogen-bond network of an enzyme active site. Proc Natl Acad Sci USA 111(52):18454–18459
41. Grisanti L, Pinotsi D, Gebauer R, Kaminski Schierle GS, Hassanali AA (2017) A computational study on how structure influences the optical properties in model crystal structures of amyloid fibrils. Phys Chem Chem Phys 19(5):4030–4040
42. Pinotsi D, Grisanti L, Mahou P, Gebauer R, Kaminski CF, Hassanali A, Kaminski Schierle GS (2016) Proton transfer and structure-specific fluorescence in hydrogen bond-rich protein structures. J Am Chem Soc 138(9):3046–3057
43. Zhao Z, Zhang H, Lam JWY, Tang BZ (2020) Aggregation-induced emission: new vistas at the aggregate level. Angew Chem 20:20
44. Wang Q, Dou X, Chen X, Zhao Z, Wang S, Wang Y, Sui K, Tan Y, Gong Y, Zhang Y, Yuan WZ (2019) Reevaluating protein photoluminescence: remarkable visible luminescence upon concentration and insight into the emission mechanism. Angew Chem Int Ed 58(36):12667–12673
45. Dou X, Zhou Q, Chen X, Tan Y, He X, Lu P, Sui K, Tang BZ, Zhang Y, Yuan WZ (2018) Clustering-triggered emission and persistent room temperature phosphorescence of sodium alginate. Biomacromol 19(6):2014–2022
46. Zhou Q, Wang Z, Dou X, Wang Y, Liu S, Zhang Y, Yuan WZ (2019) Emission mechanism understanding and tunable persistent room temperature phosphorescence of amorphous nonaromatic polymers. Mater Chem Front 3(2):257–264
47. Chen X, Liu X, Lei J, Xu L, Zhao Z, Kausar F, Xie X, Zhu X, Zhang Y, Yuan WZ (2018) Synthesis, clustering-triggered emission, explosive detection and cell imaging of nonaromatic polyurethanes. Mol Syst Des Eng 3(2):364–375
48. Zhao Z, Chen X, Wang Q, Yang T, Zhang Y, Yuan WZ (2019) Sulphur-containing nonaromatic polymers: clustering-triggered emission and luminescence regulation by oxidation. Polym Chem 10(26):3639–3646
49. Eftink MR, Selva TJ, Wasylewski Z (1987) Studies of the efficiency and mechanism of fluorescence quenching reactions using acrylamide and succinimide as quenchers. Photochem Photobiol 46(1):23–30
50. Yan J, Zheng B, Pan D, Yang R, Xu Y, Wang L, Yang M (2015) Unexpected fluorescence from polymers containing dithio/amino-succinimides. Polym Chem 6(34):6133–6139
51. Wang Y, Tang S, Wen Y, Zheng S, Yang B, Yuan WZ (2020) Nonconventional luminophores with unprecedented efficiencies and color-tunable afterglows. Mater Horizons 7(8):2105–2112

52. Zhao E, Lam JWY, Meng L, Hong Y, Deng H, Bai G, Huang X, Hao J, Tang BZ (2015) Poly[(maleic anhydride)-alt-(vinyl acetate)]: a pure oxygenic nonconjugated macromolecule with strong light emission and solvatochromic effect. Macromolecules 48(1):64–71
53. Hu C, Guo Z, Ru Y, Song W, Liu Z, Zhang X, Qiao J (2018) A new family of photoluminescent polymers with dual chromophores. Macromol Rapid Commun 39(10):1–6
54. Zhou Q, Yang T, Zhong Z, Kausar F, Wang Z, Zhang Y, Yuan WZ (2020) A Clustering-triggered emission strategy for tunable multicolor persistent phosphorescence. Chem Sci 11(11):2926–2933
55. Teki Y (2020) Excited-state dynamics of non-luminescent and luminescent π-radicals. Chem A Eur J 26(5):980–996
56. Wang Z, Zou X, Xie Y, Zhang H, Hu L, Chan CCS (2020) A nonconjugated radical polymer with stable red luminescence in solid state. ChemRxiv. https://doi.org/10.26434/chemrxiv.12924200.v1
57. Hirayama F (1965) Intramolecular excimer formation. I. Diphenyl and triphenyl alkanes. J Chem Phys 42(9):3163–3171
58. Brown CJ, Farthing AC (1949) Preparation and structure of di-p-xylylene [2]. Nature 20:915–916
59. Zhang H, Zheng X, Xie N, He Z, Liu J, Leung NLC, Niu Y, Huang X, Wong KS, Kwok RTK, Sung HHY, Williams ID, Qin A, Lam JWY, Tang BZ (2017) Why do simple molecules with "isolated" phenyl rings emit visible light? J Am Chem Soc 139(45):16264–16272
60. Han T, Deng H, Qiu Z, Zhao Z, Zhang H, Zou H, Leung NLC, Shan G, Elsegood MRJ, Lam JWY, Tang BZ (2018) Facile multicomponent polymerizations toward unconventional luminescent polymers with readily openable small heterocycles. J Am Chem Soc 140(16):5588–5598
61. Chen L, Wang YH, He B, Nie H, Hu R, Huang F, Qin A, Zhou XS, Zhao Z, Tang BZ (2015) Multichannel conductance of folded single-molecule wires aided by through-space conjugation. Angew Chem Int Ed 54(14):4231–4235
62. Robb MJ, Li W, Gergely RCR, Matthews CC, White SR, Sottos NR, Moore JS (2016) A robust damage-reporting strategy for polymeric materials enabled by aggregation-induced emission. ACS Cent Sci 2(9):598–603
63. Bartholomew GP, Bazan GC (2001) Bichromophoric paracyclophanes: models for interchromophore delocalization. Acc Chem Res 34(1):30–39
64. Zhang H, Du L, Wang L, Liu J, Wan Q, Kwok RTK, Lam JWY, Phillips DL, Tang BZ (2019) Visualization and manipulation of molecular motion in the solid state through photoinduced clusteroluminescence. J Phys Chem Lett 10(22):7077–7085
65. Yang W, Pan CY, Luo MD, Bin ZH (2010) Fluorescent mannose-functionalized hyperbranched poly(amido amine)s: synthesis and interaction with *E. coli*. Biomacromol 11(7):1840–1846
66. Yang W, Pan CY, Liu XQ, Wang J (2011) Multiple functional hyperbranched poly(amido amine) nanoparticles: synthesis and application in cell imaging. Biomacromol 12(5):1523–1531
67. Qiu L, Zhu C, Chen H, Hu M, He W, Guo Z (2014) A turn-on fluorescent Fe3+ sensor derived from an anthracene-bearing bisdiene macrocycle and its intracellular imaging application. Chem Commun 50(35):4631–4634
68. Tao S, Lu S, Geng Y, Zhu S, Redfern SAT, Song Y, Feng T, Xu W, Yang B (2018) Design of metal-free polymer carbon dots: a new class of room-temperature phosphorescent materials. Angew Chem Int Ed 57:2393–2398
69. Zhen S, Mao JC, Chen L, Ding S, Luo W, Zhou XS, Qin A, Zhao Z, Tang BZ (2018) Remarkable multichannel conductance of novel single-molecule wires built on through-space conjugated hexaphenylbenzene. Nano Lett 18:4200–4205
70. Li J, Shen P, Zhao Z, Tang BZ (2019) Through-space conjugation: a thriving alternative for optoelectronic materials. CCS Chem 1(2):181–196

REVIEW

Aggregation-Induced Emission in Electrochemiluminescence: Advances and Perspectives

Guillermo Moreno-Alcántar[1] · Alessandro Aliprandi[1] · Luisa De Cola[1,2,3]

Received: 10 March 2021 / Accepted: 25 May 2021 / Published online: 19 June 2021

Abstract

The discovery of aggregation-induced electrochemiluminescence (AIECL) in 2017 opened new research paths in the quest for novel, more efficient emitters and platforms for biological and environmental sensing applications. The great abundance of fluorophores presenting aggregation-induced emission in aqueous media renders AIECL a potentially powerful tool for future diagnostics. In the short time following this discovery, many scientists have found the phenomenon interesting, with research findings contributing to advances in the comprehension of the processes involved and in attempts to design new sensing platforms. Herein, we explore these advances and reflect on the future directions to take for the development of sensing devices based on AIECL.

Graphic abstract

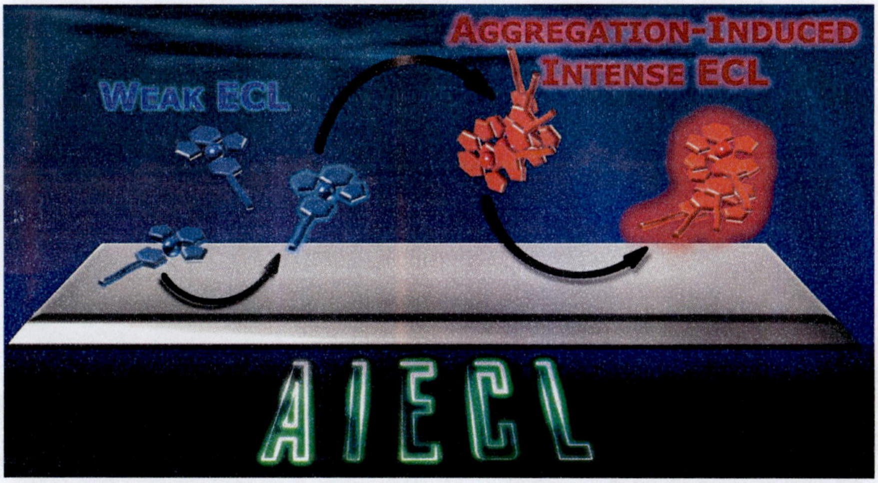

Chapter 3 was originally published as Moreno-Alcántar, G., Aliprandi, A. & Cola, L. D. Topics in Current Chemistry (2021) 379: 31. https://doi.org/10.1007/s41061-021-00343-9.

Extended author information available on the last page of the article

Keywords Aggregation-induced electrochemiluminescence · Sensing · Electrochemistry · Photophysics · Enhanced emission

1 Aggregation-Induced Electrochemiluminescence

Electrochemiluminescence (ECL) is a technique that uses an applied electrical bias to generate reactive species in the vicinity of an electrode that afterwards undergo subsequent electron-transfer reactions, leading to the formation of a luminescent excited state [1]. Although the discovery of this phenomenon was reported at the beginning of the twentieth century [2, 3], the first acknowledgeable studies were performed by Hercules [4], Visco and Chandross [5], and Santhanam and Bard [6] in the 1960s. Since then, and particularly in recent decades, the understanding of the processes and the applications of ECL have rapidly expanded [7]. In ECL, light emission can be obtained mainly by two mechanisms (see Fig. 1). If the reaction takes place between the oxidized and the reduced forms of the luminophore, both generated sequentially at the working electrode, the mechanism is defined as annihilation. In contrast, if the reaction involves the use of sacrificial reagents that upon oxidation/reduction decompose into highly energetic species able to react with the luminophore, thereby generating the excited state, during a single potential step or scan, the ECL mechanism is denoted as the co-reactant type. As opposed to chemiluminescent systems where all reagents are irreversibly consumed, in ECL, the luminophore is regenerated after each cycle; consequently, a large number of photons can be produced. Such a signal generation mechanism combined with the great spatial and temporal control derived from the electrochemical generation of reactive species are the strengths of ECL compared to other analytical techniques. Moreover, the absence of background noise, since no light for excitation is needed as in traditional photoluminescent (PL) measurements, results in an extremely high signal-to-noise ratio and high selectivity [8]. Therefore, it is not surprising that nowadays, ECL is used widely as a clinical diagnosis technique and has even become the gold standard in hospitals all over the world [9]. In addition, its applications are not limited to the biomedical field but also for food and water analysis [10, 11] as well as explosives detection [12, 13], just to cite some of the technique's many applications.

Since the luminophore is regenerated at each cycle the emission intensity is not only dictated by the intrinsic photoluminescent properties of the compound but also by the rate of the redox reactions responsible for the formation of the excited state once the voltage is applied. Therefore, an efficient ECL luminophore must possess suitable redox properties, be stable in its oxidized or reduced state and possibly have excellent PL properties. For diagnostics and biomedical applications, the luminophore must also possess a good degree of solubility in aqueous media where biological analyses are usually performed. Keeping into account also the role of the spin statistics in the formation of the exciton [14], the use of phosphorescent emitters have been found to outperform fluorescent dyes, resulting in the field being dominated by transition metal complexes, in particular ruthenium and iridium complexes [15–21]. Unfortunately, phosphorescent compounds possess relatively low photoluminescence quantum yield (PLQY), especially in aerated conditions as elemental

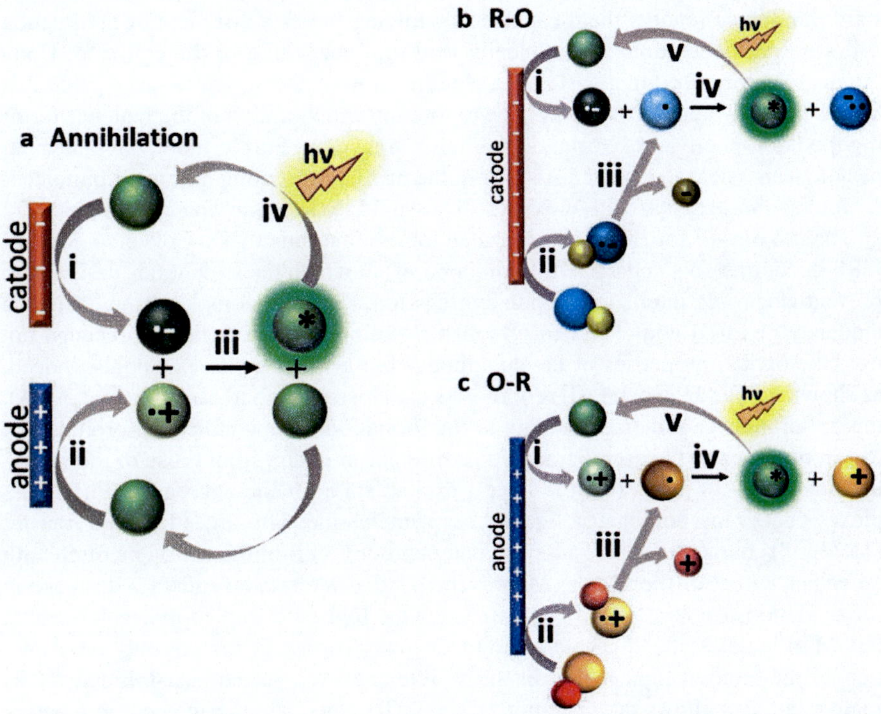

Fig. 1 Electrochemiluminescence (ECL) mechanisms. **a** Annihilation: first the reduction (*i*) and oxidation (*ii*) of the luminophore take place; then the annihilation reaction between the oxidized and reduced luminophore radicals (*iii*) produces the electronically excited form of the luminophore that relaxes to the ground state upon light emission (*iv*). **b** Co-reactant reduction–oxidation (*R–O*): the reductions of both luminophore (*i*) and co-reactant (*ii*) take place; then, after a subsequent reaction (generally bond cleavage), the reduced form of the co-reactant generates a strong oxidizing radical (*iii*) which exergonically reacts with the reduced luminophore (*iv*) to generate an excited state, which in turn relaxes to the ground state upon light emission (*v*). **c** Co-reactant oxidation–reduction (*O–R*): the oxidations of both luminophore (*i*) and co-reactant (*ii*) take place; then the co-reactant decomposes to form a strong reducing radical (*iii*) that subsequently reduces the luminogen (*iv*) to generate the electronic excited state that relaxes to the ground state upon light emission (*v*)

oxygen is an effective quencher of the long-lived excited triplet state. Even for the fluorescent probes, water and polar solvents are often not ideal media for the emission properties, compared with an organic apolar solvent [22]. An aqueous solution can detrimentally affect the emissive behavior by acting as an electron or proton acceptor through hydrogen bond formation [23] as well as by promoting the formation of excimers or exciplex [24]. Such limitations have motivated researchers to exploit new approaches to increase the performance of luminophores, such as by employing nanomaterials [25–27] and supramolecular aggregates [28, 29]. The aggregation-induced emission (AIE) [30–34] and aggregation-induced emission enhancement (AIEE) [35–37] phenomena could be considered to be an ideal tool to overcome the limitation exerted by the media. AIE systems require a careful molecular design in order to make them readily aggregate in solution but introducing steric hindrance renders them capable of avoiding the establishment of detrimental π–π stacking interactions that typically lead to a quenching of the emission. Upon the aggregation of certain systems, a rigidification of the structure takes place that blocks the non-radiative pathways due to rotation and vibration of the moieties forming the fluorophore and favoring the radiative transition. Furthermore, the molecular packing can shield the excited state from the media, preventing unwanted bimolecular interactions, such as with dioxygen, that would lead to emission quenching.

The use of AIE for ECL was described for the first time by De Cola et al. in 2017 [38]. In contrast to standard AIE luminophores, in which the structure is designed to prevent electronic interactions upon aggregation, the authors took advantage of the tendency of Pt(II) complexes to establish metallophilic interactions to change not only the optical properties of the compounds but also the redox potentials. Indeed, as shown in Fig. 2a, when Pt(II) complexes are close enough to each other (< 3.5 Å), the dz^2-orbitals can interact, leading to the formation of new molecular orbitals that can in turn lead to electronic transitions different in nature from those of the monomeric species. In particular, the lowest excited state for the aggregated Pt(II)complexes containing conjugated ligands has a metal–metal–to–ligand charge transfer (MMLCT) character. This results in a bathochromic shift of the emission, often with an enhancement of the PLQY. Moreover, this d–d interaction causes a decrease in the oxidation potential of the complex since the highest occupied molecular orbital (HOMO) is destabilized. The shift towards lower energy favors not only the detection of the emitted light, being in the visible and even in the near-infrared (NIR) region, but also allows the oxidation of the Pt(II) units, which can occur at a potential that can be reached by the common co-reactants or by applying a low voltage at the electrode. Therefore, ECL that is not possible with monomeric species becomes available upon their self-assembly in discrete structures in which electronic interactions are present. This is the reason why this new phenomenon has been defined as aggregation-induced electrochemiluminescence (AIECL), i.e. to stress that the aggregation leads to ECL-active materials. As shown in Fig. 2, the investigated Pt(II) amphiphilic complexes consist of a tridentate chromophoric ligand and an ancillary pyridine functionalized with one (**Pt-PEG**) or two (**Pt-PEG$_2$**) hydrophilic

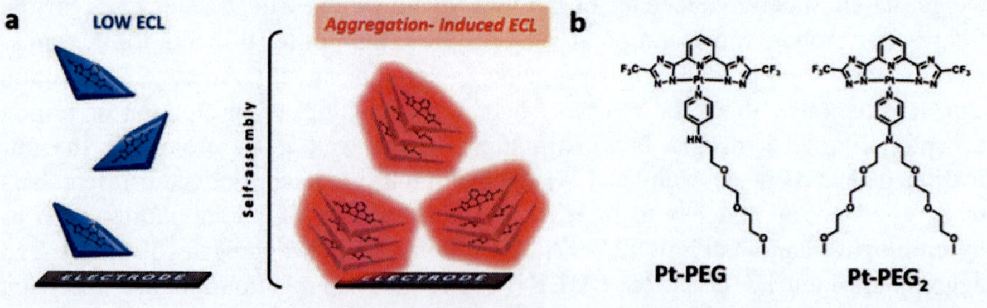

Fig. 2 **a** Aggregation-induced electrochemiluminescence (AIECL) observed upon aggregation of neutral Pt(II) amphiphilic compounds. **b** Structures of the Pt(II) amphiphilic complexes **Pt–PEG** and **Pt–PEG2** displaying AIECL. Adapted with permission from Carrara et al. [38]. Copyright 2017 American Chemical Society

triethylene glycol chains. The overall charge has been kept neutral to promote aggregation in polar media while the hydrophilic chains have been introduced to promote their solubility in water. While **Pt-PEG** is almost insoluble in water, **Pt-PEG$_2$** is able to form a stable orange colloidal dispersion in pure aqueous solutions and its ECL performance outperforms the efficiency of the classical ECL reference $[Ru(bpy)_3]^{2+}$[7, 16, 39].

2 Current Advances and Applications

Since the first report of AIECL, several papers have been published that have expanded and diversified the library of luminophores suitable for this technique, ranging from small molecules, both organic [40–42] and inorganic [43], to polymeric arrangements [44] and composites [45]. Diverse types of aggregates, such as nanoparticles (NPs), nanocrystals (NCs) [47], polymeric dots (Pdots) [44] and many others, have been shown to display AIECL. In all these cases, the mechanisms of triggering the ECL upon aggregation remain the same: changing the redox properties, introducing new excited states or changing the nature of the emissive state—and, of course, rigidification of the systems. The availability of a larger library of luminophores has incentivized the application of AIECL in analytical assays to detect diverse analytes of interest, ranging from heavy metals to biologically relevant molecules [46, 48]. AIECL has proven to be a way to extend ECL methodologies to water by harvesting the tendency of hydrophobic luminogens to aggregate in aqueous solution and has already shown superior efficacies and selectivity. Avoiding the more common aggregation-caused quenching (ACQ) is a challenge in the design of systems, but the number of new creative approaches is increasing. Regarding the scope of this review, we consider AIECL processes to be those for which the ECL measurements are performed on aggregates formed in solution, frequently by the addition of non-solvents, as well as observations of the formation and deposition of aggregates of a defined shape and size on the electrode surface prior to the ECL measurements. Some of the more remarkable results are discussed in later sections of this article.

2.1 Small Molecule-Based AIECL

Shortly after the description of our AIECL system [38], Zhang and collaborators reported the first AIECL active organic NPs [46]. Their design, based on a donor–acceptor molecule, namely 6-[4-(*N*,*N*-diphenylamino)phenyl]-3-ethoxycarbonyl coumarin (**DPA-CM**; Fig. 3a), NPs of 5.8 nm were formed by a reprecipitation method and deposited on glassy carbon electrodes (GCE). The **DPA-CM NPs** exhibit AIECL in the presence of $K_2S_2O_8$ or tripropylamine (TPrA). As the latter is far more intense, the oxidation–reduction mechanism was proposed as the most favorable (Fig. 3b). Further, the AIECL signal quenching by different biologically relevant analytes (ascorbic acid, uric acid, and dopamine) was shown to be

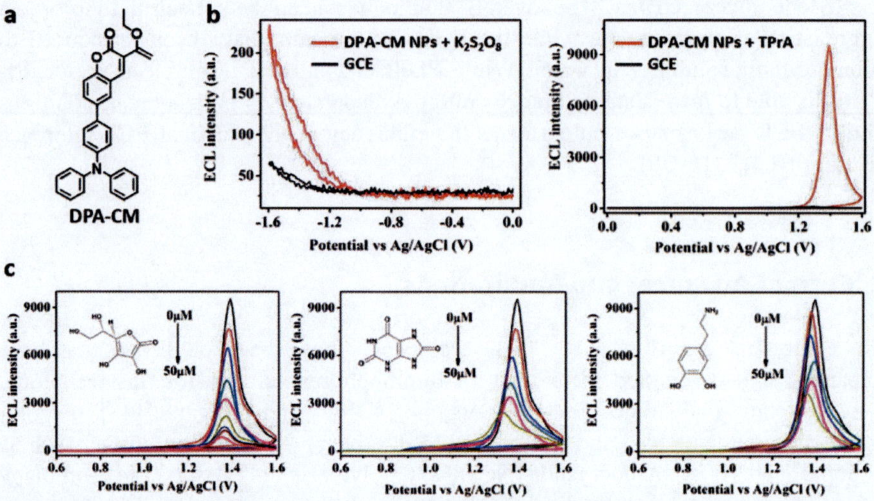

Fig. 3 a **DPA–CM** (6-[4-(*N*,*N*-diphenylamino)phenyl]-3-ethoxycarbonyl coumarin) structure. **b** ECL intensity potential observed in the presence of in 0.1 M phophate buffered saline (PBS; pH 7.40) containing 50 mM $K_2S_2O_8$ or 50 mM tripropylamine (*TPrA*). Scan rate: 0.1 V/s. for bare glassy carbon electrodes (*GCE*; black trace) and **DPA–CM** nanoparticles (**DPA–CM NPs**)-modified GCE (red trace). **c** ECL intensity potential of the **DPA–CM NPs**-modified GCE in 0.1 M PBS (pH 7.40) containing 50 mM TPrA and different concentrations (0, 0.05, 0.1, 0.5, 1.0, 5.0, 10, 25, 50 μM) of acetic acid (left), uric acid (middle) and dopamine (right). Adapted with permission from Liu et al. [46]. Copyright 2017 American Chemical Society

applicable for the detection of these analytes in the 0.05- to 50-μM linear interval, thereby also representing the first application of AIECL in sensing (Fig. 3c).

AIE luminogens result a straightforward choice in the quest for AIECL luminophores. To date, most of the existing systems proving electrochemically generated luminescence upon aggregation are based on well-known aggregation-induced emitters. Tetraphenylethylene (TPE) is among the most explored AIE luminophores. In the molecular form, TPE and its derivatives present a highly dynamic propeller motion; as a result, they are not—or poorly—emissive when dissolved in solvents. In contrast, TPE aggregation impairs free rotation of the phenyl groups and thus deactivates non-radiative relaxation pathways, promoting an enhanced emission [40–42]. AIECL-based on TPE derivatives is typically seen as an oxidation–reduction process, in which the motion restriction plays a key role, in agreement to what has been observed for photoluminescence.

In this regard, Yuan and collaborators [47, 49, 50] have shown AIECL in TPE microcrystals (**TPE MCs**) formed by surfactant-assisted self-assembly (Fig. 4a). While faint ECL was displayed by the dissolved form of TPE in tetrahydrofuran (THF) (20 mM), the microcrystalline form showed an intense red ECL emission ($\lambda_{max}=675$ nm) signal in PBS dispersion using triethylamine (TEA) as co-reactant (Fig. 4b). These MCs have been used for building a biosensor for the detection of a cancer biomarker, Mucin 1 (MUC1), a membrane protein, achieving a linearity range of 10^{-3}–10^{3} pg mL^{-1} and a limit of detection

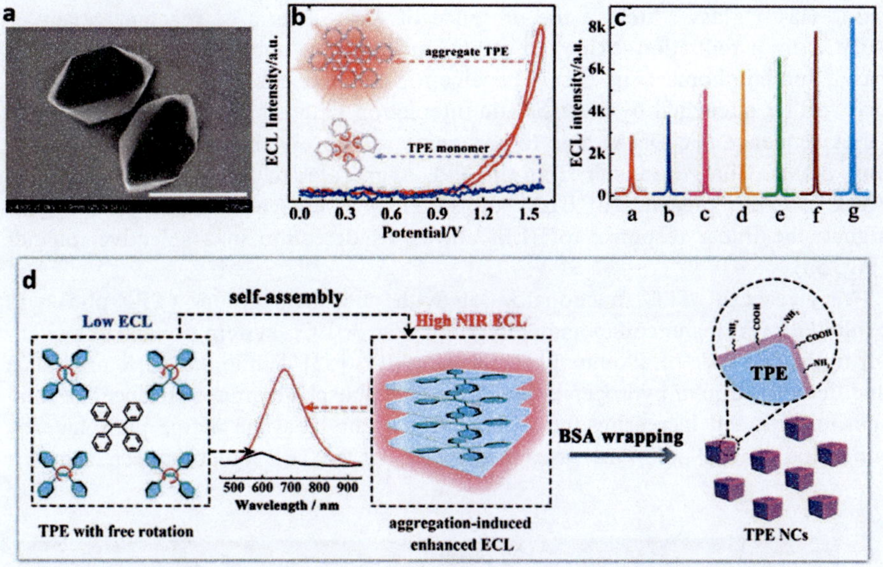

Fig. 4 **a** Scanning electron microscopy image of the tetraphenylethylene (*TPE*) microcrystals (scale bar 2 μm). **b** ECL-potential profile for bare GCE in 0.1 M $TBAPF_6$ THF solution containing 1 mg mL^{-1} TPE monomers and 20 mM triethylamine (TEA; blue trace), and in 0.1 M PBS containing 1 mg mL^{-1} TPE microcrystals (**TPE MCs**) and 20 mM TEA (red trace). **c** ECL responses of the developed biosensor incubated with different Mucin 1 concentrations: *a* 1 fg mL^{-1} *b* 10 fg mL^{-1} *c* 100 fg mL^{-1} *d* 1 pg mL^{-1} *e* 10 pg mL^{-1}, *f* 100 pg mL^{-1}, *g* 1 ng mL^{-1}. **d** Schematic AIECL NIR-emission observed in **BSA–TPE NCs**. Panels **a**–**c** Adapted with permission from Jiang et al. [49]. Panel **d** Adapted from Liu et al. [47] Article [49], copyright 2019 American Chemical Society. Article [47], with permission of The Royal Society of Chemistry

(LOD) of 0.29 fg mL^{-1} which is comparable to the best reported MUC1 sensing platforms (Fig. 4c) [49]. Further, the same group has addressed the origin of the strongly red-shifted AIECL ($\lambda_{max} = 678$ nm) in relation to the photoluminescence ($\lambda_{max} = 440$ nm) of bovine serum albumin (BSA)-coated TPE NCs (**BSA–TPE NCs**). Employing cyclic voltammetry, these authors observed a lowered energetic gap in comparison with that of the PL, suggesting the that the existence of surface states originating the ECL was different from the bulk bandgap observed in the PL; the NIR emission efficiency relative to $[Ru(bpy)_3]^{2+}$ was found to be 1.35% (Fig. 4d) [47]. These **BSA–TPE NCs** were further used for developing a highly sensitive and selective biosensor for the detection of miRNA-141 [50]

Chemical modification of TPE has shown to be a powerful strategy not only for the tuning of the aggregation and emission properties of the materials but, as shown by Lu et al. [51], because it represents a simple way to produce sensing platforms in the aqueous phase. Functionalization of TPE, i.e. introducing positively charged ammonium groups, has been recently explored by the group of Li [52]. The generated positively charged molecule (**QAU-1**) was used to produce self-assemblies with high charge density on the surface of indium tin oxide (ITO) glass which in the presence of $K_2S_2O_8$ as a co-reactant generates AIECL by a reduction–oxidation (R–O) mechanism, uncommon to other TPE-based luminophores (Fig. 5a). The electrogenerated emission of the assembly was further quenched by electrostatic interaction with a ferrocene functionalized DNA sequence (Fc-DNA) able to recognize selectively the antitumoral bleomycin (BLM). Charge transfer from the TPE aggregates to ferrocene quenches the AIECL. In the presence of BLM, the Fc-DNA is cleaved, turning on the ECL signal; the linear response to BLM allows its detection in a selective manner (Fig. 5b).

Very recently, TPE functionalization with phosphate groups (**TPE-pho**) was exploited by Yan and collaborators to produce an AIECL system capable of switch-off in the presence of alkaline phosphatase (ALP) [53]. While the steric hindrance and the formation of hydrogen bonds within the phosphate groups enhance the rotation impairment, increasing the AIECL signal, the cleavage of the phosphate by the action of ALP group increases the mobility of the TPE moieties, decreasing the

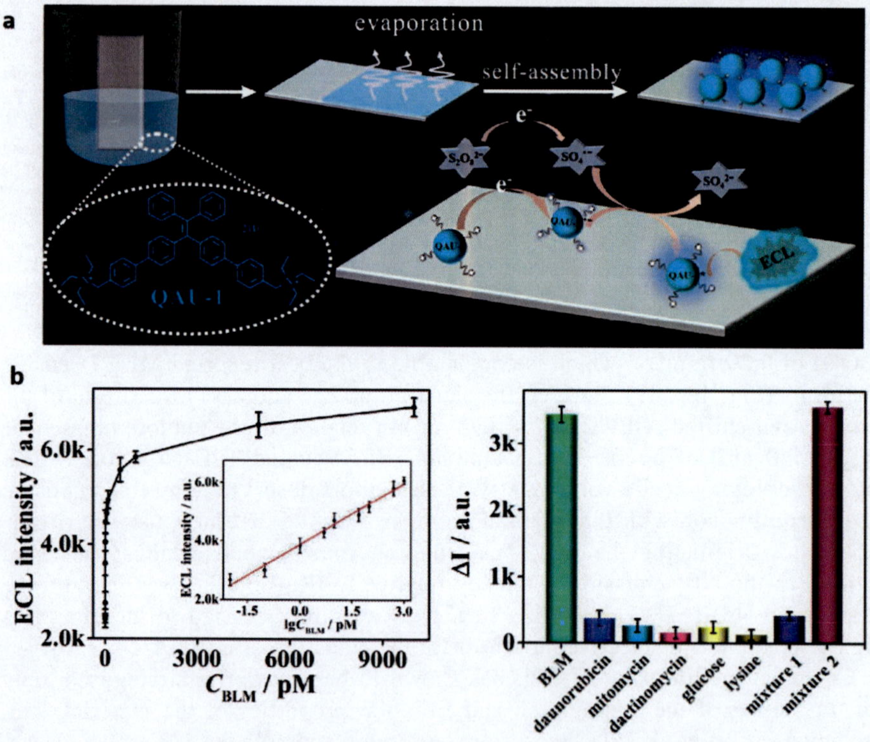

Fig. 5 **a** Schematic representation of the indium tin oxide (ITO) glass functionalization with **QAU-1** assemblies and the AIECL mechanism proposed. **b** Left: ECL response of a ferrocene functionalized DNA sequence (Fc-DNA)/**QAU-1**/ITO-based biosensor to the concentration of bleomycin (*BLM*); the inset shows the linearity of the ECL response with the logarithm of the concentration. Right: Change in the ECL intensity of Fc-DNA/**QAU-1**/ITO-based biosensor in the presence of different analytes. Adapted with permission from Lv et al. [52] Copyright 2020 American Chemical Society

response. The system was used as one-step rapid detector of ALP and showed a linear range of 0.1–6.0 U L^{-1} and a LOD of 0.037 U L^{-1}. These authors also analyzed ALP in human serum, reaching recoveries ranging from 95.12 to 113.8% [53].

Another elegant example employing TPE is the approach followed by Lu et al. [54] in which the aromatic moieties were linked to 5-(4-aminophenyl)-10,15,20-triphenylporphyrin (ATPP) (Fig. 6a). Porphyrins, despite their interesting photophysical properties, suffer from a low solubility in water and a high tendency to aggregation due to π-stacking interactions, resulting in quenching of the H-aggregates. Introducing TPE in the ATTP structure, not only changed the character of the aggregates from ACQ to AIE (Fig. 6b, c) but also enhanced the AIECL response of the system through the favorable energy transfer from the TPE moiety to the ATTP. In optimal conditions (90% water content, using $K_2S_2O_8$ as co-reactant), **ATPP–TPE** reaches a 34% efficiency relative to the benchmark $[Ru(bpy)_3]^{2+}$ and the intensity signal is stable for more than 20 cycles (Fig. 6d). The chemical functionalization including TPE units for AIECL has also been exploited in systems based on polymers (discussed in following section).

It has been reported that J-aggregates of zinc porphyrin complexes display a strong AIECL [55]. Thus, the choice of methodologies that suppress the formation

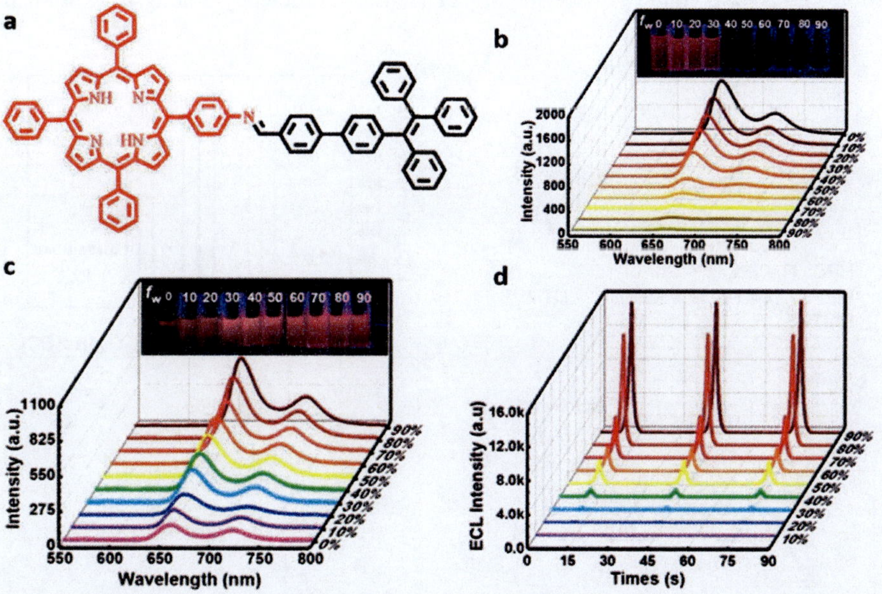

Fig. 6 **a** TPE-decorated 5-(4-aminophenyl)-10,15,20-triphenylporphyrin (ATPP). **b** Photoluminescent (PL) aggregation-induced quenching observed upon the increase of water content from 0 to 90% in solutions containing non-functionalized ATPP in DMF. **c** AIE PL emission displayed by **ATPP–TPE** upon increases in water content from 0 to 90%. Insets of **b** and **c** show the corresponding solutions under UV light (365 nm) **d** ECL-time curves of different degrees of aggregation in **ATPP–TPE**. Conditions: 0.1 M $K_2S_2O_8$ in 0.1 M PBS (pH 7.5) with 0.1 MKCl. Adapted with permission from An et al. [36]. Copyright 2020 John Wiley and Sons

of the π-stacked H-aggregation and favor the displaced J-aggregates is a strategy that can promote the AIECL in porphyrin-based luminophores. Using surfactant-assisted synthesis, Fu and collaborators [56] obtained the J-aggregates of 5,10,15,20-tetrakis (4-carboxyphenyl)porphyrin (**TCPP**) that by an R–O mechanism, using peroxydisulfate as co-reactant, displays AIECL. This system was applied in the construction of a composed platform for the sensing of Cu^{2+}. The authors discovered that the use ofL-cysteine-capped zinc oxide "nanoflowers" (**ZnO@Cys NFs**) improves both the AIECL response and the sensor sensibility through a dual effect: (1) acting as a co-reactant accelerator and (2) acting as an energy donor to populate **TCPP** excited states. Although there was good performance of the sensing platform (linear range 0.001–500 nmol L^{-1} and LOD (0.33 pmol L^{-1}), the mechanism of the dispositive emission quenching by Cu^{2+} was not fully addressed. The authors suggest Cu^{2+} strong coordination to the cysteine as the main factor causing the quenching.

Siloles are another prominent family of AIE active compounds [57, 58] that have recently been explored as AIECL luminogens. Lu et al. [59] demonstrated the potential of 1,1-disubstituted 2,3,4,5-tetraphenylsiloles (Fig. 7a) as AIECL luminophores when deposited on the electrode surface to generate what was denoted heterogeneous aggregation-induced electrochemiluminescence (HAIECL); in solutions of compounds in DMSO with 20% water content weaker ECL was observed, and the behavior in higher water content was not addressed. 1,1,2,3,4,5-Hexaphenylsilole (**HPS**) was found to have the higher HAIECL efficiency amongst the studied

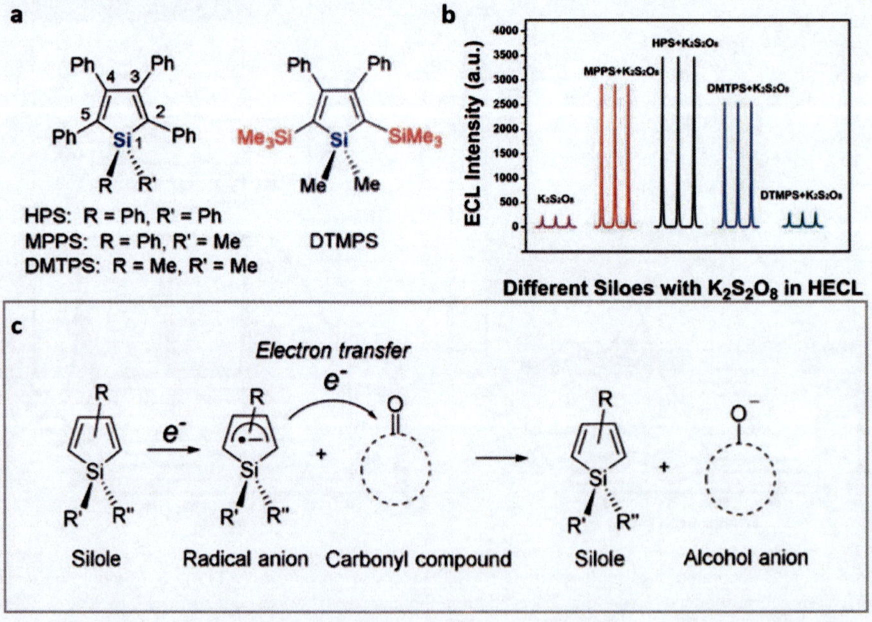

Fig. 7 **a** Small AIECL active siloles. **b** ECL intensity of the different siloles in the aggregated state at the electrode surface. Potential window: – 1.6 to 0 V (vs. Ag/Ag^+); scan rate: 0.1 V s^{-1}; 0.1 M PBS containing 0.1 M KCl and 0.1 M $K_2S_2O_8$, at pH 7.5. **c** Mechanism of the siloles ECL quenching by carbonyl compounds. Adapted with permission from Han et al. [59]. Copyright 2019 John Wiley and Sons

compounds, with 37.8% (relative to $[Ru(bpy)_3]^{2+}$) under optimal conditions, using $K_2S_2O_8$ as co-reactant (Fig. 7b). Further, the authors proposed the use of **HPS** for the detection of carbonyl compounds, taking advantage of the formation of the silole radical anion intermediary in the ECL mechanism which can react with carbonyls, thus quenching the emission (Fig. 7c). This concept was proved in the sensing of the industrial plasticizer di-*n*-butyl ortho-phthalate (DNBP). More recently, this group has also studied the effects of a 4-substitution on the phenyl groups of tetraphenylbenzosiloles (TPBS), observing the positive effect of electron withdrawing groups on the enhancement of the AIECL response [60]. The competing reduction of $Cr_2O_7^{2-}$ was exploited in order to use the system for the detection of hexavalent chromium [60].

An extension of the silole's work was described by the group of Pagenkopf [61] who extended the concept of crystallization-induced emission enhancement [62] to benzosilole systems displaying enhanced AIECL in the crystalline state rather than in the amorphous aggregates. The synthesized compounds (**4T1** and **4T2**; Fig. 8a) display weak ECL in dichloromethane solutions; in the absence of co-reactants the annihilation mechanism produces ECL signals with an efficiency of 2.3 and 2.1%, respectively, for **4T1** and **4T2** relative to $[Ru(bpy)_3]^{2+}$ in the same conditions. The addition of benzoyl peroxide (BPO) as a co-reactant reduces the relative efficiency to 0.08 and 0.4%, respectively. Upon the drop-casting formation of crystalline

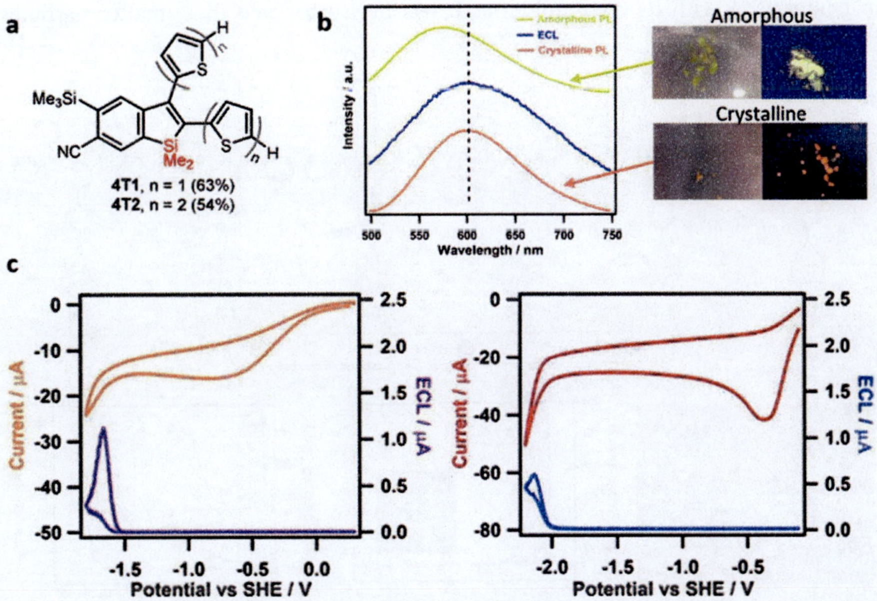

Fig. 8 **a** Structure of the benzosiloles **4T1** and **4T2**. **b** Emission spectra of **4T2**, upon photoexcitation, for the amorphous (yellow trace) and crystalline forms (orange trace), and ECL of the films (blue trace) along with photographs of the **4T2** under natural (left) and 365 nm UV (right) light. **c** Cyclic voltammograms and ECL-potential diagrams of **4T2** (left) and **4T1** (right) films with 5 mM benzoyl peroxide as co-reactant at 0.1 Vs^{-1}. Adapted with permission from Yang et al. [61]. Copyright 2020 John Wiley and Sons

films in GCEs, using a 1:1 water:acetonitrile solvent and upon addition of BPO, as co-reactant, an increase of the efficiency to 2.5% (**4T1**) and 6.5% (**4T2**) has been reported. The mechanism proposed involves a dimerization step, which enhances the emission and causes a red-shift for both solution and amorphous states (Fig. 8b).

One of the first observations of ECL involved 9,10-diphenylanthracene (DPA) as luminophore [6]; recently, DPA cubic NPs (CNPs) have been exploited for the construction of an AIECL immunosensor for the detection of the mycotoxin aflatoxin B_1 (AFB_1) [63]. The authors observed an important increase of the ECL signal when the DPA CNPs were deposited on the GCE surface in comparison with the signal observed when the NPs were dispersed in the solution. The mechanistic studies suggested an oxidation–reduction (O–R) mechanism using TPrA as co-reactant. Upon optimization, the constructed immunosensor showed a LOD of 3 fg·mL^{-1} and a linear range of 0.01–100 ng mL^{-1} for the detection of AFB_1.

Wei et al. proposed the use of carboranyl carbazoles as efficient R–O AIECL fluorophores in order to overcome the limited applications of O–R systems in biosensing [64]. In aqueous media, the synthesized molecules (**T-1–T-6**; Fig. 9a) associate in NPs that display AIE behavior, which is remarkable for compound **T-3**. The excellent properties of this compound are thought to be related to the presence of the 1,2-diphenyl-*o*-carboranyl moiety in the structure, which allows a more efficient back-donation. Although all of the reported compounds performed 10- to 40-fold-enhanced ECL emission upon aggregation, **T-3** increased the emission by 221-fold in comparison with its THF solution. It has been observed that smaller particles

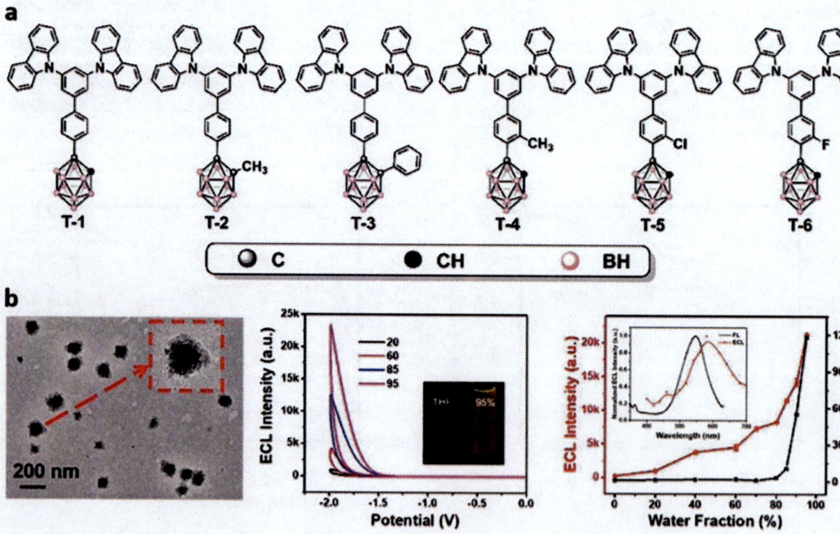

Fig. 9 **a** Structures of the carboranyl carbazoles **T-1**—**T-6**. **b** Left: Transmission electron microscopy image of **T-3** aggregates formed in 95% H_2O. Center: ECL signal enhancement on **T-3** at different water contents; the inset shows the AIE PL. Right: AIECL (red trace) and AIE (black trace) as a function of the water content of the media; the inset shows the AIECL (red) and PL (black) spectra. Adapted with permission from Wei et al. [64]. Copyright 2019 John Wiley and Sons

show a more intense ECL. The red-shifted emission of the ECL peak (582 nm) in comparison with the PL (547 nm) is ascribed to surface state transitions that are less energetic than the bulk transitions (Fig. 9b); in all the cases, $K_2S_2O_8$ was used as a co-reactant in solutions containing 95% water: 5% THF.

Recently, following up on their interest on the use of donor–acceptor AIE luminophores to generate ECL-active films [65], the group of Liang reported the AIECL behavior of a benzothiadiazole-bistriphenylamine (**BTD–TPA**) [66]. This compound displays an aggregation-induced ECL enhancement of 252-fold when compared with the aggregates formed in water with the compound in THF solution. The authors also produced films in which the use of Au wafers as electrodes boosted the observed signal by a factor of eight compared with the films formed in GCE. This effect was addressed because of the electrocatalytical role of gold in the oxidation processes of both luminogen and coreactant. In addition, the study of the mass amplifying effect of the ECL by the increase of the loading leads to the morphological characterization of the aggregates formed on the electrode, showing that gold acts as nucleating centers that promote the growing of grass-like aggregates not observed in GCE. Under optimal conditions, using triethanolamine (TEOA; 300 mM) as coreactant these grassy aggregates have shown an ECL efficiency of up to 25.6% relative to $[Ru(bpy)_3]^{2+}$. Finally, the system was used to detect dopamine with a low LOD (3.3×10^{-16} M) in a linear range from 10^{-15} to 10^{-8} M [66].

All of the examples illustrated up to this point are for organic systems, even though the first description of AIECL refers to transition-metal coordination compounds. Indeed, metal complexes have also been employed to construct AIECL systems. These include coordination compounds in both polymeric (vide infra) or molecular forms. The group of Ye reported the first molecular iridium complex displaying AIECL [43]. The synthesized **[Ir(tpy)(bbbi)]** (with bbbiH_3 = 1,3-*bis*(1*H*-benzimidazol-2-yl)benzene) (Fig. 10a) displays weak PL and ECL in the dissolved forms, which the authors attributed to the vibrational and rotational freedom. The restriction of molecular vibrations in the aggregated state activates the radiative paths giving place to enhanced PL and ECL. In 90% water:10% DMSO, **[Ir(tpy)(bbbi)]** forms monodisperse 120-nm aggregates; a further increase of the water content to 98% increases the average size to 160 nm. The increase in the size of

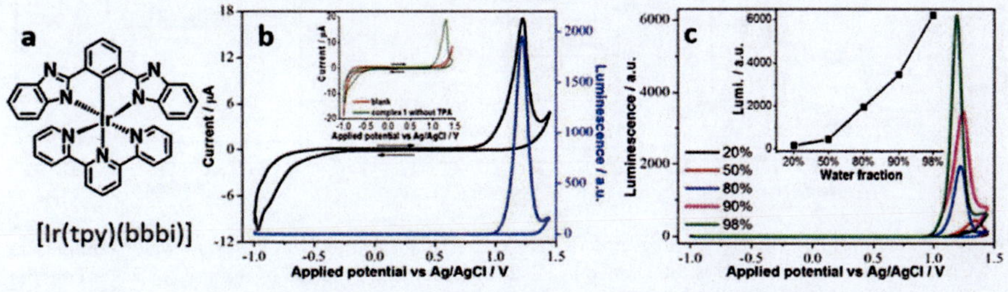

Fig. 10 **a** Chemical structure of **[Ir(tpy)(bbbi)]**. **b** Cyclic voltammogram (black trace) and ECL-potential curves of the iridium complex (c = 200 μM) in DMSO:H_2O 20:80; the inset shows the blank curve (red trace). Conditions: 1 mM TPrA and 100 mM NaCl in 10 mM PBS at pH 7.4. Adapted with permission from Gao et al. [43]. Copyright 2018 American Chemical Society

the nanoaggregates is consistent with an increase of the ECL signal using TPrA as co-reactant and applying a potential of about + 1.23 V, which the authors attributed to the oxidation of **[Ir(tpy)(bbbi)]** to $[Ir(tpy)(bbbi)]^+$ (Fig. 10b). The ECL intensity of these aggregates has shown to be approximately 39-fold more intense than that of its dissolved form and approximately fourfold more intense than that of $[Ru(bpy)_3]^{2+}$(Fig. 10c). Additionally, the complex aggregates showed a linear increase in ECL intensity upon the addition of bovine serum albumin (BSA), which binds to the surface of the nanoaggregates, diminishing the molecular vibrations; this response suggests the potential of the compound in the development of AIECL biosensors.

Very recently, Gu et al. reported the use of dichlorobis(1,10-phenanthroline) ruthenium(II) (**[Ru$_2$(phen)$_2$Cl$_2$)]** as AIECL luminophore and its application in the sensing and differentiation of nucleic acids [67]. The PL and AIECL emissions of **[Ru$_2$(phen)$_2$Cl$_2$]** aggregates, using TPrA as co-reactant, were found to be at a maximum in a 70% water:30% CH_3CN mixture; at higher water content the increase in the size of the aggregates becomes counterproductive, and a decrease in the intensity of the AIECL is observed (Fig. 11a). In 10% H_2O:90% CH_3CN, minor AIECL is observed from the pure complex. The addition of different nucleic acid types (i.e. RNA, ssDNA or dsDNA) triggers different aggregation forms that display differentiable ECL signals (Fig. 11b). Furthermore, the ECL response was affected by the abundance of different nucleobases. This approach even allows the differentiation of diverse miRNA strands (Fig. 11c).

A very innovative approach has recently been shown by Wei et al. that produced an AIECL fluorophore by confining molecules of *fac*-tris(2-phenylpyridine) iridium(III) ($Ir(ppy)_3$) [68]. The aggregation of around 44 molecules within the apoferritin (apoFt) cavity caused a 5.3-fold enhancement of the ECL emission compared to the monomer signal. The encapsulation was achieved by the pH-triggered assembly–disassembly of apoFt. The **Ir(ppy)$_3$@apoF**t bioconjugate was used to construct an immunosensor for the biomarker for squamous cell carcinomas CYFRA21, with a good linearity range (1 pg mL^{-1} – 50 ng mL^{-1}) and an interesting

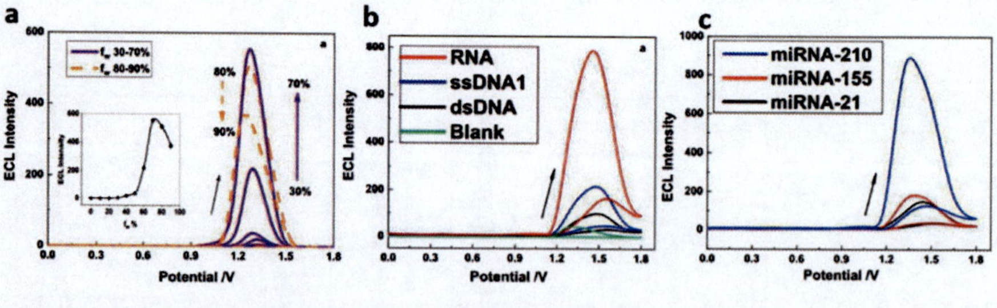

Fig. 11 **a** Potential-AIECL signal diagrams of **[Ru$_2$(phen)$_2$Cl$_2$]** (50 μM) at different water contents in CH_3CN mixtures; the inset shows the change in EICL intensity as a function of the solvent water content. Conditions: 10 mM TPrA, 0.10 M LiClO4 on a GCE, scan rate of 0.10 V/s. **c** ECL of **[Ru$_2$(phen)$_2$Cl$_2$]** aggregates (50 μM) in the presence of 2 μM ssDNA-1, dsDNA or RNA. **c** ECL of $Ru(phen)_2Cl_2$ aggregates (50 μM) in the presence of 2 μM miRNA-210, miRNA-155, and miRNA-2. Conditions for **b** and **c**: 0.10 M $LiClO_4$ in 10% (v/v) H_2O–MeCN using a GCE at a scan rate of 0.10 Vs^{-1}. Adapted with permission from Gu et al. [67]. Copyright 2020 American Chemical Society

LOD (0.43 pg mL^{-1}). This type of bioconjugate approach can help to overcome the limited colloidal stability of some strongly insoluble AIECL luminogens [68].

2.2 Macromolecules, Polymeric Structures and Metallic Clusters Matrix-Based AIECL

With the increasing interest in and production of AIECL luminophores and platforms for biosensing, not only small molecules have been used but macromolecular systems have also been explored. An example of such large structures is polymeric dots (Pdots). The robustness, tunability and versatility of Pdots allow the inclusion of active molecular moieties as monomers while preserving their properties. Concomitant to the interest in the aggregation of TPE molecules (see previous section), the TPE moiety has been included in several polymeric designs to generate AIECL. The group of Ju reported donor–acceptor terpolymers including fluorene (**P-1**) or carbazole (**P-2**) donor units, TPE as AIE agents and BODIPYs as acceptor chromophores [44] (Fig. 12a). These polymers were combined with poly(styrene-co-maleic anhydride) (PSMA) to form Pdots. The use of carbazole as a strong donor unit redshifts the emission of the Pdots, increases the ECL intensity of **P-2** in comparison with **P-1** (by approx. fourfold), and also decreases the anodic peak potential by about 553 mV in the annihilation mechanism. Further, the use of TPrA as co-reactant enhances the ECL signal by promoting an O–R mechanism, and the enhancement is more pronounced for **P-2**, giving an approximately sixfold more intense signal than **P1** (Fig. 12b). **P-1** and **P-2** achieved ECL efficiencies of 5.8 and 11.8%, respectively, when compared with $[Ru(bpy)_3]^{2+}$. A similar design was used to sense two biologically relevant catechol derivatives: epinephrine and dopamine [69]. The sensing principle relies on the quenching of Pdots emission by the formation of *o*-benzoquinone derivatives as a product of the catechol oxidation. For epinephrine and dopamine, this method has been shown to have linear ranges of 10 to 500 μM and of 10 nM to 100 μM, with detection limits of 3 nM and 7 nM, respectively.

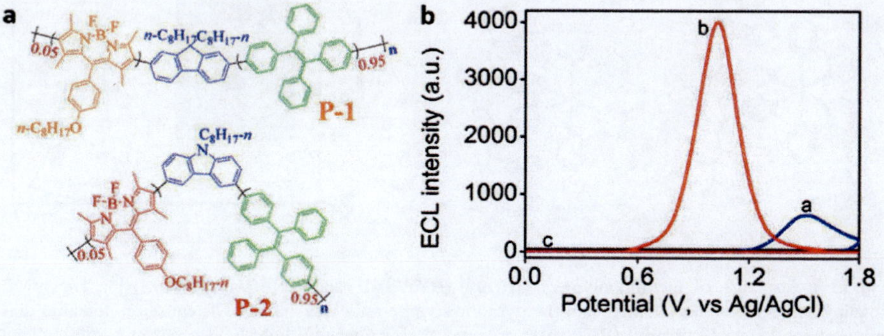

Fig. 12 **a** Structures of donor–acceptor-TPE polymers **P-1** and **P-2**. **b** ELC-potential signal of **P-1** Pdots (*a*, blue trace), **P-2** Pdots (*b*, red trace), and GCE (*c*, black trace). Conditions: 0.1 M pH PBS at pH 7.4 in the presence of 0.1 M TPrA as anodic co-reactant. Scan rate: 0.1 V s^{-1}. Adapted with permission from Wang et al. [44]. Copyright 2018 American Chemical Society

TPE was also recently used to generate conjugated microporous polymers by Suzuki cross-coupling of 1,1,2,2-tetrakis(4-bromophenyl)ethylene (TBPE) with *tris*(4-ethynylphenyl)amine. [70] The system displays an 1.72 % relative ECL efficiency vs. $[Ru(bpy)_3]^{2+}$ and was applied to sense dopamine using the *o*-benzoquinone-based quenching principle previously described, reaching linearity in the interval of 10 nM to 500 μM with a detection limit of 0.85 nM. The application of TPE-conjugated polymers for sensing has been scaled to a more practical level by the group of Hua et al. [48] who have used boron ketoiminate (BKM)-TPE polymers together with PSMA to build AIECL **BKM–TPE Pdots** of approximately 11 nm in size. Functionalization with DNA opens up the possibility of aggregate selectively with UO_2^{2+}, a common nuclear waste, which is capable of transferring energy to the Pdots, thereby increasing the ECL signal. The detection of UO_2^{2+} by this method is linear in the range of 0.05–100 nM and presents a very low limit of detection (10.6 pM) for uranyl. These excellent results were translated into a portable device to detect UO_2^{2+} in natural water sources.

In 2018, Hogan et al. reported AIECL observed in cyclometalated Iridium(III) metallopolymers **$[Ir(ppy)_2(PVP/S)_2]^+$** and **$[Ir(dfppy)_2(PVP/S)_2]^+$** (Fig. 13a) [71]. After addressing the PL and AIE behavior of the compounds by using theoretical methods, the authors showed that the synthesized polymers display ECL emissions in DMF solutions and thin layer films over GCE, following an O–R mechanism in the presence of $Na_2C_2O_4$ as co-reactant. More importantly, the authors reported the formation of small polymeric NPs (PNPs) of around 44 nm by a reprecipitation method from THF/water (Fig. 13b). The intensity of the ECL signal of **$[Ir(dfppy)_2(PVP/S)_2]^+$** is increased by a factor of 12 when compared to the thin layer signal under similar conditions (Fig. 13c).

The effects of the inclusion of iridium luminophores into polymeric TPE-conjugated matrices attracted the attention of Xu's group who found that the position of the iridium complexes capping or embedding the polymer is important in the design

Fig. 13 **a** Structure of metallopolymers **$[Ir(ppy)_2(PVP/S)_2]^+$** and **$[Ir(dfppy)_2(PVP/S)_2]^+$**. The metal loading was 1 metal:5 monomers and the pyridine:styrene ratio was 2:1. **b** ECL emission-potential diagram of solid-state **$[Ir(dfppy)_2(PVP/S)_2]^+$** polymer (lower intensity: light blue trace) and immobilized polymeric NPs (PNPs) (higher intensity: dark-blue trace) on GCE (scan rate 0.05 V s^{-1} in 0.1 M H_2SO_4 water solution using 20 mM $Na_2C_2O_4$ as co-reactant. Inset shows the PL of **$[Ir(dfppy)_2(PVP/S)_2]^+$** in THF solution and water as PNPs. Adapted with permission from Carrara et al. [71]. Copyright 2018 American Chemical Society

of these hybrid Pdots [72]. The Ir complex-capped Pdots (**Pdots1**) showed a more intense ECL signal compared to those including different quantities of embedded complex using 2,2′-(butylimino)diethanol as co-reactant. Actually, while the efficiency of the ECL for **Pdots1** (19.8%) was higher than that observed in the non-iridium functionalized Pdots0 (5.05%), the inclusion of an embedded Ir complex into **Pdots2-5** decreased the efficiency (2.92, 1.35, 0.88, 0.150). The gradual decrease in the efficiency from **Pdots2** to **Pdots5** correlates with an increase in the iridium loading of the polymer. This behavior is attributed to the disruption of the TPE conjugation by the inclusion of the Ir-complex units.

The construction of two-dimensional polymeric metal–organic layers (MOL) using the TPE-derived ligand **H_4ETTC** (4′,4‴,4‴″,4‴‴″-(ethene-1,1,2,2-tetrayl) tetrakis {([1,1'-biphenyl]-4-carboxylic acid))} coordinated to hafnium(IV) (**Hf-ETTC-MOL**; Fig. 14a) has shown to have an advantageous ECL performance not only when compared to monomeric or aggregated bare **H_4ETTC**, but also against the related three-dimensional metal–organic framework **Hf-ETTC-MOF** [73]. According to Xiao and collaborators, the main factor causing this superior performance of the MOL is the facile diffusion next to the electrode of the entities involved in the O–R ECL process (ions, electrons, co-reactants and intermediaries) through the thin porous layer. This material was then employed in the building of an aptasensor for the detection of carcinoembryonic antigen (CEA), a biomarker overexpressed in several cancer types. The proposed sensor is able to give a linear ECL response in the interval of concentration from 1 fg mL^{-1} to 1 ng mL^{-1} with a LOD

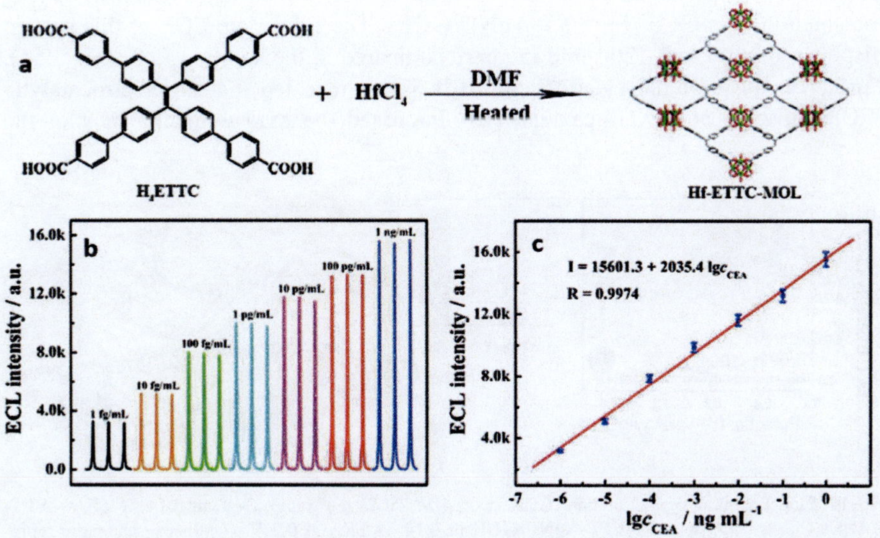

Fig. 14 **a** Synthesis of **Hf-ETTC-MOL**. **b** ECL responses of the aptasensor developed using **Hf-ETTC-MOL** when incubated with different concentrations of carcinoembryonic antigen (c_{CEA}). **c** Calibration curve for the ECL intensity vs. the logarithm of c_{CEA}. Adapted from Yang et al. [73] with permission from The Royal Society of Chemistry

of 0.63 fg mL^{-1}. This sensor also displayed excellent stability and selectivity for CEA in serum (Fig. 14b, c).

Wang and collaborators created an AIECL system based on adenosine phosphate-coated gold nanoclusters (**AuAXP**; where X stands for the number of phosphate units; M for mono; D for di; T for tri) [74]. In the presence of divalent ions, such as Ca^{2+}, Mg^{2+} or Zn^{2+}, the **AuAXP** aggregate forms a nano-hydrogel scaffold. This aggregation causes an enhancement in both the PL and ECL response of the material. The ECL response caused by Ca^{2+} has been shown to be modulated by the presence of the Ca^{2+} binding protein calmodulin (CaM). The response demonstrates a linear interval at concentrations ranging from 0.3 to 50 μg mL^{-1}, with a LOD of 0.1 μg mL^{-1}. This method also allows the assays of CaM in serum samples, with a recovery of 92–117.5%.

The application of EuSe in AIECL was explored by Yafei Li and coworkers [75]. First, they developed a method to control the size of EuSe nanocubes to the range 3 to 70 nm by changing the reaction conditions, in particular the amount of surfactant (1-dodecanethiol). They then observed a regular increase of the ECL signal upon increasing the size of the aggregates and assigned the ECL to a R–O mechanism using $K_2S_2O_8$ as coreactant. Next, they prepared core–shell structures by covering the EuSe nanocubes successively with CTAB, silica and folic acid; the system was applied as a cytosensor capable of detecting the folate receptor on HeLa cells membrane using a GCE modified with three-dimensional porous graphene-adsorbed Au NPs (3D-GR@Au NPs) [75].

An interesting double enhancement effect of ECL was reported by Liu et al. [76]. These researchers discovered that 6-aza-2-thiothymine-coated gold nanoclusters (**ATT-AuNCs**) displayed an enhancement of the ECL upon aggregation by drying the solution on a GCE (**SS-ATT-AuNC/GCE**). The solid state ECL of this system displays an important 1200-fold increase compared to the solution ECL (Fig. 15a). Further analysis of the electrochemical behavior revealed that an electrocatalytic (EC) oxidation of the co-reactant TEA improved the oxidation efficiency of the

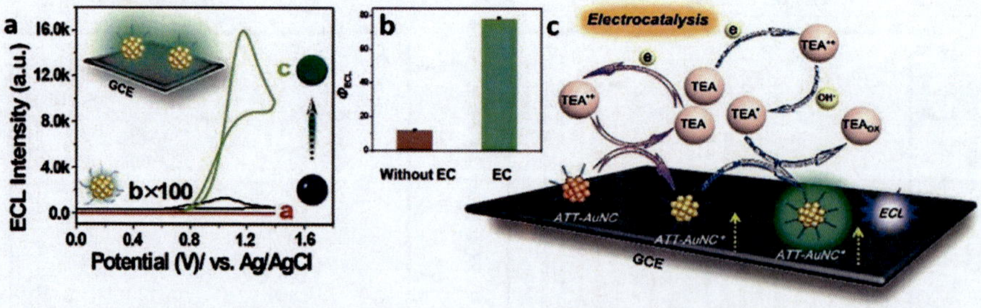

Fig. 15 ECL-Potential signals of bare GCE in 0.14 M TEA (red trace), GCE in 0.14 M TEA + **ATT-AuNCs** (black trace) and **SS-ATT-AuNC/GCE** in 0.14 M TEA at 0.2 V s^{-1}. Inset: schematic representation of the **SS-ATT-AuNC/GCE** and visual ECL of the corresponding electrodes. **b** Effect of the electrocatalysis (*EC*) on the efficiency of **SS-ATT-AuNC/GCE** at pH 11.7 in a concentration of TEA = 0.3 M (without EC) or 0.14 M (with EC). **c** Mechanism of the dual electrocatalytic/aggregation enhancement of the ECL signal. Adapted with permission from Liu et al. [76]. Copyright 2020 John Wiley and Sons

AuNCs. Increasing the TEA concentration avoids its electrocatalytical oxidation due to its concentration-dependent oxidation potential. In a higher TEA concentration, the overall ECL efficiency was reduced from 78% (with EC, $c_{TEA}=0.14$ M) to 11.8% (without EC, $c_{TEA}=0.3$ M) (Fig. 15b), showing that the dual electrocatalytic/aggregation enhancement of the ECL (Fig. 15c) is a concept that can be further explored for the improvement of other platforms' efficiency.

As summarized herein, diverse approaches towards the generation of different AIECL luminophores, with different performances and possible applications, have already been described. Table 1 summarizes the efficiency of AIECL systems compared to the standard, $Ru(bpy)_3^{2+}$that have been reported to date. Similarly, the first applications of AIECL for sensing have seen the light with some of these systems, and Table 2 recaps the sensing applications reported so far.

Table 1 Reported AIECL efficiencies for the distinct luminophores addressed herein

System	Efficiency (%)[a, b, c]	Co-reactant (c/mM)	Mechanism	Reference
Pt-PEG2	120[b]	TPrA (18)	O–R	[38]
Pt-PEG2	1400[b]	$Na_2C_2O_4$ (10)	O–R	[38]
TPE-$(NO_2)_4$/$K_2S_2O_8$	41.2[c]	TPrA (100)	O–R	[51]
ATPP-TPE	34[c]	$K_2S_2O_8$ (100)	R–O	[54]
HPS	37.8[c]	$K_2S_2O_8$ (100)	R–O	[59]
4T1	2.5[c]	BPO (5)	R–O	[61]
4T2	6.5[c]	BPO (5)	R–O	[61]
BTD-TPA@Au	26.5	TEOA (300)	O–R	[66]
[Ir(tpy)(bbbi)]	404[b]	TPrA (1)	O–R	[43]
[$RuCl_2$(phen)$_2$]	100[c]	TPrA (10)	O–R	[67]
P-1	5.8[c]	TPrA (100)	O–R	[44]
P-2	11.8[c]	TPrA (100)	O–R	[44]
TBPE-CMP-1	1.72[c]	TPrA (10)	O–R	[70]
BKM-TPE Pdots	0.5[c]	TPrA (25)	O–R	[48]
BKM-TPE Pdots + UO_2^{2+}	18.6[c]	TPrA (25)	O–R	[48]
Pdots0	5.05[c]	BDEA (25)	O–R	[72]
Pdots1	18.9[c]	BDEA (25)	O–R	[72]
Pdots2	2.92[c]	BDEA (25)	O–R	[72]
Pdots3	1.35[c]	BDEA (25)	O–R	[72]
Pdots4	0.88[c]	BDEA (25)	O–R	[72]
Pdots5	0.15[c]	BDEA (25)	O–R	[72]
H_4ETTCAgg	10.34[c]	TEA (36)	O–R	[73]
Hf-H_4ETTC-MOF	15.52[c]	TEA (36)	O–R	[73]
Hf-H_4ETTC-MOL	27.63[c]	TEA (36)	O–R	[73]
SS-ATT-AuNC	78[c]	TEA (140)	O–R	[76]

[a]Relative to $[Ru(bpy)_3]^{2+}$/TPrA (100%)

[b]Direct maximum intensity comparison

[c]Integrated intensities quotient

Table 2 AIECL-based sensing platforms

System	Analyte	Linear range	LOD	Reference
DPA–CM NPs	Uric acid	0.05–50 μM	0.2 μM	[46]
DPA–CM NPs	Dopamine	0.05–50 μM	0.04 μM	[46]
DPA–CM NPs	Ascorbic Acid	0.05–50 μM	0.04 μM	[46]
TPE MCs	Mucin 1	10^{-6}-1 ng mL^{-1}	0.29 fg mL^{-1}	[49]
BSA–TPE NCs	miRNA-141	10^{-7}-1 nM	13.6 aM	[50]
TPE–$(NO_2)_4$/$K_2S_2O_8$	Iodide	5–2000 nM	0.23 nM	[51]
QAU-1	Bleomycin	0.01–10^4 pM	4.64 fM	[52]
TPE-pho	ALP	0.1–6.0 U L^{-1}	0.037 U mL^{-1}	[53]
ZnO@Cys NFs-TCPP	Cu^{2+}	0.001–500 nM	0.33 pM	[56]
HPS	DNBP	5–2500 nM	0.15 nM	[59]
BTD-TPA@Au	Dopamine	0.001–10^3 pM	0.33 fM	[66]
$Ir(ppy)_3$@apoFt	CYFRA21	0.001–50 ng mL^{-1}	0.43 pg mL^{-1}	[68]
P-1	Catechol	2–10^6 nM	1 nM	[69]
TBPE–CMP-1	Dopamine	10^{-6}-1 mM	0.85 nM	[70]
BKM–TPE Pdots	UO_2^{2+}	0.05–100 nM	10.6 pM	[48]
Hf-ETTC-MOL	CEA	10^{-6}-1 ng mL^{-1}	0.63 fg mL^{-1}	[73]
Ca^{2+}@AuATP	CaM	0.3–50 μg mL^{-1}	0.1 μg mL^{-1}	[74]

3 Perspectives

It is clear that even in its infancy, the progress made in AIECL during the last 3 years has been brilliant, and interesting applications and promising results have demonstrated the potentialities of this tool. AIECL represents a strategy capable not only of overcoming the limitations of ECL applications in aqueous, biological and environmental media but also to increase the sensing capabilities of existing ECL platforms. The introduction of new or imported concepts that enrich the AIECL toolbox, such as RIM, [49], HAIE-ECL [59], CIEE [61] or SS-AIECL [74], is important to feed the creativity that underlies the conception of more powerful materials capable of detecting the more challenging analytes in the most diverse conditions. Until now, just a handful of the abundant collection of known AIE luminophores have been explored in AIECL, and the expansion to more diverse systems provides AIECL with an important potential that still needs to be explored; in addition, pursuing the de novo design of luminogens specifically for AIECL also has a boundless potential. The interest in AIECL is indeed related to the potential to overcome barriers that typically influence the classical luminophores. The photoexcitation of molecules is regulated by a transition that must satisfy the selection rules of spin and symmetry selection. When the excited state is generated by charge recombination, as in ECL or EL (electroluminescence), such rules can be neglected and the excited state population will be dictated by statistical quantum mechanics consideration, with 25% of the excitons representing singlets and 75% triplets. Therefore, phosphorescent compounds are a better choice for ECL and explain why $Ru(bpy)^{32+}$ has been a standard for decades. If we consider the aggregation of luminophores, this process often leads to quenching because of the formation of low-lying excited states or because of triplet–triplet annhiliation. An accurate choice of emitters can prevent such effects, avoiding π–π interactions or very long-lived species. However, aggregation can be very beneficial for molecules possessing a high number of vibrations or groups that can rotate or that can be quenched by molecules present in solution, such dioxygen. The motions in solution increase the non-radiative decays while the quenchers will both decrease the radiative constants and lower the emission quantum yields. The packing of these molecules in aggregates, where the motion is blocked or strongly hindered, contributes therefore toward increasing the emission quantum yields, reducing the non-radiative pathways but also preventing the diffusion of the quenchers so that they cannot approach the emitters. Finally, for ECL it is very important that the redox properties match the co-reactant's ability to act as a reducing or oxidizing agent and, therefore, often the oxidation or reduction of the luminophore must be accessible at rather low potentials and possibly reversible. Aggregation can strongly modify such redox behavior since electronic interactions between the molecules will stabilize (or destabilize) the HOMO and/or LUMO (lowest un-occupied molecular orbital) levels.

With such considerations in mind, the design of new luminophores that can undergo AIECL is possible keeping into account that the aggregation process in solution depends on both the concentration of the compound and the media. Indeed, the majority of the reported systems are based on the use of solvent/antisolvent to create the aggregates in the forms of colloidal suspensions; therefore, the final properties are dictated by the preparation protocol. Finally, factors such as morphology, size and shape of aggregates can influence the AIECL performance of the luminophores. Addressing these effects, which are system dependent, is an arduous task required for all the new luminophores and, as discussed above, while an increase in size can in some cases increase the efficiency [75], in others the opposite effect is observed [64]. Likewise, the crystalline and amorphous aggregates can display dissimilar behaviors [62], or even H or J aggregates can exhibit different AIECL response [55]. Also, in very diluted conditions, we expect the compounds to exist in the molecularly dissolved form, or at least in equilibrium with the monomer whose photophysical properties are typically poor. Important analytical applications, such as immunoassays, require the luminogen to be covalently bound to an antibody to form a sandwich complex in a 1:1 ratio. Therefore, the extreme dilution of the luminogen and the steric hindrance exerted by the biological macromolecule would prevent the formation of aggregates and in turn disable the AIECL phenomenon. Therefore, it is of importance to develop new systems that take advantage of the AIE and AIECL phenomena, but which are not affected by media, concentration and preparation protocol. A possible strategy would be to covalently link several luminophores to obtain a discrete system that behaves as an aggregate. Very recently we have shown [77] that by linking together three Pt(II) complexes it is possible to fully benefit from the assembly emerging properties in terms of elongation of the excited state lifetime, protection from the dioxygen quenching and increase in the emission quantum yield (AIE phenomenon). Thanks to the covalent bond, the photophysical properties are not affected by concentration or sample preparations since the "aggregation" is due to intramolecular interactions rather than intermolecular ones. While such an approach has been demonstrated successfully for the case of AIE of neutral Pt(II) complexes, there are no other examples in literature with other luminophores, therefore opening a new avenue in the development of systems with enhanced properties.

Author contributions The first draft of the manuscript was written by GM-A and all authors commented on previous versions of the manuscript. All authors read and approved the final manuscript.

Funding Open Access funding enabled and organized by Projekt DEAL. L.D.C. thanks IUF and AXA funding for financial support.

Data Availability N/A.

Code Availability N/A.

Declaration

Conflict of Interest On behalf of all authors, the corresponding author states that there is no conflict of interest.

References

1. Verhoeven JW (1996) Glossary of terms used in photochemistry (IUPAC recommendations 1996). Pure Appl Chem 68:2223–2286. https://doi.org/10.1351/pac199668122223
2. Harvey N (1929) Luminescence during electrolysis. J Phys Chem 33:1456–1459. https://doi.org/10.1021/j150304a002
3. Dufford RT, Nightingale D, Gaddum LW (1927) Luminescence of Grignard compounds in electric and magnetic fields, and related electrical phenomena. J Am Chem Soc 49:1858–1864. https://doi.org/10.1021/ja01407a002
4. Hercules DM (1964) Chemiluminescence resulting from electrochemically generated species. Science 145:808–809. https://doi.org/10.1126/science.145.3634.808
5. Visco RE, Chandross EA (1964) Electroluminescence in solutions of aromatic hydrocarbons. J Am Chem Soc 86:5350–5351. https://doi.org/10.1021/ja01077a073
6. Santhanam KSV, Bard AJ (1965) Chemiluminescence of electrogenerated 9,10-diphenylanthracene anion radical 1. J Am Chem Soc 87:139–140. https://doi.org/10.1021/ja01079a039
7. Richter MM (2004) Electrochemiluminescence (ECL). Chem Rev 104:3003–3036. https://doi.org/10.1021/cr020373d
8. Aliprandi A, DiMarco BN, De Cola L (2019) Transition metal complexes in ECL: diagnostics and biosensing. In: Albini A, Protti S (eds) Photochemistry, volume 46. The Royal Society of Chemistry, London, pp 319–351

9. Liu Z, Qi W, Xu G (2015) Recent advances in electrochemiluminescence. Chem Soc Rev 44:3117–3142. https://doi.org/10.1039/c5cs00086f
10. Hao N, Wang K (2016) Recent development of electrochemiluminescence sensors for food analysis. Anal Bioanal Chem 408:7035–7048. https://doi.org/10.1007/s00216-016-9548-2
11. Busa LSA, Mohammadi S, Maeki M, et al (2016) Advances in microfluidic paper-based analytical devices for food and water analysis. Micromachines 7(5):10.3390/mi7050086. https://doi.org/10.3390/mi7050086
12. Cui Y, Jin Y, Chen X, Wu J (2018) Two-dimensional electrochemiluminescence on porous silicon platform for explosive detection and discrimination. ACS Sensors 3:1439–1444. https://doi.org/10.1021/acssensors.8b00113
13. Li G, Yu X, Liu D et al (2015) Label-free electrochemiluminescence aptasensor for 2,4,6-trinitrotoluene based on bilayer structure of luminescence functionalized graphene hybrids. Anal Chem 87:10976–10981. https://doi.org/10.1021/acs.analchem.5b02913
14. Uoyama H, Goushi K, Shizu K et al (2012) Highly efficient organic light-emitting diodes from delayed fluorescence. Nature 492:234–238. https://doi.org/10.1038/nature11687
15. Shum J, Leung PK-K, Lo KK-W (2019) Luminescent ruthenium(II) polypyridine complexes for a wide variety of biomolecular and cellular applications. Inorg Chem 58:2231–2247. https://doi.org/10.1021/acs.inorgchem.8b02979
16. Huang B, Zhou X, Xue Z, Lu X (2013) Electrochemiluminescence quenching of tris(2,2′-bipyridyl) ruthenium. TrAC Trends Anal Chem 51:107–116. https://doi.org/10.1016/j.trac.2013.06.012
17. Laird SE, Hogan CF (2017) Electrochemiluminescence of iridium complexes. In: Zysman-Colman E (ed) Iridium(III) in optoelectronic and photonics applications. Wiley, Chichester, pp 359–414
18. Haghighatbin MA, Laird SE, Hogan CF (2018) Electrochemiluminescence of cyclometalated iridium (III) complexes. Curr Opin Electrochem 7:216–223
19. Zhang S, Ding Y, Wei H (2014) Ruthenium polypyridine complexes combined with oligonucleotides for bioanalysis: a review. Molecules 19:11933–11987. https://doi.org/10.3390/molecules190811933
20. Muegge BD, Richter MM (2005) Electrogenerated chemiluminescence from polymer-boundorthometallated iridium(III) systems. Luminescence 20:76–80. https://doi.org/10.1002/bio.807
21. Li J, Wang E (2012) Applications of tris(2,2′-bipyridyl)ruthenium(II) in electrochemiluminescence. Chem Rec 12:177–187. https://doi.org/10.1002/tcr.201100017
22. Homocianu M, Airinei A, Dorohoi DO (2011) Solvent effects on the electronic absorption and fluorescence spectra. J Adv Res Phys 2:1–9
23. Dobretsov GE, Syrejschikova TI, Smolina NV (2014) On mechanisms of fluorescence quenching by water. Mol Biophys 59:183–188. https://doi.org/10.1134/S0006350914020079
24. Lakowicz JR (2013) Principles of fluorescence spectroscopy. Springer SBM, New York
25. Chen Y, Zhou S, Li L, Zhu J jie (2017) Nanomaterials-based sensitive electrochemiluminescence biosensing. Nano Today 12:98–115
26. Bertoncello P, Forster RJ (2009) Nanostructured materials for electrochemiluminescence (ECL)-based detection methods: recent advances and future perspectives. Biosens Bioelectron 24:3191–3200
27. Deng S, Ju H (2013) Electrogenerated chemiluminescence of nanomaterials for bioanalysis. Analyst 138:43–61. https://doi.org/10.1039/c2an36122a
28. Lei Y-M, Zhao M, Wang A et al (2016) Electrochemiluminescence of supramolecular nanorods and their application in the "On-Off-On" detection of copper ions. Chem Eur J 22:8207–8214. https://doi.org/10.1002/chem.201504995
29. Sun X, Du Y, Zhang L et al (2007) Luminescent supramolecular microstructures containing Ru(bpy) 32+: Solution-based self-assembly preparation and solid-state electrochemiluminescence detection application. Anal Chem 79:2588–2592. https://doi.org/10.1021/ac062130h
30. Luo J, Xie Z, Xie Z et al (2001) Aggregation-induced emission of 1-methyl-1,2,3,4,5-pentaphenylsilole. Chem Commun 18:1740–1741. https://doi.org/10.1039/b105159h
31. Tang BZ, Zhan X, Yu G et al (2001) Efficient blue emission from siloles. J Mater Chem 11:2974–2978. https://doi.org/10.1039/b102221k
32. Hong Y, Lam JWY, Tang BZ (2009) Aggregation-induced emission: phenomenon, mechanism and applications. Chem Commun (Camb) (29):4332–4353. https://doi.org/10.1039/b904665h
33. Hong Y, Lam JWY, Tang BZ (2011) Aggregation-induced emission. Chem Soc Rev 40:5361–5388. https://doi.org/10.1039/c1cs15113d

34. Mei J, Hong Y, Lam JWY et al (2014) Aggregation-induced emission: the whole is more brilliant than the parts. Adv Mater 26:5429–5479. https://doi.org/10.1002/adma.201401356
35. Tang W, Xiang Y, Tong A (2009) Salieylaldehyde azines as fluorophores of aggregation-induced emission enhancement characteristics. J Org Chem 74:2163–2166. https://doi.org/10.1021/jo802631m
36. An BK, Gierschner J, Park SY (2012) π-conjugated cyanostilbene derivatives: a unique self-Assembly motif for molecular nanostructures with enhanced emission and transport. Acc Chem Res 45:544–554. https://doi.org/10.1021/ar2001952
37. Hong J-L (2013) Enhanced emission by restriction of molecular rotation. In: Tang AQZ (ed) Aggregation-induced emission: fundamentals. Wiley, Chichester, pp 285–305
38. Carrara S, Aliprandi A, Hogan CF, De Cola L (2017) Aggregation-induced electrochemiluminescence of platinum(II) complexes. J Am Chem Soc 139:14605–14610. https://doi.org/10.1021/jacs.7b07710
39. Chen L, Wei J, Chi Y, Zhou S (2019) Tris(2,2'-bipyridyl)ruthenium(II)-nanomaterial co-reactant electrochemiluminescence. ChemElectroChem 6:3878–3884. https://doi.org/10.1002/celc.201900693
40. Yang Z, Chi Z, Mao Z et al (2018) Recent advances in mechano-responsive luminescence of tetraphenylethylene derivatives with aggregation-induced emission properties. Mater Chem Front 2:861–890. https://doi.org/10.1039/C8QM00062J
41. Feng H-T, Yuan Y-X, Xiong J-B et al (2018) Macrocycles and cages based on tetraphenylethylene with aggregation-induced emission effect. Chem Soc Rev 47:7452–7476. https://doi.org/10.1039/C8CS00444G
42. La DD, Bhosale SV, Jones LA, Bhosale SV (2018) Tetraphenylethylene-based AIE-active probes for sensing applications. ACS Appl Mater Interfaces 10:12189–12216. https://doi.org/10.1021/acsami.7b12320
43. Gao T-B, Zhang J-J, Yan R-Q et al (2018) Aggregation-induced electrochemiluminescence from a cyclometalated iridium(III) complex. Inorg Chem 57:4310–4316. https://doi.org/10.1021/acs.inorgchem.7b03093
44. Wang Z, Feng Y, Wang N et al (2018) Donor-acceptor conjugated polymer dots for tunable electrochemiluminescence activated by aggregation-induced emission-active moieties. J Phys Chem Lett 9:5296–5302. https://doi.org/10.1021/acs.jpclett.8b02087
45. Wei X, Zhu M, Yan H, et al (2019) Recent advances in aggregation-induced electrochemiluminescence. Chem—A Eur J 25:12671–12683. https://doi.org/10.1002/chem.201902465
46. Liu H, Wang L, Gao H et al (2017) Aggregation-induced enhanced electrochemiluminescence from organic nanoparticles of donor-acceptor based coumarin derivatives. ACS Appl Mater Interfaces 9:44324–44331. https://doi.org/10.1021/acsami.7b15434
47. Liu J-L, Zhang J-Q, Tang Z-L et al (2019) Near-infrared aggregation-induced enhanced electrochemiluminescence from tetraphenylethylene nanocrystals: a new generation of ECL emitters. Chem Sci 10:4497–4501. https://doi.org/10.1039/C9SC00084D
48. Wang Z, Pan J, Li Q et al (2020) Improved AIE-active probe with high sensitivity for accurate uranyl ion monitoring in the wild using portable electrochemiluminescence system for environmental applications. Adv Funct Mater 30:2000220. https://doi.org/10.1002/adfm.202000220
49. Jiang M-H, Li S-K, Zhong X et al (2019) Electrochemiluminescence enhanced by restriction of intramolecular motions (RIM): tetraphenylethylene microcrystals as a novel emitter for mucin 1 detection. Anal Chem 91:3710–3716. https://doi.org/10.1021/acs.analchem.8b05949
50. Liu JL, Zhuo Y, Chai YQ, Yuan R (2019) BSA stabilized tetraphenylethylene nanocrystals as aggregation-induced enhanced electrochemiluminescence emitters for ultrasensitive microRNA assay. Chem Commun 55:9959–9962. https://doi.org/10.1039/c9cc04660g
51. Han Z, Zhang Y, Wu Y et al (2019) Substituent-induced aggregated state electrochemiluminescence of tetraphenylethene derivatives. Anal Chem 91:8676–8682. https://doi.org/10.1021/acs.analchem.9b02357
52. Lv W, Yang Q, Li Q et al (2020) Quaternary ammonium salt-functionalized tetraphenylethene derivative boosts electrochemiluminescence for highly sensitive aqueous-phase biosensing. Anal Chem 92:11747–11754. https://doi.org/10.1021/acs.analchem.0c01796
53. Li S, Li J, Geng B et al (2021) TPE based electrochemiluminescence for ALP selective rapid one-step detection applied in vitro. Microchem J 164:106041. https://doi.org/10.1016/j.microc.2021.106041
54. Zhang Y, Zhao Y, Han Z, et al (2020) Switching the photoluminescence and electrochemiluminescence of liposoluble porphyrin in aqueous phase by molecular regulation. Angew Chem Int Edn. https://doi.org/10.1002/anie.202010216

55. Yang Z, Pu G, Ning X et al (2019) J-Aggregates of zinc tetraphenylporphyrin: a new pathway to excellent electrochemiluminescence emitters. Phys Chem Chem Phys 21:10614–10620. https://doi.org/10.1039/c9cp01278h
56. Han Q, Wang C, Li Z et al (2020) Multifunctional zinc oxide promotes electrochemiluminescence of porphyrin aggregates for ultrasensitive detection of copper ion. Anal Chem 92:3324–3331. https://doi.org/10.1021/acs.analchem.9b05262
57. Zhao Z, He B, Tang BZ (2015) Aggregation-induced emission of siloles. Chem Sci 6:5347–5365
58. Chen J, Law CCW, Lam JWY et al (2003) Synthesis, light emission, nanoaggregation, and restricted intramolecular rotation of 1,1-substituted 2,3,4,5-tetraphenylsiloles. Chem Mater 15:1535–1546. https://doi.org/10.1021/cm021715z
59. Han Z, Yang Z, Sun H et al (2019) Electrochemiluminescence platforms based on small water-insoluble organic molecules for ultrasensitive aqueous-phase detection. Angew Chem Int Edn 58:5915–5919. https://doi.org/10.1002/anie.201814507
60. Guo J, Feng W, Du P et al (2020) Aggregation-induced electrochemiluminescence of tetraphenylbenzosilole derivatives in an aqueous phase system for ultrasensitive detection of hexavalent chromium. Anal Chem 92:14838–14845. https://doi.org/10.1021/acs.analchem.0c03709
61. Yang L, Koo D, Wu J, et al (2020) Benzosiloles with crystallization-induced emission enhancement of electrochemiluminescence: synthesis, electrochemistry, and crystallography. Chem—A Eur J 26:11715–11721. https://doi.org/10.1002/chem.202002647
62. Dong Y (2013) Crystallization-induced emission enhancement. In: Tang AQZ (ed) Aggregation-induced emission: fundamentals. Wiley, Chichester, pp 323–335
63. Lv X, Xu X, Miao T et al (2020) Aggregation-induced electrochemiluminescence immunosensor based on 9,10-diphenylanthracene cubic nanoparticles for ultrasensitive detection of aflatoxin B1. ACS Appl Bio Mater 3:8933–8942. https://doi.org/10.1021/acsabm.0c01201
64. Wei X, Zhu MJ, Cheng Z et al (2019) Aggregation-induced electrochemiluminescence of carboranyl carbazoles in aqueous media. Angew Chem—Int Edn 58:3162–3166. https://doi.org/10.1002/anie.201900283
65. Li Z, Qin W, Wu J et al (2019) Bright electrochemiluminescent films of efficient aggregation-induced emission luminogens for sensitive detection of dopamine. Mater Chem Front 3:2051–2057. https://doi.org/10.1039/c9qm00401g
66. Li Z, Qin W, Liang G (2020) A mass-amplifying electrochemiluminescence film (MAEF) for the visual detection of dopamine in aqueous media. Nanoscale 12:8828–8835. https://doi.org/10.1039/d0nr01025a
67. Lu L, Zhang L, Miao W et al (2020) Aggregation-induced electrochemiluminescence of the dichlorobis(1,10-phenanthroline)ruthenium(II) (Ru(phen)2Cl2)/Tri-n-propylamine (TPrA) system in H2O-MeCN mixtures for identification of nucleic acids. Anal Chem 92:9613–9619. https://doi.org/10.1021/acs.analchem.0c00799
68. Yang L, Sun X, Wei D et al (2021) Aggregation-induced electrochemiluminescence bioconjugates of apoferritin-encapsulated iridium(III) complexes for biosensing application. Anal Chem 93:1553–1560. https://doi.org/10.1021/acs.analchem.0c03877
69. Wang Z, Wang N, Gao H et al (2020) Amplified electrochemiluminescence signals promoted by the AIE-active moiety of D-A type polymer dots for biosensing. Analyst 145:233–239. https://doi.org/10.1039/c9an01992h
70. Cui L, Yu S, Gao W et al (2020) Tetraphenylenthene-based conjugated microporous polymer for aggregation-induced electrochemiluminescence. ACS Appl Mater Interfaces 12:7966–7973. https://doi.org/10.1021/acsami.9b21943
71. Carrara S, Stringer B, Shokouhi A et al (2018) Unusually strong electrochemiluminescence from iridium-based redox polymers immobilized as thin layers or polymer nanoparticles. ACS Appl Mater Interfaces 10:37251–37257. https://doi.org/10.1021/acsami.8b12995
72. Gao H, Zhang N, Pan J-B, et al (2020) Aggregation-induced electrochemiluminescence of conjugated pdots containing a trace Ir(III) complex: insights into structure–property relationships. ACS Appl. Mater Interfaces 12 (48):54012–54019. https://doi.org/10.1021/acsami.0c18197
73. Yang Y, Hu GB, Bin LW et al (2020) An AIEgen-based 2D ultrathin metal-organic layer as an electrochemiluminescence platform for ultrasensitive biosensing of carcinoembryonic antigen. Nanoscale 12:5932–5941. https://doi.org/10.1039/c9nr10712f
74. Jiang H, Qin Z, Zheng Y et al (2019) Aggregation-induced electrochemiluminescence by metal-binding protein responsive hydrogel scaffolds. Small 15:1901170. https://doi.org/10.1002/smll.201901170

75. Zhang J, Xie H, Shen Y et al (2018) Facile synthesis of highly monodisperse EuSe nanocubes with size-dependent optical/magnetic properties and their electrochemiluminescence performance. Nanoscale 10:13617–13625. https://doi.org/10.1039/C8NR02500B
76. Peng H, Huang Z, Deng H et al (2020) Dual enhancement of gold nanocluster electrochemiluminescence: electrocatalytic excitation and aggregation-induced emission. Angew Chemie Int Ed 59:9982–9985. https://doi.org/10.1002/anie.201913445
77. Chakraborty S, Aliprandi A, De Cola L (2020) Multinuclear Pt II complexes: Why three is better than two to enhance photophysical properties. Chem—A Eur J 26:11007–11012. https://doi.org/10.1002/chem.202001510

Authors and Affiliations

Guillermo Moreno-Alcántar[1] · Alessandro Aliprandi[1] · Luisa De Cola[1,2,3]

✉ Luisa De Cola
luisa.decola@marionegri.it

Alessandro Aliprandi
aliprandi@unistra.fr

1 Institut de Science Et D'Ingénierie Supramoléculaires (ISIS), University of Strasbourg & CNRS, 8 allée Gaspard Monge, 67083 Strasbourg, France

2 Institute for Nanotechnology (INT), Karlsruhe Institute of Technology, Hermann-von-Helmholtz-Platz 1, 76344 Eggenstein-Leopoldshafen, Germany

3 Dipartimento Di Scienze Farmaceutiche, DISFARM, and Istituto Di Ricerche Farmacologiche Mario Negri, IRCCS, University of Milan, Milan, Italy

Mechanofluorochromism with Aggregation-Induced Emission (AIE) Characteristics: A Perspective Applying Isotropic and Anisotropic Force

Vishal Kachwal[1] · Inamur Rahaman Laskar[1]

Received: 3 March 2021 / Accepted: 24 May 2021 / Published online: 9 June 2021

Abstract

Organic mechanofluorochromic (MFC) materials (that change their emission under anisotropic and isotropic pressure) have attracted a great attention in recent years due to their promising applications in sensing pressure, storage devices, security inks, three-dimensional (3D) printing, etc. Stimuli-responsive organic materials with aggregation-induced emission (AIE) characteristics would be an interesting class of materials to enrich the chemistry of MFC compounds. A diamond anvil cell (DAC) is a small tool that is employed to generate high and uniform pressure on materials over a small area. This article discusses the relationship between the chemical structure of AIE compounds and the change in emission properties under anisotropic (mechanical grinding) and isotropic (hydrostatic) pressure. The luminescent properties of such materials depend on the molecular rearrangement in the lattice, conformational changes, excited state transitions and weak intermolecular interactions. Hence, studying the change in luminescent property of these compounds under varying pressure will provide a deeper understanding of the excited-state properties of various emissive compounds with stress. The development of such materials and studies into the effect of pressure on their luminescence properties are summarized.

Keywords Mechanofluorochromic · Mechanical grinding · Hydrostatic pressure · Aggregation-induced emission

Chapter 4 was originally published as Kachwal, V. & Laskar, I. R. Topics in Current Chemistry (2021) 379: 28. https://doi.org/10.1007/s41061-021-00341-x.

✉ Inamur Rahaman Laskar
ir_laskar@pilani.bits-pilani.ac.in

1 Department of Chemistry, BITS PILANI, Pilani campus, Pilani, India

1 Introduction

Solid-state smart materials are high-performance materials that change their photophysical properties in response to external stimuli such as humidity, pH, temperature, viscosity, polarity, pressure (mechanofluorochromism (MFC)) and chemical entities [1–3]. The luminescence property of such materials depends primarily on molecular arrangements, conformational changes and intermolecular interactions. Thus, stimuli that change their molecular interaction and enable conformational changes in their molecular packing will affect the excited state and energy levels [4]. Emission intensity and tuning of emission color can be controlled by changing the intermolecular forces (π-stacking force, H-bonding, van der Waals interactions, hydrophobicity, etc.) operated within and in between the molecules, present in crystal solids of luminescent materials. The modulation of such interactions is expected to result in interesting photophysical properties of luminescent compounds [5, 6].

Mechanofluorochromic (MFC) materials are a class of smart stimuli-responsive materials in which the emission properties of the material change or respond upon the application of force/pressure, etc. MFC materials have attracted considerable attention because of their use in data storage, security inks and structural health monitoring, and their fundamental interest such as in the study of conformational changes, etc. [7–9]. In biology, pressure can direct protein folding and unfolding and control cell migration [10]. The difficulties in these cases are to monitor and measure the pressure, which makes pressure sensing of the utmost importance. High pressure reduces the volume of molecular accumulation, affecting the molecular rearrangement, the formation of different excited states (changes intermolecular π–π interaction), either improving or reducing the stability of the system and even making possible solid-state mechanical reactions [11]. At high pressure, molecules may experience a change in intermolecular distance as well as conformational changes. Generally, under high pressure, molecular geometry (distance and angle) and energy (wavelengths, bandwidths) usually differ from those found in ambient conditions. The isotropic hydrostatic pressure (HP) modifies both molecular conformation and crystal packing without the destruction of crystals [12].

To induce high pressure, scientists use the diamond anvil cell (DAC). The DAC is used extensively as an appropriate instrument to apply high pressure and for studying the behavior of luminescent materials. The DAC has a diamond of low birefringence and good spectral permeability [13]. There are many different forms of DAC, but the structures comprise mainly three parts: a diamond anvil, gasket, and a support guide device. Typical culet sizes for diamond anvils are 100–250 μm for applying high pressure over a small area. HP can be applied by filling the pressure chamber with a liquid pressure-transmitting medium, generally paraffin oil, silicone oil, 4:1 (methanol: ethanol) and cyclohexane, etc. Samples are loaded between the two flat and parallel surfaces of the diamond anvils. Using the DAC, it is possible to record characteristics in situ, including Raman spectroscopy, UV–visible absorption, fluorescence spectroscopy, lifetime, and imaging of a single crystal under high pressure. These characterizations describe the structural deformation or changes that happen in the materials under high pressure [14].

Before the discovery of aggregation-induced emission (AIE) compounds, MFC were the most common luminescent compounds. The use of traditional fluorophores was limited in MFC materials as they are non-emissive in the solid-state due to the aggregation-caused quenching (ACQ) effect. Tang and coworkers [15–17] designed a family of molecules that are emissive in the solid-state and termed this phenomenon AIE, which is contradictory to the ACQ effect. Many AIE luminogens with different properties have been developed, and the relationship between the molecular structure of AIE luminogens and their photophysical properties has been studied. AIE-active compounds are often characterized by strongly twisted conformations, which lead easily to efficient emission in the aggregated state by preventing the close stacking of molecules. Furthermore, the strongly twisted conformations often result in loose packing arrangements of molecules in the crystals, which are thus sensitive to pressure stimulation. Therefore, compounds with AIE properties are considered potential candidates prone to MFC-active materials under a given stress. [9, 18, 19]. Recently, there has been a drive to understand the mechanoflurochromic materials based on AIE molecules, but in-depth analysis of design strategies as proved inadequate. There are many reviews available on mechanofluorochromic materials changing their properties upon mechanical grinding [20, 21]. Upon mechanical grinding, AIE luminogens undergo a phase transformation from a crystalline state to an amorphous state [20]. But, as pressure in mechanical grinding is anisotropic, not uniform in direction and magnitude. It is often difficult to understand the mechanism in mechanical grinding. However, HP is isotropic and uniform in nature. Therefore, understanding the property changes of molecules with HP may help us our in-depth understanding of the mechanofluorochromic properties of molecules [22–24].

To date, three mini-reviews have summarized changes in photophysical properties with structural changes of a few organic-based materials under HP. Some materials' emissions are quenched, some have wavelength changes and some materials' emissions are enhanced under HP [25–27]. The materials have a distinct response when applying mechanical grinding and HP. Under low or high pressure, the intermolecular and π–π interactions between molecules differ, which alters the photophysical properties in accordance with the change in the excited state. Pressure-induced emission (PIE) is found in halogen perovskite materials, inorganic nanocrystals and organic materials. Owing to their different twisted forms, secondary interaction, planarized rigid structure and molecular arrangement, various organic molecules, like triphenylamine (TPA), tetraphenylethylene (TPE), anthracene and stilbene, etc., each have their own pattern and arrangements in the solid state. Each organic molecule responds to pressure according to its specific molecular arrangements and interactions. Most materials are π-conjugated organic D–A type materials that show red-shifted with emission quenching at high pressure. The emission is quenched as a result of the formation of a low-energy emission species and a non-radiative "dark" state. Very few cases have been reported with pressure-induced blue shifted and enhanced emission [26, 27].

This article provides a brief description of, and highlights the recent advances in, different AIE active organic materials that respond with emission property changes upon applying pressure. The study of MFCs under HP not only helps our understanding of structure–property relationships, but also triggers further development of new MFC materials with improved properties. We explain the distinct response of these materials based on molecular structural changes under mechanical grinding and high pressure. At different pressure, the excited state of the molecules is changed according to the molecular rearrangement and intermolecular interaction. These intermolecular interactions and molecular packing arrangements govern the photophysical properties of these compounds.

1.1 Tetraphenylethylene

TPE (**1**) act as a molecular rotor prototype for AIE materials [17, 28, 29]. It has highly twisted conformations that are locked with multiple CH–π and CH–C shorts contacts. Mechanical grinding (anisotropic force) and HP (isotropic force) modulate these weak interactions, which alters the spectroscopic properties, especially photoluminescence (PL). TPE in solid-state under ambient temperature exhibits a deep bluish emission (λ_{emi} = 448 nm) [12]. Any substitution on TPE may cause changes in its conformation, which may modulate its response under pressure. Here, we describe TPE derivatives that show distinct responses under pressure (Fig. 1). TPE itself has a different response under mechanical grinding and HP. Zou and colleagues [30, 31] carried out an in-depth analysis of TPE under high pressure and revealed that, while applying low pressure up to 1.5 GPa, the emission intensity of TPE decreases due to the detrimental effects. Under low pressure (below 1.5 GPa), intermolecular interactions lead to an additional non-radiative pathway, reducing PL efficiency. Emission intensity increases on applying pressure above 1.5–5.3 GPa. In the pressure range 1.5–5.3 GPa, the molecules are close enough for the formation and strengthening of hydrogen bonds between the aromatic rings, which restricts the aromatic rings of TPE. Thus, energy loss due to intramolecular motion can be suppressed more efficiently, which reduces the detrimental effects [32].

Applying further higher pressure (> 5.3 GPa) results in deformation of the CH–π and CH–C network, leading to an amorphous state. Usually, in the amorphous state,

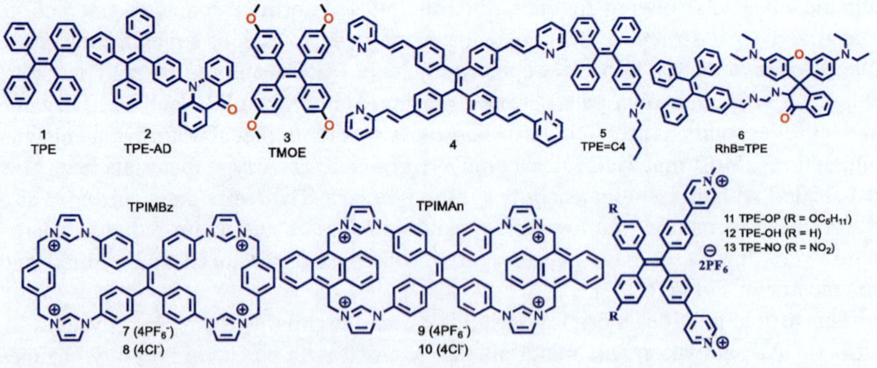

Fig. 1 Chemical structures of various tetraphenylethylene (TPE) and TPE derivatives that exhibit changes in luminescent properties when applying pressure

intermolecular forces are not as effective at limiting intramolecular rotation in crystals, allowing non-radiative decay to decrease PL efficiency [30].

To understand structural deformation in the TPE crystal, Xu and coworkers [33] studied the MFC properties of a methoxy-substituted centrosymmetric TPE molecule [tetra(4-methoxyphenyl)ethylene, TMOE]. They aimed to find a correlation between the chemical structure and MFC property changes with HP. TMOE resulted in a red-shift, with a gradual weakening emission peak (480 nm → 540 nm) under HP. The gradual weakening of C–C (ethylene core and benzene ring) and C=C (ethylene core in the molecule) bonds can be observed from the Raman spectra with increasing HP, suggesting an increase in bond strength. The observation of a red-shift in the fluorescence property of TMOE with HP, along with Raman shifts rationalizes the suggestion that the twisted benzene ring gradually turns into a planar structure with an ethylene core, hence drawing the neighboring molecules closer, triggering π–π stacking and H-bond interaction. Theoretical calculations corroborate distortion of the torsion angle, which leads to enhanced intermolecular interaction and facilitates H-bond interaction. The grinding process is unable to show the change observed by HP, indicating the limited and uncontrolled nature of the grinding process [33].

By introducing donor and acceptor nature in a molecule, interesting excited state properties are often observed [34]. The concept of switching of excited states, i.e., switching between a locally excited state and an intramolecular charge transfer (ICT) state during mechanical stimuli, is an excellent dimension for the development of new turn on (with high contrast) piezochromic materials [35]. Tian and colleagues [36] synthesized a novel molecule that integrated the polycyclic electron-accepting acridonyl (AD) with the AIE active and twisted tetraphenylethylene (TPE) (**2**). The pristine powder of AD-TPE was originally in the dark state (D-Phase); after grinding, a strong cyan color emission was observed in the bright phase (B-Phase). Thus, it exhibited a turn on and color-tunable emission. The in-depth mechanism underlying this phenomenon was studied. A weak emission arising from the local excited (LE state) was first observed that was accounted for by the orthogonal conformation of AD and TPE, which fully separates the electronic distribution. The twisted conformation was changed by the mechanical stimuli, leading to overlap of the frontier orbitals between donor and acceptors and formation of the ICT state [36].

Intermolecular interactions between molecules could play a significant role in changing the molecular conformation upon applying HP. Here, Bu and colleagues tried to explain these properties based on the change in molecular conformation of the TPE derivative. They synthesized a multi-responsive TPE derivative (**4**) by attaching the 2-vinylpyridine moiety into the four arms of the TPE matrix, where vinyl is conjugated on the p-position of the TPE rings. Crystallographic data suggested that the molecule forms a highly twisted conformation, with a distinct pore formed by intermolecular interactions between the four units of molecules and a long-range orderly arrayed channel (Fig. 2a, b) [37]. Compound (**4**) resulted in red-shifted emission (540 nm–570 nm) with a relatively weaker emission peak after grinding that was reversible under fuming conditions only. Upon applying HP, the same compound exhibits a large emission shift of 120 nm (from 540 to 660 nm). Moreover, the change is reversible and this reversibility is sustained upon pressure

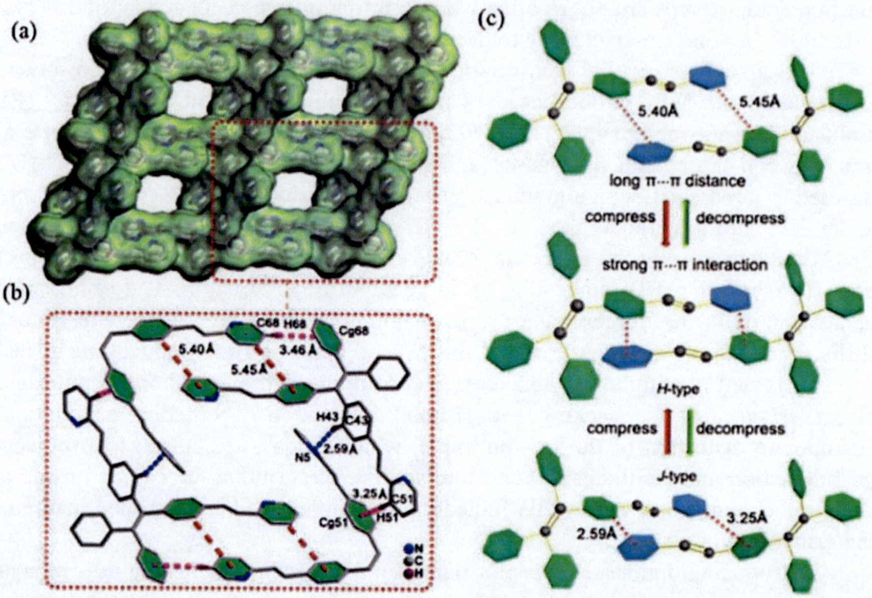

Fig. 2 **a** Porous crystal structure of TP2VPE. **b** Intermolecular supramolecular interactions, including C–H···N hydrogen bond, C–H···π interactions, and the long π···π distance (**c**) Stacking modes and proposed mechanisms under hydrostatic pressure (HP) of TP2VPE [37]. Copyright 2020, American Chemical Society

greater than 10 GPa. The emission color change was thought to be due to the transformation of J-aggregates (slipped interaction between the pyridine rings, long-distance) to H-aggregates (face-to-face π–π interaction between pyridine rings, shorter distance) (Fig. 2c). The existence of hollow space in the crystal packing provide space for the shrunken structure during the decompression process [37]. Fluorophores with a twisted D-A nature often show a distinct energy band with characteristics of both LE and ICT excited state, i.e., a 'hybridized local and charge-transfer excited state' (HLCT) [38, 39]. The HLCT state is recognized if (1) the excited state has a large transition moment from the LE state, and (2) there is total utilization of the exciton from the ICT state [40, 41].

In this context, Ma and colleagues synthesized D-A-based twisted compound by incorporating a Schiff base substituent into TPE (TPE=C4) (**5**). It is claimed that the presence of the HLCT state in TPE=C4 leads to a distinct response towards anisotropic grinding and isotropic compression. In grinding, the emission change is due to the phase transformation from the crystalline to the amorphous state, leading to polarization of the TPE=C4 structure and a red-shifted emission band (blue to bluish-green, 436 nm to 474 nm). TPE=C4 is recoverable by treating with ethanol or self-recoverable by keeping the ground sample for 2 days. Compound **5** has a low recrystallization temperature (51.2 °C), which is the driving force for the self-recovering property. HP is applied through the DAC device. It was observed that the original crystals resulted in only single emission that was mainly LE in nature while under HP the HLCT band split into dual-fluorescence (446 nm and 532 nm) with a drastic emission color change (bluish-green to sky green) because of the pressure-induced band separation (Fig. 3a,b). The ICT emission band (532 nm) is more sensitive to pressure than the LE emission band (445 nm). Once the pressure is released, the ICT band disappears (Fig. 3c) [42].

In Raman spectroscopy, most of the Raman peaks show blue shifts and are broad at high pressure due to a decrease in interatomic distance and increase in the force constant. No new peaks are observed and none disappear in Raman spectra, indicating that conformational change happens only under compression. In a similar study, Jia and colleagues synthesized TPE-rhodamine dyad (RhB = TPE) (**6**) based on a twisted D-A compound. In the same manner, RhB = TPE exhibits a distinct response under grinding and HP due to the pressure-induced separation of HLCT state and further pressure-induced quenching of emission. At high pressure, due to the compact packing and decreased intermolecular distance, the ICT state dominates, which results in red-shifted quenching of PL intensity [43].

The decrease in emission intensity with redshift could be understood by enhancement of the π–π interaction between the rings causing increased pressure-induced planarization (PIP). To have only the PIP phenomenon, Cao and colleagues designed beautiful cage-like molecules consisting of tetraphenylethene-based tetracationicdicyclophanes (TPIMBz and TPIMAn) (Fig. 1) with tetraimidazolium TPE as the central core and p-benzene (**7, 8**) and 9,10-anthracene (**9, 10**) as outer linkers. Due to its single (benzene) ring, TPIMBz can rotate freely in the bicyclic structure to induce a change in the molecular conformation, yielding two rotational conformers

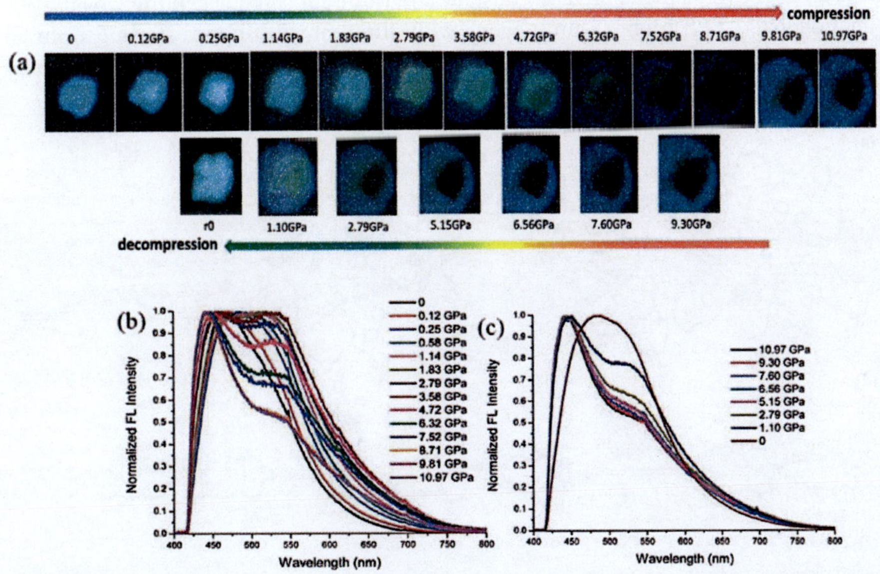

Fig. 3 (**a**) Image of TPE = C4 powder at different pressure under excitation by a UV-lamp (λ_{ex}, 365 nm). **b**, **c** Normalized photoluminescence (PL) spectra of TPE = C4 powder under compression (**b**) and decompression (**c**). [42] Copyright 2020, royal society of chemistry

[clockwise (P) and anticlockwise (M)] in the stacking structure (Fig. 4a). However, in TPIMAn, due to steric hindrance, free rotation of the bicyclic structure is restricted and there is only one clockwise conformer (P) (Fig. 4b). The presence of this slight difference between TPIMBz and TPIMAn on the outer linkers (p-benzene versus 9,10-anthracene) induces mechanochromic luminescence of TPIMBz under grinding, vaporing and HP. The compound TPIMBz undergoes a red-shifted emission under grinding force (blue to green, $\Delta\lambda \sim 12$ nm) and reverts to the original compound upon treating with water vapor. The suitability of water vapor to recover the original emission color may be because of solvation of the cationic imidazolyl rings in water. No significant difference was observed in the original and post-X-ray diffraction (XRD) peaks, suggesting that the stacking mode in the crystal lattice remains intact (Fig. 4c) and that the observed red-shift is due to partial planarization of the benzene ring of TPE with the ethylene core (Fig. 4d). The same compound exhibited a large red-shifted emission (~72 nm) under HP because of the complete planarization of benzene rings in the bicyclic structure. Furthermore, it showed decreased emission intensity which was rationalized by increased intermolecular π–π interactions among the benzene rings, which was facilitated by complete planarization (Fig. 4d). No wavelength change was observed under grinding or HP for TPIMAn. Two anthracene rings in TPIMAn are big enough to rotate freely through the small cycle under pressure, so hardly any change of emission wavelength was observed [44].

Molecules with high ICT character are highly sensitive to pressure. To understand the influence of molecular conformation on the electronic state of the molecule upon application of pressure, Tang and coworkers [46] developed TPE derivatives with different ICT character. They designed pyridinium (high accepting capability)-functionalized TPE (donor) luminogens with different substituents starting from an electron-donating *n*-pentyloxy group to electron-withdrawing nitro groups, namely

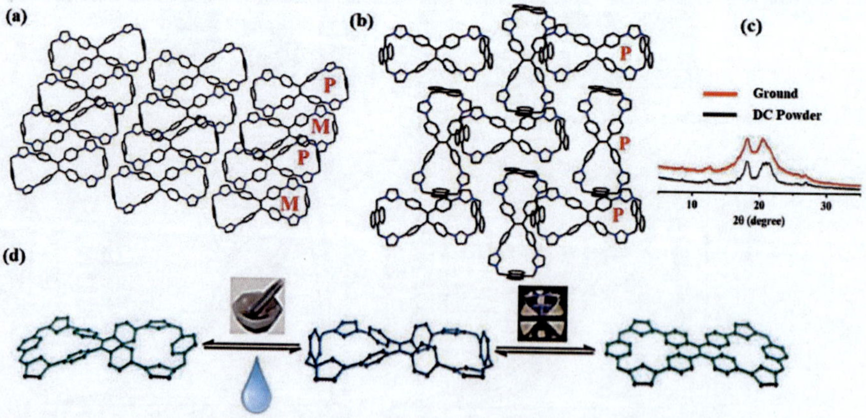

Fig. 4 **a** Stacking pattern of P (clockwise) and M (anticlockwise) rotational conformers of TPIMBz. **b** P stacking pattern of TPIMAn. **c** Powdered X-ray diffraction (XRD) pattern of TPIMBz before and after grinding. **d** Plausible mechanism of the mechanofluorochromic material (TPIMBz) [44, 45]. Copyright 2020, Royal Society of Chemistry

TPE-OP, TPE-H and TPE-NO (Fig. 1), respectively, with differences in the electronic push–pull effect. Due to the enhanced ICT strength, TPE-OP and TPE-H have high contrast emission when applying pressure. All the derivatives have a twisted conformation but the long chain in TPE-OP serves as a block to loosen molecular packing. Upon applying pressure, the TPE-H crystal cannot revert to its original emission due to the hysteresis effect. Due to this effect, two polymorphs (TPE-H–K and TPE-H-T) were produced using a compressed sample with different standing time. Molecular structures with strong ICT character AIE molecules and multiple interactions would be ideal candidates to exhibit MFC with high contrast [46].

1.2 Triphenylamine

Triphenylamine (TPA) is another molecular rotor for AIE materials [47, 48]. As the TPA molecule has a propeller-like shape and the amine nitrogen has a lone pair of electrons, derivatives of TPA forming donor–acceptor structures exhibit typical ICT characteristics. Because of the propeller-shaped structure, the TPA and its derivatives are expected to be external stimuli-responsive materials (Fig. 5) [49, 50].

TPA itself responds differently under pressure. Xu et al. [51] reported that TPA responds differently with grinding and HP. The observation of a blue-shifted emission of TPA under grinding force suggested that the original TPA molecule has better molecular coplanarity and conjugation in the crystal state. This organization leads to less vibration and reduced non-radiative transition. But, after grinding, the crystalline structure is transformed into a less packed amorphous phase; hence, the emission intensity is reduced. When HP is applied on TPA through DAC, the change in fluorescence property is similar to the grinding force up to the pressure 0.1 GPa. With further increasing external pressure, fluorescence intensity was gradually increased until the maximum pressure reaches 1.3 GPa. It was speculated that the

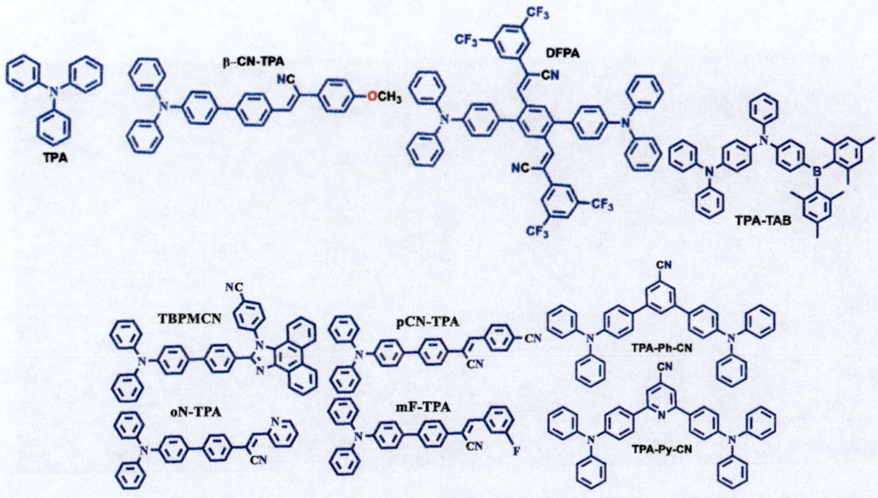

Fig. 5 Chemical structures of triphenylamine (TPA)-based derivatives

phenyl ring of TPA becomes more twisted and the TPA molecule becomes more flattened, which decreases steric hindrance between the phenyl rings with pressure and makes the molecules more closely packed. Thus, rotation of the molecule is restricted, which leads to increasing fluorescence intensity. This phenomenon is well supported by the observed Raman spectra [51]. That no phase change is observed with such high pressure indicates the good stability of this compound with the variation of pressure.

In recent years, there has been an upsurge in introducing secondary interactions (C–H–π, C–N–H etc.) by incorporating cyano stilbene into molecules. These secondary interactions cause tight packing of the crystalline state for high emission. The emission intensity of cyano-substituted oligo(para-phenylene vinylene) in the crystalline state increased due to the crystal-induced emission (CIE) phenomenon. In such a case, the presence of secondary bond interactions restricts cis–trans isomerization around the double bond, which is responsible for the resulting bright emission [52–54]. Zhang and colleagues [55] developed a TPA derivative (β–CN–TPA) in which diphenyl acetonitrile acts as an acceptor and TPA as an electron donor. β–CN–TPA holds a distorted molecular conformation in the solid-state and does not show any changes under mechanical grinding. However, under HP, it undergoes an emission color change from green to deep red with a 135 nm emission shift. Two different types of intermolecular arrangements are present in the crystalline state, one of which is a J type of arrangement, with weak intermolecular stacking interactions and multiple secondary bond interactions like C-H···π (3.482 Å) and π···π (3.473 Å) existing in two neighboring molecules. These multiple secondary bond interactions play an important role in crystal stability. The T-type intermolecular arrangement was observed with multiple secondary interactions, which makes packing tight and limits intramolecular rotation (Fig. 6). Because of this strong interaction, mechanical grinding force is not sufficient to make any change in these optical properties. When HP is applied through the DAC instrument, at initial pressure

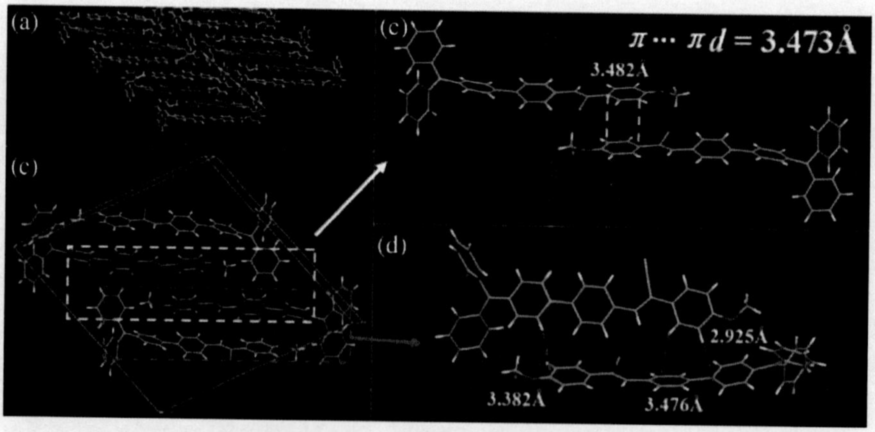

Fig. 6 **a–d** Crystal packing diagram β-CN-TPA (**a,b**) showing J-J type intermolecular stacking (**c**) and the secondary -bonds interactions (C-H···π, π···π) (**d**). Copyright 2020, Royal Society of Chemistry [55]

the emission intensity increases due to the twisted geometry of the TPA rings and decreases the intermolecular distance, which restricts rotation. On further application of pressure, the $\pi\cdots\pi$ interactions existing in the diphenyl acrylonitrile segment of the molecules was strengthened. Strengthening of the $\pi\cdots\pi$ interactions resulted in a decrease in fluorescence intensity with a red-shifted emission color. The decrease in intermolecular distance was supported by Raman spectroscopy blue-shifted frequencies. These blue shifts indicated strengthening of the $\pi\cdots\pi$ interaction with increasing pressure [55]. Zhang and colleagues reported a D-A type of cyano-substituted p-phenylenevinylene derivative (mF-TPA and pCN-TPA) [56]. The molecule has a typical HLCT character, which has both a major fraction of CT transition from TPA to DiCN (diphenylacrylonitrile) unit and a minor fraction of LE transition calculated through excited state natural transition orbitals (NTOs).

Several other stilbene-based TPA derivatives have been reported. Here, mF-TPA exhibits greater efficiency in emission than pCN-TPA due to the increase in the LE component. From the XRD pattern and differential scanning calorimetry (DSC) analysis, the crystalline nature of mF-TPA was transformed into the chaotic amorphous form (green to greenish-yellow emission color) upon grinding. Upon grinding, the XRD pattern of the compound shows a very weak and broad diffraction peak, indicating the formation of a distorted packing arrangement. Under HP, a linear relationship can be seen between pressure change and wavelength shift (500–660 nm) in the range of 0–10 GPa. Raman spectroscopy depicts a blue shift with a broad peak under high pressure, which indicates the closeness in the intermolecular interaction. The decrease in the intermolecular distance increases the polarity effect of the neighboring molecules. This results in a redshift of the PL wavelength and a decrease in PL intensity [56]. Using the same framework, Wei and colleagues also synthesized a TPA-cyanostilbene derivative named oN-TPA containing cyano and pyridine as electron-accepting units. The excited state of oN-TPA is confirmed to be HLCT state in character. Under mechanical grinding, the emission color changes from green to yellow, with a 26 nm shift due to phase transition from the crystalline to the amorphous state supported by the powder XRD pattern of the ground sample as weak and broadening of the peaks, respectively. Upon HP (1 GPa–9.3 GPa), the emission intensity decreases, with an initial marked redshifted emission, from green (552 nm) to orange (598 nm) and, with further pressure, shifting to red (642 nm; 90 nm shift). The original emission reverts upon releasing pressure to 1 atm. Because of the HLCT character, oN-TPA exhibits a huge shift under HP. Due to the increase in CT character in HLCT, the emission peak is red-shifted and fluorescence is quenched [57].

A D–A type of molecular probe containing a TPA moiety has been utilized for sensing applications because of its quick response to external stimuli due to the sensitive ICT state. TPA emission depends on the spacer between D and A and the accepting ability of the substituent attached to the rings. Yang et al. [47] developed charge separation in between the molecules by synthesizing a (D–π–A) TPA derivative (TBPMCN) composed of a rigid triphenylamine-phenanthroimidazole (TBPM) backbone as an electron donor and an orthogonally linked cyanophenyl as an electron acceptor. TBPMCN showed enhancement of emission intensity under pressure change due to unique rehybridization of the nitrogen atom of the TPA unit.

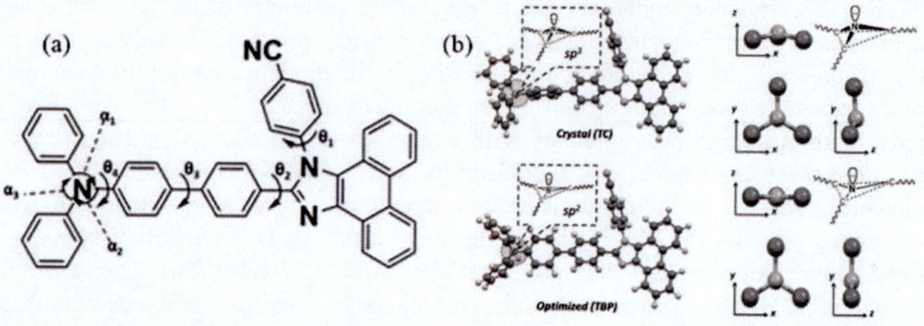

Fig. 7 **a** Chem draw structure of TBPMCN molecule. **b** Enlarged view of TBPMCN triangular-cone (TC) and three-blade-propeller (TBP) conformation. Reproduced with Permission Copyright 2020, Wiley [59]

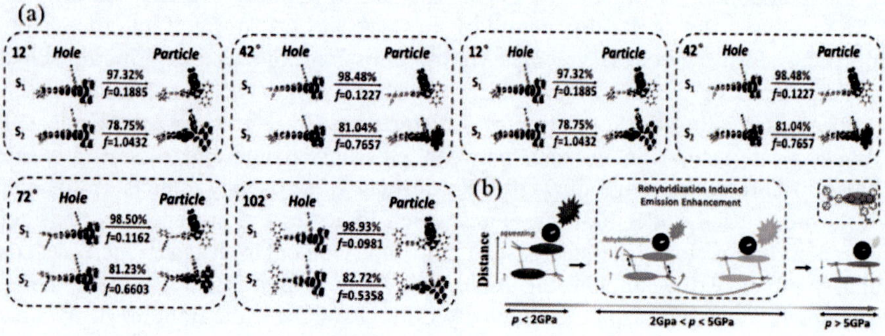

Fig. 8 **a** Natural transition orbital (NTO) analysis on twisted TC conformation. The twisted angle is chosen as the θ_4 for the TC conformation. **b** Proposed mechanism of RHIEE (rehybridization induced emission enhancement) in TBPMCN crystal. Reproduced with Permission Copyright 2020, Wiley [59]

The TBPMCN crystal emission shows redshift with a decrease in the intensity under pressure in the range of 0–2.04 GPa, while emission intensity increases when the external pressure increases gradually from 2.50 to 4.55 GPa; beyond 4.5 GPa, the rise in emission intensity followed a monotonic decline up to 13.2 GPa. The abnormal behavior of TBPMCN under pressure stimulation is due to the transformation of a CT-dominated to a HLCT state [58], because of the conformational change from metastable TC (triangular-cone) to TBP (three-blade-propeller) with rehybridization of the N atom in the TPA unit from sp^3 to sp^2 (Fig. 7a,b). At lower pressures (0–2.04 GPa), the consecutive neighboring molecules are squeezed by the enhanced π–π interaction, which results in a decrease in emission intensity. When the pressure is further enhanced (2.50–4.55 GPa), the energy barrier from the TC to the TBP conformation transition could be overcome, and the N atom is rehybridized.

The TBP conformation of TPA is understood as the planarization of TC conformation. The emission intensity increases with increasing external pressure due to rehybridization of the N atom from sp^3 to sp^2, which was named as the "planarization of donor" (Fig. 8b). To confirm the effect of rehybridization, NTO calculations were carried out depending on the twisting angle θ_4; other twisting angles θ_1–θ_3 were kept steady to maintain the TC conformation, which rules out the possibility of planarization induced under high pressure. From these calculations, it was clear that there was no portion of TPA attributable to the S_1 and the oscillator strength of the S_1 state did not show any obvious change in any of the twisting angles, which indicates that the emission enhancement is not due to a change in the twisting angle (Fig. 8a).

Thus, under the pressure, the TPA moiety with the TC conformation is compressed into a thermodynamically stable TBP conformation. For efficient PL, the TB P conformation is more suitable due to the formation of the HLCT emissive state, which essentially exhibits a pressure-induced emission phenomenon at this stage (2.50–4.55 GPa). On increasing the pressure (4.55–13.2 GPa), the emission decreases, with a broad red-shifted peak. The decreased emission can be attributed to the intermolecular formation of a new excited-state species with lower energy and lower emission as a result of the increasing π–π interaction and the CT-dominated excited state property in the solid-state of TBPMCN [59]. Molecular conformational changes in the molecules lead to a change in the excited state (Fig. 8b).

Molecules that contain the TICT state usually have a larger bathochromic shift with applying pressure due to intramolecular electron transfer accompanied by twisting around the single bond. Conformation changes around the single bond could be achieved by disturbing the propeller shape of TPA, which could enhance the response towards pressure and have the larger stoke shift. On conceptualizing this idea, Bo Zou and coworkers synthesized an X-shaped π architecture TPA derivative (DFPA). DFPA exhibits solid-state TICT (twisted state charge transfer) emission.

$$\eta = \frac{\lambda_1 - \lambda_2}{p_1 - p_2} \tag{1}$$

Crystals of DFPA quantitatively detect external pressures as high as 10 GPa, with a large red shift (162 nm) and high sensitivity of 15 nm GPa (Fig. 9a) with a color change from yellow to deep red. DEPA exhibits a linear relationship related to HP, which implies that pressure could be determined quantitatively from the PL peak [60]. The sensitivity of DEPA was calculated by Eq. (1) Where η represents the slope

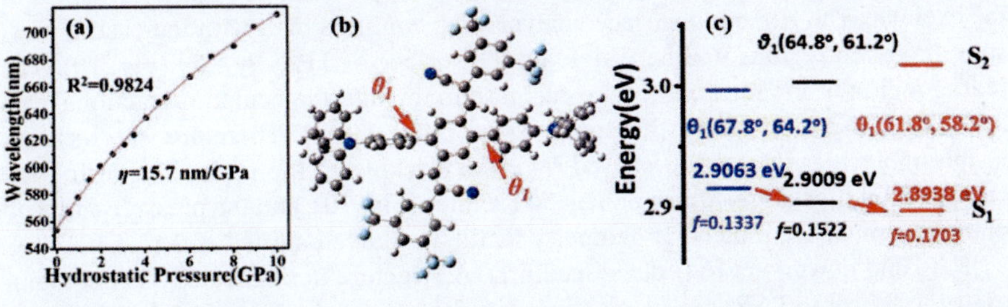

Fig. 9 **a** Linear fit plot between PL peak and HP. **b** Optimized geometry. **c** Excitation energies of DFPA (X-shaped π architecture TPA derivative) molecular conformations with different dihedral angles (θ_1). The calculations were carried out using the TD/M06-2X/6-31 g (d,p) method, and f is the oscillator strength. Reproduced with Permission [60] Copyright 2021, Wiley

of the curve and $\lambda_1$and λ_2 refer to the PL peaks at pressures of P_1 and P_2, respectively. For DEPA, the piezochromic sensitivity η was estimated to be 15.1 nm GPa^{-1} (Fig. 9a). The red shift of the TICT emission was related to molecular planarization. To confirm this, Zou and colleagues calculated the excitation energies with decreasing dihedral angles (θ_1) (Fig. 9b). As shown in Fig. 9c, the rotation of the TPA ring lessens the dihedral angle, which results in a decrease in S_1 excitation energy, which in turn corresponds to red shift in the TICT band during the HP experiment. Therefore, the planarity of the molecular conformation may be responsible for the emission redshift. Deformation of the peaks in infrared (IR) and spectroscopy under high pressure also indicate the configuration change in the molecules. This study helped to develop ratiometric pressure sensing materials [60].

In the same context, Zhiyong Ma and colleagues [61], synthesized molecules with a twisted donor–acceptor–donor (D–A–D) structure noted as TPA–Ph–CN and TPA–Py–CN. Both molecules have completely different packing styles in the crystal. TPA–Py–CN forms a molecular dimer in the crystal state, while TPA–Ph–CN does not form any dimer. So, TPA–Py–CN exhibits green emission with three bands and forms an excimeic emission (449, 477 and 538 nm) while TPA–Ph–CN results in blue emission with a single peak (435 nm). Both have distinct emission behavior under grinding and pressure. A single crystal was squeezed directly into the DAC apparatus and silicone oil was loaded as a pressure-transmitting medium. In the case of TPA–Ph–CN, the emission gradually broadens with redshift and PL intensity decreases. For TPA–Py–CN, the crystals undergo a three-stage change process in emission. At the initial pressure applied, TPA–Py–CN shows a predominantly sky-blue emission (477 nm), which then turns green with the growth of the emission band at 538 nm; the emission band gradually broadens. At elevated pressure, due to the increase in the intermolecular distance, the emission intensity of TPA–Py–CN increases from the excimer effectively (under pressure the neighboring molecules come closer to each other). Most of the Raman peak of TPA–Py–CN exhibits continuous blue shifts during the compression process due to the decrease in the intermolecular distance and an increase in the effective force constants [61]. No new Raman peak appears and no old peaks disappear, indicating that there is no phase transition [62]. A theoretical calculation was performed based on a single crystal at different pressures. According to the results obtained, pressure will decrease the distance between the neighboring molecules effectively, which enhances formation of the excimer. The Hirshfeld surface analysis also confirms that, with increasing pressure, more red regions will be visible on the surface of TPA–Py–CN (Fig. 10). The redder region is an indication of the decrease in the intermolecular interaction (from 2.771 to 2.152 nm) during the compression process [63]. Therefore, the increase in intermolecular interaction caused by close stacking of the molecules during the compression process results in enhanced excimer (Fig. 10), and the pressure-induced planarization of the molecular geometry facilitates the red-shifted emission [61].

Tang and coworkers [64] developed a D–A structure derivative of TPA containing triarylborons (TABs) as an electron-deficient molecule. In mechanical grinding (anisotropic shearing), the emission color is changed because of the phase transition. But, in isotropic HP, TPA–TAB shows three color changes from green, to yellow and finally to orange emission. In mechanical grinding, the intermolecular hydrogen

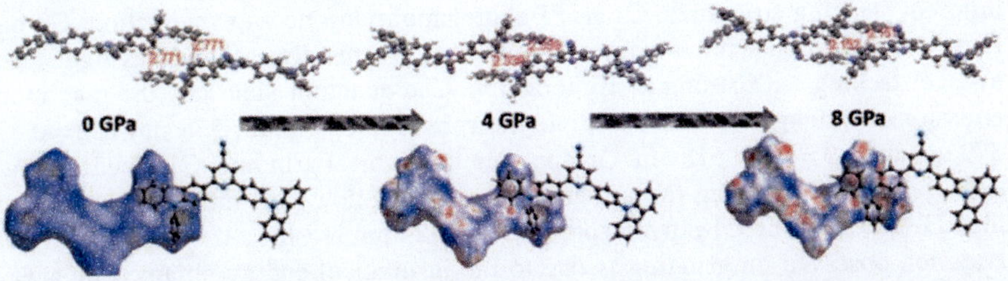

Fig. 10 Changes in dimeric distance and Hirshfeld surface analysis of TPA–Py–CN under compression at pressures of 0, 4, and 8 GPa. Red represents distance shorter than the sum of the van der Waals radii, while white and blue indicate distance longer than the sum of the van der Waals radii. Reproduced with Permission Copyright 2020, Royal Society of Chemistry [61]

bonding and ICT processes will be perturbed, also due to the conformational planarization of the molecules, resulting in a redshift with enhanced emission intensity. The HP is strong enough to squeeze the neighboring molecules, which increases π–π intermolecular interactions. The increase in π–π interactions between the molecules leads to an increase in the intermolecular charge transfer process. This causes the red-shifted decreased emission [64].

1.3 Anthracene

Anthracene (An) is a polyaromatic hydrocarbon with a planar rigid structure and wide bandgap. Due to its planar structure, π–π interactions between the molecules increase with application of pressure, which causes a decrease in emission intensity. The examples shown in Figure 11 show that, basically, the planar conformation is hampered by substituting anthracene with a twisted moiety and hetero atom, which causes fluctuation in the π–π interaction and intermolecular interaction between the molecules under pressure [65, 66]. Tian and colleagues synthesized 9,10-Bis((E)-2-(pyrid-2-yl)vinyl)anthracene (BP2VA) as an AIE active piezochromic luminescent material. BP2VA exhibits a strong green emission in its powder state. They obtained three polymorphs of BP2VA, named C1, C2 and C3. These three polymorphs show

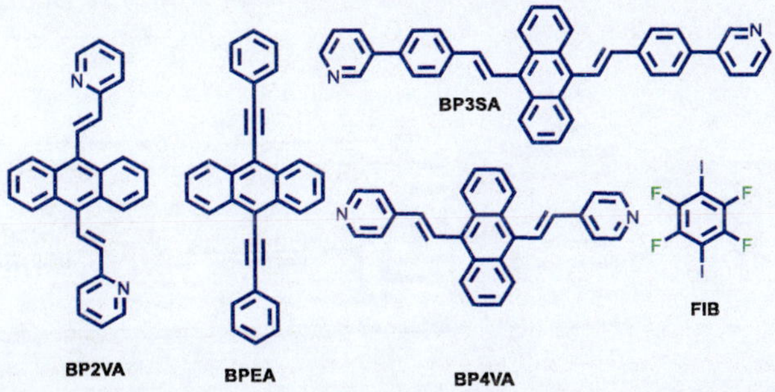

Fig. 11 Chemical structure of all anthracene (An)-based derivatives

different stacking structures. C1 has J-aggregation with no π–π interaction, C2 has H-type aggregation with π–π interaction and C3 forms dimers, and has tight face-to-face stacking with strong π–π interaction. The quantum yield and the maximum emission wavelengths of the three polymorphs are, C1=0.48, 528 nm, C2=0.37, 579 nm and C3=0.28, 618 nm. On applying mechanical grinding, C1 powder shows a sharp red shift of 33 nm from green (528 nm) to yellow (561 nm) emission. But at high HP (8 GPa), the C1 polymorphs undergo 124 nm redshift. The small redshifted emission observed on grinding is due to the insufficient energy obtained by shearing force. At the initial pressure of 2.43 GPa, the green emission turned into orange emission, indicating the C2 form, and upon applying further pressure, the orange one turned into C3. The strength of the intermolecular π–π interactions increases gradually, which induced the PL emission of BP2VA to change from a green emission (no π–π interactions) to an orange emission (weak π–π interactions) and then to a red emission (strong π–π interactions [67] (Fig. 12).

In 2019, Tian and coworkers synthesized 9,10-bis-((E)-4-(pyridin-3-yl)styryl) anthracene (BP3SA), similar toBP2VA, by introducing one more phenyl ring in between pyridine and the vinylic bond to increase the conjugation. Two kinds of crystals were obtained [G (green) and O (orange) phases]. The G-phase crystals are triclinic with J-type aggregation and the O-phase is monoclinic with H-Type aggregation. Upon mechanical grinding, the G-phase and O-phase exhibited redshifted emission spectra and were interconvertible through heating. Under HP from 0 to 5.27 GPa, the emission spectra changed from 511 to 660 nm and from 575 nm–641 nm for G-phase and O-phase, respectively. The sensitivity of the G and P phases towards pressure is 29.1 nm and 33 nm, respectively. The observed differences in sensitivity may be due to the difference in molecular packing mode of the G-phase with J-type arrangement and O-phase with H-type arrangement. For the G-phase, the intense C–H–N interaction helps with the rigidity of the molecules and locking the rotation, whereas for O-phase, the lack of C–H–N interaction increases the sensitivity of the response to external pressure [68].

Cui and colleagues stretched the intermolecular interactions by obtaining two polymorphs of 9,10-bis(phenyl ethynyl)anthracene named BPEA [orange crystal (OC) and red crystal (RC)]. BPEA consist of a phenyl ring attached to An through an acetylene group that is flexible enough to cause the conformational changes. Both polymorphs show dimer formation in the crystal state. Upon grinding, the emission intensity is enhanced with a slight blue shift (2–4 nm) in both polymorphs.

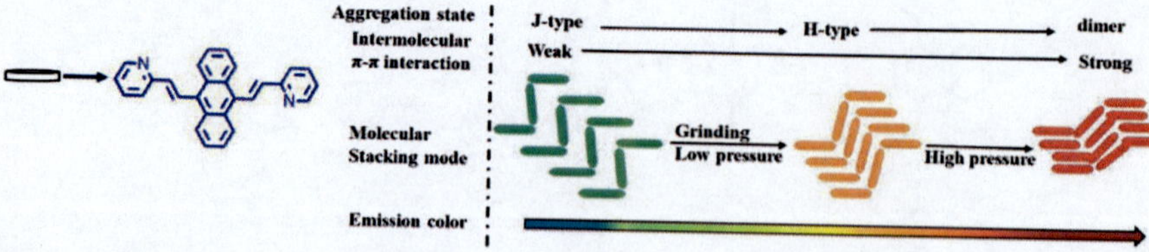

Fig. 12 Stacking mode of three polymorphs (C1, C2 and C3) and corresponding emission color with different molecular packing. Copyright: 2020, Wiley [64]

Under grinding conditions, the ordered structure of the crystal state was destroyed and an amorphous state was formed, as supported by the XRD pattern. With applying HP from 0.12 to 1.70 GPa, only BPEA-OC shows a decrease in emission intensity. Then, a continuous red shift was observed from 1.70 to 9.41 GPa. The gradual change is caused by π–π interactions between anthracene planes [69].

Co-crystallization is a unique technique in which two different molecules can crystallize in the same phase. Here, this could mean higher possibilities to form polymorphs with different intermolecular interactions and molecular packing, which could modulate the resulting photophysical properties under external stimuli. Recently, Liu et al. [70] reported a co-crystal system containing 9,10-bis((E)-2-(pyridin-4-yl)vinyl)anthracene (BP2VA) and 1,3,5-trifluoro-2,4,6-triiodobenzene (FIB). Through the simple solution evaporation method, two different polymorphs of BP4VA-FIB cocrystals were obtained (denoted I and II). Cocrystal I emits yellow emission while II emits green emission. When crystal I was fumed with the solvent or heated (160 °C), I was transformed into the green-emitting II polymorph (Fig. 13a). On applying grinding force, the fumed crystals (or heated) (I_f) returned to the yellow-emitting original polymorph I. [70]. (Fig. 13b). This is one of the rarest examples of the conversion of one form of co-crystal into another. The powder XRD pattern of the fumed and heated crystal of I_f and I_h, respectively, are different from I but both are similar to II, which suggests that I_f and I_h adopt the same packing arrangement of crystal II. The pattern of IIg was similar to that of I (Fig. 13c).The crystal-to-crystal phase transition luminescent switch behavior of BP2VA was also supported by Raman spectroscopy.

In the crystal packing structure, crystal I has several CH–F hydrogen bonds and CI–N halogen bonds between the BP4VA and FIB molecules. These intermolecular hydrogen and interhalogen bonds are the driving force for self-assembly co-crystallization. The 'I' has a J-aggregate type arrangement; therefore, it has a bright yellow emission, whereas II has an H-type self-assembled arrangement and, therefore, exhibits a blue shifted emission. Under HP, both showed a red-shifted emission

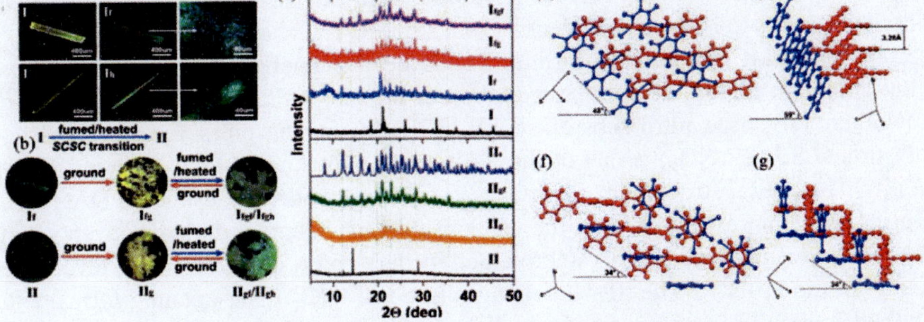

Fig. 13 **a** Images of crystal I and fumed crystal I (I_f) under irradiation with UV light (365 nm). **b** Images of crystal I and II after different treatment under irradiation with UV light (365 nm). **c** Corresponding powder X-ray diffraction (XRD) pattern. **d, e** Crystal packing showing slipped stacking mode along a-axis in the crystal I. **f, g** Crystal Packing along a-axis in crystal II. Right to permission, Copyright: 2020, Wiley [67]

When applying HP up to 3.26 GPa, the emission spectra of I showed bathochromic shifting of 92 nm (554 nm–646 nm). In sharp contrast, crystal II exhibited a bathochromic shift of 24 nm (510 nm–534 nm) with increasing pressure up to 3.38 GPa. This result indicated that crystal I was more sensitive than crystal II because of the different stacking pattern. In crystal I, BP4VA molecules adopt a slipped stacking mode with CH–π interactions along the a-axis, whereas in crystal II no such stacking mode is seen (Fig. 13d–g). Therefore, under pressure, crystal I exhibited more exciton coupling between neighboring BP4VA chromophores [70]. For planar structures like An, the pressure of mechanical grinding is not sufficient to disturb the molecular arrangement.

1.4 Spiropyrans

Spiropyrans (SP) are one of the most extensively studied families of photochromic, thermochromic and solvatochromic materials in both solution and in the crystalline state. Upon exposure to UV irradiation, SP materials change their emission from colorless to colored [open planar merocyanine (MC) form] compounds [71]. The MC form is less stable than the corresponding SP type, which is quite different from other derivatives in terms of stimulus response [72]. In this case, the response to stimuli is based on their tautomerization and orientations between their bistable state in solution or their polymer blends [73, 74]. Zhang and colleagues [75] designed and synthesized a pair of SP-based [benzo[1, 3]oxazine] compounds: OX-1 and OX-2. Interestingly, one of them undergoes tautomerization in the crystalline phase under mechanical stimulus, with OX-1 exhibiting a dynamic switching behavior with varying HP. On applying pressure from 0 to 10 GPa, a smooth modulation of absorption was seen across the whole visible light range (~430 nm to ~700 nm). The low tautomeric energy barrier and the strong environmental sensitivity of the tautomerization equilibrium are responsible for the high sensitivity of OX-1 towards pressure. The lack of dynamic switching of OX-2 can be interpreted as being due to the large tautomeric energy barrier and lower sensitivity of the tautomerization equilibrium. For SP molecular switches, the decrease in volume from ring-opening reactions makes it possible for SPs to isomerize under external stimuli. The change of sp^3 carbon to sp^2 carbon in the MC (ring open) form and subsequent planarization of the molecules plays the main role in the volume reduction of the molecules [75]. In 2015, Ma and coworkers [76] used nitro-substituted SP (SP-NO_2) for the pressure sensor. Nitro-substituted SP (SP-NO_2) is one of the most commonly studied molecules in the SP family. The crystal structure of Sp-NO_2 was examined under HP using the DAC cell. During the compression process, the yellowish crystal becomes dark and, even when the pressure is released, the crystal retained its dark color, indicating the irreversible nature of the Sp-NO_2. The SP-NO_2changes into MC–NO_2 (ring-opening form) with applied pressure. To provide support, a controlled experiment was carried out by irradiating Sp-NO_2 using a Hg lamp at 365 nm for 1 h under atmospheric pressure. The spectra reported for the isomerization in the PMMA matrix through the UV irradiation also matched well with the spectra of Sp-NO_2 under pressure.

This result indicates that Sp-NO_2 is converted to MC–NO_2 under high pressure. The entire change process is visible to the naked eye without any UV-lamp as the color of the crystal changes from colorless to a dark red [76]. Ma and coworkers [77] also used the same concept of ring-opening of SP for visible response under high pressure in crystals of phenolphthalein and its analogues. Although the phenolphthalein derivative PP has been used extensively in host–guest chemistry, its piezoceramic properties were first studied by Ma and co-workers. Ring-opening of PP occurs when the pressure reaches 7 GPa, different from the 5 GPa required for plastic deformation. Ma and coworkers [77] studied the electron-withdrawing effect on the molecules, which reduces the ring-opening threshold pressure to 5 GPa. The crystal structure of the derivative revealed a lactone structure on the sp^3 carbon. The length of the C–O bond in the lactone is longer than the normal C–O bond. The twisted molecular arrangement of PP(Fig. 14a, b, c) makes it easy to transform into a planar structure under high pressure, which causes the ring-opening reaction and helps break the C–O bond in the lactone. Under initial HP, the absorbance in the visible region increased slightly. When the pressure increased further, the absorption band at 530 nm increased dramatically, with a red-shift. The red-shifted absorbance indicates that the ring opening happens under high pressure. In situ images of the crystal sample revealed that the PP crystal changed from transparent to red and then to dark red in the compression process, which could be seen by the naked eye. In the decompression process, the absorption peak is reduced but the peak is not completely blue shifted. This indicates that the ring-closing reaction was not completed under decompression. In Raman spectra of all the molecules, there were two relatively strong peaks around 770 cm^{-1} and 1600 cm^{-1} assigned to ring C–H out-of-plane bending and ring C–C stretching, respectively. Two Raman peaks shifted to higher frequencies with increasing pressure, which indicates that, at high pressure, the intermolecular interactions are further strengthened, and the crystals are more compact (Fig. 14d, e). For the SP derivatives, HP reduces the volume of the crystals but the peak is not completely blue shifted [77].

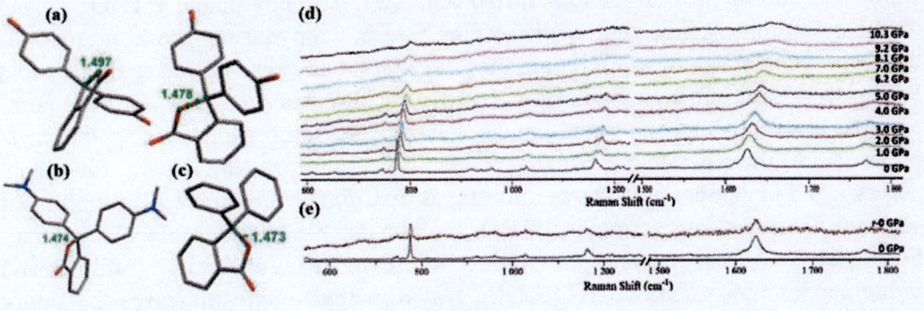

Fig. 14 **a–c** Oak Ridge thermal ellipsoid plot (ORTEP) diagram of the crystal structures of phenolphthalein derivative PP (**a**) PP-NMe_2 (**b**) and PP-H (**c**). **d, e** Change in Raman peak spectra of PP-NMe_2 under compression process in the range of 0–10.3 GPa (**d**) and before pressure treatment (**e**). Copyright Permission, Royal Society Chemistry, 2020 [74]

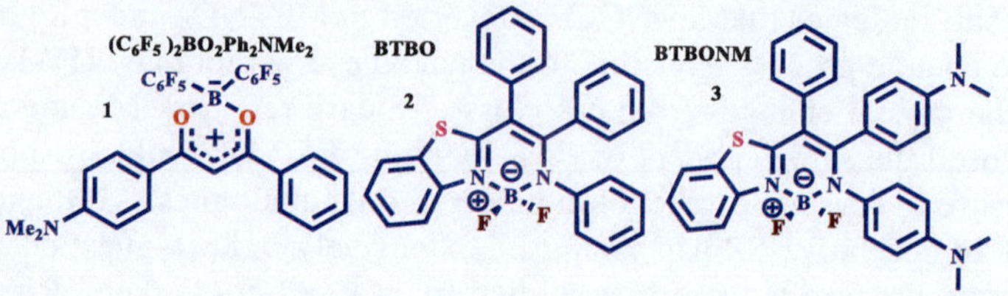

Fig. 15 Chemical structures of boron derivatives that exhibit changes in luminescent properties with applied pressure

1.5 Boron Derivatives

Apart from the organic materials, boron-containing compounds (Fig. 15) were also utilized for HP [78]. In 2015, Wang and colleagues synthesized an orange (1OC) and red (1RC) crystalline boron derivative by refluxing a mixture of 1,3-diaryl-b-diketone and fluorosis (pentafluorophenyl)borane. The molecular conformations of both these crystals are quite different. 1OC exhibits a different response in mechanical grinding and HP. Under mechanical grinding, 1OC is converted into an orange-red solid. In the mechanical grinding process, the well-ordered molecular aggregates are deformed into disordered aggregates, which is supported by the broad XRD pattern and an exothermic peak in DSC of the ground sample. When applying HP, 1OC shows a gradual shifting to red emission. The change in the Raman peak under HP indicates the change in the molecular conformation of the crystals [79].

In the same year, Beijing and coworkers [80] synthesized a propeller-shaped π-conjugated benzothiazole based on two boron compounds, namely BTBO and BTBONM. Both compounds exhibit red shift and decreased luminescence upon grinding, and the luminescence of the ground powder reverted upon decompressing. The XRD pattern shows peak weakening in BTBO, while in BTBONM it shows peak broadening too, which indicates the higher extent of amorphization for BTBONM upon grinding. It is speculated that the higher amorphization led the molecules to be more planar, thereby increasing dipole–dipole intramolecular interactions, resulting in a larger red-shifted emission. The compound BTBO showed dual-band luminescence and a distinct hysteresis loop during the compression-decompression process, while compound BTBONM always showed a single band luminescence and no hysteresis (Fig. 16a,b). By observing the crystal of the compounds, both have several short contacts and CH–π interactions. Using the crystal structure, their molecular packing modes were optimized by density functional theory (DFT) calculations. The results stated that higher pressure led to the planar structure. Under compression in BTBO, the enhanced π–π intermolecular interaction favored excimer formation, resulting in dual-band luminescence (excimer and monomer luminescence). In BTBONM, *N,N*-dimethyl amino increases the donor and acceptor nature. Under compression, planarization induced the ICT between *N,N*-dimethylamino to the deficient boron ring, which resulted in a luminescence bathochromic shift with a decrease in intensity (Fig. 16c,d) [80].

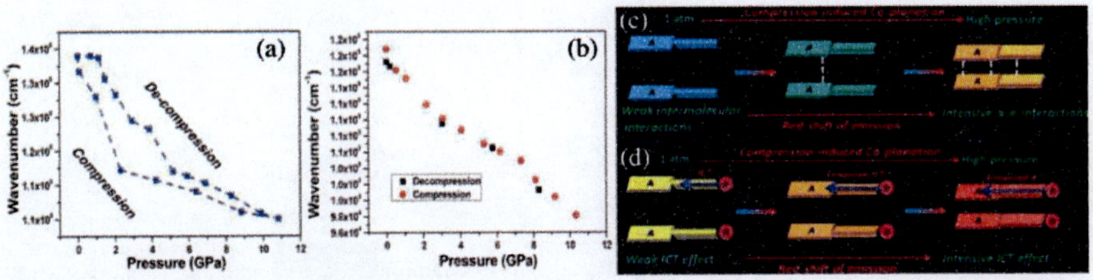

Fig. 16 **a**, **b** Hydrostatic compression–decompression profiles of the two boron compounds BTBO (**a**) and BTBONM (**b**) based on the luminescence wavenumber at different HPs. **c**, **d** Proposed mechanism under compression for BTBO (**c**) and BTBONM (**d**). *A* Electron-withdrawing motif, *D* electron-donating group. Copyright 2020, Royal Society Chemistry [80]

1.6 Schiff Base

Owing to the presence of π–π interactions at high pressure (isotropic pressure), many AIE molecules exhibit an emission peak that is redshifted with a decrease in intensity [81, 82]. To achieve turn-on emission under pressure, Han and colleagues [83] designed and synthesized a biplane molecule [2-amino-3-((E)-((2-hydroxynaphthalen-1-yl)methylene)amino)maleonitrile (AHM)]. A biplane molecule consists of electron donor (D) and electron acceptor (A) planes connected through a single bond as shown in Fig. 17. This type of molecule forms a co-plane via single-bond rotation after crystallizing, which favors D–A coupling by the pronounced overlap of the conjugated plane to suppress light emission. AHM shows turn-on emission under grinding. The powder XRD pattern shows the crystalline nature of the compound before and after grinding, some of the peaks disappeared, but a sharp peak indicating the defect in the crystalline nature is observed. In the crystals, the molecules exist in coplanar nature, favoring D–A coupling; hence, emission is weak. Upon grinding, the molecules are freed from severe D–A coupling and π–π interaction by twisting through the single bond. Pressure is applied directly through the digital trust meter, containing an indenter of 1.5 mm on the test strip containing the compound. There was no observed change of emission at initial force, up to 0.5 N, but the emission intensity increases when the force reaches 0.7 N and emission enhancement can still be noticed up to 60 N [83]. Han and colleagues synthesized a simple Schiff base compound, which is denoted as HNMB (Fig. 18). The compound is AIE active and responds to pressure sensing. In the crystalline structure, HNMB has J-type coupling, which is due to the less overlapping of π···π stacking and fixed by multiple intermolecular interactions, including mainly CH···π, CH···O and hydrogen bonding. This type of interaction is disturbed at low pressure and does not require high pressure. By applying pressure, the flexible molecular conformation is disturbed without any intermolecular interaction, which allows the non-radiative decay path to occur. At high pressure, the molecules are planarized, which increases π overlap between neighboring molecules (Fig. 17) [84]. The same group, encouraged by their previous idea in 2020, synthesized a similar Schiff base—CDBTA (Fig. 18)—containing benzothiazole and *N, N* dimethylamine as a donor. Two forms of the crystal for CDBTA were isolated by slow and fast evaporation techniques,

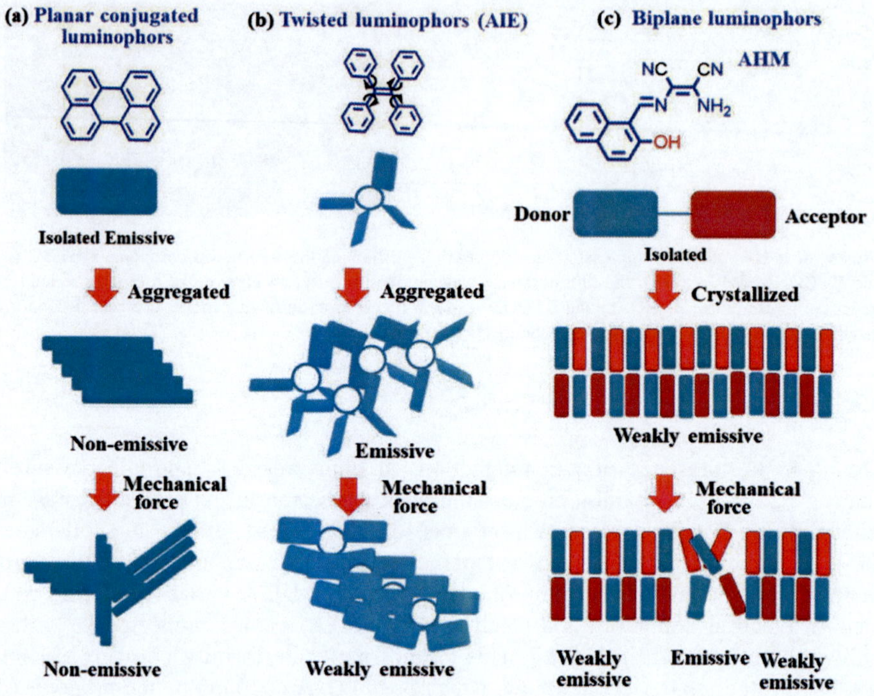

Fig. 17 The design of three molecules and the proposed mechanism for their mechanochromic responses. Copyright 2020, Royal Society Chemistry [80]

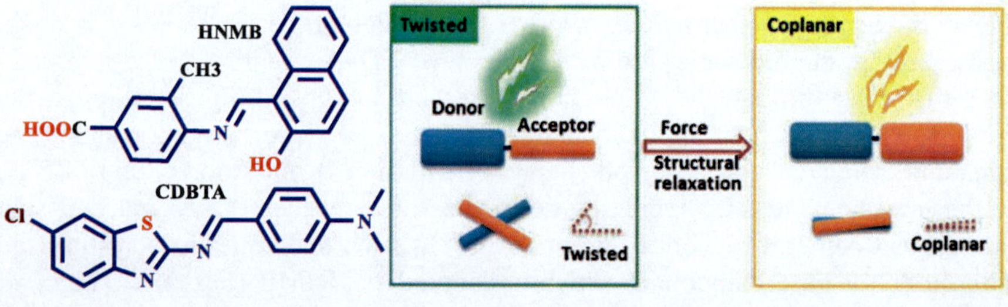

Fig. 18 Chemical structure of HNMB and CDBTA, and working mechanisms for D–A type Schiff base [82] Copyright 2020, Wiley

and are denoted as G and Y forms, respectively. The powder XRD patterns for the two forms differ. The G form emits green fluorescence and the Y form emits yellow fluorescence. The crystal structure of the G form reveals that it has an antiparallel twisted conformation, while the Y form is more planar in structure. Force applied to CDBTA by a digital thrust meter converted the G form into the Y form. Mechanical grinding destroys the intermolecular interactions to release intramolecular torsional stress, which transferred the twisted D-A biplane into a coplanar structure. Such a dihedral-angle-controlled mechanism makes Schiff-base-type materials compatible for pressure detection with high sensitivity (Fig. 17) [85]

1.7 Carbazole and Phenazine

Carbazole is a blue-emitting crystal organic fluorophore. The crystal structure of the carbazole is connected by intermolecular N–H···π hydrogen bonds and C–H···π hydrogen bonds, exhibiting herringbone motifs (Fig. 19a). By applying HP, carbazole emission intensity is increased up to 1.0 GPa. The fluorescence spectrum was red-shifted 40 nm with decreasing emission intensity after applying pressure from 1 up to 10.3 GPa (Fig. 19b). Infrared spectra at higher pressure were examined to study the changes in intermolecular interactions. The observed band at 3050 cm^{-1} was assigned to C–H stretching vibrations and gradually blue shifted with pressure, indicating that the C–H bond is shortened. The stretching of vibrations increases the non-radiation vibration process, resulting in a decrease in intensity. Simultaneously, the shape of the C–H band also changes due to changes in the C–H–π interactions. During compression, the molecules adopt a smaller volume and come closer, which increases the intermolecular interactions. Single crystals of carbazole exhibit herringbone arrangements (Fig. 19a), and the volume of crystals contract in the herringbone arrangements, increasing the intermolecular interaction between the molecules. The contraction in such herringbone motifs rearranges the molecules to make them more parallel to each other. The parallel arrangement increases π–π interactions, which results in quenching of the emission. In carbazole, the N–H frequency is observed at ~ 3420 cm^{-1} under compression. The peak is observed to initially redshift then blue shift (Fig. 19c).

These results indicate that, at the initial pressure, the bond is elongated, demonstrating that N–H–π interactions became stronger with increasing pressure, which restricts N–H vibrations thus inhibiting the non-radiative vibrational process. The

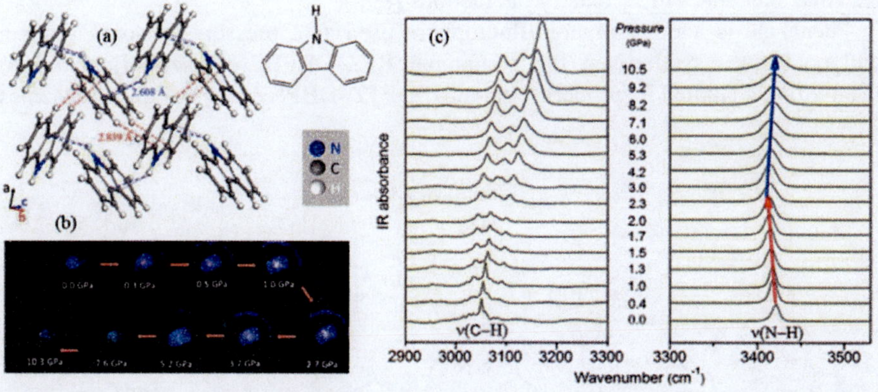

Fig. 19 **a** Herringbone arrangement of carbazole crystals, **b** In situ PL image of single crystal under compression in a diamond anvil cell (DAC). **c** Infrared (IR) spectra of carbazole with high pressure up to 10.5 GPa at room temperature in the region of ν(C–H) and ν(N–H) vibrational modes. Copyright 2020, Wiley [83]

emission intensity increases up to 1.0 GPa; beyond that, π–π stacking interactions increase, which quenches emission [86]. Ling and coworkers [87] synthesized carbazole-based diphenyl maleimide dyes with different N-substituent denoted BCPMs. All compound BCPMs are highly emissive, and their packing modes are similar to each other. The peak intensity is gradually redshifted, with a decrease in intensity with applying pressure due to a change in phase transformation in the crystal [87].

One way to have multi-colored emission under external stimuli is to incorporate a large charge separation within the molecule. The ICT emission is more sensitive to external stimuli. In TPA and TPE derivatives, we noticed that the unique HLCT state consists of both ICT and LE character, which respond differently under external stimuli. Liu et al. [88] designed a molecule Cz2CN (Fig. 20) with a twisted D–A structure using the carbazole group as the donor and a dicyanoethylene fragment as the acceptor to have two characteristic emission bands, LE and ICT Cz2CN showed AIEE properties. The crystal packing shows that it does not contain π–π interaction with twisted conformation, which makes Cz2CN emissive in the solid state. With increasing pressure, the adjacent molecules approach each other, which makes the molecular geometry of Cz2CN more planarized, leading to the redshifted ICT emission. The fluorescence spectra could not return fully to the original state because, at high pressure, its single crystal structure was damaged [88].

Nowadays, π-conjugated chromophores such as 2,1,3-benzothiadiazole (BTD) are used extensively for molecular optoelectronics and are excellent candidates for efficient solid-state emitters at longer wavelength due to several advantages over other chromophores (simpler methods of synthetic route, strong electron-withdrawing capacity, and ease of crystallization). Zhao and colleagues synthesized three derivatives of BTD from flexible to rigid structure. Under compression and decompression processes, all these crystals showed a reversible bathochromic shift from orange emission to red/near infrared (NIR) emission. Here, the molecular structure governs the secondary interaction, which leads to different emission in the crystals. These different emissions are due to the different arrangements in aggregation mode and also due to changes in secondary interactions [89].

Phenazine is another organic fluorophore used as a pressure sensor. Fukuhara and colleagues synthesized the compound PTZ–DBPHZ(dibenzo[a,j]phenazine) (Fig. 20) and studied its properties under HP. PTZ–DBPHZ has two phenothiazines

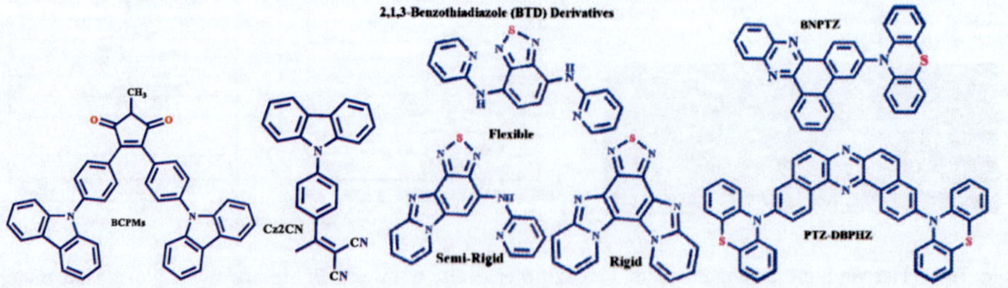

Fig. 20 Chemical structure of different organic molecules based on carbazole, benzothiadiazole and phenazine

(PTZ) units that are attached to a dibenzo[*a,j*]phenazine (DBPHZ) acceptor. PTZ–DBPHZ has four conformers and is sensitive to external stimuli. The compound responds to the luminescence property change under HP in solution as well as in different solvents (toluene or methylcyclohexane). PTZ–DBPHZ shows a ratiometric response under HP that originates from the ternary emission comprising ^{1}LE and two different ^{1}CT excited state associated with the different conformers. Upon HP, and in a high viscous solvent, the observed emission spectrum demonstrates interconversion between the conformers [90]. Jia and coworkers also synthesized dibenzo[a,c]phenazine-phenothiazine dyad (BNPTZ) linked by a single C–N bond. BNPTZ (Fig. 20) is an AIEE active compound and shows polymorphism, mechanochromism and sensitivity to HP due to the twisted D–A structure. By mechanical grinding, the crystal undergoes a color change from orange-yellow to red. When the ground powder was fumed with organic solvents, the emission was recovered. This was an example of the solid crystal to crystal phase transformation reaction under mechanical grinding. Upon HP, both the crystals show a redshifted emission with quenching. Both crystals showed high sensitivity to pressure with a large spectral shift (100 nm). These types of transformation under pressure could be assigned as emission from the locally excited state at higher energy, and emission from the ICT at lower energy in twister D–A type compounds [91].

Such types of hetropolyaromatic twisted D-A type compounds have a different type of packing mode and have two or more different conformers active in the crystal. Upon application of external stimuli, these conformers are interconvertible under mechanical grinding and have a distinct response in HP due to their molecular arrangements.

1.8 Pyrene

Pyrene is a rich π-conjugated polyaromatic hydrocarbon. It exhibits monomeric and excimeric emission in the excited state. Tang and colleagues [92] synthesized (4-(phenothiazin-10-yl)phenyl)(pyren-1-yl)-methanone (Py–BP–PTZ) (Fig. 21),

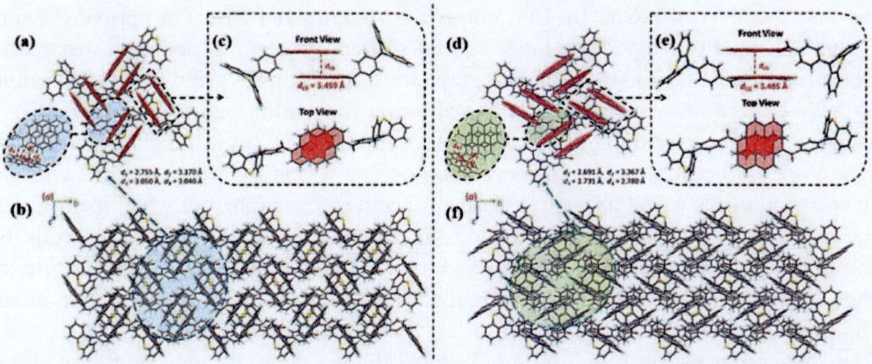

Fig. 21 **a–f** Intermolecular interactions and π–π stacking patterns of Py–BP–PTZ (**a**, **b**, **c**) CB and Py–BP–PTZ (**d**, **e**, **f**) CG. Copyright 2020, Royal Society Chemistry [87]

which exhibits AIEE, polymorphism and mechanochromic properties. Py–BP–PTZ has two conformers in the crystalline state by the slow evaporation technique: Py–BP–PTZ (CB) (blue emission) and Py–BP–PTZ (CG) (green emission). Many intermolecular weak interactions (such as C–H–π) are observed in both conformers. The extent of overlapping and inter-plane distances present in CB and CG crystals differs. Both possess a herringbone arrangement in the packing diagram (Fig. 21). With high pressure, the emission of CB is red-shifted with decreasing emission intensity. A new broad peak appears under decompression. Upon grinding of Py–BP–PTZ (CB), the original emission color changed to bright yellow, and upon fuming with DCM for 10 min, the ground sample returned to its original emission. The planar pyrene derivative changes emission mainly due to planarization under high pressure and change of molecular packing mode with changing intermolecular interactions. With the application of adequate pressure (1.83 GPa), the molecules of pyrene derivatives come closer and restrict the motion of the rings, which increases the radiative pathway; hence, the emission intensity increases. But if high pressure is applied, then, due to the closeness and pyrene ring slippage, the planarization of the molecule increases, which indeed enhances the π–π stacking, ultimately resulting in quenching of emission with red-shift [92].

1.9 Combined Organic Moieties

Conformational changes affect molecular packing in the crystal state [93]. In the above section, we discussed details of the relationship between the structure of different derivatives, their packing modes and the effect of isotropic pressure on them. The packing modes in the crystal lattice and the subsequent optical responses of various derivatives of TPE, TPA and An change on applying HP. Many scientists have attempted to integrate the properties of different derivatives such as ratiometric fluorescence property change, colorimetric change, reversibility, high stroke shift and high quantum yield under isotropic pressure, in a single compound [25].

Ma and coworkers [94] integrated the property of isomerization of SPs at high pressure and dual emission of pyrene (excimer and monomeric emission) in a single entity and synthesized Sp-Pyr, combining pyrene and SPs. On applying force (grinding), Sp-Pyr (Fig. 22) exhibited a red-shifted, sequential tricolored mechanofluorochromic behavior (blue–bluish–green–red). Under mild grinding, the different aggregated fragment of pyrene, Sp-Pyr shows a shift from blue to green emission, while, with heavy grinding, the change in luminescence from bluish-green to red was observed because of the isomerization of SPs. Upon light grinding the change in emission is due to the phase transformation and is reversible, but when the force is increased (heavy grinding), the change from green to red does not fully revert to its initial state. The initial change in Sp-Pyr is due to transfer of the crystalline nature to the amorphous phase. Further grinding increases the force on Sp-Pyr, which induces the ring-opening reaction of the SP moiety with planarized zwitterionic structure, thus this conformation leads to more reddish fluorescence. It could be rationalized that the multi-luminescence of Sp-Pyr arises from the highly crystalline nature of the pristine sample and the large change in the chromophore caused by the chemical

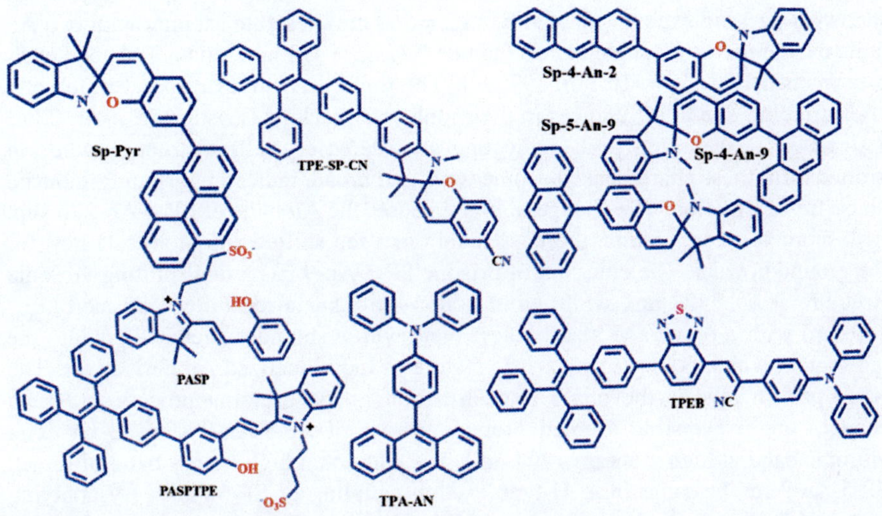

Fig. 22 Chemical structures of combined organic molecules that exhibit changes in luminescent property with applying pressure

reaction, which is not fully reversible. Under HP, the emission change is reversible if the pressure falls in the range of 0 GPa–8.3 GPa. The change in emission was observed to be linear with increasing pressure. Here, a reversible shift arises due to the conformational changes, and, as molecules come closer due to the pressure, the pyrene may exist in the excimer form, which also results in the red-shifted emission. As the pressure increases beyond 11.1 GPa, there were 100 nm shifts with quenched emission. After decompression, a new peak arises due to the isomerization of the SPs. The emission of pyrene can be divided into two parts—one reverts to its original position (424 nm) and another shifts to 467 nm as a result of irreversible changes in pyrene stacking after isomerization of the SPs [94]. The same group, using a similar approach, replaced pyrene with TPE and found the same result with multicolored emission under isotropic pressure [95].

In 2017, Jia and colleagues [96] synthesized three molecules composed of anthracene and SPs, abbreviated as SP-4-An-2, SP-4-An-9 and SP-5-An-9 (Fig. 22). These synthesized molecules have small variations in structure but show relatively different responsive behavior to anisotropic grinding and isotropic HP. The compounds are interrelated through linking of the SP moiety to the α or γ position of an An unit, and also through connecting an An unit to the 5-position of a SP moiety. The change in molecular geometry is certainly affected by the packing in the crystals. In SP-4-An-9, adjacent An units in single crystals are highly overlapped, while in SP-4-An-2, there is only weak π–π interaction between them. No interaction exists in An units in SP-5-An-9. The emission color of An in the crystal state was correlated to the extent of the degree of π–π interaction between the adjacent moieties. In An derivatives, the bathochromic shifted emission change indicates the increased π–π overlapping between An moieties. In the literature, based on the geometry arrangements of An, there are three kinds of anthracene excimers, which include sandwich-packing excimer (named E1, $\lambda_{em}=560$ nm), a stable excimer with two An units overlapped at an angle of 55° (named E2, $\lambda_{em}=470$ nm), and a T-shaped excimer (named E3, $\lambda_{em}=510$ nm) [97–99]. The three derivatives respond differently after grinding due to differences in their molecular packing. Pristine SP-4-An-2 has a sharp XRD scattering peak, indicating an ordered crystalline structure while in ground form these sharp peaks become weak and broad, indicating grinding-induced phase transformation. The change of phase causes the An units of SP-4-An-2 to slip, with more π–π conjugation, thereby leading to a red-shifted emission (505 nm) for the ground powder. The emission of pristine SP-5-An-9 has a fine splitting vibronic structure at λ_{max} 422 nm, while ground SP-5-An-9 has a redshifted emission (λ_{max} 443 nm) with a broad and structureless band with a shoulder stretched far into the red region, which is the characteristic feature of individualized An derivatives. The XRD pattern reveals the phase transformation in the grinding process and such changes are irreversible. Fumed ground powder of SP-5-An-9 showed an extra vibronic band at higher energy (404 nm). The additional high energy band of fumed SP-5-An-9 might come from H-type exciton coupling of SP-5-An-9. By applying isotropic pressure, similar changes could be observed. Sp-4-An-2 exhibited a high contrast change with a multicolor switch by HP, and isomerization of the SP caused more red-shift at high pressure [96].

A couple of similar works have been carried out by Zhang and colleagues [101], who reported on new AIE fluorescents, with similar prototype, 1',3',3'-trimethyl-5'-(4-(1,2,2-triphenylvinyl)phenyl)spiro[chromene-2,2'-indoline]-6-carbonitrile (TPE–Sp–CN) (Fig. 22), and applied these to the quantitative detection of three parameters—pH value, sulfite anion and HP—with different molecular states. Here, in the presence of an acid, the TPE–Sp–CN (ring-closed form) was transformed into TPE–MCH–CN (ring-open form). The solid-state of TPE–MCH–CN is more sensitive to pressure in the range of 0 to 6 GPa than TPE–Sp–CN [100] and a cyanine-based photoacid-PS derivative named PASPTPE was also developed using TPE as the molecular rotor and PASP without TPE (Fig. 22). The isotropic pressure effect depends on the molecular arrangement in the crystal state as PASPTPE emission intensity decreases with bathochromic shift due to the strong π–π interaction under pressure. Conversely, for PASP at mild pressure, emission intensity increases due to the twisted and antiparallel arrangement of the ring, which increases the intermolecular interaction. Applying high pressure (above 4 GPa) resulted in a bathochromic shift with quenched emission because of strong π–π interaction [101].

In 2018, Xu and coworkers integrated TPA and An units through a C–C single bond (TPA–AN). In the crystal structure of TPA–AN (Fig. 22), the presence of weak C–H–N and C–H–π interactions is observed. The TPA–AN compound exhibits a typical ICT character. Upon grinding, a slight blue shift of 6 nm from the original emission (476 nm–470 nm) was observed, with increasing emission intensity. In the powder XRD pattern, the ground form is amorphous in nature, i.e., the peak is broadened with weaker intensity. At initial pressure up to 0.77 GPa, there was no obvious change, but when the applied pressure increases up to 2.67 GPa, the blue emission of the crystals changes to a cyan-colored emission. Upon adequate HP (>2.69 GPa), TPA–AN exhibits HLCT character. Xu and colleagues came to this conclusion from the observation of a new emission band at longer wavelength, which originates from the ICT state, and these results are in good agreement with previous works [38, 102]. The blue-shifted peak in the Raman spectrum also indicates the closeness and the decrease in the intermolecular distance, which increase the interactions resulting in the bathochromic shift [103]. A similar concept was utilized by Li and colleagues to develop piezochromic materials by replacing An with pyrene [102].

In 2020, Yang and coworkers combined a π-rich conjugated planar chromophore with a highly twisted molecular rotor. In this case, TPE was conjugated with An, linked at a different position (mTPE-AN and pTPE-AN). In the crystal structure of mTPE-AN, the TPE substituent is oriented asymmetrically along one side of the An plane in a molecular conformation allowing An dimer formation with antiparallel π–π stacking. In pTPE–AN, the TPE molecular conformation was divided equally by the An plane, which results in the complete deflection of π–π An dimer formation (Fig. 23a). The emission of mTPE–AN is red-shifted in comparison to that of pTPE–AN. Upon mechanical grinding, there is no change in the emission observed, indicating a stable crystal state. By applying HP, three steps are observed in mTPE-AN. As pressure increased on mTPE-AN from 1–1.23 GPa, the emission is redshifted, with a decrease in the intensity indicating the formation of An dimers. When the pressure was increased above 1.23 GPa, a new emission band appeared at 438 nm. The new emission band increases gradually as pressure increases up to 4.28 GPa with a blue shift. With further compression of 4.28 GPa, high- and low-energy emission showed a red-shift, with quenching of emission intensity. Hirshfeld surface analysis supports a decrease in the π–π distance of the An-dimer and the TPE dimer with pressure that corresponds to the enhancement of the intermolecular interactions (Fig. 23b) [104]. The effect of low temperature on the decrease in the π–π distance was the same as that observed at high pressure. At room temperature, mTPE-AN shows a single exponential decay corresponding to the pure excited state of the An-excimer. Moreover, the emission of mTPE–AN depends on the excitation,

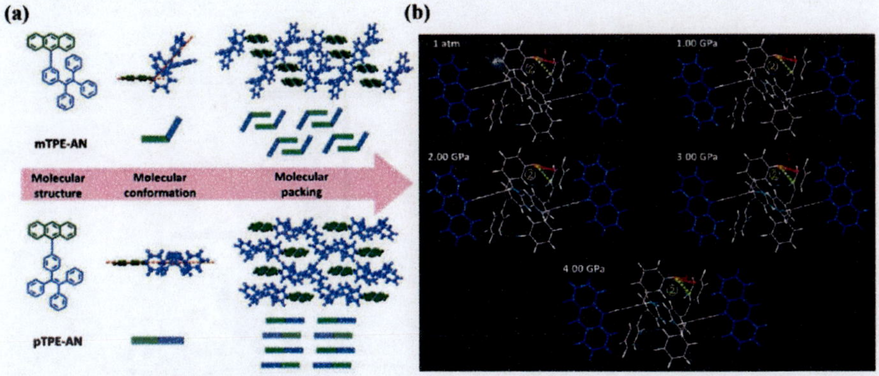

Fig. 23 **a** Design of molecular structure with the formation of An-dimer. Crystal packing and molecular conformation are obtained from the single-crystal XRD. **b** From the Hirshfeld analysis, the strength of the C–H–C interaction is seen to increase with increasing pressure (at 1 atm 1 = 2.694 Å& 2 = 2.993 Å, at 1 GPa = 2.610 Å& 2 = 2.895 Å, at 2 GPa = 2.539 Å& 2 = 2.833 Å, at 3 GPa = 2.466 Å& 2 = 2.719 Å, 2, at 4 GPa = 2.429 Å& 2 = 2.644 Å). Copyright 2020, American Chemical Society [98]

indicating a fast internal conversion process. The abnormal behavior of mTPE-AN was explained as follows. From 1 atm pressure to 1.23 GPa, there is a complete energy transfer from TPE to An excimer in mTPE-AN, which produces one emission band from the An excimer because of the increasing spectral overlap between the emission of TPE and absorption of the An dimer. As the pressure increases on mTPE–AN, the absorption from the AN dimer weakens, while the emission spectra of TPE remain quite unchanged, which gradually decreases the overlap between TPE and An dimer, leading to the suppression of energy transfer (ET). The suppression of ET is responsible for the high energy band from TPE. As the volume decreases during pressure, the molecules come closer and, due to this, the TPE units aggregate into the tighter packs and motion is restricted; therefore, emission intensity increases with higher energy bands.

Once the pressure increases beyond 4.3 GPa, π–π stacking between TPE units dominates, and emission intensity is quenched. Upon the decompression process, the An-dimer reverts, but a small spectral residue is visible due to the hysteresis effect of TPE under high pressure. The hysteresis effect of TPE is the large structural deformation under high pressure (Fig. 24) [104].

When a flat π-conjugated chromophore is combined with a twisted conformation ring such as TPA and TPE, one can expect dimer formation in the crystal packing that leads to excimeric emission. Formation of the dimer is responsible for the different responses to pressure. In 2020, Luo et al. synthesized the novel molecule TPEB (Fig. 22) by combining TPA and TPE together linked by the π-conjugated electron-accepting chromophore [benzothiadiazole (BTA)]. The charge separation creates a different excited state, such as LE, HLCT or ICT. Two kinds of polymorphs were formed: TPEB-γ, which has a dihedral angle of 16.3° and 9.3° between the plane of the acrylonitrile and its aromatic rings, respectively, while TPEB-α has a higher dihedral angle (21.4° and 14.6°). Due to this, TPEB-γ adopted a more planar conformation. Both polymorphs have antiparallel π-dimers (head-to-tail π-dimers) with

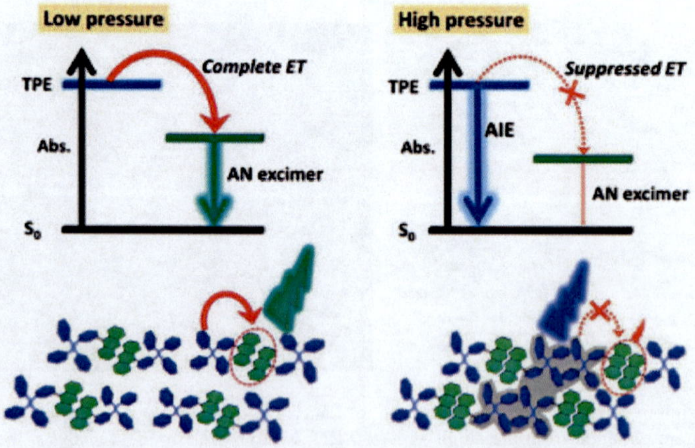

Fig. 24 Schematic representation of the proposed mechanism of pressure-induced emission blue-shift and enhancement in mTPE-AN crystals. Copyright 2020, American Chemical Society [98]

a separation of nearly 4 Å due to the twisted TPE units. With HP, TPEB-α shows a huge bathochromic emission shift from deep red (658 nm) to NIR (870 nm).The observed bathochromic shift is due to the decrease in the intermolecular distance under compression. The dimers in the molecular packing come closer due to the molecular planarization during compression, which enhances the extent of electron delocalization, resulting in redshifted emission [105].

1.10 Other Materials

Yamaguchi and colleagues [107] synthesized the new material tetrathiazolylthiophene, which has a distinct response to mechanical grinding and HP. Here, the molecular arrangements with the four-thiazole ring play a crucial role in the distinct response. Two thiazole rings are coplanar to each other and the other two are in a twisted form. In the crystal form, weak intermolecular interaction and hydrogen bonding play an important role in the molecular rearrangements. Under mechanical grinding, the crystals are disordered, which suppressed the excimer formation results in the blue-shifted emission, while at high pressure the emission is red-shifted with quenching. By applying high pressure, the molecules come close to each other and form excimers due to the face-to-face molecular arrangements [106]. Wang and coworkers [106] demonstrated that pressure can induce single electron transfer in viologen species (viologens are organic compounds with the formula $[©_5H_4NR)_2{}^{n+}]$. The change could also be triggered by electricity and light. Applying HP, the absorption of the viologen is red-shifted in compression and blue-shifted in the decompression process. The powder XRD and IR data of viologen with pressure and after removal of pressure do not show any significant change from the original yellow state. This indicates that there is no phase change or structural deformation. In the Raman spectra, the peaks shifted to higher frequencies as expected under high pressure, but no new peak was formed and all peaks were reverted by the decompression process. These results suggest that the compression and decompression process does not involve any molecular arrangement (related to the viologen, and triggered by electricity and light [107]).

1.11 Summary and Future Perspectives

AIE mechanofluorochromic materials are advantageous because they have enhanced fluorescence intensity in the solid or aggregated state. Moreover, their twisted geometry and the presence of void space in the crystal lattice is inherently sensitive and prone to stress. These properties identified classes of AIE active compounds as a perfect platform to study and explore mechanisms for change in the photophysical properties of materials with variations of pressure. This article provides an exhaustive overview of recent progress in the study of AIE active organic materials that exhibit distinct responses in emission properties upon mechanical grinding (anisotropic) and HP (isotropic). The isotropic force is applied uniformly by the DAC tool. In mechanical grinding, pressure is not applied isotopically (comparable to shearing force) and is also insufficient to have complete structural change while HP is uniform; high pressure can be generated by the DAC instrument. In the mechanical grinding process (basically, shear force), the phase transformation of the solid crystals takes place from crystalline to amorphous, and this phase transformation is irreversible. Conversely, HP increases intermolecular interactions by decreasing intermolecular distance and is able to change the molecular arrangements/ conformational changes in the crystals reversibly.

In the case of twisted molecular structures (such as TPE and TPA, etc.), the twisting ability of the ring, conformational changes and intermolecular interactions depend on the linker moiety and the position at which it is linked. Polyaromatic compounds (e.g., pyrene, An, etc.) containing rigid and planar π-conjugated molecules support excimeric emission. But when these are attached to twisted moieties, the twisted ring results in an increased intermolecular distance, which precludes dimer formation. However, at higher pressure, due to the planarization of the rings, dimer formation is allowed, leading to excimeric emission creating a distinct response in low and high pressure. Usually, the twisted rings rotate with gradually increasing pressure, which triggers emission property changes through the perturbation of intermolecular forces. It would be easy to investigate such a study with the help of Raman spectroscopy, which rationalizes structural changes with pressure. One of the observed mechanisms in such twisted structures is that at, relatively low pressure, the intermolecular distance is decreased, which restricts motion, resulting in emission enhancement. But, at higher pressure, planarized moieties approach closer, which leads to the enhancement of π–π stacking, resulting in emission quenching. By introducing rigid π-chromophores, mechanisms such as energy transfer and excimeric emission play a significant role in HP, which leads to greater bathochromic shift. Incorporation of heterocycles such as pyridine, thiophene, benzothiadiazole, etc., will make compounds more sensitive towards pressure and gives the possibility to have multi-color emission properties under stress. These heterocycles create more short contacts between the molecules, and help stabilize the molecular structure. Changes under pressure lead to conformational changes, which result in changes in emission. Co-crystallization, formed by the growing of different components (both luminophores/one luminophores and one non-luminophore) with variable geometries, may lead to interesting responses with pressure through molecular rearrangement/conformational changes/varying intermolecular interactions. Co-crystallization is the best and easiest way to study properties under HP due to the combined and distinct effect of each component. Very few studies have reported co-crystallized solids and this need to be explored further. We have also reported on different combinations of organic entities with distinct responsive properties utilized for pressure sensing. Different excited states, such as LE, HLCT and ICT states, play a significant and interesting role at variable pressure. When a π-rigid electron-withdrawing moiety is attached to a twisted donor, the HLCT state rises under pressure, which enhances emission intensity. The ICT state is more sensitive to pressure than the HLCT state.

This review provides an understanding of chemical structures with changes in emission properties in AIE active compounds under mechanical stress. Undoubtedly, the knowledge to be gained will support and inspire the scientists to explore new mechanofluorochromic materials showing exciting properties with pressure.

Although some AIE-based MFC materials result in exceptionally good performances, these are still in their infancy in terms of industrial applications. The present review will not only inspire scientists to explore new AIE-based MFC materials but also given a thrust towards real-world applications.

Acknowledgements We thank DST, Government of India, for financial assistance under the following project (SB/S1/IC-13/2014). V.K. thanks CSIR India for the SRF fellowship [09/719(0082)/2018EMR-I]. BITS Pilani, Pilani Campus, are also acknowledged for infrastructure and facilities.

Declarations

Conflict of interest The authors declare no conflict of interest.

References

1. Guo F, Guo Z (2016) Inspired smart materials with external stimuli responsive wettability: a review. Rsc Adv 6(43):36623–36641. https://doi.org/10.1039/C6RA04079A
2. Wang H, Zhao E, Lam JWY, Tang BZ (2015) AIE luminogens: emission brightened by aggregation. Mater Today 18(7):365–377. https://doi.org/10.1016/j.mattod.2015.03.004
3. Han T, Liu L, Wang D, Yang J, Tang BZ (2021) Mechanochromic fluorescent polymers enabled by AIE processes 42(1):2000311. https://doi.org/10.1002/marc.202000311
4. Roy E, Nagar A, Chaudhary S, Pal S (2020) Advanced properties and applications of AIEgens-inspired smart materials. Ind Eng Chem Res 59(23):10721–10736. https://doi.org/10.1021/acs.iecr.0c01869
5. Zhang X, Chi Z, Zhang Y, Liu S, Xu J (2013) Recent advances in mechanochromic luminescent metal complexes. J Mater ChemC 1(21):3376–3390. https://doi.org/10.1039/C3TC30316K
6. Gundu S, Kim M, Mergu N, Son Y-A (2017) AIE-active and reversible mechanochromic tetraphenylethene-tetradiphenylacrylonitrile hybrid luminogens with re-writable optical data storage application. Dyes Pigm 146:7–13. https://doi.org/10.1016/j.dyepig.2017.06.043
7. Feng A, Smet PF (2018) A review of mechanoluminescence in inorganic solids: compounds. Mech Models Appl 11(4):484
8. La DD, Bhosale SV, Jones LA, Bhosale SV (2018) Tetraphenylethylene-based AIE-active probes for sensing applications. ACS Appl Mater Interfaces 10(15):12189–12216. https://doi.org/10.1021/acsami.7b12320
9. Di B-H, Chen Y-L (2018) Recent progress in organic mechanoluminescent materials. Chin Chem Lett 29(2):245–251. https://doi.org/10.1016/j.cclet.2017.08.043
10. Bustamante C, Alexander L, Maciuba K, Kaiser CM (2020) Single-molecule studies of protein folding with optical tweezers. Annu Rev Biochem 89(1):443–470. https://doi.org/10.1146/annurev-biochem-013118-111442
11. Xie Y, Li Z (2020) The development of mechanoluminescence from organic compounds: breakthrough and deep insight. Mater Chem Front 4(2):317–331. https://doi.org/10.1039/C9QM00580C
12. Zhou X, Li H, Chi Z, Zhang X, Zhang J, Xu B, Zhang Y, Liu S, Xu J (2012) Piezofluorochromism and morphology of a new aggregation-induced emission compound derived from tetraphenylethylene and carbazole. New J Chem 36(3):685–693. https://doi.org/10.1039/C1NJ20782B
13. Moggach SA, Allan DR, Parsons S, Warren JE (2008) Incorporation of a new design of backing seat and anvil in a Merrill-Bassett diamond anvil cell. J Appl Crystallogr 41(2):249–251. https://doi.org/10.1107/S0021889808000514
14. Evans WJ, Yoo C-S, Lee GW, Cynn H, Lipp MJ, Visbeck K (2007) Dynamic diamond anvil cell (dDAC): a novel device for studying the dynamic-pressure properties of materials. Rev Sci Instrum 78(7):073904. https://doi.org/10.1063/1.2751409
15. Zhao F, Chen Z, Fan C, Liu G, Pu S (2019) Aggregation-induced emission (AIE)-active highly emissive novel carbazole-based dyes with various solid-state fluorescence and reversible mechanofluorochromism characteristics. Dyes Pigm 164:390–397. https://doi.org/10.1016/j.dyepig.2019.01.057

16. Zhang G, Singer JP, Kooi SE, Evans RE, Thomas EL, Fraser CL (2011) Reversible solid-state mechanochromic fluorescence from a boron lipid dye. J Mater Chem 21(23):8295–8299. https://doi.org/10.1039/C0JM03871G
17. Hong Y, Lam JWY, Tang BZ (2011) Aggregation-induced emission. Chem Soc Rev 40(11):5361–5388. https://doi.org/10.1039/C1CS15113D
18. Wu W, Duan Y, Liu B (2020) Mechanoluminescence: quantitative pressure-brightness relationship enables new applications. Matter 2(2):291–293. https://doi.org/10.1016/j.matt.2020.01.007
19. Ubba E, Tao Y, Yang Z, Zhao J, Wang L, Chi Z (2018) Organic mechanoluminescence aggregation-induced emission. Chemistry 13(21):3106–3121. https://doi.org/10.1002/asia.201800926
20. Suman GR, Pandey M, Jeevan Chakravarthy AS (2021) Review on new horizons of aggregation induced emission: from design to development. Materials Chem Front 5(4):1541–1584. https://doi.org/10.1039/D0QM00825G
21. Clough JM, Weder C, Schrettl S (2021) Mechanochromism in structurally colored polymeric materials. Macromol Rapid Commun 42(1):2000528. https://doi.org/10.1002/marc.202000528
22. Pucci A (2019) Mechanochromic fluorescent polymers aggregation-induced emission features. Sensors 19(22):4969
23. Liu J-J, Yang J, Wang J-L, Chang Z-F, Li B, Song W-T, Zhao Z, Lou X, Dai J, Xia F (2018) Tetrathienylethene based red aggregation-enhanced emission probes: super red-shifted mechanochromic behavior and highly photostable cell membrane imaging. Mater Chem Front 2(6):1126–1136. https://doi.org/10.1039/C8QM00008E
24. Hao J, Xu C-N (2018) Piezophotonics: From fundamentals and materials to applications. MRS Bull 43(12):965–969. https://doi.org/10.1557/mrs.2018.296
25. Wang L, Ye K-Q, Zhang H-Y (2016) Organic materials with hydrostatic pressure induced mechanochromic properties. Chin Chem Lett 27(8):1367–1375. https://doi.org/10.1016/j.cclet.2016.06.049
26. Fu Z, Wang K, Zou B (2019) Recent advances in organic pressure-responsive luminescent materials. Chin Chem Lett 30(11):1883–1894
27. Wang L, Liu L, Xu B, Tian W (2021) Recent advances in mechanism of AIE mechanochromic materials. Chem Res Chin Univ 37(1):100–109. https://doi.org/10.1007/s40242-021-0431-0
28. Mei J, Hong Y, Lam JWY, Qin A, Tang Y, Tang BZ (2014) Aggregation-induced emission. Adv Mater 26(31):5429–5479. https://doi.org/10.1002/adma.201401356
29. Valuer B (2001) Molecular fluorescence: principles and applications. In: Digital encyclopedia of applied physics. Wiley, New York, pp 477–531. https://doi.org/10.1002/3527600434.eap684
30. Yuan H, Wang K, Yang K, Liu B, Zou B (2014) Luminescence properties of compressed tetraphenylethene: the role of intermolecular interactions. J Phys Chem Lett 5(17):2968–2973. https://doi.org/10.1021/jz501371k
31. Li N, Gu Y, Chen Y, Zhang L, Zeng Q, Geng T, Wu L, Jiang L, Xiao G, Wang K, Zou B (2019) Pressure-induced emission enhancement and piezochromism of triphenylethylene. J Phys Chem C 123(11):6763–6767. https://doi.org/10.1021/acs.jpcc.9b00670
32. Schmidtke JP, Kim J-S, Gierschner J, Silva C, Friend RH (2007) Optical spectroscopy of a polyfluorene copolymer at high pressure: intra- and intermolecular interactions. Phys Rev Lett 99(16):167401. https://doi.org/10.1103/PhysRevLett.99.167401
33. Wu J, Tang J, Wang H, Qi Q, Fang X, Liu Y, Xu S, Zhang SX-A, Zhang H, Xu W (2015) Reversible piezofluorochromic property and intrinsic structure changes of tetra(4-methoxyphenyl)ethylene under high pressure. J Phys Chem A 119(35):9218–9224. https://doi.org/10.1021/acs.jpca.5b02362
34. Kwon JE, Park SY (2011) Advanced organic optoelectronic materials: harnessing excited-state intramolecular proton transfer (ESIPT) process. Adv Mater 23(32):3615–3642. https://doi.org/10.1002/adma.201102046
35. Guo Z-H, Jin Z-X, Wang J-Y, Pei J (2014) A donor–acceptor–donor conjugated molecule: twist intramolecular charge transfer and piezochromic luminescent properties. Chem Commun 50(46):6088–6090. https://doi.org/10.1039/C3CC48980A
36. Qi Q, Qian J, Tan X, Zhang J, Wang L, Xu B, Zou B, Tian W (2015) Remarkable turn-on color-tuned piezochromic luminescence. Adv Funct Mater 25(26):4005–4010. https://doi.org/10.1002/adfm.201501224
37. Xiong J, Wang K, Yao Z, Zou B, Xu J, Bu X-H (2018) Multi-stimuli-responsive fluorescence switching from a pyridine-functionalized tetraphenylethene AIEgen. ACS Appl Mater Interfaces 10(6):5819–5827. https://doi.org/10.1021/acsami.7b18718

38. Huang L, Liu J, Liu L, Yang Q, Ma Z, Jia X (2020) A D-A-D' type organic molecule with persistent phosphorescence exhibiting dual-mode mechanochromism. Dyes Pigm 173:107963. https://doi.org/10.1016/j.dyepig.2019.107963
39. Qiu X, Ying S, Wang C, Hanif M, Xu Y, Li Y, Zhao R, Hu D, Ma D, Ma Y (2019) Novel 9,9-dimethylfluorene-bridged D–π–A-type fluorophores with a hybridized local and charge-transfer excited state for deep-blue electroluminescence with CIE_y ~ 0.05. J Mater Chem C 7(3):592–600. https://doi.org/10.1039/C8TC05469J
40. Li W, Pan Y, Yao L, Liu H, Zhang S, Wang C, Shen F, Lu P, Yang B, Ma Y (2014) A hybridized local and charge-transfer excited state for highly efficient fluorescent OLEDs: molecular design, spectral character, and full exciton utilization. Adv Opt Mater 2(9):892–901. https://doi.org/10.1002/adom.201400154
41. Li W, Pan Y, Xiao R, Peng Q, Zhang S, Ma D, Li F, Shen F, Wang Y, Yang B, Ma Y (2014) Employing ~100% Excitons in OLEDs by utilizing a fluorescent molecule with hybridized local and charge-transfer excited state. Adv Funct Mater 24(11):1609–1614. https://doi.org/10.1002/adfm.201301750
42. Liu X, Li A, Xu W, Ma Z, Jia X (2018) Pressure-induced emission band separation of the hybridized local and charge transfer excited state in a TPE-based crystal. Phys Chem Chem Phys 20(19):13249–13254. https://doi.org/10.1039/C8CP02096E
43. Liu X, Ma C, Li A, Xu W, Ma Z, Jia X (2019) Schiff base-bridged TPE-rhodamine dyad: facile synthesis, distinct response to shearing and hydrostatic pressure, and sequential multicolored acidichromism. J Mater Chem C 7(27):8398–8403. https://doi.org/10.1039/C9TC02113B
44. Nian H, Li A, Li Y, Cheng L, Wang L, Xu W, Cao L (2020) Tetraphenylethene-based tetracationic dicyclophanes: synthesis, mechanochromic luminescence, and photochemical reactions. Chem Commun 56(21):3195–3198. https://doi.org/10.1039/D0CC00860E
45. Gao YW, Jun; Zhang, Chao-Feng; Xu, Xiang-Hong; Zhang, Mian; Kong, Ling-Yi, (2014) CCDC 940012: experimental crystal structure determination. Cambridge Crystallographic Data Centre. https://doi.org/10.5517/CC10K4ZP
46. Wang X, Qi C, Fu Z, Zhang H, Wang J, Feng H-T, Wang K, Zou B, Lam JWY, Tang BZ (2021) A synergy between the push–pull electronic effect and twisted conformation for high-contrast mechanochromic AIEgens. Mater Horiz 8(2):630–638. https://doi.org/10.1039/D0MH01251C
47. Yang W, Li C, Zhang M, Zhou W, Xue R, Liu H, Li Y (2016) Aggregation-induced emission and intermolecular charge transfer effect in triphenylamine fluorophores containing diphenylhydrazone structures. Phys Chem Chem Phys 18(40):28052–28060. https://doi.org/10.1039/C6CP04755F
48. Pham HD, Hu H, Feron K, Manzhos S, Wang H, Lam YM, Sonar P (2017) Thienylvinylenethienyl and naphthalene core substituted with triphenylamines—highly efficient hole transporting materials and their comparative study for inverted perovskite solar cells. Sol RRL 1(8):1700105. https://doi.org/10.1002/solr.201700105
49. Jia J, Zhao H (2019) Structure-dependent reversible mechanochromism of D-π-A triphenylamine derivatives. Tetrahedron Lett 60(3):252–259. https://doi.org/10.1016/j.tetlet.2018.12.024
50. Fang M, Yang J, Liao Q, Gong Y, Xie Z, Chi Z, Peng Q, Li Q, Li Z (2017) Triphenylamine derivatives: different molecular packing and the corresponding mechanoluminescent or mechanochromism property. J Mater ChemC 5(38):9879–9885. https://doi.org/10.1039/C7TC03641H
51. Wu J, Wang H, Xu S, Xu W (2015) Comparison of Shearing force and hydrostatic pressure on molecular structures of triphenylamine by fluorescence and Raman spectroscopies. J Phys Chem A 119(8):1303–1308. https://doi.org/10.1021/jp511380a
52. Yue B, Xie Z, Lu P, Ma Y (2013) The organic photoelectric functions and applications in organic light-emitting diodes of aggregation-induced emission molecules. Sci China 43(1674–7224):1065. https://doi.org/10.1360/032013-197
53. Zhang Y, Zhuang G, Ouyang M, Hu B, Song Q, Sun J, Zhang C, Gu C, Xu Y, Ma Y (2013) Mechanochromic and thermochromic fluorescent properties of cyanostilbene derivatives. Dyes Pigm 98(3):486–492. https://doi.org/10.1016/j.dyepig.2013.03.017
54. Zhao H, Wang Y, Harrington S, Ma L, Hu S, Wu X, Tang H, Xue M, Wang Y (2016) Remarkable substitution influence on the mechanochromism of cyanostilbene derivatives. Rsc Adv 6(71):66477–66483. https://doi.org/10.1039/C6RA14707K
55. Ouyang M, Zhan L, Lv X, Cao F, Li W, Zhang Y, Wang K, Zhang C (2016) Clear piezochromic behaviors of AIE-active organic powders under hydrostatic pressure. Rsc Adv 6(2):1188–1193. https://doi.org/10.1039/C5RA21218A

56. Zhang Y, Qile M, Sun J, Xu M, Wang K, Cao F, Li W, Song Q, Zou B, Zhang C (2016) Ratiometric pressure sensors based on cyano-substituted oligo(p-phenylene vinylene) derivatives in the hybridized local and charge-transfer excited state. J Mater ChemC 4(42):9954–9960. https://doi.org/10.1039/C6TC03157A
57. Wang B, Wei C (2018) Stimuli-responsive fluorescence switching of cyanostilbene derivatives: ultrasensitive water, acidochromism and mechanochromism. Rsc Adv 8(40):22806–22812. https://doi.org/10.1039/C8RA03598A
58. Zhang Y, Wang K, Zhuang G, Xie Z, Zhang C, Cao F, Pan G, Chen H, Zou B, Ma Y (2015) Multicolored-fluorescence switching ICT-type organic solids clear color difference: mechanically controlled excited state. Chemistry 21(6):2474–2479. https://doi.org/10.1002/chem.201405348
59. Zhang S, Dai Y, Luo S, Gao Y, Gao N, Wang K, Zou B, Yang B, Ma Y (2017) Rehybridization of nitrogen atom induced photoluminescence enhancement pressure stimulation. Adv Funct Mater 27(1):1602276. https://doi.org/10.1002/adfm.201602276
60. Zhang Y, Zhang J, Shen J, Sun J, Wang K, Xie Z, Gao H, Zou B (2018) Solid-state TICT-emissive cruciform: aggregation-enhanced emission, deep-red to near-infrared piezochromism and imaging in vivo. Adv Opt Mater 6(22):1800956. https://doi.org/10.1002/adom.201800956
61. Li A, Chu N, Liu J, Liu H, Wang J, Xu S, Cui H, Zhang H, Xu W, Ma Z (2019) Pressure-induced remarkable luminescence switch of a dimer form of donor–acceptor–donor triphenylamine (TPA) derivative. Mater Chem Front 3(12):2768–2774. https://doi.org/10.1039/C9QM00529C
62. Mosca S, Milani A, Peña-Álvarez M, Yamaguchi S, Hernández V, Ruiz Delgado MC, Castiglioni C (2018) Mechanochromic luminescent tetrathiazolylthiophenes: evaluating the role of intermolecular interactions through pressure and temperature-dependent Raman spectroscopy. J Phys Chem C 122(30):17537–17543. https://doi.org/10.1021/acs.jpcc.8b05423
63. Spackman MA, Jayatilaka D (2009) Hirshfeld surface analysis. CrystEngComm 11(1):19–32. https://doi.org/10.1039/B818330A
64. Zhang J, Li A, Zou H, Peng J, Guo J, Wu W, Zhang H, Zhang J, Gu X, Xu W, Xu S, Liu SH, Qin A, Lam JWY, Tang BZ (2020) A "simple" donor–acceptor AIEgen with multi-stimuli responsive behavior. Mater Horiz 7(1):135–142. https://doi.org/10.1039/C9MH01041F
65. Yoshizawa M, Klosterman JK (2014) Molecular architectures of multi-anthracene assemblies. Chem Soc Rev 43(6):1885–1898. https://doi.org/10.1039/C3CS60315F
66. Sun B, Yang X, Ma L, Niu C, Wang F, Na N, Wen J, Ouyang J (2013) Design and application of anthracene derivative with aggregation-induced emission charateristics for visualization and monitoring of erythropoietin unfolding. Langmuir 29(6):1956–1962. https://doi.org/10.1021/la3048278
67. Dong Y, Xu B, Zhang J, Tan X, Wang L, Chen J, Lv H, Wen S, Li B, Ye L, Zou B, Tian W (2012) Piezochromic luminescence based on the molecular aggregation of 9,10-bis((E)-2-(pyrid-2-yl) vinyl)anthracene. Angew Chem Int Ed 51(43):10782–10785. https://doi.org/10.1002/anie.201204660
68. Shao B, Jin R, Li A, Liu Y, Li B, Xu S, Xu W, Xu B, Tian W (2019) Luminescent switching and structural transition through multiple external stimuli based on organic molecular polymorphs. J Mater ChemC 7(11):3263–3268. https://doi.org/10.1039/C9TC00051H
69. Li A, Wang J, Xu S, Huo Z, Geng Y, Xu W, Cui H (2019) Distinct stimuli-responsive behavior for two polymorphs of 9,10-bis(phenylethynyl)anthracene under pressure based on intermolecular interactions. Dyes Pigm 170:107603. https://doi.org/10.1016/j.dyepig.2019.107603
70. Liu Y, Li A, Xu S, Xu W, Liu Y, Tian W, Xu B (2020) Reversible luminescent switching in an organic cocrystal: multi-stimuli-induced crystal-to-crystal phase transformation. Angew Chem Int Ed 59(35):15098–15103. https://doi.org/10.1002/anie.202002220
71. Bénard S, Yu P (2000) New spiropyrans showing crystalline-state photochromism. Adv Mater 12(1):48–50. https://doi.org/10.1002/(SICI)1521-4095(200001)12:1%3c48::AID-ADMA48%3e3.0.CO;2-G
72. Klajn R (2014) Spiropyran-based dynamic materials. Chem Soc Rev 43(1):148–184. https://doi.org/10.1039/C3CS60181A
73. Kortekaas L, Browne WR (2019) The evolution of spiropyran: fundamentals and progress of an extraordinarily versatile photochrome. Chem Soc Rev 48(12):3406–3424. https://doi.org/10.1039/C9CS00203K
74. Such G, Evans RA, Yee LH, Davis TP (2003) Factors influencing photochromism of spiro-compounds within polymeric matrices. J Macromol Sci Part C 43(4):547–579. https://doi.org/10.1081/MC-120025978

75. Wang Y, Tan X, Zhang Y-M, Zhu S, Zhang I, Yu B, Wang K, Yang B, Li M, Zou B, Zhang SX-A (2015) Dynamic behavior of molecular switches in crystal under pressure and its reflection on tactile sensing. J Am Chem Soc 137(2):931–939. https://doi.org/10.1021/ja511499p
76. Meng X, Qi G, Zhang C, Wang K, Zou B, Ma Y (2015) Visible mechanochromic responses of spiropyrans in crystals via pressure-induced isomerization. Chem Commun 51(45):9320–9323. https://doi.org/10.1039/C5CC01064K
77. Deng X, Guo H, Meng X, Wang K, Zou B, Ma Y (2019) Visible responses under high pressure in crystals: phenolphthalein and its analogues with adjustable ring-opening threshold pressures. Chem Commun 55(32):4663–4666. https://doi.org/10.1039/C9CC01145E
78. Parris MD, MacKay BA, Rathke JW, Klingler RJ, Gerald RE (2008) Influence of pressure on boron cross-linked polymer gels. Macromolecules 41(21):8181–8186. https://doi.org/10.1021/ma801187q
79. Wang L, Wang K, Zhang H, Jiao C, Zou B, Ye K, Zhang H, Wang Y (2015) The facile realization of RGB luminescence based on one yellow emissive four-coordinate organoboron material. Chem Commun 51(36):7701–7704. https://doi.org/10.1039/C5CC01113B
80. Wang X, Liu Q, Yan H, Liu Z, Yao M, Zhang Q, Gong S, He W (2015) Piezochromic luminescence behaviors of two new benzothiazole–enamido boron difluoride complexes: intra- and intermolecular effects induced by hydrostatic compression. Chem Commun 51(35):7497–7500. https://doi.org/10.1039/C5CC01902H
81. Pasha SS, Yadav HR, Choudhury AR, Laskar IR (2017) Synthesis of an aggregation-induced emission (AIE) active salicylaldehyde based Schiff base: study of mechanoluminescence and sensitive Zn(ii) sensing. J Mater ChemC 5(37):9651–9658. https://doi.org/10.1039/C7TC03046K
82. Yadav P, Singh AK, Upadhyay C, Singh VP (2019) Photoluminescence behaviour of a stimuli responsive schiff base: aggregation induced emission and piezochromism. Dyes Pigm 160:731–739. https://doi.org/10.1016/j.dyepig.2018.08.065
83. Shi P, Duan Y, Wei W, Xu Z, Li Z, Han T (2018) A turn-on type mechanochromic fluorescent material based on defect-induced emission: implication for pressure sensing and mechanical printing. J Mater ChemC 6(10):2476–2482. https://doi.org/10.1039/C7TC05683D
84. Zhang M, Zhao L, Zhao R, Li Z, Liu Y, Duan Y, Han T (2019) A mechanochromic luminescent material with aggregation-induced emission: application for pressure sensing and mapping. Spectrochim Acta Pt A 220:117125. https://doi.org/10.1016/j.saa.2019.05.030
85. Hou Y, Li Y, Zhang M, Zhang X, Xu Z, Li Y, Duan Y, Han T (2020) A dihedral-angle-controlled mechanochromic luminescent material: application for pressure sensing. Dyes Pigm 180:108505. https://doi.org/10.1016/j.dyepig.2020.108505
86. Gu Y, Wang K, Dai Y, Xiao G, Ma Y, Qiao Y, Zou B (2017) Pressure-induced emission enhancement of carbazole: the restriction of intramolecular vibration. J PhysChemLett 8(17):4191–4196. https://doi.org/10.1021/acs.jpclett.7b01796
87. Mei X, Wei K, Wen G, Liu Z, Lin Z, Zhou Z, Huang L, Yang E, Ling Q (2016) Carbazole-based diphenyl maleimides: multi-functional smart fluorescent materials for data process and sensing for pressure, explosive and pH. Dyes Pigm 133:345–353. https://doi.org/10.1016/j.dyepig.2016.06.015
88. Liu Y, Li A, Ma Z, Xu W, Ma Z, Jia X (2020) Multicolored fluorescence variation of a new carbazole-based AIEE molecule by external stimuli. Phys Chem Chem Phys 22(34):19195–19201. https://doi.org/10.1039/D0CP02783A
89. Zhao Y, Ding J, Han X, Geng T, Zhou X, Hu C, Wang Y, Xiao G, Zou B, Hou H (2020) Tuning the optical properties of N-aryl benzothiadiazole via Cu(ii)-catalyzed intramolecular C-H amination: the impact of the molecular structure on aggregation and solid state luminescence. Organic Chem Front 7(23):3853–3861. https://doi.org/10.1039/D0QO00915F
90. Takeda Y, Mizuno H, Okada Y, Okazaki M, Minakata S, Penfold T, Fukuhara G (2019) Hydrostatic pressure-controlled ratiometric luminescence responses of a dibenzo[a,j]phenazine-cored mechanoluminophore. ChemPhotoChem 3(12):1203–1211. https://doi.org/10.1002/cptc.201900190
91. Yang Y, Li A, Ma Z, Liu J, Xu W, Ma Z, Jia X (2020) Dibenzo[a, c]phenazine-phenothiazine dyad: AIEE, polymorphism, distinctive mechanochromism, high sensitivity to pressure. Dyes Pigm 181:108575. https://doi.org/10.1016/j.dyepig.2020.108575
92. Xiong Y, Huang J, Liu Y, Xiao B, Xu B, Zhao Z, Tang BZ (2020) High-contrast luminescence dependent on polymorphism and mechanochromism of AIE-active (4-(phenothiazin-10-yl)phenyl)(pyren-1-yl)methanone. J Mater ChemC 8(7):2460–2466. https://doi.org/10.1039/C9TC05064G

93. Wang C, Li Z (2017) Molecular conformation and packing: their critical roles in the emission performance of mechanochromic fluorescence materials. Mater Chem Front 1(11):2174–2194. https://doi.org/10.1039/C7QM00201G
94. Meng X, Qi G, Li X, Wang Z, Wang K, Zou B, Ma Y (2016) Spiropyran-based multi-colored switching tuned by pressure and mechanical grinding. J Mater ChemC 4(32):7584–7588. https://doi.org/10.1039/C6TC02578A
95. Meng X, Chen C, Qi G, Li X, Wang K, Zou B, Ma Y (2017) From two, to three, to multi-color switches: developing AIEgen-based mechanochromic materials. ChemNanoMat 3(8):569–574. https://doi.org/10.1002/cnma.201700120
96. Ma Z, Meng X, Ji Y, Li A, Qi G, Xu W, Zou B, Ma Y, Kuang G-C, Jia X (2019) Pressure induced the largest emission wavelength change in a single crystal. Dyes Pigm 162:136–144. https://doi.org/10.1016/j.dyepig.2017.11.036
97. Zhang Z, Zhang Y, Yao D, Bi H, Javed I, Fan Y, Zhang H, Wang Y (2009) Anthracene-arrangement-dependent emissions of crystals of 9-anthrylpyrazole derivatives. Cryst Growth Des 9(12):5069–5076. https://doi.org/10.1021/cg9008569
98. Zhang HY, Zhang ZL, Ye KQ, Zhang JY, Wang Y (2006) Organic crystals with tunable emission colors based on a single organic molecule and different molecular packing structures. Adv Mater 18(18):2369–2372. https://doi.org/10.1002/adma.200600704
99. Lekha PK, Prasad E (2010) Aggregation-controlled excimer emission from anthracene-containing polyamidoamine dendrimers. Chem Eur J 16(12):3699–3706. https://doi.org/10.1002/chem.200902391
100. Lin T, Su X, Wang K, Li M, Guo H, Liu L, Zou B, Zhang Y-M, Liu Y, Zhang SX-A (2019) An AIE fluorescent switch with multi-stimuli responsive properties and applications for quantitatively detecting pH value, sulfite anion and hydrostatic pressure. Mater Chem Front 3(6):1052–1061. https://doi.org/10.1039/C8QM00544C
101. Liu L, Su X, Yu Q, Guo H, Wang K, Yu B, Li M, Zou B, Liu Y, Xiao-An Zhang S (2019) Photoacid-spiropyran exhibits different mechanofluorochromism before and after modification of tetraphenylethene under grinding and hydrostatic pressure. J Phys Chem C 123(41):25366–25372. https://doi.org/10.1021/acs.jpcc.9b06928
102. Li A, Liu Y, Bi C, Xu W, Ma Z, Cui H, Xu S (2020) Pressure-dependent distinct luminescent evolutions of pyrene and TPA-Py single crystals. Spectrochim Acta Part A Mol Biomol Spectrosc 237:118390. https://doi.org/10.1016/j.saa.2020.118390
103. Li A, Ma Z, Wu J, Li P, Wang H, Geng Y, Xu S, Yang B, Zhang H, Cui H, Xu W (2018) Pressure-induced wide-range reversible emission shift of triphenylamine-substituted anthracene via hybridized local and charge transfer (HLCT) excited state. Adv Opt Mater 6(3):1700647. https://doi.org/10.1002/adom.201700647
104. Liu H, Gu Y, Dai Y, Wang K, Zhang S, Chen G, Zou B, Yang B (2020) Pressure-induced blue-shifted and enhanced emission: a cooperative effect between aggregation-induced emission and energy-transfer suppression. J Am Chem Soc 142(3):1153–1158. https://doi.org/10.1021/jacs.9b11080
105. Luo Q, Lv C, Sheng H, Cao F, Sun J, Zhang C, Ouyang M, Zou B, Zhang Y (2020) Highly bright fluorescence from dispersed dimers: deep-red polymorphs and wide-range piezochromism. Adv Opt Mater 8(7):1901836. https://doi.org/10.1002/adom.201901836
106. Nagura K, Saito S, Yusa H, Yamawaki H, Fujihisa H, Sato H, Shimoikeda Y, Yamaguchi S (2013) Distinct responses to mechanical grinding and hydrostatic pressure in luminescent chromism of tetrathiazolylthiophene. J Am Chem Soc 135(28):10322–10325. https://doi.org/10.1021/ja4055228
107. Sui Q, Ren X-T, Dai Y-X, Wang K, Li W-T, Gong T, Fang J-J, Zou B, Gao E-Q, Wang L (2017) Piezochromism and hydrochromism through electron transfer: new stories for viologen materials. Chem Sci 8(4):2758–2768. https://doi.org/10.1039/C6SC04579K

REVIEW

Check for updates

Recent Trends in the Design, Synthesis, Spectroscopic Behavior, and Applications of Benzazole-Based Molecules with Solid-State Luminescence Enhancement Properties

Suzanne Fery-Forgues[1] · Corinne Vanucci-Bacqué[1]

Received: 3 February 2021 / Accepted: 10 July 2021

Abstract

Molecules that exhibit solid-state luminescence enhancement, i.e. the rare property to be more strongly emissive in the solid state than in solution, find an increasing number of applications in the fields of optoelectronic and nanophotonic devices, sensors, security papers, imaging, and theranostics. Benzazole (BZ) heterocycles are of particular value in this context. The simple enlargement of their π-electron system using a –C=C–Ar or –N=C–Ar moiety is enough for intrinsic solid-state luminescence enhancement (SLE) properties to appear. Their association with a variety of polyaromatic motifs leads to SLE-active molecules that frequently display attractive electroluminescent properties and are sensitive to mechanical stimuli. The excited-state intramolecular proton transfer (ESIPT) process that takes place in some hydroxy derivatives reinforces the SLE effect and enables the development of new sensors based on a protection/deprotection strategy. BZ may also be incorporated into frameworks that are prototypical aggregation-induced enhancement (AIE) luminogens, such as the popular tetraphenylethene (TPE), leading to materials with excellent optical and electroluminescent performance. This review encompasses the various ways to use BZ units in SLE systems. It underlines the significant progresses recently made in the understanding of the photophysical mechanisms involved. A brief overview of the synthesis shows that BZ units are robust building blocks, easily incorporated into a variety of structures. Generally speaking, we try to show how these small heterocycles may offer advantages for the design of increasingly efficient luminescent materials.

Keywords Benzoxazole · Benzothiazole · Benzimidazole · Solid-state luminescence · Aggregation-induced emission

Chapter 5 was originally published as Fery-Forgues, S. & Vanucci-Bacqué, C. Topics in Current Chemistry (2021) 379: 32. https://doi.org/10.1007/s41061-021-00344-8.

✉ Suzanne Fery-Forgues
sff@chimie.ups-tlse.fr

Extended author information available on the last page of the article

Published online: 03 August 2021

Abbreviations

ACQ	Aggregation-caused fluorescence quenching
AIE	Aggregation-induced emission
AIE-gen	AIE-luminogen
BI	Benzimidazole
BO	Benzoxazole
BT	Benzothiazole
BZ	Benzazole
CI	Conical intersection
DMF	Dimethylformamide
D-π-A	Electron donor-π-bridge-electron acceptor
ESIPT	Excited-state intramolecular charge transfer
HBO	Hydroxybenzoxazole
HBT	Hydroxybenzothiazole
HOMO	Highest occupied molecular orbital
ICT	Intramolecular charge transfer
LUMO	Lowest unoccupied molecular orbital
MFC	Mechanofluorochromism
NIR	Near infrared
OLED	Organic light-emitting diode
PL	Photoluminescence
PLQY	Photoluminescence quantum yield
PBI	Phenylbenzimidazole
PBO	Phenylbenzoxazole
PBT	Phenylbenzothiazole
RIR	Restriction of intramolecular rotations
RIM	Restriction of intramolecular motions
RACI	Restricted access to a conical intersection
SLE	Solid-state luminescence enhancement
TICT	Twisted internal charge transfer
THF	Tetrahydrofuran
TD-DFT	Time-dependent density functional theory
TPE	Tetraphylethene/tetraphenylethylene

1 Introduction

Whether they contain one oxygen, nitrogen, or a sulfur bridging atom, benzazoles are among the most relevant biologically active heterocycles, often incorporated as building blocks in ligands to target a variety of receptors and enzymes in medicinal chemistry studies [1–4]. They are also the basic units of an important class of fluorescent organic dyes, particularly attractive due to their high quantum yield, ease of synthesis, and excellent thermal, chemical, and photochemical stability. The three isologs, namely benzoxazole (BO), benzimidazole (BI), and benzothiazole (BT), have rather close spectroscopic properties governed by a stiff aromatic framework and by the presence of the heteroatoms [5]. These electron-accepting building blocks are generally connected to a π-conjugated substituent in the 2-position, and then they lead to compounds that exhibit superior luminescence properties. For this reason, all benzazole (BZ) units are very popular fluorescent tags of bio-conjugated molecules. They have also led to a variety of probes for the efficient detection of numerous analytes such as ions, biomolecules, and explosives [6]. Distinct advantages of the BI heterocycle are its chromogenic pH sensitivity and the possibility to introduce a variety of substituents on the nitrogen atom, for example to modulate the solubility or to bring specific bio-targeting ability. As regards BO, 2-phenylbenzoxazole derivatives (PBO) have been used for a long time as near-UV dyes in scintillation counters, lasers, textile brighteners, liquid crystals, and high-temperature probes [7]. Like all organic dyes, BZ derivatives have been traditionally used at small concentrations in dilute solution or after dispersion in solid matrices such as polymers. For the last two decades, they have contributed to the development of fluorescent solids made of pure organic dye molecules or containing a large amount of them. They are present in molecules that show aggregation-induced emission (AIE) properties [8, 9], and more widely speaking, solid-state luminescence enhancement (SLE) [10], i.e. the rare property to be more strongly emissive in the solid state than in solution. Hence, they find new applications in the fields of optoelectronic and nanophotonic devices, organic solar cells, switches, sensors, security papers, biological imaging, and theranostics.

The current review, which does not pretend to be exhaustive, is only focused on metal-free BZ derivatives. It begins by recalling the fundamentals of luminescence enhancement in the various solid states. Then, it explains how proper substitutions of BZ enable the emergence of intrinsic SLE properties. The association of BZ units with organic motifs chosen for their electroluminescent properties, gelation capacity, or for their response to mechanical stimuli, is described. It is shown that the hydroxyl derivatives, which undergo an excited-state intramolecular proton transfer (ESIPT) process, are of particular interest. This is also the case for the strongly luminescent borate complexes. Finally, the incorporation of BZ units into frameworks already well known for their AIE properties is addressed, underlining the value of these new systems. Emphasis is put on the significant progresses recently made in the understanding of the photophysical mechanisms involved. Figures gather the molecular structures together with the corresponding spectroscopic properties in a given solid state, when specified in the publications. In particular, the term "aggregates" (agg) refers to aggregates in suspension in an organic solvent/water mixture, and "powder" generally indicates a microcrystalline as-synthesized powder.

Our aim is to show the value of benzazoles in the field of SLE, with the hope that this walk through multiple structures will be a source of inspiration to move towards increasingly efficient luminescent materials.

2 From emissive solids to solid-state luminescence enhancement: a brief overview of a long journey

The development of organic molecules that emit light efficiently in the solid state has long been very challenging [11, 12]. In fact, conventional organic fluorophores rarely exhibit this property. Even though they are excellent emitters in solution or after dispersion in a matrix, most of them become very weakly emissive at high concentrations and in the condensed state. This frequently encountered phenomenon is called aggregation-caused fluorescence quenching (ACQ). To understand its origin, it is important to consider that solid-state luminescence results from the contribution of a variety of intra- and intermolecular effects (i.e. geometrical factors affecting the molecular planarity, polarizability issues, exciton coupling, and excimer formation [13, 14]) related to the molecular arrangement and to morphological features. A recent and very inspiring review from Gierschner et al. [15] identifies these effects and shows how difficult it is to disentangle them. Reasons for fluorescence quenching in the solid state are therefore of high complexity. Intuitively, it can be thought that the large delocalized π-conjugated moieties of organic fluorophores are prone to stacking, thus promoting non-radiative decay. Indeed, H-aggregates formed by the perfect "side-by-side" alignment of two molecules have been predicted to be non-emissive, but this simplified model hardly reflects the complexity of interactions between neighboring molecules in a solid. According to Gierschner et al., H-aggregation is not the cause of fluorescence quenching in most molecular solids. It is rather the trap sites, generally due to impurities and structural defects, that would play the major role in the deactivation of excited states [16].

Whatever its origin, ACQ may be circumvented by creating space between the fluorophores. To do so, fluorophores may be hosted by cyclodextrins and zeolites, or borne by macromolecules (e.g. dendrimers). According to their nature, they may be associated to bulky counterions [17–19], co-crystallized with neutral species [20–22], or chemically modified by the insertion of sterically demanding substituents that prevent them from stacking [23]. Another strategy is to organize the aromatic moieties around an sp^3-hybridized carbon or nitrogen atom that impedes planar geometry and hence prevents π-π stacking. This strategy has proved to be very successful in the design of a new generation of dyes, the behavior of which is precisely at the opposite of that induced by ACQ. These molecules do not emit, or emit very weakly in solution, and become luminescent upon aggregation and in the solid state [8, 9]. This effect, first reported by Stokes in 1853, was rediscovered in 2001 by Tang et al., who coined the term "aggregation-induced emission (AIE)" with the explicit goal of exploring the tremendous range of applications offered by such fluorophores in the fields of fluorescent materials and probes for biology. Typical AIE luminogens (AIE-gens) contain aromatic groups connected by rotatable C–C, C–N, or N–N single bonds. They are often propeller- or spiro-shaped compounds. It was first accepted that fluorescence quenching in solution results from the waste of excitation energy via thermal processes linked to rotations and large molecular motions. The restriction of intramolecular rotations (RIR) and motions (RIM) that occurs in the solid state was then assumed to be responsible for AIE. Indeed, the photophysical mechanisms involved are much more complex, although molecular stiffening is still at the basis of fluorescence revival [10]. Advances from theoretical chemistry have recently given another perspective on the AIE phenomenon. They have shown that the geometry of typical AIE-gens in a fluid medium may vary considerably under excitation, taking the molecule very far from the ground-state equilibrium structure. The potential energy surfaces of the ground and excited states that correspond to this geometry are close to each other. They are degenerate and almost intersect in a region called conical intersection (CI). The probability to pass from the excited state to the ground state via non-radiative internal conversion is therefore maximal in this region. In contrast, in a rigid medium where the molecular motions are impeded, the CI is no longer thermally accessible, and deactivation via the emission of light is strongly enhanced (Fig. 1). This restricted access to a conical intersection (RACI) model, introduced by Blancafort, Crespo-Otero et al., is now widely accepted to account for the behavior of AIE-gens [24–26]. Furthermore, the group of Ciofini and Adamo has shown theoretically that considering the electrostatic field induced by the environment at the excited state can be enough to explain the brightness enhancement and shift of the emission wavelength observed for an AIE-gen in the crystalline state, in comparison with dilute solutions [27].

The stiffening of the structure is not the only prerequisite for emission to be observed in the solid state. Fluorophores must adopt a molecular geometry that promotes the emission of light, and arrange between them in a favorable way. The initial AIE label paid little attention to these interactions because it mainly referred to phenomena observed with the formation of aggregates. Consequently, the term "solid state luminescence enhancement" (SLE) now seems to be more appropriate to consider the whole set of intra- and intermolecular factors involved in solid-state emission, and to encompass the various solid states, which range from highly viscous or solid solutions to amorphous powders and crystals [10].

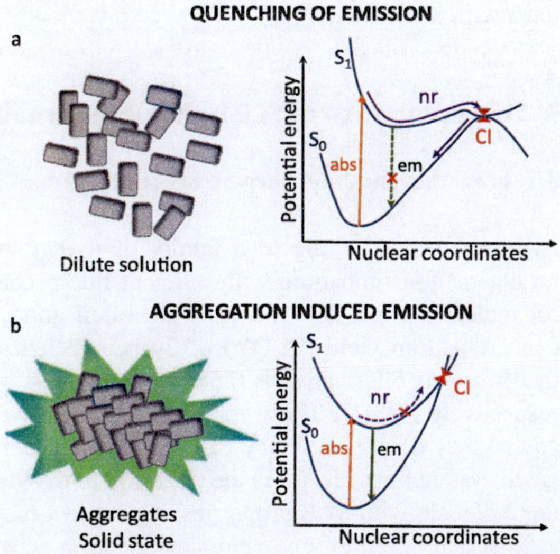

Fig. 1 Left: Representation of typical AIE-active fluorophore molecules in dilute fluid solution (**a**) and in the aggregate and solid states (**b**). Right: For each case, schematic description of the absorption (abs), emission (em), and non-radiative (nr) processes along the potential energy surface curves of the ground state (S_0) and first excited singlet state (S_1). Conical intersections (CI) are indicated in red ink

Generally speaking, it must be kept in mind that any change in the molecular conformation and in the supramolecular organization may induce variations in the photoluminescence (PL) color and/or intensity [28, 29]. It is possible to play on these factors by non-covalent approaches, the merit of which is to reduce the effort in synthesis and to make the best from a given fluorophore [11, 30]. For instance, polymorphism, i.e. the ability of a molecule to adopt different crystal structures, provides many opportunities to tune the optical properties just by implementing various conditions of crystallization [31–33]. Some fluorescent organic molecules also show variations in their PL properties in response to a physical stimulus, in particular mechanical stress, in other words they exhibit mechanofluorochromism (MFC) [14, 28, 34–38]. Typically, emission is changed by grinding, rubbing, crushing, pressing, shearing, or smearing, and the molecules are restored to their original state by annealing or fuming by solvent vapors, which increase the molecular mobility. The chemical structure of the molecules is not affected. This phenomenon is attributed to subtle and reversible changes taking place in molecular conformation and planarity, or in the crystal packing mode. It often results from transition between crystalline phase to amorphous phase, under mechanical forces. A prerequisite is that the molecular stacking in the crystal state is loose and can be easily destroyed under mechanical stress, resulting in a transformation of the emission properties. This is the reason why MFC is frequently associated to SLE materials. From a general viewpoint, the development of such mechanofluorochromic compounds is an emerging field of research that leads to smart materials for the detection of stress in polymers, security inks, data storage, and memory devices. Finally, it must be underlined that the modifications undergone by one molecule or one group of molecules may impact a large number of surrounding molecules. The response to a stimulus is then amplified in the solid state with respect to dispersed molecules [39, 40], a collective behavior that may be of great value for sensing materials.

Benzazole derivatives have taken part in the epic development of photoluminescent materials. However, some molecular engineering is necessary to bestow the BZ units with SLE behavior.

3 The emergence of SLE behavior in small benzazole derivatives

3.1 From dicyanovinyl derivatives to styryl dyes

Aryl-benzazoles belong to a family of dyes that spontaneously adopt solid-state arrangements compatible with efficient fluorescence emission. Even for the smallest molecules, the PL properties are often good. For example, the photoluminescence quantum yield (PLQY) of 2-phenylbenzoxazole (PBO) is 0.26, and that of the 4-methoxy derivative is 0.38, with emission maximum at 360 nm and 382 nm, respectively [7, 41]. These molecules are also very strongly emissive in solution, and so they do not show any SLE behavior. The adequate extension of the π-conjugated system not only shifts the emission to the visible range, but frequently endows the molecule with SLE properties. This effect may be enhanced by the presence of at least one cyano group, a classical electron acceptor that has been identified for a

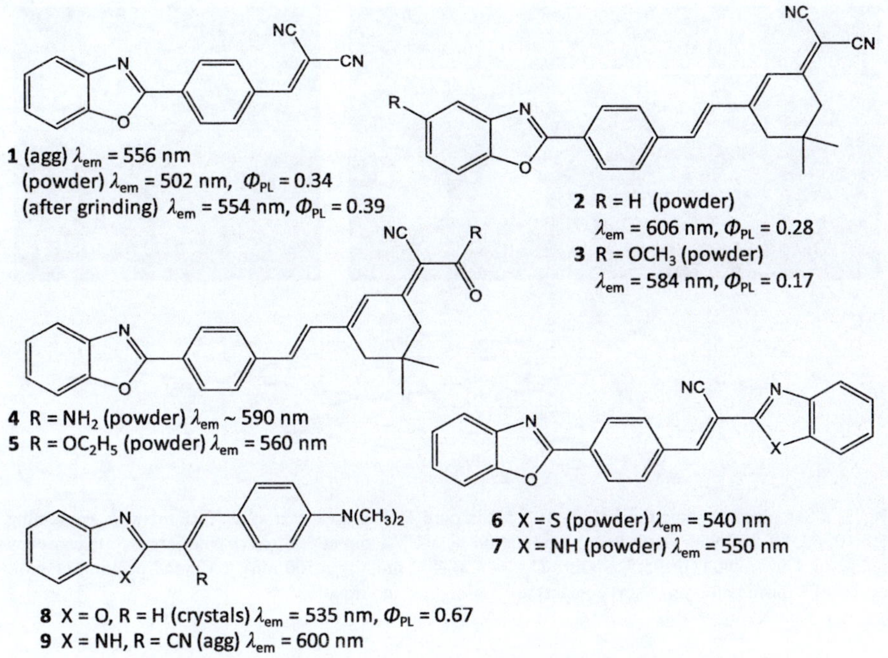

Fig. 2 Molecular structure of compounds **1** [43], **2**, and **3** [45], **4–7** [44], **8** [49], and **9** [50, 52] with solid-state PL maxima (λ_{em}) and photoluminescence quantum yields (Φ_{PL})

long time as a key factor in the creation of supramolecular interactions, leading to strongly fluorescent SLE-active organic materials (Fig. 2) [42].

The dicyanovinyl-PBO derivative **1** developed by our team is probably the smallest benzazole-based molecule endowed with SLE properties [43]. This molecule is weakly emissive in the blue region in pure THF solution. By increasing the proportion of water in a THF/water solution, the emission is first shifted from blue to turquoise. Then, above 80% water in the medium, opacification of the medium occurs with emission of strong yellow light with a maximum at 556 nm (Fig. 3a). Observation with the fluorescence microscope of the suspension with 95% water shows that emission arises from small fluorescent microcrystals (Fig. 3b). The PLQY of the microcrystalline powder is 0.34, from 11- to 16-fold higher than in organic solutions. It is noteworthy that the relatively small chemical modification brought to PBO has allowed the PL spectrum to be shifted from the near UV to the first half of the visible range.

In microcrystals, strong N···H–C hydrogen bonds are formed between the cyano groups and the hydrogen atoms of the vinylene unit and BO ring of adjacent molecules. As a result, molecules form layers (Fig. 3c), prone to slip with respect to each other upon mechanical stimuli. A mechanofluorochromic behavior is indeed observed. The pristine microcrystals emit yellow-green light (502 nm), which becomes golden yellow (554 nm) upon grinding, and vice versa by heating or solvent fuming. This phenomenon is accompanied by a reversible evolution

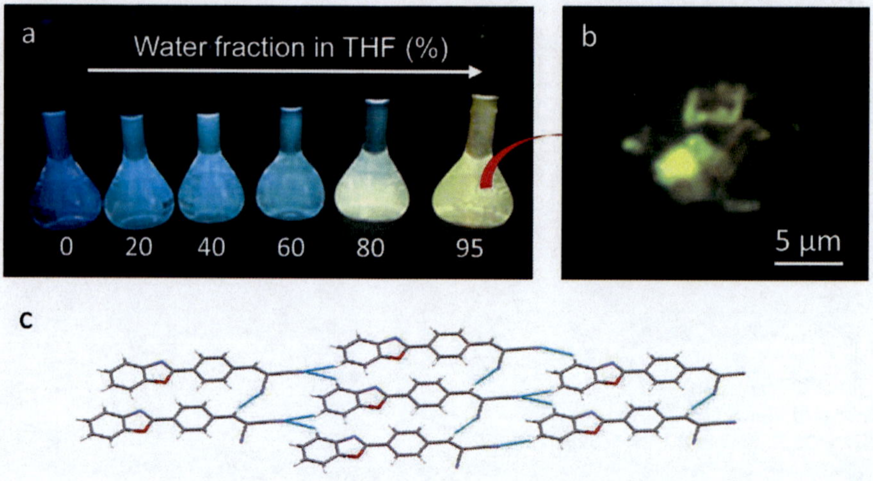

Fig. 3 **a** Samples of compound **1** (2.6×10^{-5} M) in pure THF and various water/THF mixtures containing 20, 40, 60, 80, and 95% water upon illumination by a UV lamp at 365 nm. **b** Fluorescence microscopy image of **1** in water/THF 95:5 mixture (λ_{ex} = 450–490 nm, λ_{em} > 500 nm). **c** Crystal packing of molecules of **1** forming a layer. Short contacts are highlighted in turquoise

of the photoluminescence quantum yields and lifetimes. These variations are attributed to the formation of different types of emissive species in the crystalline phases and in the ground material, which is a mixture of very small microcrystals and amorphous matter. Remarkably, evidence is given for additional spectroscopic effects, which may be easily confused with MFC, but actually depend on the size, compactness, and heterogeneity of the solid samples. These effects are presumably linked to the generation of regions rich in low-energy exciton traps in polycrystalline and pressed samples, and to reabsorption of light.

In compound **2** introduced by Yadav et al. [44] and subsequently studied by our group, the dicyanomethylene group is integrated into a rigid and bulky dimethylcyclohexene moiety. The enlargement of the π-electron system results in a significant redshift of the excitation and emission spectra with respect to **1**, and in the enhancement of the SLE effect. This compound is indeed very weakly fluorescent in fluid organic solvents, but fluorescence intensity increases with increasing viscosity, and the PLQY reaches 0.28 in the solid state (Fig. 4a) [45]. When increasing the proportion of water into a THF solution of **2**, the emission intensity at the PL maximum wavelength (584 nm) is strongly enhanced up to 45-fold (Fig. 4b, c). Scanning electron microscopy shows the presence of micrometer-sized amorphous particles, some of them resembling empty rice grains, when the water fraction in the mixture (f_w) reaches 60%, while the sample at f_w = 95% contains agglomerates of cubic microcrystals measuring several hundreds of nanometers (Fig. 4d, e). The presence of a methoxy group on the BO ring (compound **3**) does not enhance the SLE behavior, while the substitution by a hydroxy group is strongly detrimental to PL emission in

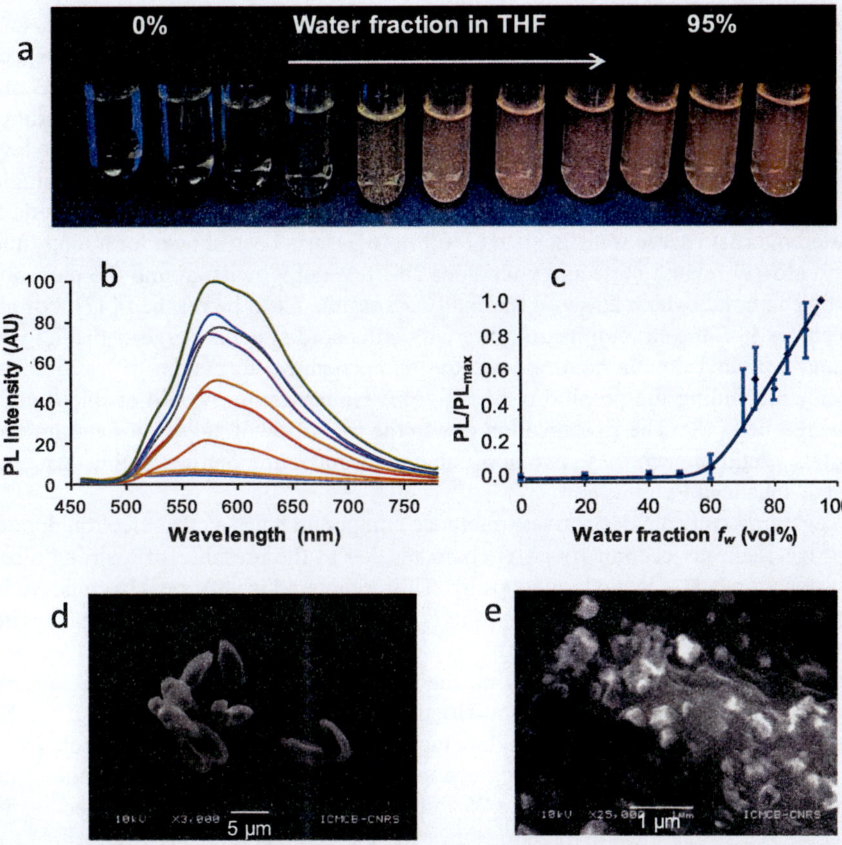

Fig. 4 **a** Samples of dye **2** (~7×10^{-6} M) in various THF/water mixtures, illuminated by a hand-held UV lamps (365 nm). **b** Corresponding PL spectra with λ_{ex} = 440 nm. **c** Plot of the PL intensity ratio at 586 nm versus the water content in the solvent mixture (average of three measurements). Scanning electron microscopy images of the suspensions at f_w = 60% (**d**) and 95% (**e**)

the solid state. No clear MFC effect is detected for molecules **2** and **3**, which arrange in densely packed crystal networks [45].

Yadav et al. [44] have also reported compounds **4** and **5** with only one terminal electron-withdrawing cyano group, as well as the stilbenic derivatives (**6** and **7**) that combine PBO with BI and BT, with the cyano group placed at the center of the molecule. Theoretical calculations show that, for all these molecules, the highest occupied molecular orbitals (HOMO) are mainly located on the benzazole moieties, and the lowest unoccupied molecular orbitals (LUMO) are centered on the cyano groups. All these molecules display SLE properties with intense emission in the solid state, varying from yellow to orange according to the position and strength of the electron-withdrawing group. These compounds are extremely stable, from a thermal and photochemical point of view.

As for the mechanism, the PBO group is quite rigid at room temperature, and rotations between the phenyl and BO moieties do not play a major role in the spectroscopic behavior. However, it may be considered that molecules **1**, **2**, and **3** are molecular rotors in fluid solution, like many compounds bearing a terminal dicyanovinyl group. Indeed, the dicyanovinyl moiety of **1** and the 5,5-(dimethylcyclohex-2-en-1-ylidene)malononitrile moiety of **2** may rotate around the single bond, thus generating large molecular motions. This type of rotation leads to a dark twisted internal charge transfer state (TICT) [46]. It has been shown for a long time that a closely related molecule undergoes an irreversible twist around the dicyanomethylene bond, which allows a thermally accessible CI to be reached [47]. For all molecules **1–7**, the structural similarity with stilbenoid systems suggests that a twist around the vinyl double bond, due to the photochemical formation of a bi-radical species precluding the possibility of *cis–trans* isomerization, would enable the CI to be reached [48]. The rigidification due to the formation of aggregates and microcrystals would prevent these two non-radiative deactivation pathways, thus explaining the enhanced PL emission.

Regarding stilbenic derivatives, reference compound **8** has a clear electron donor-π-bridge-electron acceptor (D–π–A) character due to the presence of a strong electron-donating *N,N*-dimethylamino group. This compound is very weakly emissive in organic solvents, whereas crystals strongly emit in the yellow green (535 nm) with PLQY as high as 0.67. A crystal film responds reversibly to volatile acid vapors, protonation occurring successively on the two nitrogen atoms of this PBO derivative [49]. In compound **9** studied by Horak et al., the cyano group close to the BI moiety reinforces the electron-withdrawing effect on this side of the molecule [50]. In water, this compound forms aggregates that emit orange-red fluorescence, the intensity of which is approximately 35-fold the emission of dissolved molecules in ethanol. The strong push–pull character of the molecule is probably responsible for the emission at long wavelength (600 nm). On alternate exposure of **9** in solid state to the vapors of HCl and NH_3, the emission is switched on and off [51]. The luminescence of the aggregates formed in aqueous organic solvent is quenched in the presence of picric acid, and so the compound has been proposed for the sensitive detection of nitroaromatic explosives [52]. Interestingly, the derivative in which the BI nitrogen atom is substituted by a phenyl ring does not show any SLE property. It seems that for stilbenic derivatives, the cyano group acts more as an auxochrome rather than as an SLE promoter. RIR and planarization have been evoked to explain the SLE effect. Besides, as shown with a BT analogue of **8**, the main deactivation mechanism of these compounds in fluid solvents is a photoisomerization, the restriction of which in viscous solvents leads to the increase in fluorescence quantum yield [53]. It is likely that the same process takes place in the solid state.

3.2 Schiff base derivatives

The Schiff bases **10–12** (Fig. 5) studied by the group of Zhou are isomers, differing by the position of the nitrogen atom in the distal pyridine group. They have a twisted

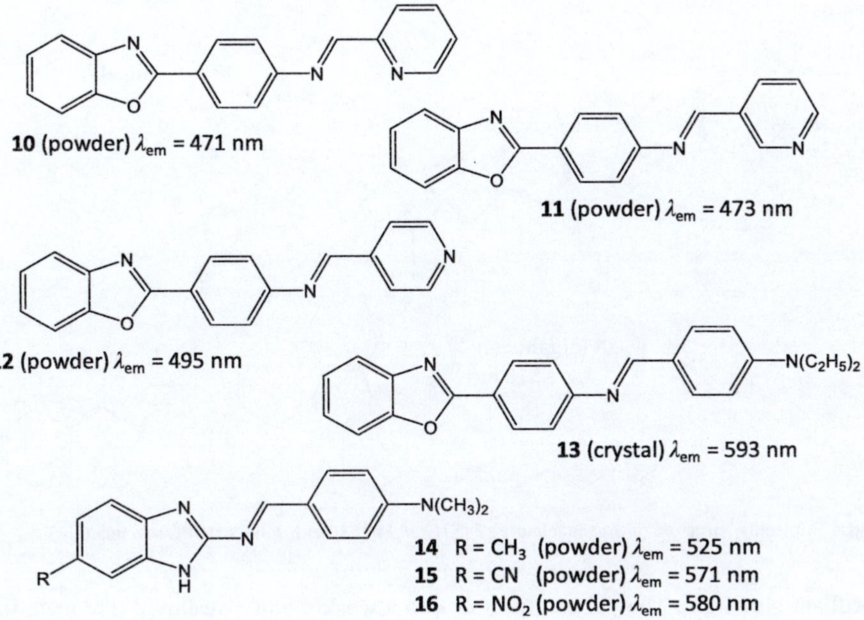

Fig. 5 Molecular structure of Schiff base derivatives **10–12** [54], **13** [55], and **14–16** [56] with solid-state PL maxima (λ_{em})

conformation and do not present significant push–pull character. The emission maximum of powders is centered between 471 and 495 nm, according to the compound. These dyes are much more emissive as aggregates and amorphous powders than in the crystalline state. This behavior is attributed to a detrimental crystal packing mode in which the molecules stack together face-to-face, with extensive overlap of their aromatic systems [54]. In contrast, the presence of a *N,N*-diethylamino group in **13** prevents π-π stacking and induces cross-packing arrangement, resulting in particularly emissive crystals. This electron-donating group also contributes to the push–pull effect, so that the emission maximum is around 593 nm in crystals and 552 nm in a powder of low crystallinity [55].

In comparison with **13**, the Schiff bases **14–16** of Horak et al. have a small π–conjugated system [56]. The intramolecular charge transfer (ICT) is reinforced by replacing the methyl group at the BI acceptor moiety by the strong electron-withdrawing CN and NO_2 groups. The solid-state emission is then tuned from 525 nm to 571 and 580 nm, respectively.

In molecules **17** and **18**, the PBI moieties are connected to the Schiff base by an aliphatic linker (Fig. 6) [57, 58]. Like many molecules that contain amide, amino acid, or urea moieties, these compounds form organogels with low minimum gelation concentration. They show strong SLE effect. For instance, in a mixed solvent of dimethylsulfoxide and ethylene glycol, **17** produces a dense, blue-emitting 3D network of entangled fibers stabilized by hydrogen bonds. The incorporation of Cd^{2+} cations leads to a metallogel that involves coordination interactions and emits

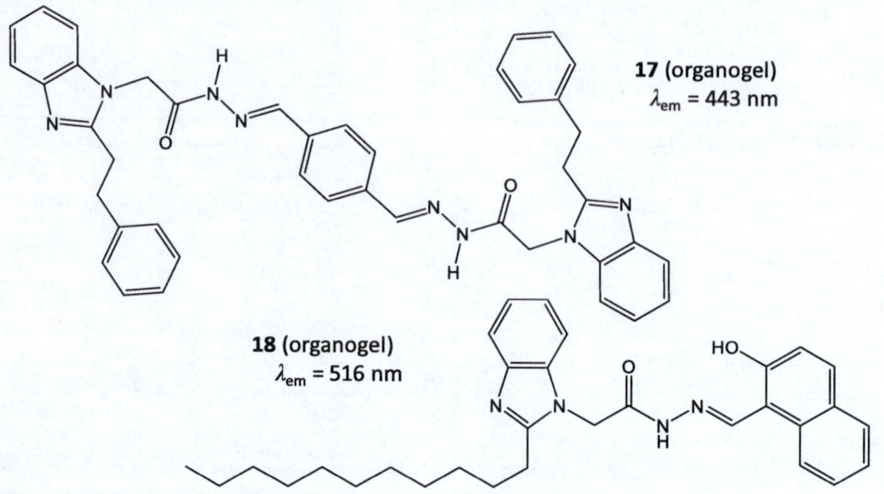

Fig. 6 Molecular structure of organogelators **17** [57] and **18** [58] with solid-state PL maxima (λ_{em})

brilliant blue light. The gel can turn back to a weakly emissive liquid-like material after heating, and vice versa [57].

The formation of the organogel from compound **18** in a glycerol/ethylene glycol mixture is accompanied by the emission of strong brilliant yellow-green fluorescence. This gel is stabilized by numerous interactions, i.e. hydrogen bonds between the acylhydrazone and the hydroxyl groups, π-π stacking interactions between the naphthyl moieties, and strong van der Waals interactions taking place between the long alkyl chains. Addition of CN^- anions and Al^{3+} cations leads to a fluorescent and colorimetric response, with good selectivity with respect to other ions. The presence of Fe^{3+} quenches fluorescence that may be recovered by addition of L-cysteine.

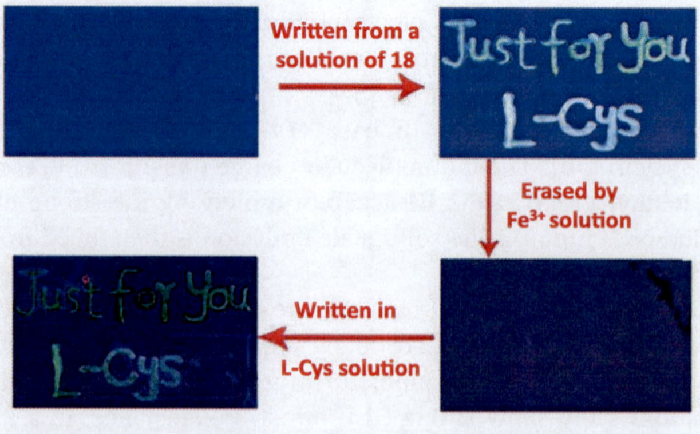

Fig. 7 Photographs of **18** on test film taken under exposure with UV light (365 nm). A solution of **18** (2.0×10^{-2} M) was used to write on a silica gel plate and let to dry in the open air. The test plate was then brushed with a solution of Fe^{3+}, then dipped into a water solution of L-cysteine. Adapted from Ref. 58 with permission from the Royal Society of Chemistry

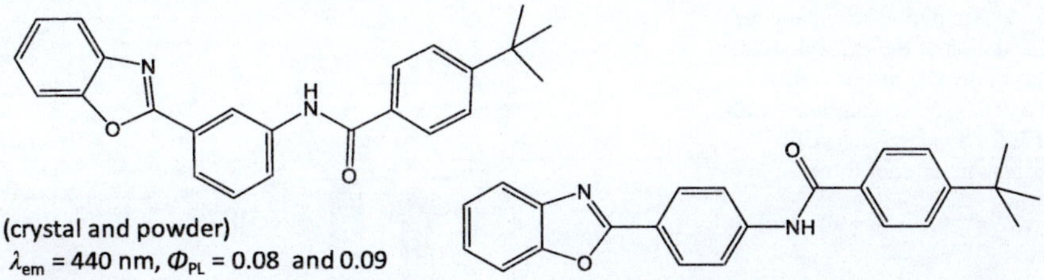

19 (crystal and powder)
λ_{em} = 440 nm, Φ_{PL} = 0.08 and 0.09
(agg and film) λ_{em} = 446 nm, Φ_{PL} = 0.02

20 (crystal and powder) λ_{em} = 410 nm, Φ_{PL} = 0.50
(amorphous film) λ_{em} = 450 nm

Fig. 8 Molecular structure of carboxamide derivatives **19** [59] and **20** [60] with solid-state PL maxima (λ_{em}) and photoluminescence quantum yields (Φ_{PL})

This "smart" gel can then act as a "on–off–on" fluorescence switch, potentially usable as an erasdeprived of the hydroxyl group, does not exhibit this behaviorable fluorescent display (Fig. 7) [58].

3.3 Carboxamide derivatives

There is no π-conjugation between the PBO unit and the rest of the molecule in the carboxamide derivatives **19** and **20**, which are two positional isomers (Fig. 8). The molecular backbone is very flexible. Both compounds mainly emit in the blue region in the condensed state, with small color changes depending on the degree of crystallinity. Isomer **19**, in which the 4-*tert*-butylbenzamide group is located in the 3' position of the PBO moiety, is almost not emissive in fluid organic solvent and only weakly emissive as a powder ($\Phi_{PL} = 0.09$). SLE is attributed to RIR that locks the molecule in an emissive quasi-TICT exited state, formed through incomplete transition of the local excited state to a dark TICT state. The RIR effect is supposed to increase with the order of the molecular arrangement, i.e. amorphous film < nanosheet < powder < crystalline state, and this could explain that the PL efficiency varies similarly [59]. In contrast, isomer **20**, in which the 4-*tert*-butylbenzamide group is placed in the 4′ position of the PBO moiety, is strongly emissive. The PLQY of the powder is close to 0.50, almost 230 times higher than in THF solution. In this case, the mechanism was attributed to restriction of non-radiative TICT. A major contribution of the crystal packing mode, in which molecules adopt the cross-dipole stacking typical of PBO [7], is also evoked to explain the spectacular SLE effect [60].

3.4 2′-Hydroxy-derivatives: exploiting the ESIPT effect for ultra-sensitive sensing

When a proton-donating hydroxyl group is in close proximity to a heteroatom that acts as a proton acceptor, an intramolecular hydrogen bond may take place. Such molecules exhibit conformational tautomerism and lead to a special photophysical

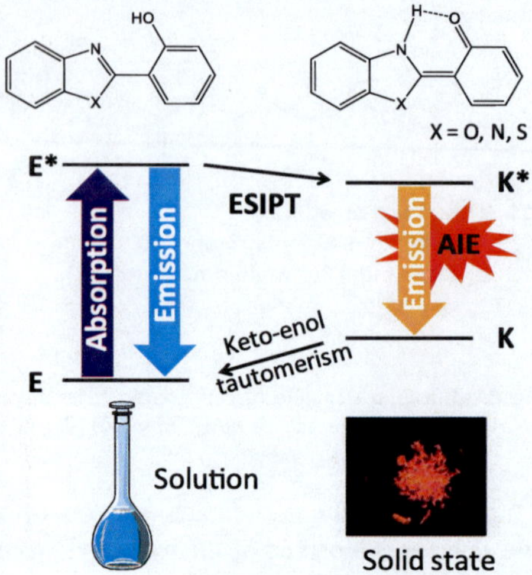

Fig. 9 Simplified schematic representation of the excited-state intramolecular proton transfer (ESIPT) process combined with SLE for a phenyl-benzazole derivative. E: enol form, K: keto form

process, i.e. excited-state intramolecular proton transfer (ESIPT) [61]. ESIPT is characterized by the presence, in addition to fluorescence at short wavelengths due to the enol form, of an intense fluorescence emission at long wavelengths due to the keto form, with an anomalously large Stokes shift. It is an extremely fast process, able to appear even in rigid glass at very low temperature and in the solid state, and

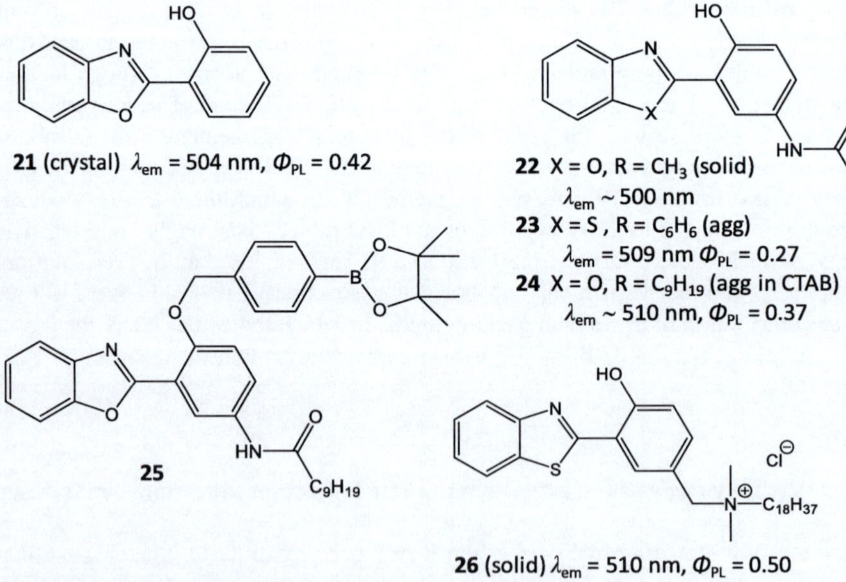

Fig. 10 Molecular structure of compounds **21** [64], **22** [65], **23** [66], **24** and **25** [65], and **26** [67], with solid-state PL maxima (λ_{em}) and photoluminescence quantum yields (Φ_{PL})

it leads to remarkable solid-state emitters [62]. As recently reviewed by Sedgwick et al., ESIPT and SLE processes are closely related [63]. The RIM and molecular planarization that result from transition to the solid state facilitate the formation of the intramolecular hydrogen bond and generally promotes the ESIPT process. Conversely, the strong emission of the excited keto form enhances the solid-state emission with respect to solutions (Fig. 9).

The 2-(2′-hydroxyphenyl)benzazole framework is a conventional ESIPT pattern. Its smallest derivatives are AIE fluorogens. For instance, 2-(2′-hydroxyphenyl)benzoxazole (HBO) **21** emits at 504 nm in its crystalline form, with a PLQY of 0.42, a value multiplied by 23 with respect to solutions [64] (Fig. 10). Its derivative bearing a small amido substituent **22** also emits near 500 nm [65]. The hydroxybenzothiazole (HBT) analogue **23** emits at 509 nm, aggregates have a PLQY of 0.27, nearly 15-fold higher than in cyclohexane. Ultrafast transient absorption spectroscopy showed that for **23**, the observed SLE effect results from the combination of intermolecular interactions and RIM, which play opposite roles on PL emission [66].

The HBO fragment has enabled the construction of ratiometric fluorescent probes using a phenolic hydroxyl protection/deprotection strategy. Compound **24**, which differs from **22** by the presence of a long alkyl chain, emits weak violet emission in dimethylsulfoxide solutions arising from the enol form, while the strong emission around 500 nm in the solid state comes from the keto form. Caged compound **25**, where the hydroxyl group is protected with a boronate-benzyl group, only emits in the violet region (405 nm) when dispersed in cationic surfactant cetyltrimethylammonium bromide (CTAB) below the critical micelle concentration (CMC). The reaction with hydrogen peroxide H_2O_2 regenerates compound **24**, and a yellow emission with a PLQY of 0.37 attributed to ESIPT appears. The following system has been proposed as a ratiometric sensor of H_2O_2 [65].

Another interesting application has been reported for the cationic molecule **26**, which interacts with anionic surfactants in water to form highly emissive catanionic aggregates with an increased keto/enol emission ratio. Owing to the affinity of the probe for negatively charged bacterial cell walls, the light-up fluorescence is also used for imaging *E. coli* bacteria according to a wash-free procedure [67, 68].

Due to the simultaneous presence of the HBO unit, urea group, and two long alkyl chains, molecule **27** establishes a variety of weak intermolecular interactions (Fig. 11). In a toluene-rich organic medium, this molecule forms a colorless gel, strongly emissive in the green region under UV light, with fluorescence intensity 26 times greater than in solution. Upon addition of fluoride anions, the gel turns green, probably because of direct interaction between fluoride anions and the hydroxyl group of HBO, and it is disrupted, which is attributed to concomitant deprotonation of the urea groups. The system is then a naked-eye fluoride sensor that shows both gel-to-sol transition and a visibly noticeable color change, with good selectivity with respect to other anions [69].

The dendritic compound **28** is also able to participate to multiple weak interactions. In a variety of organic solvents, it forms gels constituted of fibers and ribbons of different sizes. The SLE effect upon gelation is clear. The presence of a fluoride anion causes the gel disruption, due to interactions with the hydroxyl group that play a crucial role in the gelation process. Other external stimuli (i.e. addition

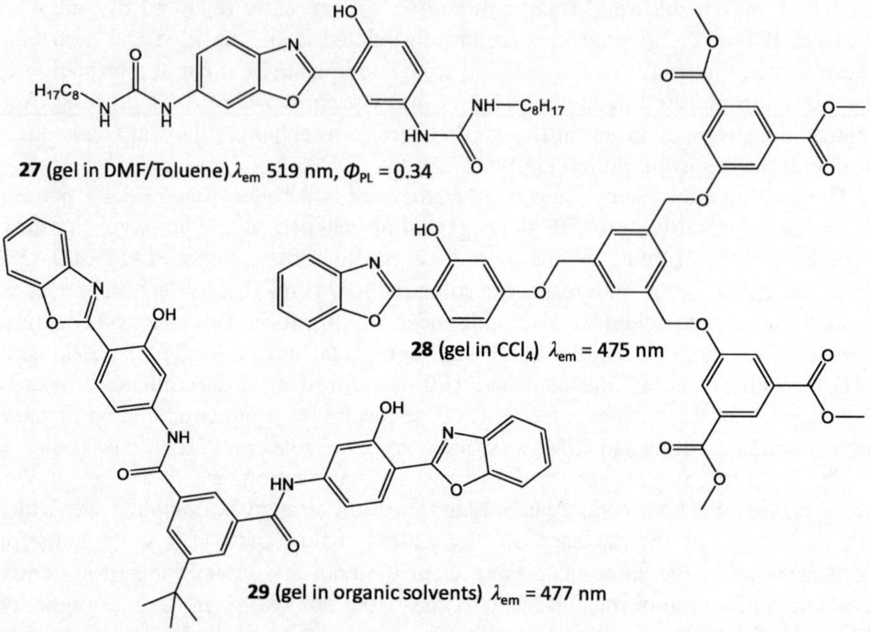

Fig. 11 Molecular structure of ESIPT-active organogelators **27** [69], **28** [70], and **29** [71] with solid-state PL maxima (λ_{em}) and photoluminescence quantum yields (Φ_{PL})

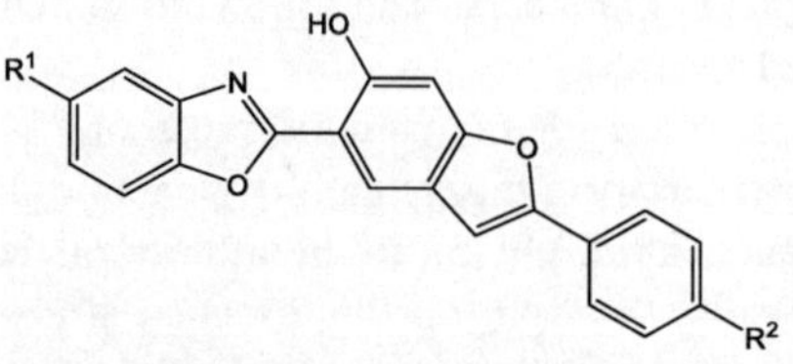

30 $R^1 = R^2 = tBu$ λ_{em} = 535 nm, Φ_{PL} = 0.54
31 $R^1 = tBu$, $R^2 = OCH_3$ λ_{em} = 551 nm, Φ_{PL} = 0.31
32 $R^1 = tBu$, $R^2 = N(C_4H_9)_2$ λ_{em} = 482/570 nm, Φ_{PL} = 0.21
33 $R^1 = F$, $R^2 = N(C_4H_9)_2$ λ_{em} = 476/571 nm, Φ_{PL} = 0.10
34 $R^1 = CF_3$, $R^2 = N(C_4H_9)_2$ λ_{em} = 501/580 nm, Φ_{PL} = 0.19
35 $R^1 = CN$, $R^2 = N(C_4H_9)_2$ λ_{em} = 524 nm, Φ_{PL} = 0.08
36 $R^1 = CO_2CH_3$, $R^2 = N(C_4H_9)_2$ λ_{em} = 574 nm, Φ_{PL} = 0.21
(solids in KBr pellets)

Fig. 12 Molecular structure of ESIPT-active compounds **30–36** [72] with solid-state PL maxima (λ_{em}) and photoluminescence quantum yields (Φ_{PL})

of zinc cations, pH changes, shear stress, and sonication) have been used to trigger the phase transition of the gel, in a reversible way. It is proposed that these variations could be monitored by fluorescence spectroscopy and that the system could evolve towards a multiresponsive fluorescent sensor [70].

Although it does not contain any fatty chain or dendritic structure, compound **29** also forms very stable gels in THF/cyclohexane mixed solvents. The examination of xerogels shows that molecules are arranged as nanobelts [71]. In all these gelator systems, the SLE phenomenon is attributed to the combination of *J*-aggregation, a fast ESIPT process, and restriction of TICT.

The rigidified π-extended HBO derivatives **30–36** (Fig. 12) have been prepared by the team of Ulrich and Ziessel. The proton donor is borne by a benzofuran

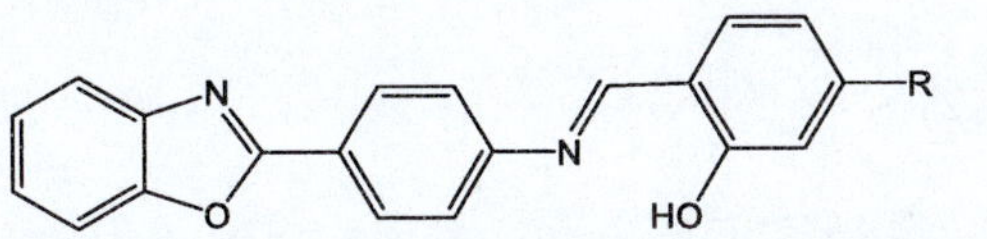

37 R = H (powder) λ_{em} = 581 nm, (crystal) 588 nm
38 R = $N(C_2H_5)_2$ (powder and crystal) λ_{em} = 564 nm

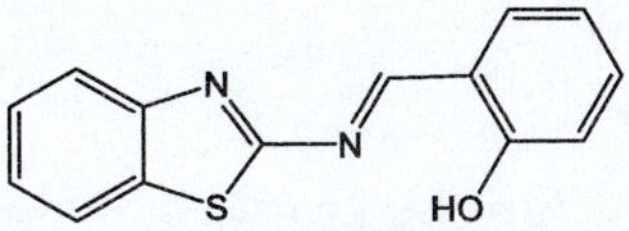

39 λ_{em} = 540 nm

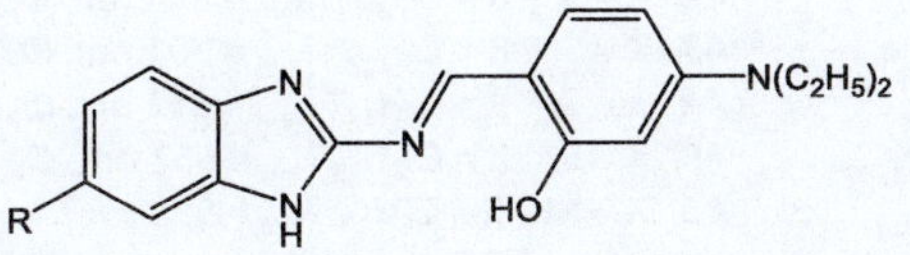

40 R = CN (powder) λ_{em} = 536 nm
41 R = NO_2 (powder) λ_{em} = 620 nm

Fig. 13 Molecular structure of ESIPT-active Schiff base derivatives **37** and **38** [55], **39** [73], and **40** and **41** [56] with solid-state PL maxima (λ_{em}) and photoluminescence quantum yields (Φ_{PL})

fragment. The PLQY of these compounds is between 0.08 and 0.54, systematically higher than in solutions. The *E*/K** ratio is fine-tuned by the substitution pattern of the molecule, with maximum emission between 476 to 580 nm. Dual emission is observed in amorphous powders, resulting in the production of white light, of great interest for electroluminescent devices [72].

In the following molecules, the proton acceptor group is the nitrogen atom of the Schiff base. The PBO derivatives **37** and **38** (Fig. 13) emit faint blue light in solutions, and strong yellow light in crystals. Compound **37** gives fibers and needle-like crystals, with cross-packing layer structures. Of course, the parent compound **13** (Fig. 5), deprived of the hydroxyl group, does not exhibit this behavior. In the solid state, the yellow emission of **37** turns blue upon exposure to vapors of dichloromethane, chloroform, and tetrahydrofuran, and vice versa after removing the vapor was removed. The molecule could then be used as a sensor to detect volatile organic solvents [55]. The BT derivative **39** is virtually not emissive in common organic solvents, and it emits yellow light at 540 nm in the solid state [73].

With their extended π-conjugated system and D–π–A structure, the BI derivatives **40** and **41** have been designed so that emission is shifted to the red region. However, the ESIPT process does not reinforce systematically the expected effect. By comparison with their respective parent compounds **15** and **16** (Fig. 5), the ESIPT-active cyano-substituted derivative **40** emits at shorter wavelengths (536 vs. 571 nm), while the nitro-substituted derivative **41** emits at much longer wavelengths (620 vs. 580 nm). The latter compound also shows neat MFC behavior [56].

It will be seen below that advantage may be taken from the ESIPT process in other types of molecules.

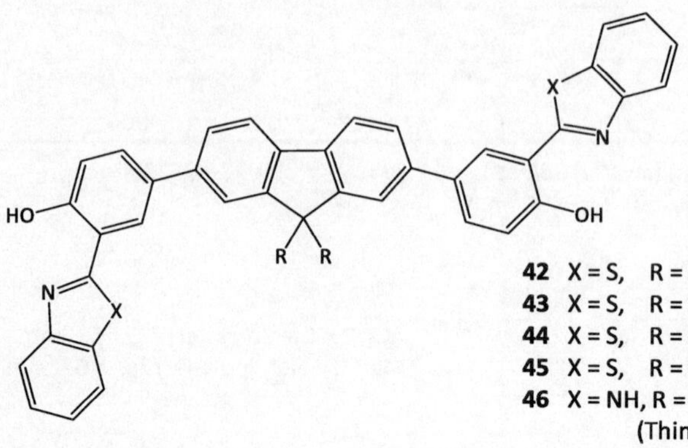

42 X = S, R = CH_3 λ_{em} = 544 nm, Φ_{PL} = 0.60
43 X = S, R = C_6H_{13} λ_{em} = 551 nm, Φ_{PL} = 0.67
44 X = S, R = C_8H_{17} λ_{em} = 552 nm, Φ_{PL} = 0.55
45 X = S, R = $C_{12}H_{25}$ λ_{em} = 552 nm, Φ_{PL} = 0.63
46 X = NH, R = C_8H_{17} λ_{em} = 590 nm
(Thin solid films)

Fig. 14 Molecular structure of fluorene derivatives **42–46** [74] with solid-state PL maxima (λ_{em}) and photoluminescence quantum yields (Φ_{PL})

3.5 Fluorene derivatives: en route to applications as electroluminescent materials

The combination of BZ moieties with polyaromatic compounds that have special physicochemical properties is expected to give new functional materials. For example, fluorene and its derivatives have desirable electron-transporting properties, and they are among the most studied electroluminescent materials for applications in optoelectronic devices. In molecules **42–46** (Fig. 14) developed by Padalkar et al., the BZ unit acts as an electron-withdrawing group, controls the crystal packing mode, and brings efficient ESIPT properties. The study of the AIE behavior performed on compound **43** shows that aggregates emit sevenfold more than solutions. For all these molecules, thin films obtained by spin casting from dichloromethane solution have excellent PLQY, up to 0.67. The size of the alkyl chain of the BT derivatives **42–45** has relatively small effect on the optical properties [74]. These fluorene derivatives are good candidates for electroluminescence applications.

3.6 Carbazole derivatives: electroluminescence and gels

This is also the case for the carbazole derivatives **47–49**. Carbazoles are known for their hole-transporting properties. They are strong electron donors, with good thermal stability. Like most of organic dyes, their fluorescence is quenched upon aggregation. However, the association with HBT units leads to D-π-A compounds that display excellent SLE properties, boosted by the ESIPT effect. For instance, molecules **47–49** (Fig. 15) are three positional isomers based on carbazole and HBT. They are strongly emissive in the green region. The PLQY reaches 0.38 for **48**, but is lower for the other two compounds, due to conformational differences [75].

Of course, carbazoles may be used only as an electron-donating substituent. By this regard, molecule **50** behaves similarly to the *N,N*-dimethylaminophenyl analogue **9** (Fig. 2) and is also suitable for the detection of picric acid [52].

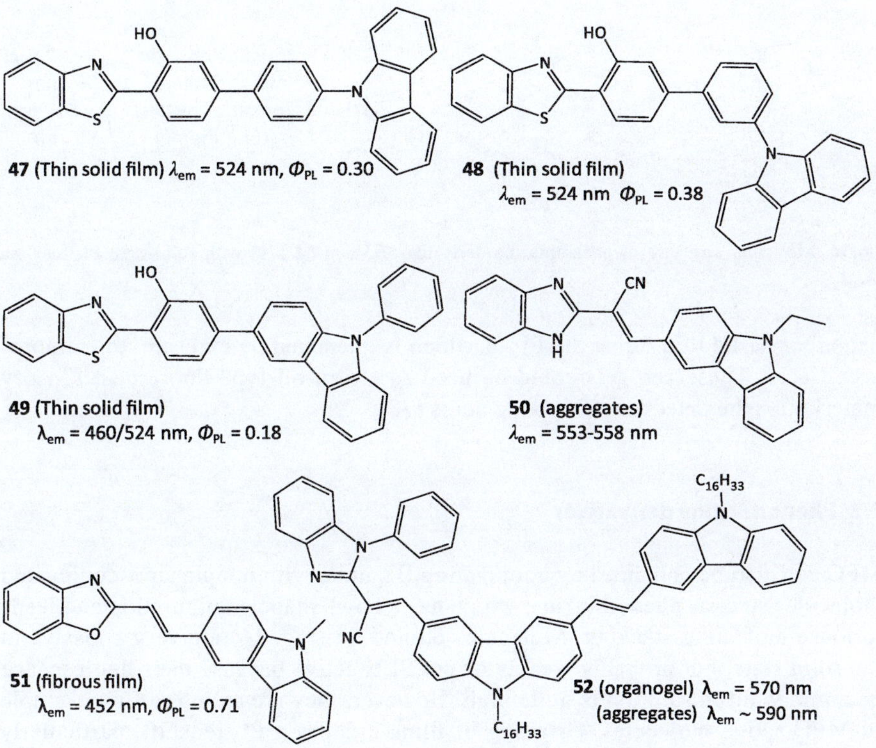

Fig. 15 Molecular structure of carbazole derivatives **47–49** [75], **50** [52], **51** [77], and **52** [78], with solid-state PL maxima (λ_{em}) and photoluminescence quantum yields (Φ_{PL})

However, the benzoxazole and carbazole conjugates present attractive additional properties. Compound **51** investigated by Xue et al. self-assembles into strongly emissive long fibers. Upon application of mechanical shearing to the fibrous film, the photoluminescence passes from bright blue to blue-green, and the intensity of emission is decreased. The initial signal is progressively restored at room temperature. In contrast, the exposure to volatile acetic acid induces the emission of orange light, and a heat treatment is required to regenerate the film. The acidic vapors selectively act as a stabilizer and developer to retain the information imparted by mechanical forces [76]. Molecule **51** has also been reported to be a gelator in organic solvents. The emission intensity of gel is increased 16-fold with respect to solutions, with a PLQY reaching 0.71 [77]. In a similar way, molecule **52** reported by Wu et al. gelates *n*-butanol with a critical gel concentration as low as 1.8×10^{-3} M, making it a super gelator. Beside the acid-sensitive PBI unit, this molecule contains two carbazole units, one of them bearing a long alkyl chain to favor gelation, and a cyanovinyl linker for enhancing the AIE properties. Indeed, aggregates and gels emit strongly in the yellow region, while fluid solutions are almost not emissive. After drying, the gels give

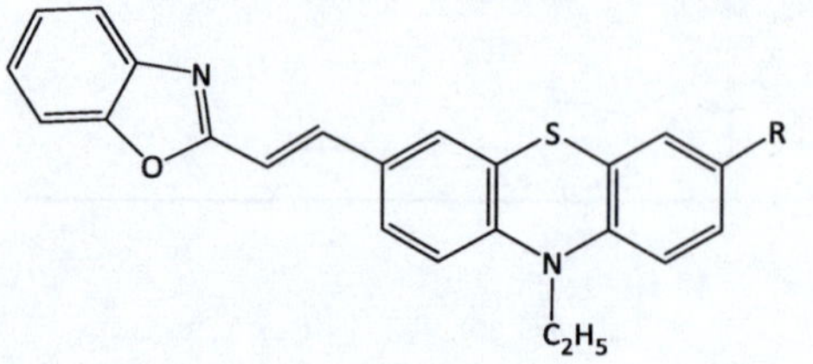

Fig. 16 Molecular structure of phenothiazine derivatives **53** and **54** [79] with solid-state PL maxima (λ_{em})

nanofiber-based thin films, the PL of which is quenched by exposure to trifluoroacetic acid. These xerogels could be used as a turn-off type fluorescent sensory material for the detection of volatile acids [78].

3.7 Phenothiazine derivatives

MFC can also be enhanced by conjugating BZ units with nonplanar π-conjugated molecules such as phenothiazine, which has a bowl-shape configuration and leads to loose molecular stacking. Molecules **53** and **54** (Fig. 16) are very emissive in the solid state, but probably weakly or not SLE-active because their fluorescence quantum yield in solution is quite high. However, they are an interesting example of MFC-active molecules. Grinding of films induces a PL redshift, particularly significant for the bromine derivative. More interestingly, the ground film of **53** can rapidly self-heal at ambient temperature, while heating or solvent fuming is necessary for the ground film of **54** to revert into the original one [79].

3.8 Triphenylamine derivatives

Owing to its good hole-transporting capabilities, triphenylamine (TPA) has frequently been incorporated into dyes designed for use in electroluminescent devices such as dye-sensitized solar cells and organic light-emitting diodes (OLEDs). It is a popular electron donor. Its fluorescent derivatives are generally well emissive in solution, but suffer from the ACQ effect, despite the fact that TPA has a propeller shape that can prevent intermolecular π-stacking. For the last decade, significant efforts have been made to transform TPA derivatives into AIE-gens [8]. The association with BZ is a simple and convenient way to reach this aim.

Compound **55** (Fig. 17) is moderately emissive in organic solution, and its pristine crystals emit strongly in the blue-green region. Upon grinding, the compound exhibits low-contrast PL color change, with a modest spectral shift of 29 nm towards long wavelengths. Interestingly, the protonated compound displays high-contrast piezofluorochromism with a large spectral change of 75 nm, and this finding shows the advantage of combining an external chemical stimulus with MFC [80].

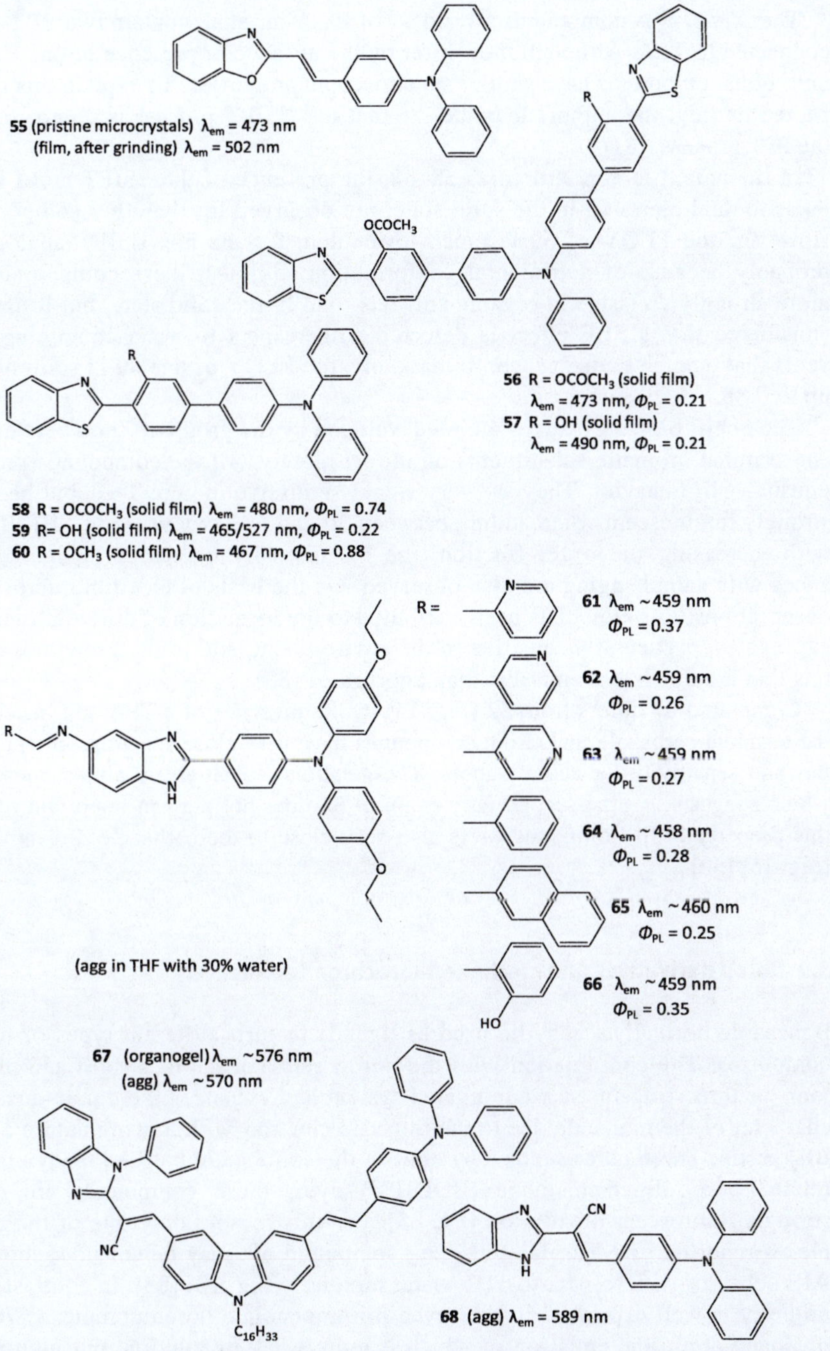

Fig. 17 Molecular structure of the triphenylamine derivatives **55** [80], **56** and **57** [81], **58–60** [82], **61–66** [83], **67** [78], and **68** [52] with solid-state PL maxima (λ_{em}) and photoluminescence quantum yields (Φ_{PL})

The A-π-D-π-A compounds **56** and **57** of Padalkar et al. contain two PBT units connected to TPA. Although they differ by the absence or presence of one ESIPT unit, both compounds have similar spectroscopic properties. To explain this unexpected finding, the authors hypothesize that the ESIPT process is dominated by the ICT process. [81].

In the parent D-π-A structures **58–60**, the presence of the ESIPT motif in **59** leads to dual emission in the solid state, not observed for the other compounds. However, the PLQY of **59** is much lower than for the non-ESIPT analogues, probably because of detrimental conformation [82]. All these compounds are more strongly emissive in organic solvents than in the solid state, but it may be considered that a SLE effect is detected with respect to water-containing solvents that quench fluorescence. Remarkably, the PLQY of dye **60** in solid film is up to 0.88.

The Schiff base derivatives **61–66** developed by the group of Zhou bear different terminal aromatic substituents on the BI moiety. All the compounds present unusual AIE behavior. They are very weakly emissive in pure THF, but become strongly luminescent when adding between 30 and 60% v/v water in THF. Then, with increasing the water fraction, the PL intensity decreases again. Differences with sample aging are also observed. On the basis of electron microscopy observations, this effect has been attributed to the formation of different types of aggregates, particularly sensitive to the environment, and to the growth of crystals that are much less emissive than aggregates [83].

Compound **67** differs from **52** (Fig. 15) by the presence of a TPA unit instead of the terminal carbazole unit. Both compounds have very close spectroscopic properties and sensitivity for acidic vapors. The gelation capacities are almost identical, which suggests that the second fatty chain of **52** does not play an important role in this property [78]. Compound **68** is also very close to the carbazole derivative **50** (Fig. 15) [52].

3.9 Boron derivatives and mechanofluorochromic effect

Benzazole derivatives may be used as ligands to form different types of boron complexes. The coordination with the boron center leads to a rigid and planar four- or three-ring-fused π-conjugated system and enhances the donor–acceptor character of the molecule due to the intrinsic electron-withdrawing nature of the BF_2 group. These compounds may emit in the solid state, but like many closely related boron-dipyrromethene (BODIPY) dyes, these compounds are often strongly fluorescent in solution [84, 85]. For this reason, only one of the complexes reported by Meesala et al., i.e., compound **69**, may be consider showing AIE behavior with respect to THF/water mixtures (Fig. 18) [85]. In contrast, this property is well expressed in the seven β-iminoenolate boron complexes **70–76** of Zhang et al., the emission of which is quite weak in solution, but multiplied up to 91-fold for aggregates of **71**. Notably, for **73**, the color of the emitted light changes significantly with the nature of the aggregates formed. This compound

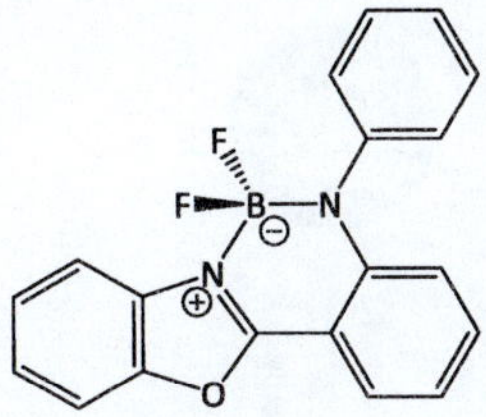

69 (agg) $\lambda_{em} \sim 500$ nm
(solid state) $\lambda_{em} = 493$ nm, $\Phi_{PL} = 0.20$

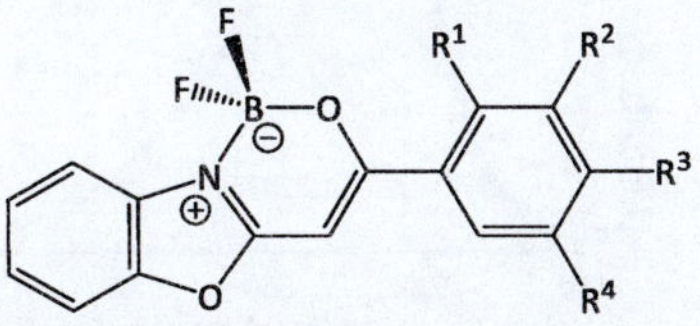

70 $R^1 = R^2 = R^3 = R^4 = H$ (agg) $\lambda_{em} = 468$ nm; (powder) $\lambda_{em} = 478$ nm, $\Phi_{PL} = 0.49$
71 $R^1 = OCH_3$, $R^2 = R^3 = R^4 = H$ (agg) $\lambda_{em} = 436/459$ nm; (powder) $\lambda_{em} = 439$ nm, $\Phi_{PL} = 0.90$
72 $R^1 = R^3 = R^4 = H$, $R^2 = OCH_3$ (agg) $\lambda_{em} = 471$ nm; (powder) $\lambda_{em} = 462$ nm, $\Phi_{PL} = 0.86$
73 $R^1 = R^2 = R^4 = H$, $R^3 = OCH_3$ (agg) $\lambda_{em} = 480/550$ nm; (powder) $\lambda_{em} = 480/550$ nm, $\Phi_{PL} = 0.92$
74 $R^1 = R^3 = OCH_3$, $R^2 = R^4 = H$ (agg) $\lambda_{em} = 476$ nm; (powder) $\lambda_{em} = 471$ nm, $\Phi_{PL} = 0.80$
75 $R^1 = R^4 = OCH_3$, $R^2 = R^3 = H$ (agg) $\lambda_{em} = 476$ nm; (powder) $\lambda_{em} = 466$ nm, $\Phi_{PL} = 0.26$
76 $R^1 = H$, $R^2 = R^3 = R^4 = OCH_3$ (agg) $\lambda_{em} = 500$ nm; (powder) $\lambda_{em} = 505$ nm, $\Phi_{PL} = 0.27$

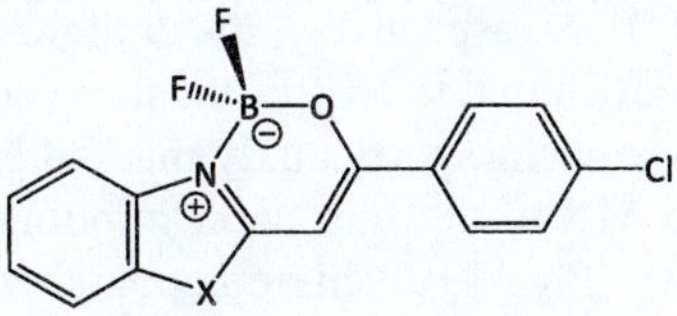

77 X = O (agg) $\lambda_{em} = 492$ nm, $\Phi_{PL} = 0.24$
(pristine powder) $\lambda_{em} = 447/472$ nm
(after grinding) $\lambda_{em} = 490$ nm
78 X = S (solid state) $\lambda_{em} = 522$ nm
(pristine powder) $\lambda_{em} = 461/490$ nm
(after grinding) $\lambda_{em} = 517$ nm

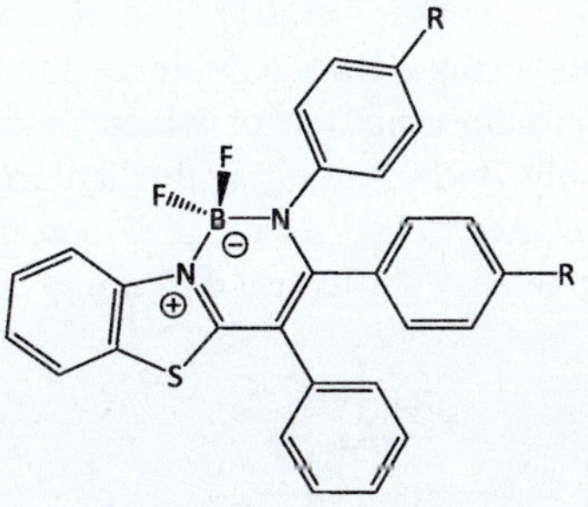

79 R = H (solid state) $\lambda_{em} = 473$ nm, $\Phi_{PL} = 0.60$
80 R = $N(CH_3)_2$ (solid state) $\lambda_{em} = 518$ nm, $\Phi_{PL} = 0.27$

Fig. 18 Molecular structure of molecules **69** [85], **70–76** [86], **77** and **78** [87], **79** and **80** [88] with solid-state PL maxima (λ_{em}) and photoluminescence quantum yields (Φ_{PL})

also exhibits unusual multicolor emission upon grinding, reversible after solvent fuming [86].

The PBO-based complex **77** reported by Zhao et al. also exhibits distinct AIE behavior, while this is not the case for the PBT analogue **78**. However, both complexes are well emissive in the solid state and show marked MFC. Their pristine crystals emit azure light, grinding turns their emitting color into cyan and green, respectively, and the initial emitting color is recovered under fuming or heating [87]. This type of MFC-active material may find applications in anti-fake labels and mechanical strength sensors.

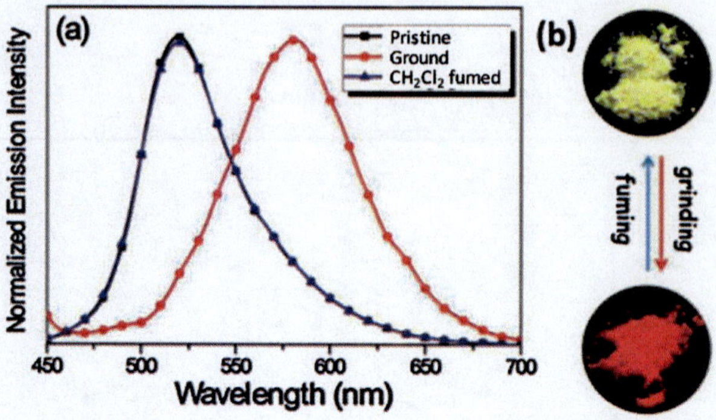

Fig. 19 **a** Normalized luminescence spectra of solid compound **80**, after being ground and CH_2Cl_2-fumed; λ_{ex} 414 nm. **b** Photographs of the ground powder of **80** and the CH_2Cl_2-fumed ground powder under UV light (365 nm). Adapted from Ref. [88] with permission from the Royal Society of Chemistry

The geometry of the propeller-shaped BT-enamido boron difluoride complexes **79** and **80** is reminiscent to that of archetypical AIE-gens. The two solid compounds are strongly emissive, with PLQY of 0.60 and 0.27, respectively, much higher than in organic solvents (<0.01). The SLE effect is attributed to RIR of the three peripheral phenyl rings. Both complexes show MFC behavior, particularly marked for **80**, the maximum emission of which passes from 518 to 582 nm upon grinding, and conversely upon fuming with dichloromethane (Fig. 19). This compound is also very sensitive to hydrostatic pressure, the maximum emission of the pristine powder being gradually shifted till 645 nm at around 10 GPa. This effect has been attributed

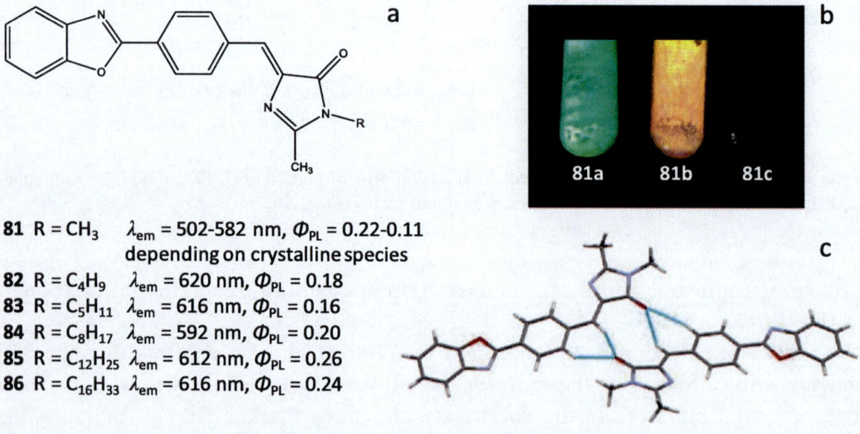

81 R = CH_3 λ_{em} = 502-582 nm, Φ_{PL} = 0.22-0.11 depending on crystalline species
82 R = C_4H_9 λ_{em} = 620 nm, Φ_{PL} = 0.18
83 R = C_5H_{11} λ_{em} = 616 nm, Φ_{PL} = 0.16
84 R = C_8H_{17} λ_{em} = 592 nm, Φ_{PL} = 0.20
85 R = $C_{12}H_{25}$ λ_{em} = 612 nm, Φ_{PL} = 0.26
86 R = $C_{16}H_{33}$ λ_{em} = 616 nm, Φ_{PL} = 0.24

Fig. 20 **a** Molecular structure of molecules **81** [91], **82** and **83** [89], **84** [90], **85** [89], and **86** [90] with solid-state PL maxima (λ_{em}) and photoluminescence quantum yields (Φ_{PL}) of the microcrystalline powders. **b** Image of the microcrystalline powders of two polymorph forms (**81a** and **81c**) and one CH_3CN solvate (**81b**) of compound **81** excited at 365 nm. **c** Example of dimer formed in the crystal structure of a non-emissive polymorph of **81**. Strong hydrogen bonds are drawn in blue ink to co-planation of the propeller structure upon compression. Compound **80** has been proposed for use as a new piezochromic luminescence material [88].

4 Insertion of benzazole moieties in AIE-gen frameworks: tuning and amplification of the AIE properties

4.1 Benzilideneimidazolinone derivatives

As seen above, BZ units may easily be converted into SLE fluorogens. However, another strategy is to merge these units with a well-known AIE-active fagment. In molecules **81–86**, introduced by our team [89, 90], a PBO unit has been combined with a synthetic derivative of the chromophore of the green fluorescent protein (GFP), i.e., *p*-hydroxybenzylideneimidazolinone (HBDI) (Fig. 20a). The HBDI chromophore is virtually not fluorescent in fluid medium, and emits blue-green light when constrained in a protein matrix, for example. However, the PLQY and photochemical stability of its derivatives are not always very good. The association with PBO strongly improves these properties. All compounds are weakly emissive in the blue region in fluid solvents. Time-dependent density functional theory (TD-DFT) and spin-flip TD-DFT calculations show that the relaxation pathway in the lowest singlet excited state leads to a thermally accessible conical intersection (CI), which favors non-radiative deactivation. A strong twist of the imidazolinone moiety with respect to the rest of the molecule is necessary to access this CI [91]. The long-chain PBO-BDI conjugates **82–86** display neat SLE effect with PLQY reaching 0.26 for the microcrystalline powders. The yellow-orange emission is shown to originate from long-distance excimers formed in the crystals [89]. Using a simple solvent-exchange method, these compounds easily form luminescent microfibers and non-doped nanoparticles [90]. The methyl derivative **81** was first assumed to be non-emissive in the solid state. However, crystallizing this molecule under precise experimental conditions allows four distinct polymorphs and one solvate to be obtained. Three of these crystalline forms are indeed virtually not emissive, while the other two exhibit strong photoluminescence (PL), close to that of the long-chain analogs (Fig. 20b). Recrystallization allows for passing reversibly from one crystalline species to another. The non-emissive polymorphs are characterized by the presence of a centrosymmetric dimer structured by hydrogen bonds, in which paired molecules vibrate with the same mode and are in turn subjected to maximum deformation, according to DFT calculations (Fig. 20c). Vibrational dispersion of the excitation energy in the dimer, together with the easy access to CI permitted by a loose crystal structure, could explain the behavior of the three non-emissive polymorphs. In contrast, RACI may account for the PL emission of the other two species, which have tight crystal packing mode. The emergence of polymorphism and of a rare on/off PL mechanism may be attributed to the presence of the BDI unit in the molecule.

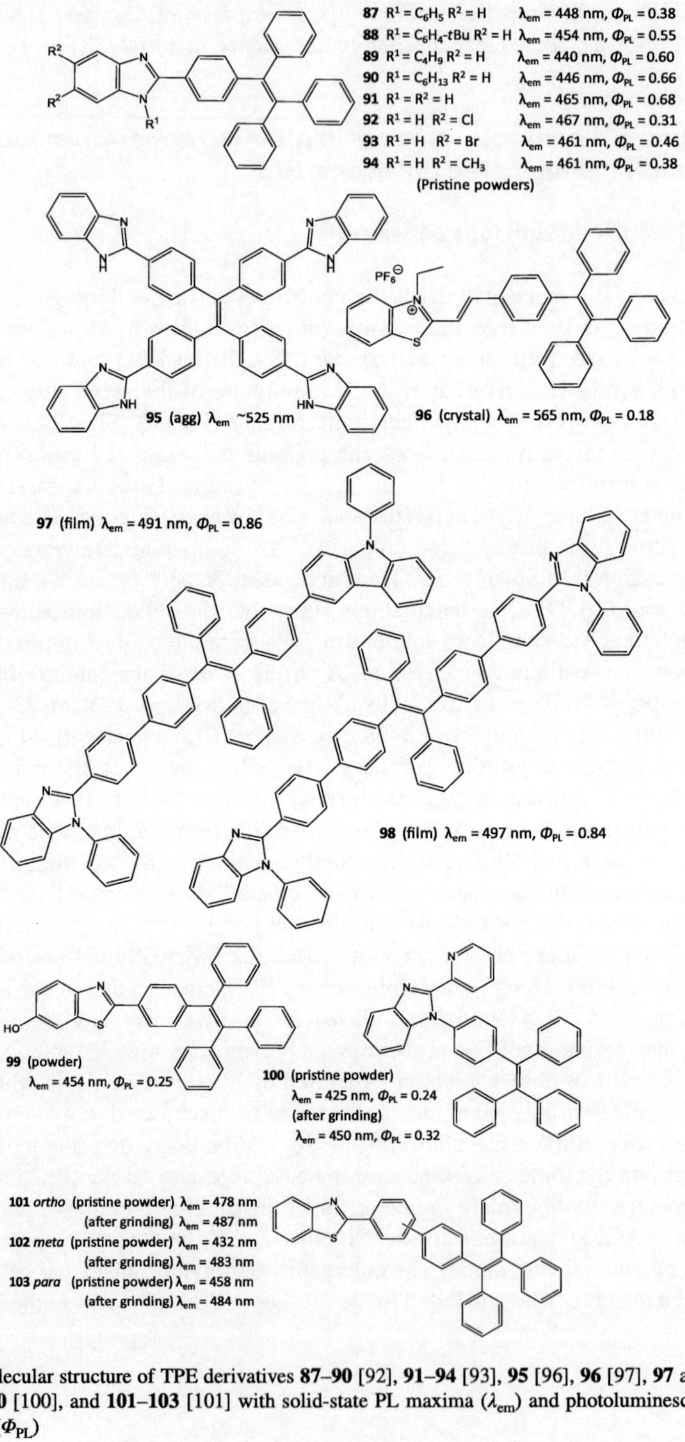

Fig. 21 Molecular structure of TPE derivatives **87–90** [92], **91–94** [93], **95** [96], **96** [97], **97** and **98** [98], **99** [99], **100** [100], and **101–103** [101] with solid-state PL maxima (λ_{em}) and photoluminescence quantum yields (Φ_{PL})

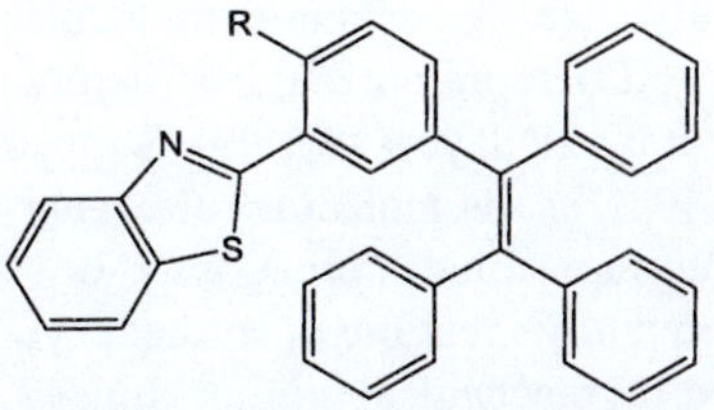

104 R = OH (agg) λ_{em} = 589 nm, Φ_{PL} = 0.36
(solid state) λ_{em} = 589 nm, Φ_{PL} = 0.97
105 R = H (agg) λ_{em} = 488 nm
106 R = OCH_3 (agg) λ_{em} = 490 nm

107 R = 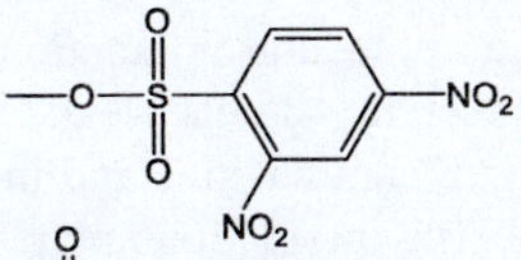

108 R = 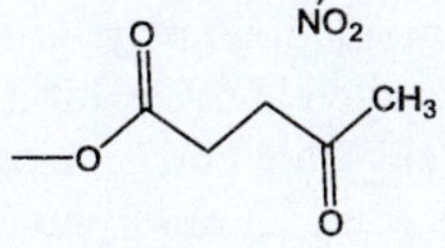

(agg) λ_{em} = 465 nm, Φ_{PL} = 0.23

Fig. 22 Molecular structure of TPE derivatives **104–106** [102] and **105–108** [103] with solid-state PL maxima (λ_{em}) and photoluminescence quantum yields (Φ_{PL})

4.2 Tetraphenylethene derivatives

Owing to its very simple structure, tetraphylethene or tetraphenylethylene (TPE) is the most popular representative of the anti-ACQ effect molecules. The modification of this building block has proven to be an efficient way to access new AIE-gens with valuable solid-state properties. The TPE molecule has been decorated with BZ units with the aim to tune its spectral properties. The design consists in the peripheral insertion of a BZ motif on one, two, or even on each of the four phenyl groups of TPE. The BZ heterocycle is frequently attached to TPE by the carbon atom in the 2-position. More rarely, it is connected via a vinyl bond or a phenyl group. The BI can also be fixed to TPE by its nitrogen atom (Figs. 21, 22).

From a general point of view, the TPE-BZ derivatives have a very good thermal stability, as illustrated by the molecules **87–90** and **91–94** developed by Zhang et al., the decomposition temperature of which is above 319 and 340 °C, respectively [92, 93]. All TPE-BZ derivatives show weak emission in organic solution, and exhibit clear fluorescence enhancement in the aggregate state. In the solid state, TPE is a blue emitter (444–453 nm) [94, 95]. Molecules **87–90** and **91–94**, which bear only one BI unit, also emit in the blue region at 440–454 nm [92] and ~ 461–467 nm [93], respectively. Molecule **95**, a highly conjugated π–system with four BI units, has a yellowish green emission at 525 nm [96]. For the benzothiazolium derivative **96**, with strong ICT, the emission maximum is 565 for the pure dye and 591 nm for a solvate [97]. However, the most stunning characteristic of some TPE-BZ derivatives is their very high PLQY, reaching for example 0.68 for **91** [93], 0.84 for **97**, and even 0.86 for **98** [98], while it is only 0.22 for TPE.

As may be expected, X-ray diffraction analyses of single crystals have shown that all the TPE-BZ derivatives have highly twisted paddle-shaped conformations, considering the position of the four phenyl rings with respect to the central ethylene segment. In most of the molecules, the BI moiety and the adjacent phenyl ring of TPE are nearly planar, as is commonly the case in aryl-substituted

BZ. This is not the case in sterically hindered compounds **87–90**, where the torsion angle between the BI heterocycle and the phenyl ring is between 34° and 44° [92]. The crystal packing mode is different in every case. For example, molecules **91**, **92**, and **94** form chain-like one-dimensional (1D) supramolecular structures [93], while **87–90** and **99** adopt a fashion stacking by 2D layers [92, 99]. Slipped π-π intermolecular interactions may be observed [93], as the molecules are generally linked by affluent weak C-H-π and C-H-N intermolecular interactions. Above all, significant cavities and interface gaps are frequently mentioned, so that crystal packing interactions may be readily destroyed by mechanical stimuli. Indeed, the molecules cited above exhibit clear MFC effect. Upon grinding by a spatula or a pestle, the sky-blue emissive microcrystals of **87–90** and **91–94** convert to yellowish green-emitting amorphous solids, and vice versa upon heating or fuming with dichloromethane vapors [92, 93]. The solid-state emission of **96** is reversibly tuned from yellow or orange to red by grinding-fuming and grinding-heating processes [97]. In the same conditions, molecule **100** reported by Cui et al. exhibits a color change from deep blue to turquoise [100]. Molecule **99** reported by Ma et al. exhibits remarkable four-color switching, by combining grinding/solvent-fuming cycles with protonation/deprotonation cycles based on exposure to HCl and NH_3 vapor [99]. Finally, the three isomers **101–103** of Jadhav et al. show that the AIE and MFC properties depend on the linkage between BT and the TPE unit. The *meta* isomer shows the strongest bathochromic spectral shift (51 nm) upon grinding, while this effect was weaker for the *para* and *ortho* isomers (26 and 9 nm, respectively). It is noteworthy that molecules of the *ortho* isomer are more tightly twisted and packed in the crystals than those of the *meta* isomer [101]. For all these MFC-active compounds, applications in the fields of chemo- and mechanosensors, optical displays, and rewritable optical media, are envisaged.

The electroluminescence properties of **88** are attractive for use in OLEDs, for which the development of blue-emitting luminogens is still a challenge [92]. In this context, molecule **97** has been especially designed to combine three functional units, i.e., the AIE-gen building block TPE, the electron-transporting PBI, and the hole-transporting carbazole. Indeed, **97** acts as a bipolar emitter that exhibits good SLE properties, together with well-balanced electron–hole transport and recombination properties. The resulting non-doped OLED shows excellent performance, with maximum luminance, current efficiency, power efficiency, and external maximum efficiency values that reach the theoretical limit for fluorescent OLEDs. This performance is much better than that of the molecules that bear two identical PBI moieties (**98**) or two carbazole units on the central TPE unit [98].

Finally, applications in the field of sensors are also possible. The tetrasubstituted nitrogen-rich BI-TPE derivative **95** possesses high affinity for Ag(I) cations, with which it forms complexes having a 1:2 stoichiometry. In methanol solution, such complexes assemble in aggregates that emit strong yellow light. The method is quite selective with respect to other cations, and it has been proposed for the facile determination of Ag(I) ions in environmental samples, with the aim of pollution control [96].

In molecules of **104** (Fig. 22), merging TPE with HTZ allows the ESIPT process to be implemented. The SLE effect is dramatic. The molecule is weakly emissive in acetone, but its PLQY is 0.36 as aggregates in aqueous acetone and 0.97 in the solid state. It exhibits a large Stokes shift, with emission from the keto form at 589 nm. Of course, this effect is not observed with the parent compound **105** and the methoxy derivative **106** that cannot show any ESIPT effect. A protection/deprotection strategy is implemented for the detection of biothiols such as cysteine, homocysteine, and glutathione that play critical roles in cells as antioxidants and signaling agents [102]. In compound **107**, the hydroxyl function is protected by a 2,4-dinitrobenzenesulfonyl group, which results in fluorescence quenching. Thiols induce the stoichiometric conversion of **107** into **104**, whose aggregation in the medium turns on the PL response. The compound is usable on test strips with good specificity with respect to other analytes, and can even be used for the detection of biothiols in living cells [102]. Similarly, Liu et al. have developed the closely related molecule **108**, in which the hydroxyl group is protected by a levulinate unit. This molecule has been successfully used to conduct the ratiometric imaging of endogenous sulfur dioxide SO_2, a signal transduction agent involved in the regulation of the cardiovascular function in mammalian cells, mouse embryonic fibroblasts, and zebrafish [103].

The AIE mechanism of TPE-based fluorophores has been well studied. Initially, the π-bond is destroyed upon excitation, and a diradical species is generated. Then, two main processes may take place, depending on the molecular structure. The first one follows the RACI model, associated with double bond torsion. The second one involves the formation of a C–C bond between adjacent phenyl rings, leading to a photocyclized intermediate. The latter process requires large motion of the phenyl rings and is blocked in the aggregate phase [24–26].

5 A glance at synthesis

The demanding and complex synthesis of AIE molecular systems (commonly large rotor molecules with luminescence in the blue and green spectral regions) is known to be a major drawback of this approach. In comparison, the synthesis of small BZ derivatives is much simpler. Through the examples given below, it can be seen that molecules exhibiting an effective SLE effect can be obtained in a few steps with satisfactory yields.

The 2-arylbenzazole moieties are generally obtained by condensation of 2-aminophenol, 2-aminothiophenols, or 1,2-diaminobenzene derivatives with carboxylic acids in the presence of polyphosphoric acid (PPA) or aldehydes. For example, Yadav et al. prepared cyano derivatives **2–7** by classical Knovenaegel condensation of 4-(benzo[d]oxazol-2-yl)benzaldehyde with substituted isophorones, benzothiazole, or benzimidazoles in the presence of piperidine (Scheme 1). The 4′-formyl derivative of BPO was obtained following a two-step procedure with 2-(*p*-tolyl)benzo[d]oxazole intermediate formation and its subsequent low yield oxidation by toxic selenium dioxide or chromium trioxide [44].

Recently, our group has reported an improved procedure to access formyl-BPO precursor by condensation of 2-aminophenol with terephtalaldehyde and subsequent

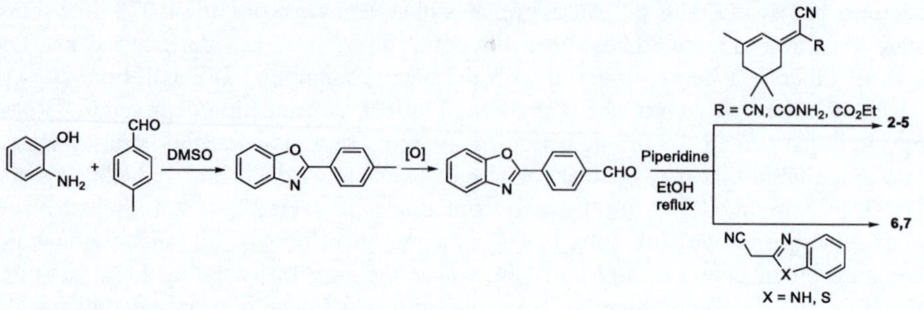

Scheme 1 Synthesis of cyano derivatives **2–7** [44]

Scheme 2 Alternative synthesis of 4-(benzo[d]oxazol-2-yl) benzaldehyde [45]

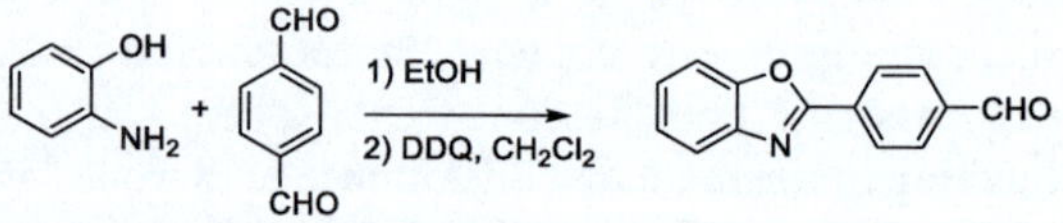

treatment with 2,3-dichloro-5,6-dicyano-1,4-benzoquinone (DDQ) in 45% overall yield [45] (Scheme 2).

The Schiff bases **10–12** were prepared in good yields (60 to 84%) from the condensation of *p*-(benzoxalzolyl)anilines (obtained from 2-aminophenol with *p*-aminobenzoic acid in PPA) with the required aromatic aldehydes in refluxing ethanol [54] (Scheme 3).

Similarly, the benzimidazole-based Schiff bases **14–16** were prepared from 2-aminobenzimidazole derivatives, readily obtained from the corresponding *o*-phenylenediamine and cyanogen bromide [56] (Scheme 4).

Organogelator **17**, which contains two BI moieties and an aromatic Schiff-base unit, was prepared by condensation in refluxing dimethylformamide (DMF) of 1,4-phtalaldehyde with an acylhydrazide derivative obtained in two steps from 2-phenethylbenzimidazole [57]. Similarly, gelator **18**, based on an acylhydrazone

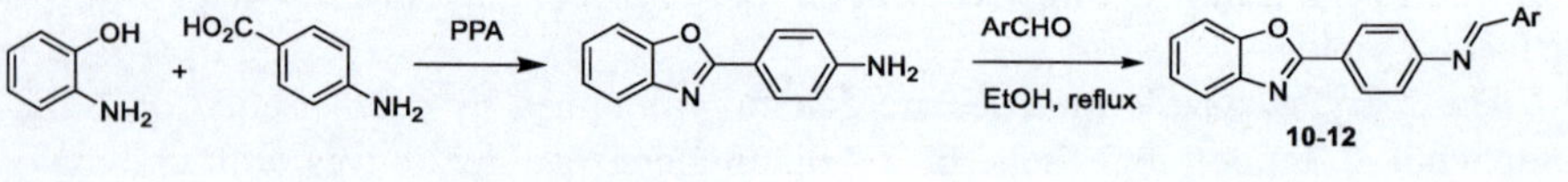

Scheme 3 Synthesis of Schiff bases **10–12** [54]

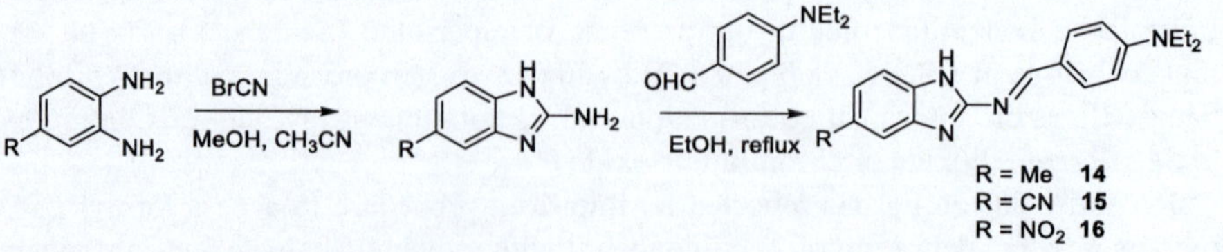

Scheme 4 Synthesis of Schiff bases **14–16** [56]

naphtol moiety, resulted from the reaction of 2-hydroxy naphthalene formaldehyde with the required benzimidazole acetylhydrazide derivative in refluxing ethanol [58] (Scheme 5).

Carboxamide BO derivatives **19** and **20** were classically obtained by coupling (benzoxalzolyl)anilines with 4-*tert*-butylbenzoic acid in the presence of *N,N′*-carbonyldiimidazole (CDI) in refluxing toluene [59, 60] (Scheme 6).

2′-Hydroxyphenyl-benzoxazole and benzothiazole derivatives come from the condensation of 2-aminophenol and 2-aminothiophenol, respectively, with an *o*-hydroxybenzoic acid reagent in the presence of PPA. Padalkar et al. developed an access to 2′-HBT derivatives by Pd-catalyzed Suzuki coupling reaction [74, 75]. The 2′-HBT framework was initially obtained from acid-catalyzed cyclization of 2-aminothiophenol and 4-bromo-2-hydroxybenzoic acid leading to intermediate 2-(benzo[*d*]thiazol-2-yl)-5-bromophenol in 52% yield. Coupling with various boronic acids in the presence of $Pd(Ph_3)_4$ and K_2CO_3 afforded a variety of carbazole, triphenylamine, and fluorene derivatives in moderate to good yields, as illustrated below for carbazole derivative **47** in Scheme 7.

A one-step Knoevenagel reaction was enough for Xue et al. to prepare the stilbene derivatives of carbazole-based (**51**) and phenothiazine benzoxazoles (**53, 54**)

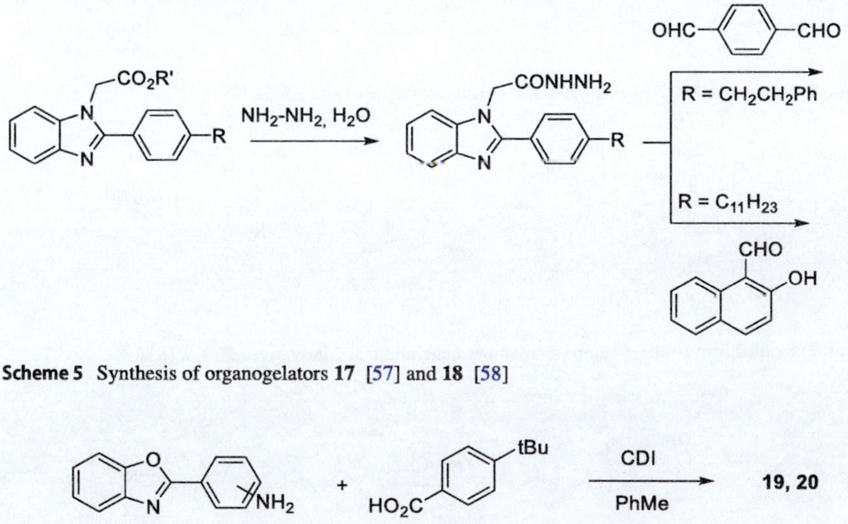

Scheme 5 Synthesis of organogelators **17** [57] and **18** [58]

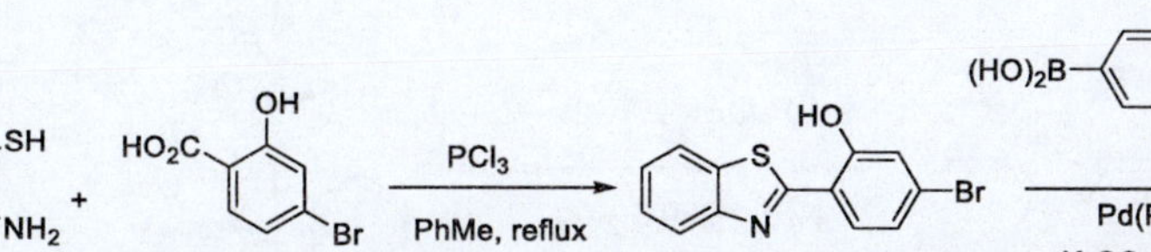

Scheme 6 Synthesis of carboxamide benzoxazole derivatives **19** [59] and **20** [60]

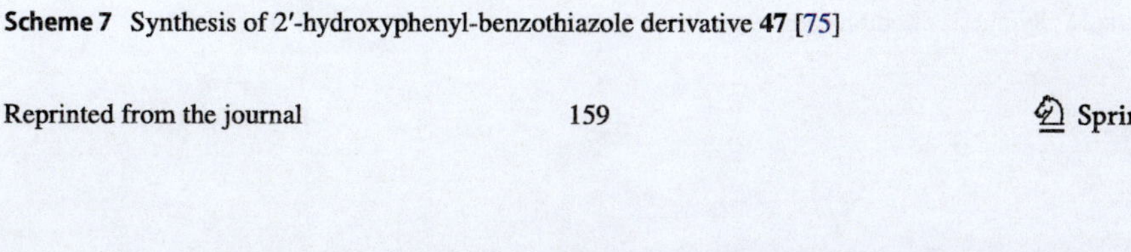

Scheme 7 Synthesis of 2′-hydroxyphenyl-benzothiazole derivative **47** [75]

in high 82 to 86% yield, starting from commercially available 2-methylbenzoxazole and carbazole or phenothiazine carbaldehydes in the presence of potassium *tert*-butoxide in THF [76, 79] (Scheme 8).

The triphenylamine-containing benzimidazole derivatives **61–66** were prepared by Cao et al. by condensation of 4-[*bis*-(4-ethoxyphenyl)-amino]-benzaldehyde with 4-nitro-benzene-1,2-diamine in refluxing DMF. Subsequent reduction of the nitro function and condensation with the requisite aldehydes provided the target Schiff bases [83] (Scheme 9).

In the same way, triphenylamine benzoxazole **55** was efficiently obtained by reaction of 2-methylbenzoxazole with 4-(diphenylamino)benzaldehyde in the presence of potassium *tert*-butoxide in THF with a yield of 86% [80] (Scheme 10).

The borate complexes were efficiently obtained by boron complexation at the chelating site of 2′-aminophenyl benzoxazole (for **69**) or β-ketoimidates derivatives after treatment with boron trifluoride diethyl etherate as exemplified for derivative **70–76** in Scheme 11 [86].

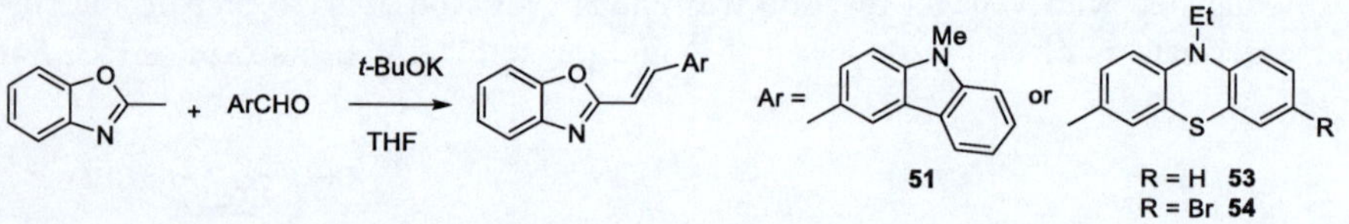

Scheme 8 Synthesis of stylbene benzoxazole derivatives **51** [76] and **53–54** [79]

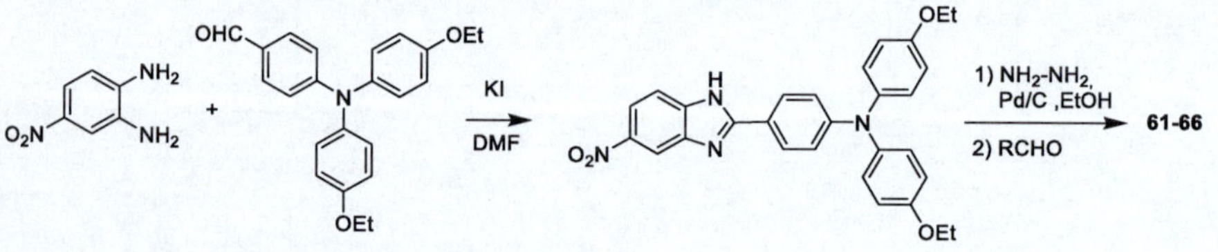

Scheme 9 Synthesis of triphenylamine-containing benzimidazole derivatives **61–66** [83]

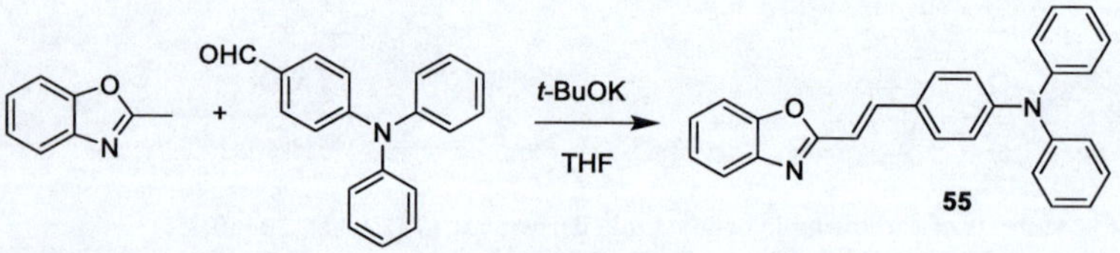

Scheme 10 Synthesis of triphenylamine benzoxazole **55** [80]

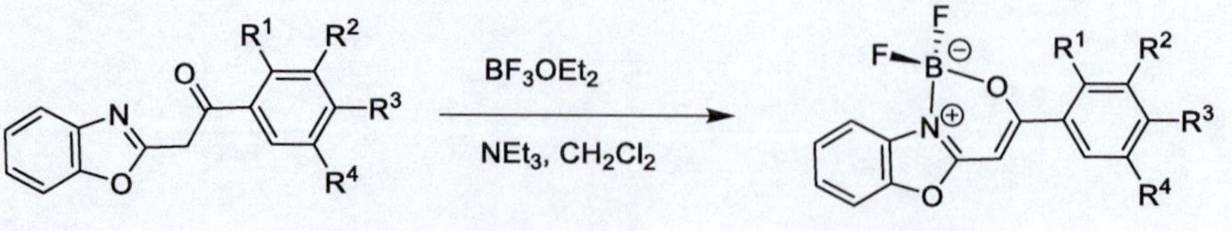

Scheme 11 Synthesis of boron complexes **70–76** [86]

Merging a BZ unit with a typical AIE-gen can also be achieved very simply. Ghodbane et al. efficiently synthesized a series of benzilideneimidazolinone derivatives **81–86** in a two-step procedure. Condensation of 2-(4′-formylphenyl)benzoxazole with a primary alkylamine led to the corresponding imino derivatives, and subsequent [2+3] cycloaddition reaction with methyl[(1-ethoxyethylidene)amino] acetate gave the target compounds in high yields [89–91] (Scheme 12).

Tetraphenylethene derivatives are generally obtained from functionalized bromo or formyl derivatives of TPE. For example, Zhang et al. synthesized compounds **91–94** in high yields starting from tetraphenylethene carbaldehydes by typical cyclization with phenylenediamine derivatives [93] (Scheme 13).

Likewise, TPE-based benzothiazoles such as **99** and **104–106** were prepared by reaction of 4-phenylethenealdehyde with the appropriate 2-aminobenzenethiols [99, 102]. The Suzuki Pd-catalyzed cross-coupling reaction was also well exploited to access TPE-BZ derivatives. For instance, compound **98** was obtained by Zhang et al. in 40% yield via a Suzuki coupling reaction between 1,2-bis(4-bromophenyl)-1,2-diphenylethene and adequate commercially available phenylbenzimidazole boronic acid [98] (Scheme 14).

Finally, it must be stressed that access to tetra(hetero)arylethylene derivatives embedding BO substituents has recently been developed by Tan et al. involving a *trans*-selective 1,2-diheteroarylation of diarylalkyne with BO via rhodium/copper co-catalyzed C–H addition/oxidative coupling process. The optimized conditions using [R Cp* $(MeCN)_3][SbF_6]_2$ as the catalyst, $Cu(OAc)_2$ as the oxidant, and pivalic

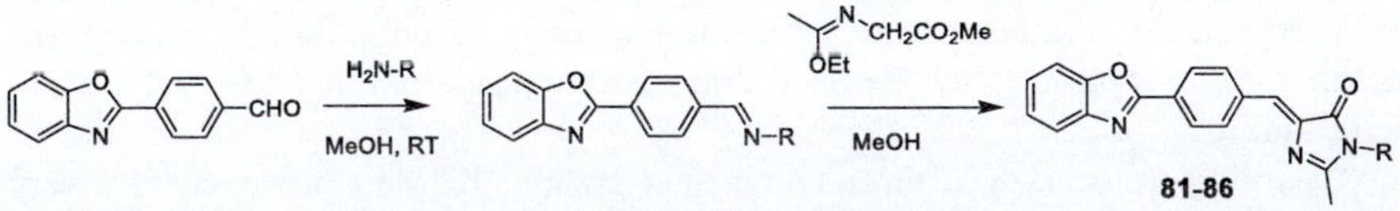

Scheme 12 Synthesis of benzilideneimidazolinone derivatives **81–86** [89–91]

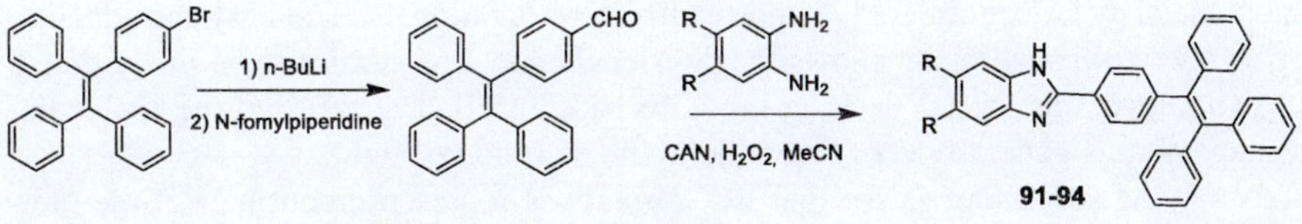

Scheme 13 Synthesis of TPE-BI derivatives **91–94** [93]

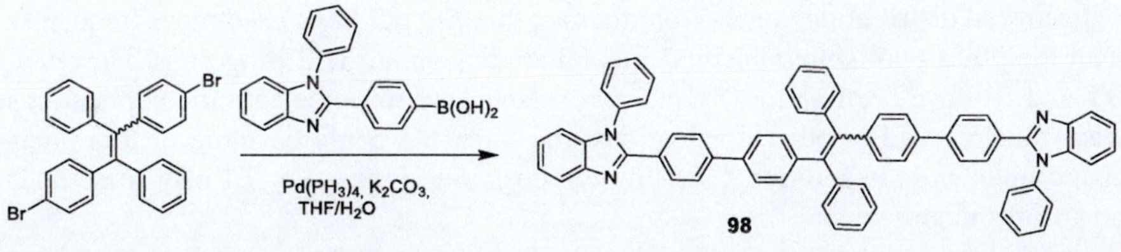

Scheme 14 Synthesis of derivative **98** [98]

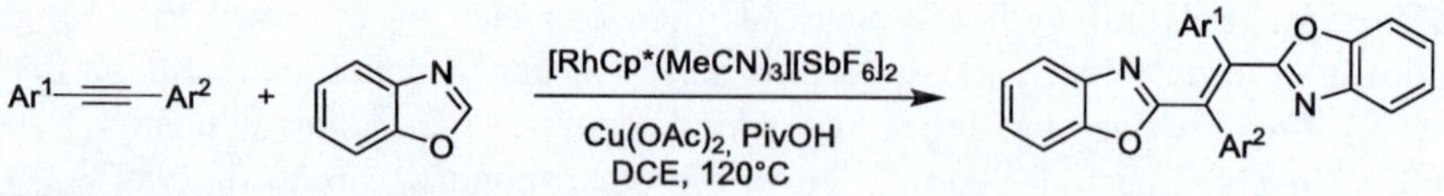

Scheme 15 Synthesis of tetra(hetero)arylethylene derivatives by addition/oxidative coupling of diaryl alkynes with benzoxazole [95]

acid as additive in dichloroethane, efficiently yielded a large variety of compounds depending on the diaryl alkyne and azole derivatives used as starting materials [95] (Scheme 15). This synthetic procedure opens wide perspectives in the development of a new generation of SLE-active compounds.

6 Conclusion and perspectives

So far, BZ units do not belong to the small family of archetypical AIE-gens. However, they are at the basis of very efficient SLE-active molecules. Substituting a BZ heterocycle with a –C=C–Ar or –N=C–Ar moiety, or connecting a $C{=}CR^1R^2$, –C=NR, or $-N{=}CR^1R^2$ group to an aryl-benzazole is sufficient to generate SLE properties. Benzazole moieties bring these properties to classical ACQ-active molecules, which in turn enhance the electroluminescence and MFC capacity of the conjugate. Such molecules are attractive for use in OLEDs and as multi-responsive sensors. The ESIPT properties of the hydroxy derivatives enable the development of sensors based on a protection/deprotection strategy. Decoration of TPE with BZ motifs leads to molecules whose electroluminescent properties are close to the theoretical limit.

According to the size of the π-conjugated system, BZ derivatives cover a very broad range of wavelengths. They range from blue emitters, in high demand for OLEDs, to red emitters, very useful for biological imaging. Various types of solids may be obtained, from microcrystals to thin films. Many BZ derivatives also arrange spontaneously in nanofibers [7] and readily give organogels. The easy introduction of a variety of substituents allows the physicochemical properties to be finely tuned. It also allows the crystal packing mode to be controlled, given that the molecular arrangement is generally very favorable to the emission of light.

It may be surprising to see that BO derivatives represent about half of the molecules reviewed above, while umpteen numbers of BT and BI derivatives have been reported for excellent fluorescent, thermal, and mechanical properties that make them potential candidates for applications in the field of optically active materials. This uneven distribution comes from the fact that BO has been used more frequently than the other two isologs in fundamental studies about SLE properties. However, BT and BI have been almost exclusively selected when optoelectronic applications were researched. For other applications, the three BZ could be more or less interchangeable, and the additional substitution possibility offered by BI may sometimes be an advantage.

Many developments are possible. SLE-active materials can be obtained by coupling BZ with other molecules, with the aim of generating new properties. It is known, for example, that crystals of some BZ-naphthalene derivatives exhibit photomechanical effect [104]. The conjugation with coumarins gives access to compounds with valuable nonlinear optic (NLO) properties [105], while cyanines derivatives are NIR-emitting bio-probes [106]. There are many things to explore in the field of closely related compounds, such as benzothiadiazole derivatives, which exhibit reversible mechanofluorochromism [107] and multi-stimuli responsiveness [108]. It is noteworthy that since benzothiadiazole cannot be substituted in the 2-position, the synthesis and architecture of the derivatives obtained are different from those of the BZ derivatives described above.

This article is focused on organic compounds. However, due to their very good metal-ion chelating properties, BZ units have also been used successfully as ligands in transition metal complexes that exhibit SLE behavior [109, 110]. Furthermore, BZ units may be included into the organic ligand, without direct participation to the coordination process, and they contribute to enhancing the SLE properties. This is the case, for example, for the SLE-active and mechanoresponsive rhenium(I) complexes developed by our group [111, 112]. A new avenue is opening up in the field of coordination chemistry. This topic surely deserves to be further developed.

At the moment, it is our hope that this review article has convinced the reader that benzazoles are precious building blocks for the design of very efficient luminescent materials based on small molecules obtained from facile and economic synthesis.

Acknowledgements This publication is part of a project that has received funding from Agence Nationale pour la Recherche (ANR), SUPERFON project # ANR 17-CE07-0029-03.

Declarations

Conflict of interest We declare no conflict of interest.

References

1. Kamal U, Javed NM, Arun K (2020) Biological potential of benzoxazole derivatives: an updated review. Asian J Pharm Clin Res 13:28–41. https://doi.org/10.22159/ajpcr.2020.v13i8.37958
2. Demmer CS, Bunch L (2015) Benzoxazoles and oxazolopyridines in medicinal chemistry studies. Eur J Med Chem 97:778–785. https://doi.org/10.1016/j.ejmech.2014.11.064
3. Keri RS, Patil MR, Patil SA, Budagumpi S (2015) A comprehensive review in current developments of benzothiazole-based molecules in medicinal chemistry. Eur J Med Chem 89:207–251. https://doi.org/10.1016/j.ejmech.2014.10.059
4. Keri RS, Hiremathad A, Budagumpi S, Nagaraja BM (2015) Comprehensive review in current developments of benzimidazole-based medicinal chemistry. Chem Biol Drug Des 86:799–845. https://doi.org/10.1111/cbdd.12462
5. Krasovitskii BM, Bolotin BM (1988) Organic luminescent materials. Wiley-VCH Verlag GmbH, Weinheim (**ISBN-13: 978-3527267286**)
6. Chen SH, Jiang K, Xiao Y, Cao XY, Arulkumar M, Wang ZY (2020) Recent endeavors on design, synthesis, fluorescence mechanisms and applications of benzazole-based molecular probes toward miscellaneous species. Dyes Pigment 175:108157. https://doi.org/10.1016/j.dyepig.2019.108157

7. Carayon C, Fery-Forgues S (2017) 2-Phenylbenzoxazole derivatives: a family of robust emitters of solid-state fluorescence. Photochem Photobiol Sci 16:1020–1035. https://doi.org/10.1039/c7pp00112f
8. Mei J, Leung NLC, Kwok RTK, Lam JWY, Tang BZ (2015) Aggregation-induced emission: together we shine, United We Soar! Chem Rev 115:11718–11940. https://doi.org/10.1021/acs.chemrev.5b00263
9. Zhao Z, Zhang H, Lam JWY, Tang BZ (2020) Aggregation-induced emission: new vistas at the aggregate level. Angew Chem Int Ed 59:9888–9907. https://doi.org/10.1002/anie.201916729
10. Shi J, Aguilar Suarez LE, Yoon SJ, Varghese S, Serpa C, Park SY, Lüer D, Roca-Sanjuán D, Milián-Medina B, Gierschner J (2017) Solid state luminescence enhancement in π-conjugated materials: unraveling the mechanism beyond the framework of AIE/AIEE. J Phys Chem C 121:23166–23183. https://doi.org/10.1021/acs.jpcc.7b08060
11. Anthony SP (2012) Organic solid-state fluorescence: strategies for generating switchable and tunable fluorescent materials. ChemPlusChem 77:518–531. https://doi.org/10.1002/cplu.201200073
12. Bera MK, Pal P, Malik S (2020) Solid-state emissive organic chromophores: design, strategy and building blocks. J Mater Chem C 8:788–802. https://doi.org/10.1039/c9tc04239c
13. Dimitriev OP, Piryatinski YP, Yuri L, Slominskii YL (2018) Excimer emission in J-aggregates. J Phys Chem Lett 9:2138–2143. https://doi.org/10.1021/acs.jpclett.8b00481
14. Wilbraham L, Louis M, Alberga D, Brosseau A, Guillot R, Ito F, Labat F, Métivier R, Allain C, Ciofini I (2018) Revealing the origins of mechanically induced fluorescence changes in organic molecular crystals. Adv Mater 30:1800817. https://doi.org/10.1002/adma.201800817
15. Gierschner J, Shi J, Milián-Medina B, Roca-Sanjuán D, Varghese S, Park SY (2021) Luminescence in crystalline organic materials: from molecules to molecular solids. Adv Opt Mater. https://doi.org/10.1002/adom.202002251
16. Gierschner J, Lüer L, Milián-Medina B, Oelkrug D, Egelhaaf HJ (2013) Highly Emissive H-aggregates or aggregation-induced emission quenching? The photophysics of all-trans para-distyrylbenzene. J Phys Chem Lett 4:2686–2697. https://doi.org/10.1021/jz400985t
17. De Silva TPD, Sahasrabudhe G, Yang B, Wang CH, Chhotaray PK, Nesterov EE, Warner IM (2019) Influence of anion variations on morphological, spectral, and physical properties of the propidium luminophore. J Phys Chem A 123:111–119. https://doi.org/10.1021/acs.jpca.8b06948
18. Andreiuk BA, Reisch A, Bernhardt E, Klymchenko AS (2019) Fighting aggregation caused quenching and leakage of dyes in fluorescent polymer nanoparticles: universal role of counterion. Chem Asian J 14:836–846. https://doi.org/10.1002/asia.201801592
19. Soulié M, Carayon C, Saffon N, Blanc S, Fery-Forgues S (2016) A comparative study of nine berberine salts in the solid state: optimization of the photoluminescence and self-association properties through the choice of the anion. Phys Chem Chem Phys 18:29999–30008. https://doi.org/10.1039/c6cp05848e
20. Lu B, Fang X, Yan D (2020) Luminescent polymorphic co-crystals: a promising way to the diversity of molecular assembly, fluorescence polarization, and optical waveguide. ACS Appl Mater Interfaces 12:31940–31951. https://doi.org/10.1021/acsami.0c06794
21. Bhowal R, Biswas S, Saseendran DPA, Koner AL, Chopra D (2019) Tuning the solid-state emission by co-crystallization through σ- and π-hole directed intermolecular interactions. CrystEngCommun 21:1940–1947. https://doi.org/10.1039/c8ce02118j
22. Jimbo T, Tsuji M, Taniguchi R, Sada K, Kokado K (2018) Control of aggregation-induced emission from a tetraphenylethene derivative through the components in the co-crystal. Cryst Growth Des 18:3863–3869. https://doi.org/10.1021/acs.cgd.8b00141
23. Kumar S, Singh M, Gaur P, Jou JH, Ghosh S (2017) Role of voluminous substituents in controlling the optical properties of disc/planar-like small organic molecules: toward molecular emission in solid state. ACS Omega 2:5348–5356. https://doi.org/10.1021/acsomega.7b00832
24. Crespo-Otero R, Li Q, Blancafort L (2019) Exploring potential energy surfaces for aggregation-induced emission-from solution to crystal. Chem Asian J 14:700–714. https://doi.org/10.1002/asia.201801649
25. Chen Y, Lam JWY, Kwok RTK, Liu B, Tang BZ (2019) Aggregation-induced emission: fundamental understanding and future developments. Mater Horiz 6:428–433. https://doi.org/10.1039/c8mh01331d
26. Rivera M, Stojanović L, Crespo-Otero R (2021) Role of conical intersections on the efficiency of fluorescent organic molecular crystals. J Phys Chem A 125:1012–1024. https://doi.org/10.1021/acs.jpca.0c11072

27. Presti D, Wilbraham L, Targa C, Labat F, Pedone A, Menziani MC, Ciofini I, Adamo C (2017) Understanding aggregation-induced emission in molecular crystals: insights from theory. J Phys Chem C 121:5747–5752. https://doi.org/10.1021/acs.jpcc.7b00488
28. Wang C, Li Z (2017) Molecular conformation and packing: their critical roles in the emission performance of mechanochromic fluorescence materials. Mater Chem Front 1:2174–2194. https://doi.org/10.1039/c7qM00201g
29. Majumdar P, Tharammal F, Gierschner J, Varghese S (2020) Tuning Solid-State Luminescence in conjugated organic materials: control of excitonic and excimeric contributions through π stacking and halogen bond driven self-assembly. ChemPhysChem 21:616–624. https://doi.org/10.1002/cphc.201901223
30. Varughese S (2014) Non-covalent routes to tune the optical properties of molecular materials. J Mater Chem C 2:3499–3516. https://doi.org/10.1039/c3tc32414a
31. Lu B, Liu S, Yan D (2019) Recent advances in photofunctional polymorphs of molecular materials. Chin Chem Lett 30:1908–1922. https://doi.org/10.1016/j.cclet.2019.09.012
32. Tan R, Wang S, Lan H, Xiao S (2017) Polymorphism-dependent and mechanochromic luminescent molecules. Curr Org Chem 21:236–248. https://doi.org/10.2174/1385272820666160905104014
33. Carayon C, André-Barrès C, Leygue N, Saffon-Merceron N, Boggio-Pasqua M, Fery-Forgues S (2021) The role of the synthetic chromophore of GFP in generating polymorphism-dependent on/off photoluminescence. Dyes Pigment 187:109119. https://doi.org/10.1016/j.dyepig.2020.109119
34. Sagara Y, Yamane S, Mitani M, Weder C, Kato T (2016) Adv Mater 28:1073–1095. https://doi.org/10.1002/adma.201502589
35. Ma Z, Wang Z, Teng M, Xu Z, Jia X (2016) ChemPhysChem 16:1811–1828. https://doi.org/10.1002/cphc.201500181
36. Khan F, Ekbote A, Misra R (2019) Reversible mechanochromism and aggregation induced enhanced emission in phenothiazine substituted tetraphenylethylene. New J Chem 43:16156–16163. https://doi.org/10.1039/c9nj03290h
37. Khan F, Ekbote A, Mobin SM, Misra R (2021) Mechanochromism and aggregation-induced emission in phenanthroimidazole derivatives: role of positional change of different donors in a multichromophoric assembly. J Org Chem 86:1560–1574. https://doi.org/10.1021/acs.joc.0c02404
38. Barman D, Gogoi R, Narang K, Iyer PK (2020) Recent developments on multi-functional metal-free mechanochromic luminescence and thermally activated delayed fluorescence organic materials. Front Chem 8:483. https://doi.org/10.3389/fchem.2020.00483
39. Su J, Fukaminato T, Placial JP, Onodera T, Suzuki R, Oikawa H, Brosseau A, Brisset F, Pansu R, Nakatani K, Métivier R (2016) Giant amplification of photoswitching by a few photons in fluorescent photochromic organic nanoparticles. Angew Chem Int Ed Engl 55:3662–3666. https://doi.org/10.1002/anie.201510600
40. Melnychuk N, Klymchenko AS (2018) DNA-functionalized dye-loaded polymeric nanoparticles: ultrabright FRET platform for amplified detection of nucleic acids. J Am Chem Soc 140:10856–10865. https://doi.org/10.1021/jacs.8b05840
41. Ghodbane A, D'Altério S, Saffon N, McClenaghan ND, Scarpantonio L, Jolinat P, Fery-Forgues S (2012) Facile access to highly fluorescent nanofibers and microcrystals via reprecipitation of 2-phenyl-benzoxazole derivatives. Langmuir 28:855–863. https://doi.org/10.1021/la2036554
42. An BK, Gierschner J, Park SY (2012) π-Conjugated cyanostilbene derivatives: a unique self-assembly motif for molecular nanostructures with enhanced emission and transport. Acc Chem Res 45:544–554. https://doi.org/10.1021/ar2001952
43. Carayon C, Ghodbane A, Leygue N, Wang J, Saffon-Merceron N, Brown R, Fery-Forgues S (2019) Mechanofluorochromic properties of an AIEE-active 2-phenylbenzoxazole derivative: more than meets the eye? ChemPhotoChem 3:545–553. https://doi.org/10.1002/cptc.201800261
44. Yadav UN, Kumbhar HS, Deshpande SS, Sahoo SK, Shankarling GS (2015) Photophysical and thermal properties of novel solid state fluorescent benzoxazole based styryl dyes from a DFT study. RSC Adv 5:42971–42977. https://doi.org/10.1039/c4ra12908c
45. Bremond E, Leygue N, Jaouhari T, Saffon-Merceron N, Erriguible A, Fery-Forgues S (2021) Effect of substitution on the solid-state fluorescence properties of styrylbenzoxazole derivatives with terminal dicyanomethylene group. J Photochem Photobiol A 404:112857. https://doi.org/10.1016/j.jphotochem.2020.112857
46. Suhina T, Amirjalayer S, Mennucci B, Woutersen S, Hilbers M, Bonn D, Brouwer AM (2016) Excited-state decay pathways of molecular rotors: twisted intermediate or conical intersection? J Phys Chem Lett 7:4285–4290. https://doi.org/10.1021/acs.jpclett.6b02277

47. Willets KA, Callis PR, Moerner WE (2004) Experimental and theoretical investigations of environmentally sensitive single-molecule fluorophores. J Phys Chem B 108:10465–10473. https://doi.org/10.1021/jp049684d
48. Levine BG, Martínez TJ (2007) Isomerization through conical intersections. Annu Rev Phys Chem 58:613–634. https://doi.org/10.1146/annurev.physchem.57.032905.104612
49. Xue P, Yao B, Sun J, Zhang Z, Li K, Liu B, Lu R (2015) Crystallization-induced emission of styrylbenzoxazole derivate with response to proton. Dyes Pigment 112:255–261. https://doi.org/10.1016/j.dyepig.2014.07.026
50. Horak E, Vianello R, Hranjec M, Krištafor S, Karminski Zamola G, Steinberg IM (2017) Benzimidazole acrylonitriles as multifunctional push–pull chromophores: spectral characterisation, protonation equilibria and nanoaggregation in aqueous solutions. Spectrochim Acta A 178:225–233. https://doi.org/10.1016/j.saa.2017.02.011
51. Horak E, Hranjec M, Vianello R, Steinberg IM (2017) Reversible pH switchable aggregation-induced emission of self-assembled benzimidazole-based acrylonitrile dye in aqueous solution. Dyes Pigment 142:108–115. https://doi.org/10.1016/j.dyepig.2017.03.021
52. Jana P, Yadav M, Kumar T, Kanvah S (2021) Benzimidazole-acrylonitriles as chemosensors for picric acid detection. J Photochem Photobiol A 404:112874. https://doi.org/10.1016/j.jphotochem.2020.112874
53. Mishra A, Thangamani A, Chatterjee S, Chipem FAS, Krishnamoorthy G (2013) Photoisomerization of trans-2-[4′-(dimethylamino)styryl]benzothiazole. Photochem Photobiol 89:247–252. https://doi.org/10.1111/j.1751-1097.2012.01227.x
54. Wang L, Zheng Z, Yu Z, Zheng J, Fang M, Wu J, Tian Y, Zhou H (2013) Schiff base particles with aggregation-induced enhanced emission: random aggregation preventing π–π stacking. J Mater Chem C 1:6952–6959. https://doi.org/10.1039/c3tc31626b
55. Wang L, Shen Y, Yang M, Zhang X, Xu W, Zhu Q, Wu J, Tian Y, Zhou H (2014) Novel highly emissive H-aggregates with aggregate fluorescence change in a phenylbenzoxazole-based system. Chem Commun 50:8723–8726. https://doi.org/10.1039/c4cc02564d
56. Horak E, Robić M, Šimanović A, Mandić V, Vianello R, Hranjec M, Steinberg IM (2019) Tuneable solid-state emitters based on benzimidazole derivatives: Aggregation induced red emission and mechanochromism of D-π-A fluorophores. Dyes Pigment 162:688–696. https://doi.org/10.1016/j.dyepig.2018.10.069
57. Ma X, Xie J, Tanga N, Wu J (2016) AIE-caused luminescence of a thermally-responsive supramolecular organogel. New J Chem 40:6584–658710. https://doi.org/10.1039/c6nj01211f
58. Yao H, Wang J, Song SS, Fan YQ, Guan XW, Zhou Q, Wei TB, Lin Q, Zhang YM (2018) A novel supramolecular AIE gel acts as a multi-analyte sensor array. New J Chem 42:18059–18065. https://doi.org/10.1039/c8nj04160a
59. Li J, Qian Y, Xie L, Yi Y, Li W, Huang W (2015) From dark TICT state to emissive *quasi*-TICT state: the AIE mechanism of *N*-(3-(benzo[*d*]oxazol-2-yl)phenyl)-4-*tert*-butylbenzamide. J Phys Chem C 119:2133–2141. https://doi.org/10.1021/jp5089433
60. Qian Y, Cai M, Zhou X, Gao Z, Wang X, Zhao Y, Yan X, Wei W, Xie L, Huang W (2012) More than restriction of twisted intramolecular charge transfer: three-dimensional expanded #-shaped cross-molecular packing for emission enhancement in aggregates. J Phys Chem C 116:12187–12195. https://doi.org/10.1021/jp212257f
61. Zhao J, Ji S, Chen Y, Guo H, Yang P (2012) Excited state intramolecular proton transfer (ESIPT): from principal photophysics to the development of new chromophores and applications in fluorescent molecular probes and luminescent materials. Phys Chem Chem Phys 14:8803–8817. https://doi.org/10.1039/c2cp23144a
62. Padalkar VS, Seki S (2016) Excited-state intramolecular proton-transfer (ESIPT)-inspired solid state emitters. Chem Soc Rev 45:169–202. https://doi.org/10.1039/C5CS00543D
63. Sedgwick AC, Wu L, Han HH, Bull SD, He XP, James TD, Sessler JL, Tang BZ, Tian H, Yoon J (2018) Excited-state intramolecular proton-transfer (ESIPT) based fluorescence sensors and imaging agents. Chem Soc Rev 47:8842–8880. https://doi.org/10.1039/c8cs00185e
64. Syetov Y (2013) Luminescence spectrum of 2-(2′-hydroxyphenyl)benzoxazole in the solid state. Ukr J Phys Opt 14:1–5. https://doi.org/10.3116/16091833/14/1/1/2013
65. Li G, Zhu D, Liu Q, Xue L, Jiang H (2013) Rapid detection of hydrogen peroxide based on aggregation induced ratiometric fluorescence change. Org Lett 15:924–927. https://doi.org/10.1021/ol4000845

66. Hu R, Li S, Zeng Y, Chen J, Wang S, Li Y, Yang G (2011) Understanding the aggregation induced emission enhancement for a compound with excited state intramolecular proton transfer character. Phys Chem Chem Phys 13:2044–2051. https://doi.org/10.1039/c0cp01181a
67. Gao M, Wang L, Chen J, Li S, Lu G, Wang L, Wang Y, Ren L, Qin A, Tang BZ (2016) Aggregation-induced emission active probe for light-up detection of anionic surfactants and wash-free bacterial imaging. Chem Eur J 22:5107–5112. https://doi.org/10.1002/chem.201505202
68. Feng X, Tong B, Shi J, Zhao C, Cai Z, Dong Y (2021) Recent progress of aggregation-induced emission luminogens (AIEgens) for bacterial detection and theranostics. Mater Chem Front 5:1164–1184. https://doi.org/10.1039/d0qm00753f
69. Kim TH, Choi MS, Sohn BH, Park SY, Lyood WS, Lee TS (2008) Gelation-induced fluorescence enhancement of benzoxazole-based organogel and its naked-eye fluoride detection. Chem Commun. https://doi.org/10.1039/b800813b
70. Chen H, Feng Y, Deng GJ, Liu ZX, He YM, Fan QH (2015) Fluorescent dendritic organogels based on 2-(2′-hydroxyphenyl)benzoxazole: emission enhancement and multiple stimuli-responsive properties. Chem Eur J 21:11018–11028. https://doi.org/10.1002/chem.201500849
71. Qian Y, Li S, Wang Q, Sheng X, Wu S, Wang S, Lia J, Yang G (2012) A nonpolymeric highly emissive ESIPT organogelator with neither dendritic structures nor long alkyl/alkoxy chains. Soft Matter 8:757–764. https://doi.org/10.1039/c1sm06358h
72. Benelhadj K, Muzuzu W, Massue J, Retailleau P, Charaf-Eddin A, Laurent AD, Jacquemin D, Ulrich G, Ziessel R (2014) White emitters by tuning the excited-state intramolecular proton-transfer fluorescence emission in 2-(2′-hydroxybenzofuran)benzoxazole dyes. Chem Eur J 20:12843–12857. https://doi.org/10.1002/chem.201402717
73. Kachwal V, Krishna ISV, Fageria L, Chaudhary J, Roy RK, Chowdhury R, Laskar IR (2018) Exploring the hidden potential of a benzothiazolebased Schiff-base exhibiting AIE and ESIPT and its activity in pH sensing, intracellular imaging and ultrasensitive & selective detection of aluminium (Al3+). Analyst 143:3741–3748. https://doi.org/10.1039/c8an00349a
74. Padalkar VS, Sakamaki D, Tohnai N, Akutagawa T, Sakai K, Seki S (2015) Highly emissive excited-state intramolecular proton transfer (ESIPT) inspired 2-(2′-hydroxy)benzothiazole-fluorene motifs: spectroscopic and photophysical properties investigation. RSC Adv 5:80283–80296. https://doi.org/10.1039/c5ra17980g
75. Padalkar VS, Kuwada K, Sakamaki D, Tohnai N, Akutagawa T, Sakai K, Sakurai T, Seki S (2017) AIE active carbazole-benzothiazole based ESIPT Motifs: positional isomers directing the optical and electronic properties. ChemSelect 2:1959–1966. https://doi.org/10.1002/slct.201602044
76. Xue P, Yao B, Wang P, Sun J, Zhang Z, Lu R (2014) Response of strongly fluorescent carbazole-based benzoxazole derivatives to external force and acidic vapors. RSC Adv 4:58732–58739. https://doi.org/10.1039/c4ra10330k
77. Xue P, Yao B, Sun J, Zhang Z, Lu R (2014) Emission enhancement of a coplanar π-conjugated gelator without any auxiliary substituents. Chem Commun 50:10284–10286. https://doi.org/10.1039/c4cc04869e
78. Wu Z, Sun J, Zhang Z, Gong P, Xue P, Lu R (2016) Organogelation of cyanovinylcarbazole with terminal benzimidazole: AIE and response for gaseous acid. RSC Adv 6:97293–97301. https://doi.org/10.1039/c6ra20910f
79. Xue P, Yao B, Sun J, Xu Q, Chen P, Zhang Z, Lu R (2014) Phenothiazine-based benzoxazole derivates exhibiting mechanochromic luminescence: the effect of a bromine atom. J Mater Chem C 2:3942–3950. https://doi.org/10.1039/c4tc00061g
80. Xue P, Chen P, Jia J, Xu Q, Sun J, Yao B, Zhang Z, Lu R (2014) A triphenylamine-based benzoxazole derivative as a high-contrast piezofluorochromic material induced by protonation. Chem Commun 50:2569–2571. https://doi.org/10.1039/c3cc49208g
81. Padalkar VS, Sakamaki D, Kuwada K, Tohnai N, Akutagawa T, Sakai K, Seki S (2016) AIE active triphenylamine–benzothiazole based motifs: ESIPT or ICT emission. RSC Adv 6:26941–26949. https://doi.org/10.1039/c6ra02417c
82. Padalkar VS, Sakamaki D, Kuwada K, Horio A, Okamoto H, Tohnai N, Akutagawa T, Sakai K, Seki S (2016) π–π Interactions: influence on molecular packing and solid-state emission of ESIPT and non-ESIPT motifs. Asian J Org Chem 5:938–945. https://doi.org/10.1002/ajoc.201600159
83. Cao Y, Yang M, Wang Y, Zhou HP, Zheng J, Zhang X, Wu J, Tian Y, Wu Z (2014) Aggregation-induced and crystallization-enhanced emissions with time-dependence of a new Schiff-base family based on benzimidazole. J Mater Chem C 2:3686–3694. https://doi.org/10.1039/c3tc32551b

84. Massue J, Frath D, Retailleau P, Ulrich G, Ziessel R (2013) Synthesis of luminescent ethynyl-extended regioisomers of borate complexes based on 2-(2′-Hydroxyphenyl)benzoxazole. Chem Eur J 19:5375–5386. https://doi.org/10.1002/chem.201203625
85. Meesala Y, Kavala V, Chang HC, Kuo TS, Yao CF, Lee WZ (2015) Synthesis, structures and electrochemical and photophysical properties of anilido-benzoxazole boron difluoride (ABB) complexes. Dalton Trans 44:1120–1129. https://doi.org/10.1039/c4dt03052d
86. Zhang Z, Wu Z, Sun J, Xue P, Lu R (2016) Multi-color solid-state emission of β-iminoenolate boron complexes tuned by methoxyl groups: aggregation-induced emission and mechanofluorochromism. RSC Adv 6:43755–43766. https://doi.org/10.1039/c6ra03722d
87. Zhao J, Peng J, Chen P, Wang H, Xue P, Lu R (2018) Mechanofluorochromism of difluoroboron β-ketoiminate boron complexes functionalized with benzoxazole and benzothiazole. Dyes Pigment 149:276–283. https://doi.org/10.1016/j.dyepig.2017.10.007
88. Wang X, Liu Q, Yan H, Liu Z, Yao M, Zhang Q, Gong S, He W (2015) Piezochromic luminescence behaviors of two new benzothiazole–enamido boron difluoride complexes: intra- and intermolecular effects induced by hydrostatic compression. Chem Commun 51:7497–7500. https://doi.org/10.1039/c5cc01902h
89. Ghodbane A, Fellows WB, Bright J, Ghosh D, Saffon N, Tolbert LM, Fery-Forgues S, Solntsev KM (2016) Effects of the benzoxazole group on green fluorescent protein chromophore crystal structure and solid state photophysics. J Mater Chem C 14:2793–2801. https://doi.org/10.1039/c5tc03776j
90. Carayon C, Ghodbane A, Gibot L, Dumur R, Wang J, Saffon N, Rols MP, Solntsev KM, Fery-Forgues S (2016) Conjugates of benzoxazole and GFP chromophore with aggregation-induced enhanced emission: influence of the chain length on the formation of particles and on the dye uptake by living cells. Small 12:6602–6612. https://doi.org/10.1002/smll.201602799
91. Carayon C, André-Barrès C, Leygue N, Saffon-Merceron N, Boggio-Pasqua M, Fery-Forgues S (2021) The role of the synthetic chromophore of GFP in generating polymorphism-dependent on/off photoluminescence. Dyes Pigment (On Press). https://doi.org/10.1016/j.dyepig.2020.109119
92. Zhang T, Zhang R, Zhao Y, Ni Z (2018) A new series of N-substituted tetraphenylethene-based benzimidazoles: aggregation-induced emission, fast-reversible mechanochromism and blue electroluminescence. Dyes Pigment 148:276–285. https://doi.org/10.1016/j.dyepig.2017.09.018
93. Zhang T, Zhang R, Zhang Z, Ni Z (2016) A series of tetraphenylethene-based benzimidazoles: syntheses, structures, aggregation-induced emission and reversible mechanochromism. RSC Adv 6:79871–79878. https://doi.org/10.1039/c6ra16514a
94. Shi J, Chang N, Li C, Mei J, Deng C, Luo X, Liu Z, Bo Z, Dong YQ, Tang BZ (2012) Locking the phenyl rings of tetraphenylethene step by step: understanding the mechanism of aggregation-induced emission. Chem Commun 48:10675–10677. https://doi.org/10.1039/C2CC35641D
95. Tan G, Zhu L, Liao X, Lan Y, You J (2017) Rhodium/Copper cocatalyzed highly trans-selective 1,2-diheteroarylation of alkynes with azoles via C–H addition/oxidative cross-coupling: a combined experimental and theoretical study. J Am Chem Soc 139:15724–15737. https://doi.org/10.1021/jacs.7b07242
96. Lu Z, Liu Y, Lu S, Li Y, Liu X, Qin Y, Zheng L (2018) A highly selective TPE-based AIE fluorescent probe is developed for the detection of Ag^{+}. RSC Adv 8:19701–19706. https://doi.org/10.1039/c8ra03591a
97. Zhao N, Yang Z, Lam JWY, Sung HHY, Xie N, Chen S, Su H, Gao M, Williams ID, Wong KS, Tang BZ (2012) Benzothiazolium-functionalized tetraphenylethene: an AIE luminogen with tunable solid-state emission. Chem Commun 48:8637–8639. https://doi.org/10.1039/c2cc33780k
98. Zhang J, Bai Y, Wei Q, Cao L, Wang T, Ge Z (2020) Efficient bipolar AIE emitters for high-performance nondoped OLEDs. J Mater Chem C 8:11771–11777. https://doi.org/10.1039/d0tc02566f
99. Ma C, Xu B, Xie G, He J, Zhou X, Peng B, Jiang L, Xu B, Tian W, Chi Z, Liu S, Zhang Y, Xu J (2014) An AIE-active luminophore with tunable and remarkable fluorescence switching based on the piezo and protonation–deprotonation control. Chem Commun 50:7374–7377. https://doi.org/10.1039/c4cc01012d
100. Cui Y, Yin YM, Cao HT, Zhang M, Shan GG, Sun HZ, Wu Y, Su ZM, Xie WF (2015) Efficient piezochromic luminescence from tetraphenylethene functionalized pyridine-azole derivatives exhibiting aggregation-induced emission. Dyes Pigment 119:62e69. https://doi.org/10.1016/j.dyepig.2015.03.024

101. Jadhav T, Dhokale B, Mobin SM, Misra R (2015) Mechanochromism and aggregation induced emission in benzothiazole substituted tetraphenylethylenes: a structure function correlation. RSC Adv 5:29878–29884. https://doi.org/10.1039/c5ra04881h
102. Chen Q, Jia C, Zhang Y, Du W, Wang Y, Huang Y, Yang Q, Zhang Q (2017) A novel fluorophore based on the coupling of AIE and ESIPT mechanisms and its application in biothiol imaging. J Mater Chem B 5:7736–7742. https://doi.org/10.1039/c7tb02076g
103. Liu Y, Nie J, Niu J, Wang W, Lin W (2018) An AIE + ESIPT ratiometric fluorescent probe for monitoring sulfur dioxide with distinct ratiometric fluorescence signals in mammalian cells, mouse embryonic fibroblast and zebrafish. J Mater Chem B 6:1973–1983. https://doi.org/10.1039/c8tb00075a
104. Peng J, Ye K, Liu C, Sun J, Lu R (2019) The photomechanic effects of the molecular crystals based on 5-chloro-2-(naphthalenylvinyl)benzoxazols fueled by topo-photochemical reactions. J Mater Chem C 7:5433–5544. https://doi.org/10.1039/c9tc01084j
105. Sun J, Zheng M, Jia J, Wang W, Cui Y, Gao J (2019) New Coumarin-benzoxazole derivatives: synthesis, photophysical and NLO properties. Dyes Pigment 164:287–295. https://doi.org/10.1016/j.dyepig.2019.01.010
106. Dahal D, McDonald L, Bi XM, Abeywickrama C, Gombedza F, Konopka M, Paruchuri S, Pang Y (2017) An NIR-emitting lysosome-targeting probe with large Stokes shift via coupling cyanine and excited-state intramolecular proton transfer. Chem Commun 53:3697–3700. https://doi.org/10.1039/c7cc00700k
107. Gautam P, Maragani R, Mobina SM, Misra R (2014) Reversible mechanochromism in dipyridylamine-substituted unsymmetrical benzothiadiazoles. RSC Adv 4:52526–52529. https://doi.org/10.1039/C4RA09921D
108. Jadhav T, Dhokale B, Patil Y, Mobin SM, Misra R (2016) Multi-stimuli responsive Donor–Acceptor tetraphenylethylene substituted benzothiadiazoles. J Phys Chem C 120:24030–24040. https://doi.org/10.1021/acs.jpcc.6b09015
109. Mao S, Han X, Li C, Huang G, Shen K, Shi X, Wu H (2018) Synthesis, crystal structure, fluorescence and electrochemical properties of two Ag(I) complexes based on 2-(4′-pyridyl)-benzoxazole/$SPPh_3$ ligands. J Coord Chem 71:3330–3341. https://doi.org/10.1080/00958972.2018.1514116
110. Chai W, Hong M, Song L, Jia G, Shi H, Guo J, Shu K, Guo B, Zhang Y, You W, Chen X (2015) Three reversible polymorphic Copper(I) complexes triggered by ligand conformation: insights into polymorphic crystal habit and luminescent properties. Inorg Chem 54:4200–4207. https://doi.org/10.1021/ic502709b
111. Wang J, Poirot A, Delavaux-Nicot B, Wolff M, Mallet-Ladeira S, Calupitan JP, Allain C, Benoist E, Fery-Forgues S (2019) Optimization of aggregation-induced phosphorescence enhancement in mononuclear tricarbonyl Rhenium(I) complexes: the influence of steric hindrance and isomerism. Dalton Trans 48:15906–15916. https://doi.org/10.1039/c9dt02786f
112. Calupitan JP, Poirot A, Wang J, Delavaux-Nicot B, Wolff M, Jaworska M, Métivier R, Benoist E, Allain C, Fery-Forgues S (2021) Mechanical modulation of the solid-state luminescence of tricarbonyl rhenium(I) complexes through the interplay between two triplet excited states. Chem Eur J 27:4191–4196. https://doi.org/10.1002/chem.202005245

Authors and Affiliations

Suzanne Fery-Forgues[1] · Corinne Vanucci-Bacqué[1]

[1] SPCMIB, CNRS UMR 5068, Université de Toulouse III Paul Sabatier, 118 route de Narbonne, 31062 Toulouse cedex 9, France

REVIEW

Status and Prospects of Aggregation-Induced Emission Materials in Organic Optoelectronic Devices

Dongge Ma[1]

Received: 30 December 2020 / Accepted: 19 February 2021 / Published online: 16 March 2021

Abstract

Aggregation induced emission (AIE) luminogens (AIEgens) have great potential in the field of organic optoelectronic devices because of their highly efficient emission property in solid state. For example, high efficiency organic light-emitting diodes (OLEDs) based on AIEgens have been developed successfully. Some AIEgens also show good photovoltaic response properties for organic solar cells (OSCs) and organic photodetectors (OPDs), and lasing properties for optically pumping organic lasers (OLs). The review will cover the status and prospects of AIEgens in OLEDs, OLs, OSCs and OPDs. It is expected that AIEgens will become an important organic optoelectronic material system in the future.

Keywords Aggregation induced emission · Organic optoelectronic devices · OLEDs · OSCs · OPDs · OLs

1 Introduction

Organic semiconductors are a new kind of semiconductor material system. With their advantages of a wide range of materials, easy synthesis, large area, flexibility, and good optical, electric and magnetic properties, organic semiconductors have been used widely in research into optoelectronic devices and have made remarkable progress. Organic semiconductors have been used successfully to fabricate optoelectronic devices such as organic light-emitting diodes (OLEDs), organic solar cells (OSCs), organic photodetectors (OPDs) and optically pumping organic lasers (OLs), among others [1]. As we know, OLEDs have achieved wide applications in display

Chapter 6 was originally published as Ma, D. Topics in Current Chemistry (2021) 379: 16. https://doi.org/10.1007/s41061-021-00328-8.

✉ Dongge Ma
msdgma@scut.edu.cn

1 Center for Aggregation-Induced Emission, State Key Laboratory of Luminescent Materials and Devices, Guangdong Provincial Key Laboratory of Luminescence from Molecular Aggregates, South China University of Technology, Guangzhou 510640, China

and lighting, and the optical conversion efficiency of OSCs has also exceeded 17% [2], which has greatly promoted the development of the organic optoelectronics industry.

In organic optoelectronic devices, the organic materials used play a very important role in the performance of organic optoelectronic devices. For example, OLEDs need high efficiency, long lifetime and low cost electroluminescent materials that can make good use of excitons. For this, the first generation of fluorescent materials [3], the second generation of phosphorescent materials [4] and the third generation of thermally activated delayed fluorescence (TADF) materials [5] have been developed successfully, and show good electroluminescent properties in OLEDs. In comparison, fluorescent materials possess the advantages of long lifetime and low cost, but their efficiency is low in OLEDs. Although phosphorescent and TADF materials can realize the utilization of 100% singlet and triplet excitons, there are still the issues of severe efficiency roll-off at high luminance and blue emitters with short lifetime. Furthermore, these organic emitters have to be doped in proper hosts at precisely controlled concentrations to avoid the aggregation-caused quenching (ACQ) effects when fabricating high efficiency OLEDs. This undoubtedly increases the difficulty of the preparation process. For the application of OLs, high efficiency organic luminescence materials are more important, and these materials cannot rapidly degrade under the intense excitation found in laser cavity because the realization of laser requires high energy pumping [6]; this therefore also needs more stable organic luminescence materials. Moreover, OSCs and OPDs also need to develop more efficient organic material systems to further improve their photovoltaic performance and accelerate their journey towards industrialization [7, 8].

Aggregation-induced emission (AIE) luminogens (AIEgens) display a photophysical phenomenon in which molecular aggregates exhibit stronger emission than the single molecules, and have become an attractive alternative as organic optoelectronic materials because of their important characteristic of strong emission in solid state, which has been attributed to their highly twisted conformation and weak intermolecular interactions [9, 10]. This means that AIEgens are able to efficiently suppress exciton annihilation, thus exhibiting high photoluminescence (PL) quantum efficiency of near 100% in films, which is a very important characteristic for OLEDs to realize high efficiency electroluminescent performance in non-doped device structures, including blue and white OLEDs (WOLEDs). Not only was high efficiency obtained, efficiency roll-off at high luminance was also greatly improved in OLEDs [11–13]. In addition to their excellent application in high efficiency OLEDs, AIEgens as a gain medium have recently exhibited good lasing characteristics [14], and AIEgens also have been used successfully to fabricate OSCs [15] and OPDs [16], fully demonstrating the multi-functional optoelectronic characteristics of AIEgens [9, 13, 17].

In this review, we summarize the application of AIEgens in OLEDs, OLs, OSCs and OPDs, and highlight possible developments in the field of organic optoelectronic devices based on AIEgens with the continuous improvement in the synthesis of AIEgens and the design of related device structures.

2 Operating Principles of Organic Optoelectronic Devices

2.1 OLED Principle and Performance Parameters

An OLED is a light-emitting diode in which a thin film of organic semiconductor is deposited on a supporting substrate that emits light via the processes of electroluminescence (EL). The emission originates from the radiative recombination of electrons and holes injected from cathode and anode contacts located on opposite sides of the organic semiconductor emissive medium. Therefore, an OLED operates by bipolar injection. Obviously, the emissive color and EL efficiency are determined by the emitting molecules and device structures used.

The light emission of OLEDs includes four basic processes, shown in Fig. 1 for a three-layer device composed of an emission layer (EML) sandwiched between an electron-transporting layer (ETL) and a hole-transporting layer (HTL) [18]: (1) the injection of charge carriers (holes and electrons) from the respective electrodes, (2) the transport of charge carriers into the EML, (3) the recombination of charge carriers on emission molecules with the creation of an excited state (excitons) and possible diffusion of these excitons, and (4) the decay of excitons accompanied by the emission of photons that emerges via the transparent anode through the substrate. Therefore, EL efficiency inside the OLEDs is decided fully by these four processes.

Efficiency is an important parameter to characterize the EL performance of OLEDs. OLEDs are obviously expected to have high EL efficiency, which is key not only for energy consumption, but also for its effect on lifetime, since the ability to operate the devices at a lower input power at a given luminance decreases the ohmic heating and increases the device lifetime. One of the important efficiency parameters is the external quantum efficiency, η_{ext} (or EQE), defined by the ratio of the

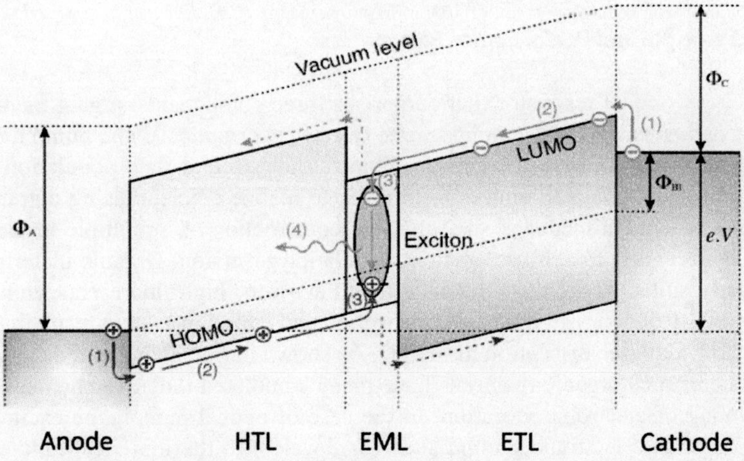

Fig. 1 Schematic of four basic processes involved in the light emission of an organic light-emitting diode (OLED). *EML* Emission layer, *ETL* electron-transporting layer, *HTL* hole-transporting layer, *LUMO* lowest unoccupied molecular orbital, *HOMO* highest occupied molecular orbital. Reproduced from [18]

number of photons emitted from OLED to the number of electrons injected. EQE is the product of the efficiencies of several processes that contribute to light emission, as follows:

$$\eta_{ext} = \eta_{\text{int}}\eta_{out} = \gamma\chi_{ST}\eta_{PL}\eta_{out} \tag{1}$$

where η_{int} is the internal quantum efficiency (i.e., the ratio of the number of photons generated to the number of injected electrons), η_{out} is the out-coupling efficiency (i.e., the fraction of the photons emitted from the front surface of the device), γ is the ratio of the number of exciton-forming events to the electrons flowing in the external circuit (basically a measure of the fraction of electrons and holes that recombine with each other), χ_{ST} is the ratio of luminescent states formed from the recombining charge carriers, and η_{PL} is the PL quantum yield.

Another parameter is the current efficiency (CE), which refers to the ratio of the luminance of the light emitted from device to the current density, and has the unit of candela per ampere (cd/A). It is simply calculated as

$$CE = \frac{LS \times 10^{-4}}{JS \times 10^{-3}} = \frac{L}{10 \times J} \tag{2}$$

Here, L is the luminance emitted from OLEDs, S is the emission area of OLEDs, J is the injection current density in OLEDs.

There is also a parameter defined as the power efficiency (PE)—the ratio of the output optical power to the electrical input with unit of lm/W—which is approximated as

$$PE = \frac{\pi LS}{IV} \tag{3}$$

Here, I is the injection current in OLEDs, and V is the applied voltage on OLEDs.

2.2 OL Principle and Performance Parameters

A laser is an optical resonator that comprises three components: a gain medium, a low loss optical cavity, and a pump source (electrical or optical). The pump intensity must be sufficient to drive the gain medium into an excited state population inversion, leading to stimulated emission. In order to realize efficient lasing output from an integrated optical feedback structure, the construction of a multiple level lasing system is necessary to achieve efficient population inversion. Organic materials are particularly suited as gain media since they have very high fluorescent yields and their Franck–Condon shifts between absorption and emission spectra result in exceptionally low material absorption losses [1]. As shown in Fig. 2, due to the quasi four-level system, such organic materials have many stimulated radiative channels under the coupling of electronic vibration. In the case of optical pumps, the excitons are excited from the electronic ground state (S_0, level 1) to the first electronic excited state (S_1, level 2) followed by a rapid non-radiative relaxation process to the lowest vibronic level of S_1 (level 3). Because of the short lifetime from level 2 to level 3 (less than 1 ps), the population inversion usually occurs from level 3 to one of the

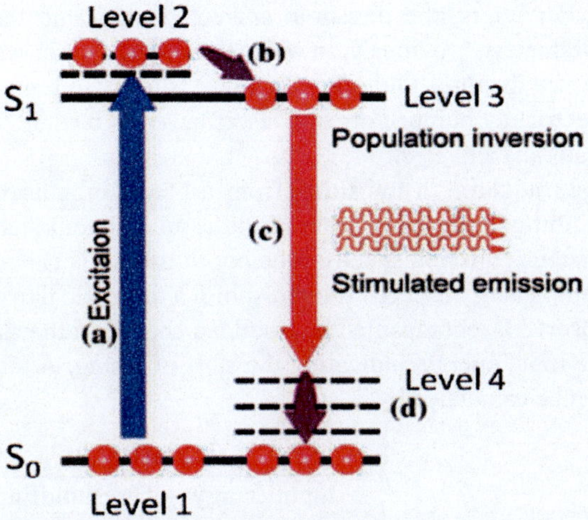

Fig. 2 **a–d** Process diagram of an organic laser with four-level system, including the corresponding optical and thermal transitions. **a** Excitation from the ground state into excited state. **b** Fast non-radiative relaxation. **c** Stimulated emission from S_1 to S_0. **d** Fast non-radiative decay to the lowest vibronic ground state

higher vibronic levels of S_0 (level 4), which is very fast for vibronic relaxation back to level 1, thus preventing the construction of significant exciton population in level 4. Considering the inevitable optical losses in gain medium and device, the inherent four-level system of organic materials is thus much more conductive to ensure initial sufficient optical gain to emit efficient lasing.

The condition for a laser to oscillate is remarkably simple: the gain due to the amplifying medium has to overcome the losses in cavity. The gain and absorption of radiation intensity per unit length are expressed by G and α, respectively, and the radiation intensity is represented by I, so

$$\frac{dI}{dx} = GI \, , \; -\frac{dI}{dx} = \alpha I \tag{4}$$

Here, G is called the gain coefficient and α the absorption coefficient. Obviously, α is related to defects in the amplifying medium (diffusion), re-absorption, reflections on optical elements inside the resonator (including the gain medium itself), whereas the value of G is dependent on the pump energy. G increases with pump energy and there is no laser emission until G is greater than α. When the gain equals the loss, the injection current density is called the threshold current density J_t, and the gain at this time is called the threshold gain G_t.

If the resonant cavity length is L and the reflection coefficient of the reflection surface is R, when the threshold is reached, then

$$G_t L = \alpha L + \ln \frac{1}{R} \tag{5}$$

Then, the threshold gain can be written as:

$$G_t = \alpha + \frac{1}{L} \ln \frac{1}{R} \tag{6}$$

For lasers, the threshold current density and threshold gain are two important parameters. Obviously, in order to make the laser work effectively, it is necessary to reduce the threshold. The main way is to try to reduce the various losses. At present, electrically pumped organic lasers have not been realized due to the serious losses of materials and electrodes, etc.

In addition, a transition from a broad spontaneous PL spectrum to a single or multimode spectrum at threshold is an essential feature of lasing. Clearly, the spontaneous emission spectrum below threshold is characteristically broad, but once the pump energy exceeds the threshold, a dramatic narrowing spectrum ($\Delta \approx 0.1$ nm) is observed, accompanied by a sudden change in the slope of the output energy against the input energy, indicating the start of lasing. A slope efficiency above threshold ρ can be calculated by:

$$\rho = \frac{\text{OuputEnergy}}{\text{InputEnergy} - \text{ThresholdEnergy}} \times 100\%. \tag{7}$$

2.3 OSC Principle and Performance Parameters

A solar cell is an optoelectronic device that converts solar irradiance into electrical energy. When the active medium used is based on organic semiconductors, then the device is called an organic solar cell.

The overall charge-generation processes of a planar heterojunction organic solar cell can be divided into four steps, as illustrated in Fig. 3 [1]: (1) absorption of light with photon energy above the optical gap and generation of excitons, η_{Abs}; (2) exciton diffusion from the bulk of the film to the D–A interface, where the exciton diffusion length determines the fraction of the layer that is actually relevant, η_{ED}; (3) exciton dissociation and formation of a CT state at the interface,

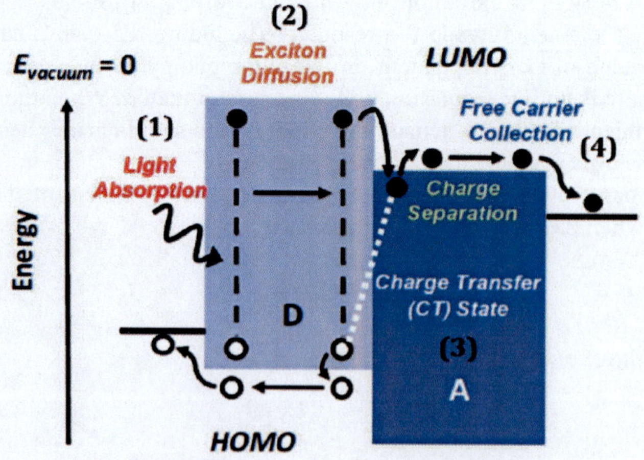

Fig. 3 Basic photoconversion processes of a planar heterojunction organic solar cell

which is still coulombically bound, η_{CT}; (4) charge separation and carrier collection at the electrodes to produce electrical power in the external circuit, η_{CC}.

Accordingly, the internal quantum efficiency is given by,

$$\eta_{int} = \eta_{Abs}\eta_{ED}\eta_{CT}\eta_{CC} \tag{8}$$

If the reflection losses for coupling light into the cell are taken into account, one obtains the EQE, which is basically the number of collected electrons with respect to the number of incident photons,

$$\eta_{ext} = (1 - R)\eta_{int} \tag{9}$$

Here, R is the reflectivity of the device. η_{ext} is determined experimentally by the measured current density at short-circuit conditions divided by the incident light intensity at a given wavelength and is also termed the incident photon-to-current efficiency (IPCE).

Organic solar cell operation can be understood from the J–V characteristics, where the open circuit voltage (V_{OC}) and short circuit current density (J_{SC}) are defined by the concepts of the *J–V* characteristics with *V* and *J* axes, respectively. Thus, the overall power conversion efficiency of a solar cell is given by:

$$\eta_P = \frac{FF\left(J_{SC}V_{OC}\right)}{P_{in}} \tag{10}$$

where FF is the fill factor, and P_{in} is the incident optical power density, preferably measured under AM1.5 sunlight conditions.

2.4 OPD Principle and Performance Parameters

A photodetector—one of the most important types of optoelectronic devices—is a transducer that converts photons to an electric current. OPDs are usually based on photovoltaic effect, so their working process is similar to that of OSCs (Fig. 3). In applications, it is required that the sensitivity of photodetectors is sufficiently high to detect weak light signal. There are two routes to improvs sensitivity, one of which is to improve the EQE and the other is to reduce the dark current density. Using photomultiplication (PM) effects to improve EQE is one of the most important approaches to achieve high-sensitivity photodetection.

An effective method for PM OPDs is based on the trap-assisted carrier tunneling effect, which consists of four steps as shown in Fig. 4 [19]: (1) energy band bending without bias, which produces a build-in electric field, thereby hindering the diffusion of holes in the organic semiconductor to the electrode; (2) after light illumination, photo-generated electrons are captured by the electron traps; (3) the trapped electrons transport toward the junction once a bias is applied; (4) the trapped electrons arriving at the junction cause the hole tunneling from the circuit into the semiconductor, finally producing the current multiplication effect.

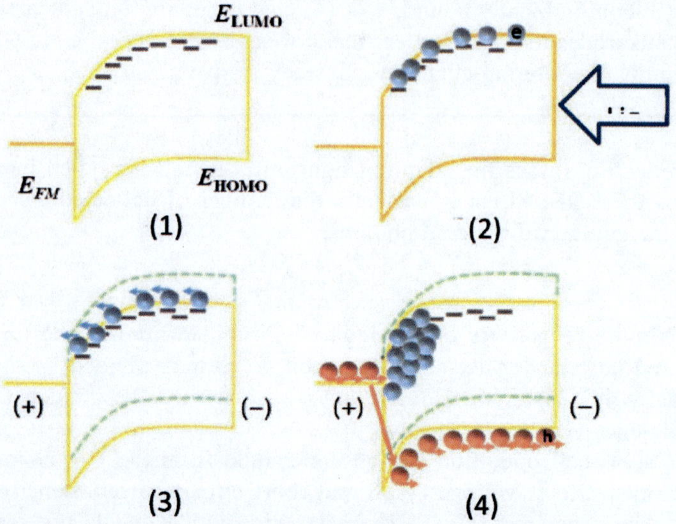

Fig. 4 Working mechanism of photomultiplication (PM) organic photodetectors (OPDs) due to the electron trap-assisted hole tunneling effect. Reproduced from [19]

To better characterize the device performance of the fabricated OPDs, a set of figures of merit are defined, including EQE, photoresponsivity, detectivity, linear dynamic range, and response time, which are listed as follows.

The EQE is defined as the electron number detected per incident photon, which can be presented by Eq. (11):

$$EQE = \frac{N_e}{N_p} = \frac{I_{ph}/q}{P_{in}/hv} \tag{11}$$

where N_e and N_p are the number of detected electrons and incident photons, respectively, h is the Planck's constant, n is the frequency of light, q is the electronic charge, and P_{in} is the incident light intensity. The absorption properties of the selected materials, the structural design of the device, and the electrical properties of the materials used are all key factors affecting EQE. For general OPDs, EQE is less than unity. However, if current gain exists, such as PM OPDs, then EQE can be greater than 1.

The photoresponsivity (R) is defined as the ratio of photocurrent to the power intensity of incident light, which characterizes the sensitivity of photodetectors to incident light. The greater the photoresponsivity is, the better the sensitivity to incident light is for PM OPDs. R can be expressed by Eq. (12):

$$R = \frac{I_{ph}}{P_{in}} = \frac{I_l - I_d}{P_{in}} = \frac{EQE \times q \times \lambda}{hc} \tag{12}$$

where I_{ph} is the photocurrent, I_l is the current under light illumination, I_d is dark current, λ is the wavelength, h is the Planck's constant, and c is the speed of light.

The detectivity (D^*) is one of the most important physical parameters to characterize the capability of weak light detection for photodetectors. It can be calculated from the noise density and the photoresponsivity R by Eq. (13):

$$D^* = \frac{R\sqrt{A\Delta f}}{i_{noise}} \tag{13}$$

where A is the active area of the detector, Δf is the electrical bandwidth, and i_{noise} is the measured total noise current. Considering that the noise current under dark is dominated by shot noise, the detectivity can be calculated by Eq. (14):

$$D^* = \frac{R}{\sqrt{2qJ_d}} \tag{14}$$

The linear dynamic range (LDR) is defined as the range of incident optical power for which the photodetectors can respond linearly. It can be calculated through Eq. (15):

$$LDR = 20\log\left(\frac{P_{max}}{P_{min}}\right) \tag{15}$$

where P_{max} and P_{min} are the maximum and minimum incident light intensities in the linear response range.

The response time presents the response speed of the photodetector in time domain, which includes the rise time (t_{raise}) and the fall time (t_{fall}). The t_{raise} and t_{fall} is defined as the time for the photocurrent to rise from 10 to 90% or fall from 90 to 10% during the on and off cycles of light illumination. Their sum is counted as the response time of the photodetector.

3 AIEgens for Organic Optoelectronic Devices

3.1 AIEgens for Blue OLEDs and as Host for Green/Orange/Red Phosphorescent OLEDs

AIEgens, which were first proposed by Tang et al. in 2001 [10], become an attractive alternative because of the superiority in terms of solid-state emission, exhibiting very high quantum yield in neat films due to their AIEgen properties. As known, the highly twisted conformation and weak intermolecular interaction of AIEgens are able to efficiently suppress the exciton annihilation in solid-state film. Therefore, AIEgens exhibit great potential as efficient pure emitter to fabricate non-doped OLEDs with high efficiency and low efficiency roll-off and will be extraordinarily prominent in OLEDs with deep blue emission due to their excellent molecular structure design and efficient solid film properties in blue light [20–23]. As we know, efficient and stable blue organic emitting light materials for OLEDs are urgently needed [12].

A typical deep blue AIEgen from triphenylethene, abbreviated as BTPE-PI, was reported by Tang et al. in 2015 [20]. It can be seen that this material showed a very high PL quantum yield of 93.8% in film, and, when used as an emitter in a simple device structure of ITO/NPB (60 nm)/BTPE-PI (20 nm)/TPBi(40 nm)/LiF (1 nm)/Al (100 nm), a pure blue emission at 463 nm with Commission Internationale de I'Eclairage (CIE) coordinates of (0.149, 0.147) was observed, and the EQE value was as high as 4.4% with low efficiency roll-off, which remained 3.4% at a luminance of 1000 cd/m^2, predicting that AIEgens are indeed promising systems for the fabrication of high efficiency blue OLEDs. The structures of the fabricated blue OLEDs and the used BTPE-PI molecule and the EQE-current density characteristic are shown in Fig. 5a, b, respectively.

Zhao et al. synthesized two novel blue AIEgens named 9-(4-(2,6-di-tert-butyl-10-(4-(1,2,2-triphenylvinyl)-phenyl)anthracen-9-yl)phenyl)-9H-carbazole (TPE-TAC) and 9-(4-(2,6-di-tert-butyl-10-(4-(1,2,2-triphenylvinyl)phenyl)anthracen-9-yl)) 1,3-di(9H-carbazol-9-l)benzene (TPE-TADC), which are shown in Fig. 6a [24]. It can be seen that they also exhibited good blue emission in an aggregated state with high quantum yields of 76.7% and 72.8%, respectively. Based on their use as emitters, high efficiency blue non-doped OLEDs were successfully fabricated by optimizing the electron-transporting layers (ETLs) (Fig. 6b). It was found that the ETLs affect the EL performance of the resulting blue OLEDs. After optimization, as shown in Fig. 6c,d, the TPE-TAC-based OLEDs emitted blue light at 452 nm with CIE coordinates of (0.159, 0.127), and efficiencies reached a maximum values of 5.11% and 3.89% at the luminance of 1000 cd/m^2, whereas the TPE-TADC-based OLEDs also showed blue emission at 451 nm with CIE coordinates of (0.165, 0.141), and efficiencies were high with a maximum of 5.71% and 4.41% at the luminance of 1000 cd/m^2. High EL efficiencies have been attributed to their good AIEgen properties.

Furthermore, using the same device structure, Ma et al. [25] fabricated high efficiency and low efficiency roll-off non-doped blue OLEDs based on another

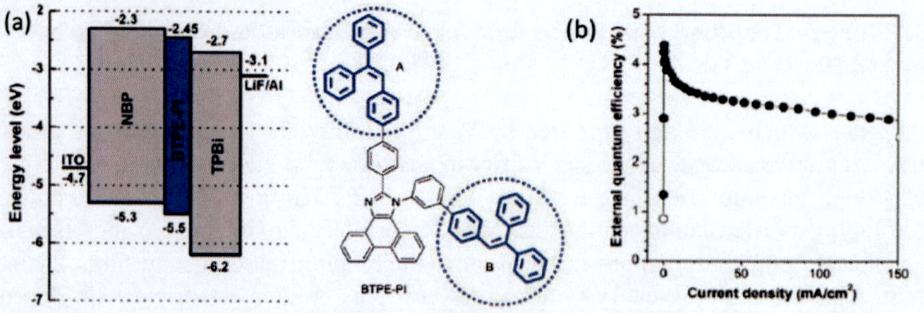

Fig. 5 **a** Structural diagrams of the fabricated blue OLED and BTPE-PI (typical deep blue AIEgen from triphenylethene) molecule. **b** External quantum efficiency (EQE)-current density characteristic of the fabricated blue OLED based on BTPE-PI. Reproduced from [20]

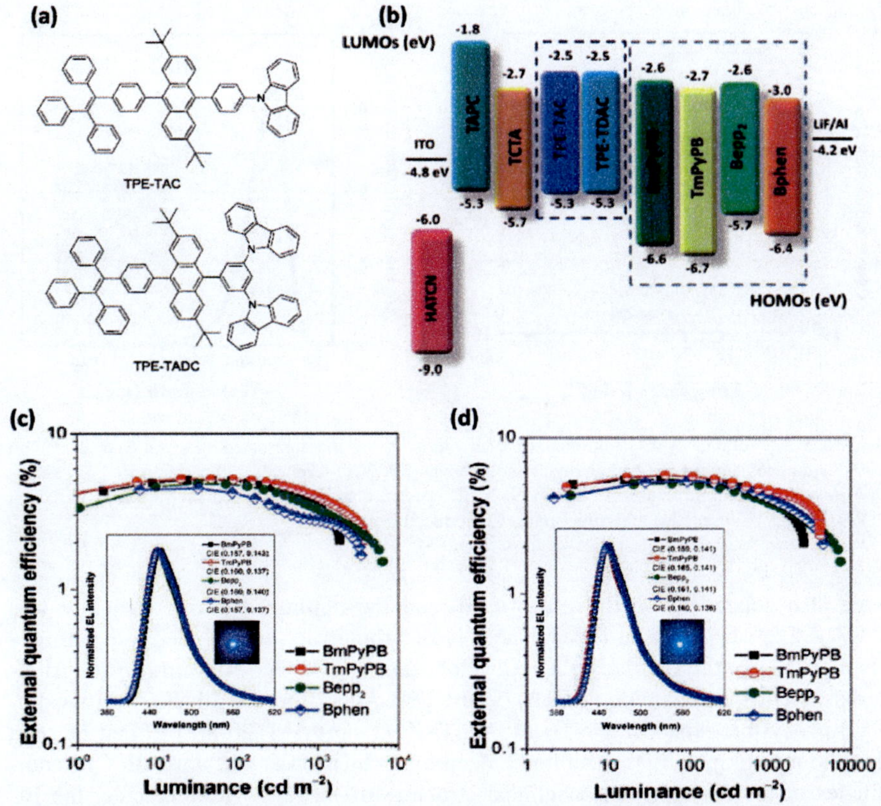

Fig. 6 **a** Chemical structures of 9-(4-(2,6-di-tert-butyl-10-(4-(1,2,2-triphenylvinyl)-phenyl)anthracen-9-yl)phenyl)-9H-carbazole (TPE-TAC) and 9-(4-(2,6-di-tert-butyl-10-(4-(1,2,2-triphenylvinyl)phenyl) anthracen-9-yl)) 1,3-di(9H-carbazol-9-l)benzene (TPE-TADC) molecules. **b** Structural diagram of the fabricated blue OLEDs. **c** EQE-luminance characteristics of the fabricated blue OLEDs based on TPE-TAC. **d** EQE-luminance characteristics of the fabricated blue OLEDs based on TPE-TADC. The insets show the electroluminescence (EL) spectra at different electron-transporting layers (ETLs) and the blue emission images. Reproduced from [24]

blue AIEgen 4′,4″-bis(diphenylamino)-5′-(4-(diphenylamino)phenyl)-[1,1′,3′,1″-terphenyl]-20-carbonitrile (3TPA-CN), shown in the inset of Fig. 7a. It can be seen that the resulting blue OLEDs not only exhibited the properties of high efficiency and low efficiency roll-off at high luminance, where the maximum efficiency reached 6.3% and remained 5.7% at the luminance of 1000 cd/m^2 (Fig. 7a), but also realized a good blue emission with CIE coordinates of (0.14, 0.11), as shown in Fig. 7b. This also indicates the potential application value of 3TPA-CN as an emitter in the fabrication of high-performance blue fluorescent OLEDs.

Since then, the maximum efficiency of AIEgen-based blue OLEDs has been enhanced to 7.0% and maintained 6.3% at the luminance of 1000 cd/m^2 using another AIEgen material 40-(4-(diphenylamino)phenyl)-50-phenyl-[1,10:20,100-terphenyl]-4-carbonitrile (TPB-AC) [11]. Similarly, the device performances

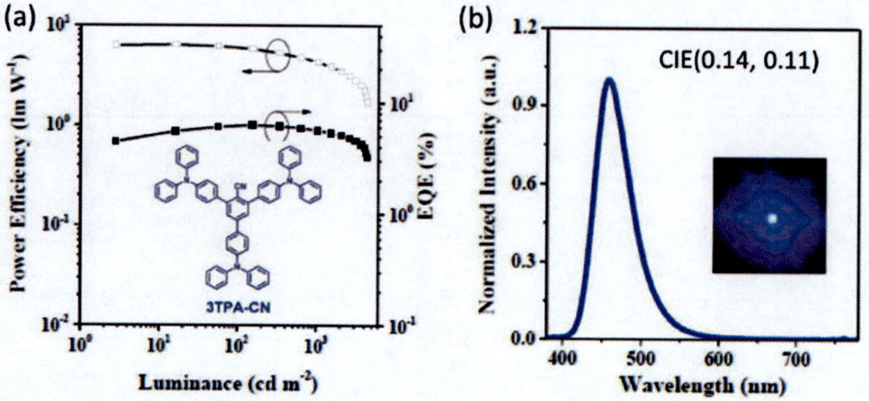

Fig. 7 **a** Power efficiency–EQE luminance characteristics of the fabricated blue OLEDs based on 4′,4″-bis(diphenylamino)-5′-(4-(diphenylamino)phenyl)-[1,1′,3′,1″-terphenyl]-20-carbonitrile (3TPA-CN). *Inset* Chemical structure of 3TPA-CN. **b** EL spectrum of the fabricated blue OLEDs based on 3TPA-CN at 5 V. *Inset* Blue emission image. Reproduced from [25]

were also dependent on the ETLs used, and the optimized device structure was ITO/1,4,5,8,9,11-hexaazatriphenylene hexacarbonitrile (HAT-CN, 5 nm)/1-bis[4-[*N*,*N*-di(4-tolyl)amino]-phenyl]cyclohexane (TAPC, 50 nm)/4,4′,4″-tri(*N*-carbazolyl)-triphenylamine (TCTA, 5 nm)/TPB-AC (20 nm)/1,3-bis[3,5-di(pyridin-3-yl)-phenyl]benzene (BmPyPB) (40 nm)/LiF (1 nm)/Al (120 nm). It can be seen that the resulting OLEDs exhibited a deep blue emission at 448 nm with CIE coordinates of (0.15, 0.08). The chemical structure of the TPB-AC molecule, the EL performances of the resulting blue OLEDs and the molecular orientation dipole

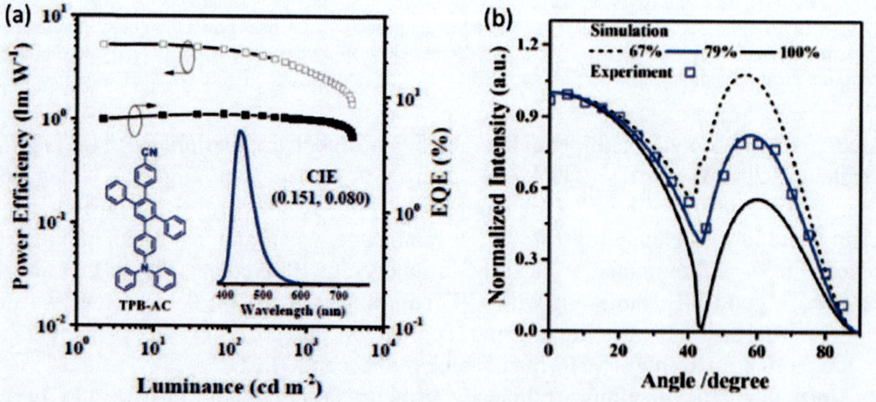

Fig. 8 **a** Power efficiency–EQE luminance characteristics of the fabricated blue OLEDs based on 40-(4-(diphenylamino)phenyl)-50-phenyl-[1,10:20,100-terphenyl]-4-carbonitrile (TPB-AC). *Inset* Chemical structure of 3TPA-CN and the EL spectrum at 4 V. **b** Measured p-polarized PL intensity at a PL peak wavelength as a function of the emission angle for TPB-AC film. Reproduced from [11]

properties of the TPB-AC film are shown in Fig. 8a, b. As we see, the EQE of the fabricated blue OLEDs exceeded the limited efficiency of fluorescence emission by 5%. Besides that, the TPB-AC film exhibited an extremely high PL quantum efficiency of 98.6%, and, more importantly, it also achieved a high horizontal dipole ratio of 79%, which should be very helpful in enhancing the light output coupling efficiency, and thus high device efficiency.

In order to further clarify the improving mechanism in EL efficiency of TPB-AC-based blue OLEDs, Ma et al. [26] studied the magneto-electroluminescence (MEL) responses of the blue OLEDs based on TPB-AC, and found a specific line shape MELs, with rise at low field and reduction at high field, which indicates a change in population between singlet and triplet excitons, thus influencing device efficiency, as shown Fig. 9a. The fact that the triplet–triplet annihilation (TTA) effect was excluded, indicates that the decrease of MELs at high magnetic field should originate from other mechanisms, and was related with the high efficiency of devices. Therefore, the singlet (S) and triplet (T) energy levels of TPB-AC were calculated and the results are plotted in Fig. 9b. Similar to traditional fluorescence molecules, the energy gap between S_1 and T_1 of TPB-AC is large enough (0.37 eV) to inhibit the reverse intersystem crossing (RISC) process from T_1 to S_1, leading to negligible contribution of T_1 excitons to the high efficiency. However, it was found that a near-zero energy gap between S_1 and T_2 exists in this material, indicating that the process of $T_2 \rightarrow S_1$ transition leads to the extra exciton utilization. According to the calculated results in Fig. 9b, the small energy gap between T_2 and S_1 in the TPB-AC molecule supplies an extra conversion process, leading more excitons on T_2 to luminescence. Generally, the MEL responses of traditional fluorescence always show monotonous positive Lorentz line shapes, it is naturally concluded that the decreased component of MELs at high magnetic field in TPB-AC-based devices is induced by the $T_2 \rightarrow S_1$ conversion.

Recently, Tang and his coworkers [27] used a novel tetraphenylbenzene (TPB)-based deep blue AIEgen CN-TPB-TPA as emitter to fabricate non-doped blue OLEDs with the structure of ITO/HAT-CN (5 nm)/TAPC (50 nm)/TCTA (5 nm)/CN-TPB-TPA (20 nm)/TmPyPB (40 nm)/LiF (1 nm)/Al. It can be seen that

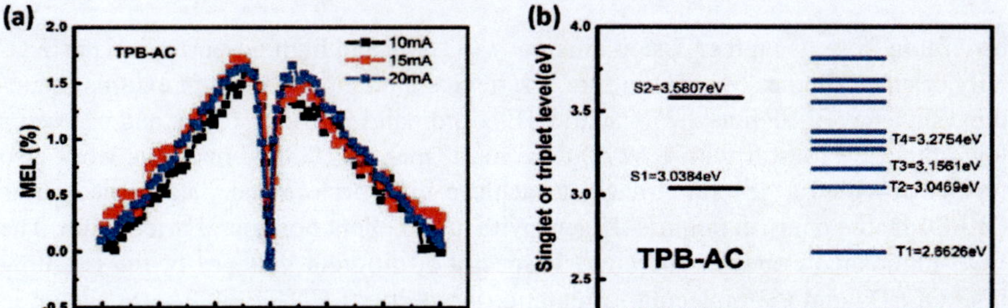

Fig. 9 **a** Magneto-electroluminescence (MEL) responses of the TPB-AC-based blue OLEDs under different applied currents. **b** The calculated energy levels of singlet and triplet states of TPB-AC by TDDFT. Reproduced from [26]

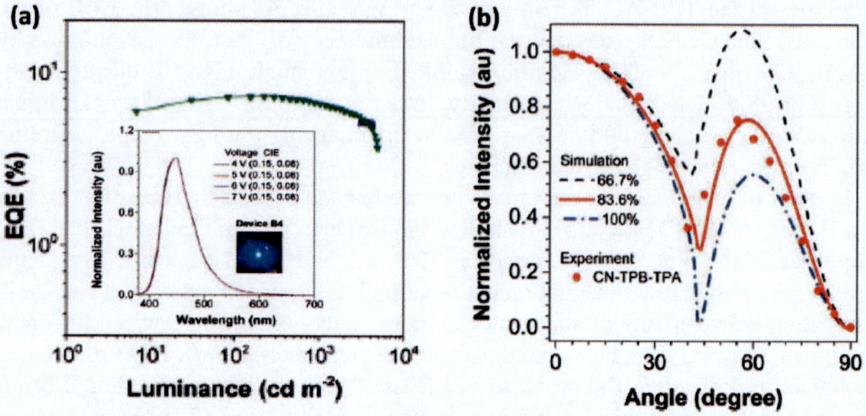

Fig. 10 **a** EQE-luminance characteristic of the fabricated blue OLEDs based on CN-TPB-TPA. *Inset* EL spectra at different voltages and EL emission image. **b** The measured p-polarized PL intensity at a PL peak wavelength as a function of the emission angle of CN-TPB-TPA film. Reproduced from [27]

Table 1 Summary of the main electroluminescence (EL) performances of the reported blue organic light-emitting diodes (OLEDs) based on aggregation induced emission (AIE) luminogens (AIEgens)

AIEgens	CE[a] (cd/A)	PE[a] (lm/W)	EQE[a] (%)	λ_{max}[b] (nm)	CIE[b] (x, y)	References
BTPE-PI	5.90/4.60	5.30/2.50	4.40/3.40	463	(0.149, 0.147)	[20]
TPE-TAC	6.19/4.29	5.81/2.32	5.11/3.89	452	(0.159, 0.127)	[24]
TPE-TADC	6.81/5.13	6.57/3.36	5.71/4.41	451	(0.165, 0.141)	[24]
3TPA-CN	6.30/5.50	6.40/4.30	6.30/5.70	457	(0.140, 0.110)	[25]
TPB-AC	5.20/4.60	5.30/3.00	7.00/6.30	448	(0.150, 0.080)	[11]
CN-TPB-TPA	5.59/5.03	5.35/2.92	7.27/6.62	448	(0.150, 0.080)	[27]

CE Current efficiency, *PE* power efficiency, *EQE* external quantum efficiency, *CIE* Commission Internationale de I'Eclairage

[a]Order of maximum, then values at 1000 cd/m^2

[b]Measured at 1000 cd/m^2

benefiting from its high emission efficiency (93.2%) and high proportion of horizontally oriented dipoles (83.6%) in film, the device emitted a maximum external quantum efficiency as high as 7.27% with CIE coordinates of (0.15, 0.08), and showed a low efficiency roll-off with 6.62% at the luminance of 1000 cd/m^2. This work also further provided a general strategy to achieve high performance, stable deep blue OLEDs by the construction of AIEgens with an excellent horizontal orientation. The EQE-luminance characteristic and EL spectra at different voltages of the resulting blue OLEDs and the molecular orientation property of CN-TPB-TPA are shown in Fig. 10.

The main EL performances of the reported blue OLEDs based on AIEgens are summarized in Table 1.

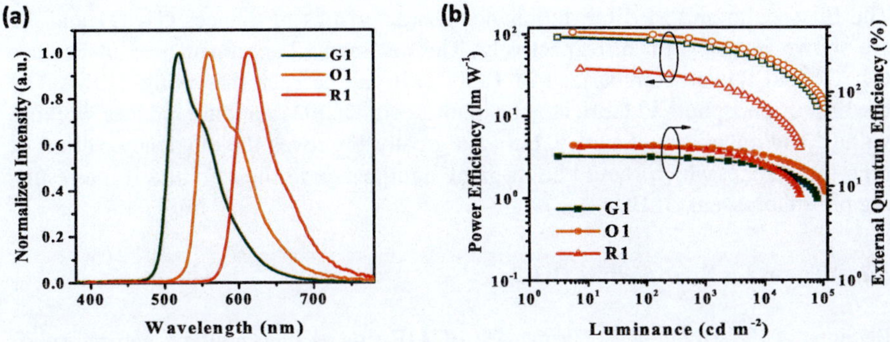

Fig. 11 **a** EL spectra and **b** PE–EQE luminance characteristics of devices G1, O1 and R1 based on TPB-AC as host. Reproduced from [11]

Table 2 Summary of the EL performances of the resulting devices G1, O1 and R1 based on TPB-AC as host. Reproduced from [11]

Device	V_{on} (V)	L_{max} (cd/m^2)	CE^a (cd/A)	PE^a (lm/W)	EQE^a (%)	λ_{max}^{b} (nm)	CIE^b (x, y)
G1	2.6	81,953	76.1/70.8	92.0/71.1	21.0/19.5	520	(0.326, 0.627)
O1	2.6	110,908	86.7/82.2	104.7/83.0	27.3/25.9	560	(0.486, 0.508)
R1	2.6	40,533	30.5/28.1	36.8/26.2	26.1/24.1	614	(0.662, 0.336)

[a]Order of maximum, then values at 1000 cd/m^2

[b]Measured at 1000 cd/m^2

As we can see, these blue AIEgens generally possess the advantages of simultaneous high PL efficiency in film, high T_1 level and good electric transport properties. This also means that these blue AIEgens should also be good host of phosphors. Recent experiments have shown that blue AIE materials are indeed very good phosphorescent hosts. Taking TPB-AC as an example, Ma and co-workers [11] examined the EL performances of green/orange/red phosphorescent OLEDs with the structure of ITO/HAT-CN (5 nm)/TAPC (50 nm)/TCTA (5 nm)/TPB-AC:phosphors (20 nm)/BmPyPB (40 nm)/LiF (1 nm)/Al (120 nm). For green device G1, 5 wt% r(ppy)$_2$(acac) doped TPB-AC was used as the emitting layer, and its maximum forward-viewing CE, PE and EQE reached 76.1 cd/A, 92.0 lm/W and 21.0%, respectively. Moreover, device G1 also showed a low EQE roll-off of 7.1% at the luminance of 1000 cd/m^2. For orange device O1, 3 wt % Ir(tptpy)$_2$(acac) doped TPB-AC was used as the emitting layer. Impressively, it showed the maximum forward-viewing CE, PE and EQE of 86.7 cd/A, 104.7 lm/W, and 27.3%, respectively, and its maximum luminance exceeded 110,000 cd/m^2 and the efficiency roll-off was as low as 5.1% at the luminance of 1000 cd/m^2. For red device R1, 3 wt % Ir(dmdpprdmp)$_2$(divm) doped TPB-AC was used as the emitting layer. A very good red emission with CIE coordinates of (0.66, 0.34) was obtained and its maximum forward-viewing CE, PE, and EQE reached, respectively, 30.5 cd/A, 36.8 lm/W, and 26.1%, with a low EQE roll-off of 7.7% at the luminance of 1000 cd/m^2, fully exhibiting the good performance of TPB-AC material as host in phosphorescent OLEDs.

The EL spectra and PE-EQE-luminance characteristics of devices G1, O1 and R1 are shown in Fig. 11a, b, respectively. The detained EL performances of devices G1, O1 and R1 are summarized in Table 2. It can be seen that using TPB-AC as the host of phosphors to fabricate phosphorescent OLEDs not only reduces working voltage and enhances efficiency, but also greatly improves the efficiency roll-off at high luminance, which should be of great significance in the stability of the resulting phosphorescent OLEDs.

3.2 AIEgens for Hybrid White OLEDs

Because of the excellent EL properties of AIEgens as blue emitters and phosphor hosts, high efficiency WOLEDs have also been manufactured and have made significant progress [24, 25, 27–32]. In 2016, Liu et al. [28] first used a blue AIEgen 9,10-bis[4-(1,2,2-triphenylvinyl)phenyl]anthracene (BTPEAn) as a fluorescent emission layer and a yellow phosphor bis(2-phenyl-4,5-dimethylpyridinato)[2-(biphenyl-3-yl)pyridinato] iridium(III) (Ir(dmppy)$_2$(dpp)) doped *N*′-di(naphthalene-1-yl)-*N*, *N*′-diphenyl-benzidine (NPB):bis[2-(2-hydroxyphenyl)-pyridine] beryllium (Bepp$_2$) bipolar host as phosphorescent emission layer to successfully fabricate two-color hybrid WOLEDs based on a multi-layer structure with spacer. The optimized device structure was ITO/HAT-CN (100 nm)/TAPC (15 nm)/NPB (5 nm)/NPB:Bepp$_2$:

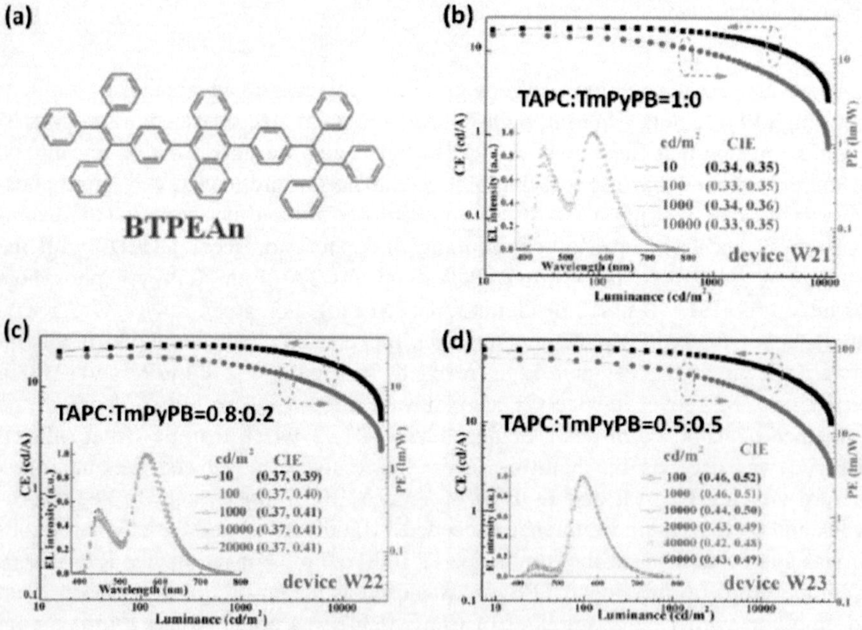

Fig. 12 **a–d** Molecular structure of BTPEAn (**a**) and EL characteristics of the resulting hybrid white organic light-emitting diodes (WOLEDs) based on BTPEAn with different ratios of TAPC:TmPyPB for **b** 1:0, **c** 0.8:0.2 and **d** 0.5:0.5. Reproduced from [28]

$Ir(dmppy)_2(dpp)$ (25 nm, 0.6:0.4:0.1)/TAPC:1,3,5-tri(*m*-pyrid-3-yl-phenyl)benzene (TmPyPB) (4 nm)/BTPEAn (10 nm)/$Bepp_2$ (40 nm)/LiF (1 nm)/Al (200 nm). Figure 12 gives the molecular structure of the blue BTPEAn and EL characteristics of the resulting hybrid WOLEDs with different ratios of TAPC:TmPyPB. It can be seen that all the devices emitted white light with a very stable spectrum over the whole luminance range, and the color-correlated temperature (CCT) was well adjusted depending on the doping ratio of TAPC:TmPyPB. When the co-doping ratio of TAPC:TmPyPB was 1:0, as shown in Fig. 12b, a hybrid WOLED (device W21) with pure white emission was well obtained, and its maximum forward-viewing CE and PE reached 19.0 cd/A and 18.8 lm/W, respectively, which correspond to the maximum total PE of 32.0 lm/W, and 29.6 and 19.4 lm/W, respectively, at 100 and 1000 cd/m^2. The achievement of pure white light emission is extremely desirable for display applications. When the co-doping ratio was changed to 0.8:0.2, as shown in Fig. 12c, the blue emission in the resulting hybrid WOLED (device W22) was reduced, resulting in a warm white emission. The maximum forward-viewing CE and PE arrived at 24.8 cd/A and 22.9 lm/W, respectively, corresponding to the maximum total PE of 38.9 lm/W. It can be seen that device W22 showed very low efficiency roll-off, reaching 38.6 lm/W and 30.4 lm/W at 100 and 1000 cd/m^2, respectively. Furthermore, when the co-doping ratio became 0.5:0.5, as shown in Fig. 12d, the warmer hybrid WOLED (device W23) was obtained and the maximum forward-viewing CE and PE reached 64.6 cd/A and 75.3 lm/W, respectively. Interestingly, at 10,000 cd/m^2, the forward-viewing CE and PE were as high as 41.3 cd/A and 25.5 lm/W, corresponding to a total CE and PE of 70.2 cd/A and 43.4 lm/W, respectively. This is an unprecedented phenomenon since no hybrid WOLED can exhibit such high efficiency at such high luminance, which should provide a new opportunity to achieve extremely high efficiency WOLEDs.

Later, they used the same AIEgen material of BTPEAn as a blue emitter and simply inserted 0.9 nm $Ir(dmppy)_2(dpp)$ and 0.3 nm $Ir(piq)_3$ ultra-thin phosphorescent layers in the TAPC or TmPyPB layer to fabricate two-color and three-color hybrid WOLEDs [29]. The device structure and the emission materials used are shown in Fig. 13. It was found that the white spectra of the resulting hybrid WOLEDs changed with the thickness of blue AIEgen emitter (0.4 nm, 1 nm, 5 nm, 10 nm and 15 nm), as shown in Fig. 14a, from pure white light to cold white light. Remarkably, their color rendering index (CRI) values were as high as 85, 84, 83, 82, and 82, respectively, which is not easy to achieve for two-color WOLEDs. These devices also emitted high efficiency, reaching CEs of 9.3, 10.0, 15.0, 18.1 and 18.4 cd/A and PEs of 10.5, 11.2, 16.8, 20.3 and 20.6 lm/W, respectively, which correspond to the total PEs of 17.9, 19.0, 28.6, 34.5 and 35.0 lm/W. It can be seen that, regardless of thin or thick film of BTPEAn, good hybrid WOLEDs can be obtained, which should contribute to the good AIEgen properties of BTPEAn material and an efficient device structure. Furthermore, the CRI can be greatly enhanced in three-color hybrid WOLEDs, and its size was dependent on whether the position of ultra-thin red and yellow phosphorescent layers in TAPC or TmPyPB and the position of blue AIEgen BTPEAn. As shown in Fig. 14, the sequence of red, yellow and blue from cathode to anode further enhanced the CRI as high as 90.3–92.8 at the high luminance of $\geq$3000 cd/m^2, and the maximum total efficiency reached 14.2 lm/W, which

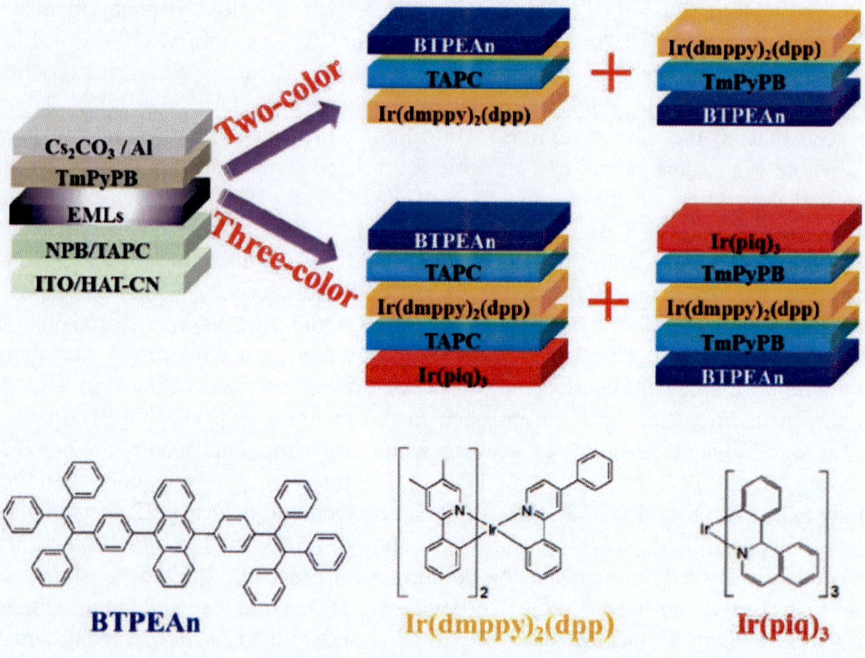

Fig. 13 Device and molecular structures of the resulting two-color and three-color hybrid WOLEDs based on BTPEAn blue AIE material with ultra-thin yellow and red phosphorescent layers. Reproduced from [29]

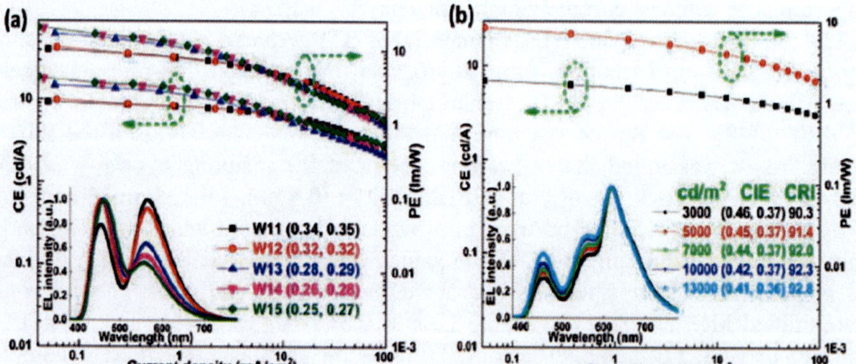

Fig. 14 **a** CE-PE-current density characteristics of the resulting two-color hybrid WOLEDs with different blue BTPEAn layer thicknesses. *Inset* EL spectra of the corresponding devices at 1000 cd/m^2 luminance. **b** CE-PE-current density characteristics of the optimized three-color hybrid WOLED based on BTPEAn. The inset gives the EL spectra, Commission Internationale de l'Eclairage (CIE) and color rendering index (CRI) of the device at different luminance. Reproduced from [29]

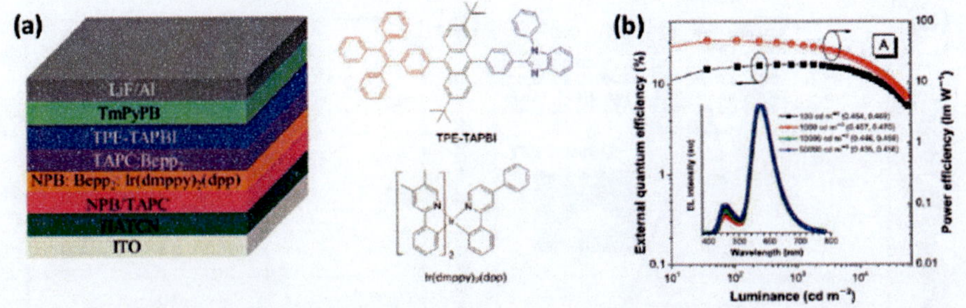

Fig. 15 **a** Device structure of the resulting hybrid WOLEDs based on PE-TAPBI and molecular structures of TPE-TAPBI and Ir(dmppy)$_2$(dpp). **b** EQE-PE-luminance characteristics of the resulting hybrid WOLEDs based on PE-TAPBI. Reproduced from [30]

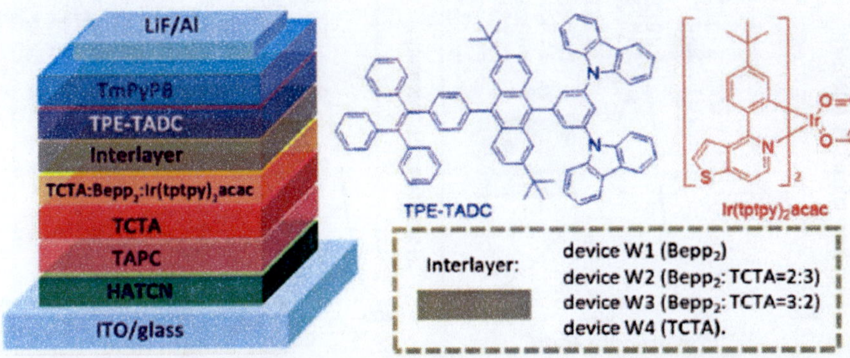

Fig. 16 Schematic of the resulting two-color hybrid WOLEDs based on TPE-TADC with different interfaces and the chemical structures of the two emitters. Reproduced from [24]

should be a higher value in hybrid WOLEDs with comprehensive performances. The CE-PE-current density characteristics and EL spectra of the three-color hybrid WOLEds at different luminance are shown in Fig. 14b.

They used still another bipolar blue AIEgen 2-(4-(2,6-di-tert-butyl -10-(4-(1,2,2-triphenylvinyl)phenyl) anthracen-9-yl)phenyl)-1-phenyl-1H-benzo[d] imidazole (TPE-TAPBI) as a blue emitter to successfully fabricate high efficiency two-color hybrid WOLEDs based on multilayer structure with spacer [30]. Among the spacer layer used the blend consisted of TAPC and Bepp$_2$, and the yellow layer was composed of a highly efficient yellow phosphorescent molecule Ir (dmppy)$_2$(DPP) doped with an NPB: Bepp$_2$ host. The device structure and the molecular structures of TPE-TAPBI and Ir (dmppy)$_2$(DPP) are displayed in Fig. 15a. It can be seen that the device exhibited stable two-color white light emission with CIE coordinates (0.457, 0.470), and also showed high efficiency properties. The maximum CE, PE and EQE efficiencies reached 76.0 cd/A, 70.5 lm/W and 28%, respectively, arriving at the efficiency of a fluorescent lamp.

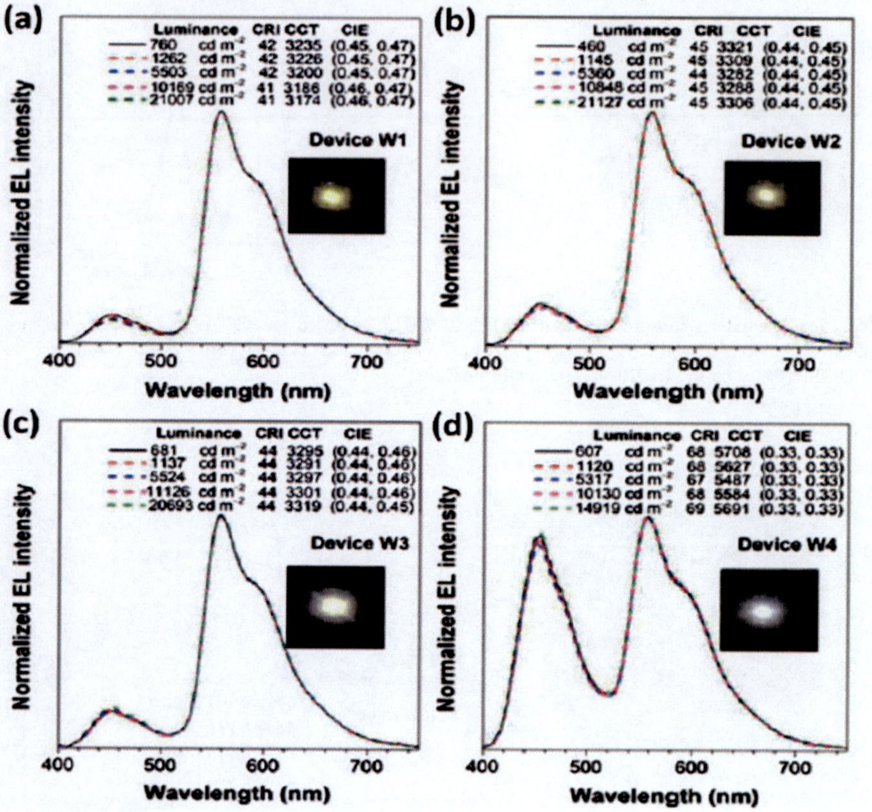

Fig. 17 EL spectra of the resulting hybrid WOLEDs (devices W1, W2, W3 and W4) based on TPE-TADC as blue emitter at different luminance. *Insets* CRI, color correlated temperature (CCT) and CIE of corresponding devices at different luminance and EL emission images. Reproduced from [24]

Zhao and coworkers have also successfully fabricated hybrid WOLEDs based on a blue AIEgen of TPE-TADC molecule by a multilayer structure with spacer [24]. The device structure and the emission molecules used are shown in Fig. 16. As shown in Fig. 17, by changing the interface between AIEgen blue fluorescent and red phosphorescent layers, the two-color hybrid WOLEDs from pure white (CIEx,y = 0.33, 0.33) to warm white (CIEx,y = 0.44, 0.46) were well obtained, and the warm hybrid WOLED emitted EL efficiencies as high as 56.7 cd/A, 55.2 lm/W, and 19.2%. More importantly, these WOLEDs all displayed ultrahigh color stability at high luminance, indicating the great potential of blue AIEgens for applications in OLED display and lighting.

Ma and coworkers used another AIEgen material 3TPA-CN as blue fluorescent emitter and Ir(tptpy)$_2$acac doped 3TPA-CN as yellow phosphorescent layer to successfully fabricate high efficiency two-color hybrid WOLEDs (devices W1 and W2) with a multilayer structure without spacer [25]. The device structure and the

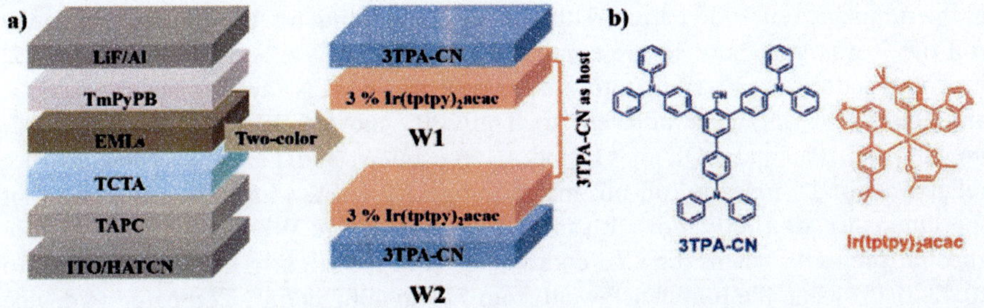

Fig. 18 **a** Device and emission layer structure scheme of the fabricated two-color hybrid WOLEDs based on 3TPA-CN. **b** Chemical structures of 3TPA-CN and $Ir(tptpy)_2acac$. Reproduced from [25]

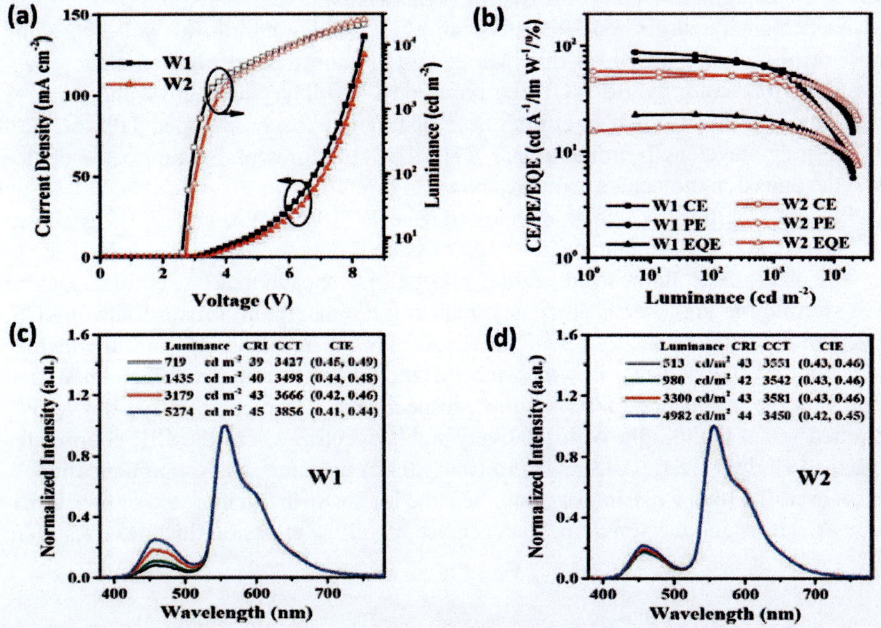

Fig. 19 **a** Current density–luminance–voltage characteristics of the fabricated two-color hybrid WOLEDs based on 3TPA-CN (devices W1 and W2). **b** PEs and EQEs as a function of luminance for devices W1 and W2. **c** Normalized EL spectra of device W1 at different luminance. **d** Normalized EL spectra of device W2 at different luminance. Reproduced from [25]

molecular structures of the emitters used are shown in Fig. 18. It was found that, although the relative position of the blue and yellow layers has little effect on the electrical characteristics, it still greatly affects the device efficiency and spectral stability. As shown in Fig. 19, by constructing the 3TPA-CN layer adjacent to TCTA and the 3 wt% $Ir(tptpy)_2acac$ doped 3TPA-CN layer adjacent to TmPyPB, device W1 exhibited a higher forward-viewing CE, PE, and EQE of 72.0 cd/A, 86.7 lm/W, and 22.3%, respectively, and the values remained 61.6 cd/A, 59.2 lm/W, and 19.6%

at the luminance of 1000 cd/m^2. Although also exhibiting a relatively low efficiency roll-off, it was still more severe compared with device W2. As shown, device W2, by constructing the 3 wt% Ir(tptpy)$_2$acac doped 3TPA-CN layer adjacent to TCTA and the 3TPA-CN layer adjacent to TmPyPB, showed relatively low forward-viewing CE, PE, and EQE of 51.8 cd/A, 56.6 lm/W, and 17.2%, respectively, but relatively good efficiency roll-off, maintaining 51.0 cd/A, 44.5 lm/W, and 16.8% at the luminance of 1000 cd/m^2. It can be seen that device W2 also showed a stable spectral property, where the CIE coordinates changed slightly from (0.43, 0.46) to (0.42, 0.45) when the luminance went from 513 to 4982 cd/m^2. Compared to device W2, device W1 exhibited a changeable spectral property with luminance going from (0.45, 0.49) at 719 cd/m^2 to (0.41, 0.44) at 5274 cd/m^2. Obviously, the exciton recombination processes in devices W1 and W2 should be different, which becomes key to the design structure of the hybrid WOLEDs fabricated here.

As demonstrated above, TPB-AC is an efficient blue emitter as well as a good host of phosphors. Based on this, Ma et al. [11] constructed high efficiency two-color and four-color hybrid WOLEDs based on a multilayer structure without spacer by using TPB-AC as the blue emitter near HTL, and phosphors-doped TPB-AC host as the long-wavelength emitters near ETL. The structures of the fabricated device and the emission molecules used are shown in Fig. 20.

For the resulting two-color devices (device W1, W2, W3, and W4), as shown in Figs. 21 and 22, the efficiencies and emission colors are dependent on the thicknesses of the blue fluorescent emitter and yellow phosphorescent emitter. Device W1 showed the highest EL efficiency, reaching the maximum forward-viewing CE, PE, and EQE of 83.5 cd/A, 99.9 lm/W, and 25.6%, respectively, and also exhibited a rather low roll-off, in which the PE and EQE yet remained 72.1 lm/W and 22.1% at the luminance of 1000 cd/m^2, respectively. It can be seen that device W1 emitted warm white light with relatively stable spectra, where the CIE coordinates changed slightly from (0.43, 0.44) to (0.41, 0.42) over the whole luminance range. Moreover, the blue emission intensity became higher with the increase of blue emission thickness and the simultaneous decrease of yellow emission thickness. As seen,

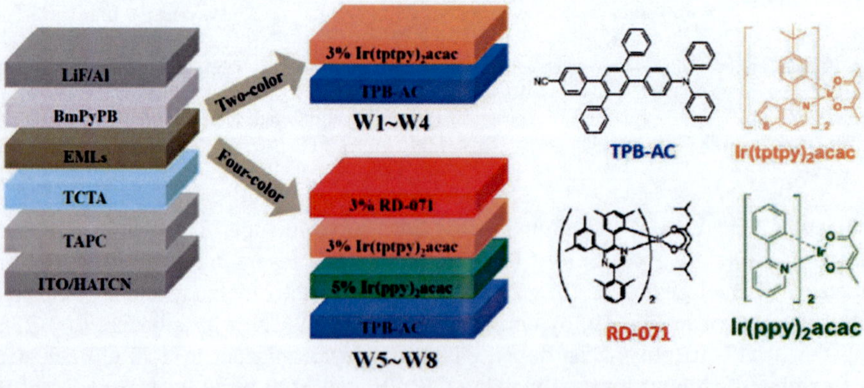

Fig. 20 Schematic diagram of the resulting two-color and four-color hybrid WOLEDs based on TPB-AC and molecular structures of TPB-AC, Ir(ppy)$_2$acac, Ir(tptpy)$_2$acac and RD-071. Reproduced from [11]

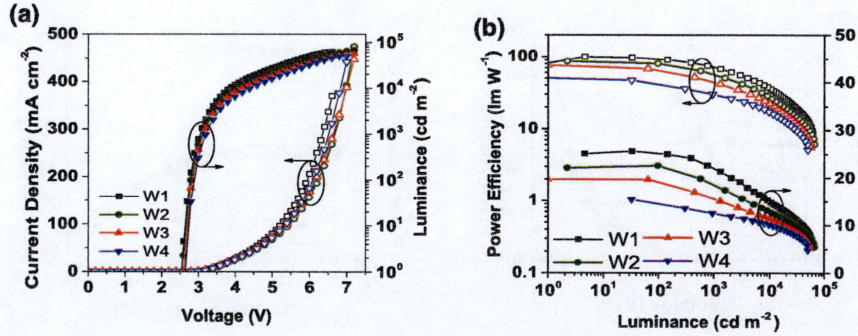

Fig. 21 **a** Current density-luminance-voltage and **b** PE–EQE-luminance characteristics of the resulting two-color hybrid WOLEDs (devices W1, W2, W3 and W4) based on TPB-AC. Reproduced from [11]

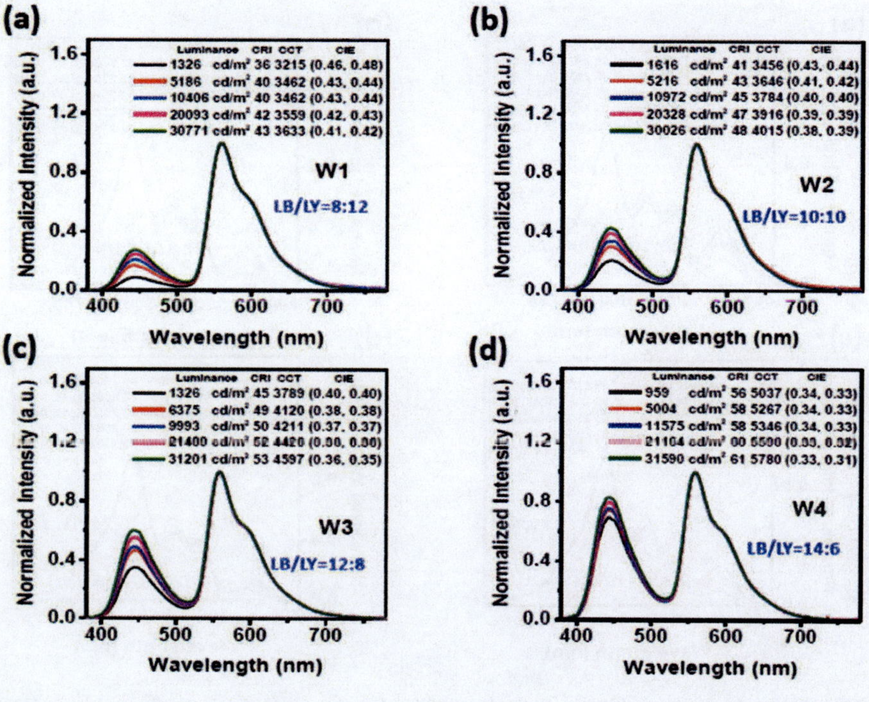

Fig. 22 EL spectra of the resulting two-color hybrid WOLEDs (devices W1 (**a**), W2 (**b**), W3 (**c**) and W4 (**d**)) based on TPB-AC for different blue and yellow layer thicknesses at different luminance. Reproduced from [11]

device W4 exhibited pure white emission with CIE coordinates of (0.34, 0.33) at the luminance of 1000 cd/m^2, which is approaching the white equivalent point of (0.33, 0.33). And device W4 also achieved a high forward-viewing PE and EQE of 47.0 lm/W and 15.4%, respectively, and fairly stable spectral profile with a small CIE coordinate variation of (0.01, 0.01) over the whole luminance range.

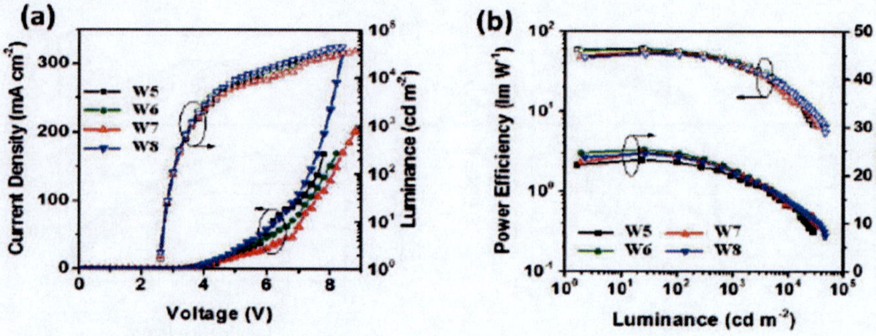

Fig. 23 **a** Current density-luminance-voltage and **b** PE–EQE-luminance characteristics of the resulting four-color hybrid WOLEDs (devices W5, W6, W7 and W8) based on TPB-AC. Reproduced from [11]

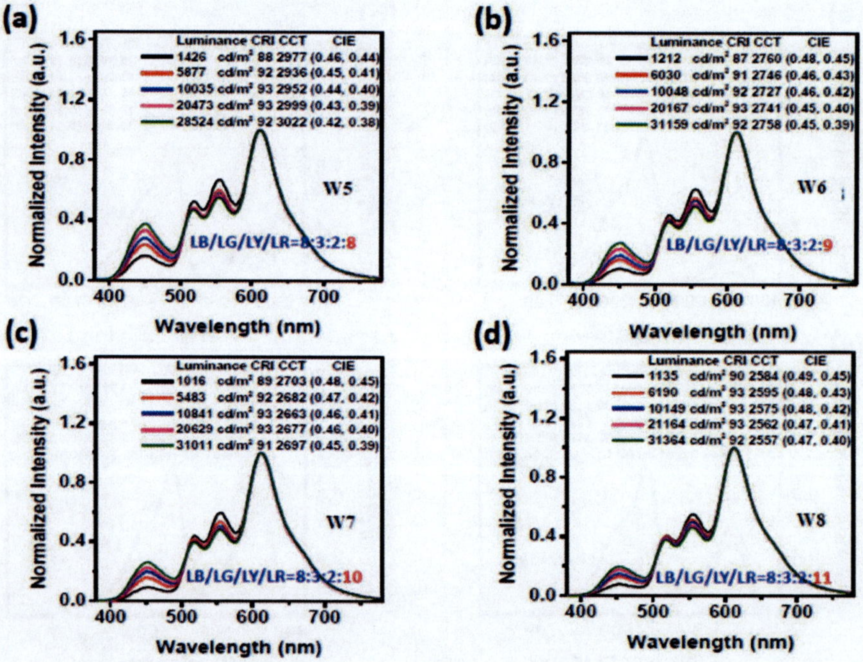

Fig. 24 EL spectra of the resulting four-color hybrid WOLEDs (devices W5 (**a**), W6 (**b**), W7 (**c**) and W8 (**d**)) based on TPB-AC for different red layer thicknesses at different luminance. Reproduced from [11]

The fabricated four-color devices (device W5, W6, W7 and W8) shown in Figs. 23 and 24 all emitted high quality warm white light with CRI of over 90 and showed high efficiency, which was changed slightly by the thickness of the red emitter to a certain extent after determining the thicknesses of blue, green and yellow layers. Comparatively, device W6 emitted relatively the best EL efficiency with CRI as high as 92 when the luminance was higher than 4000 cd/m^2. The maximum forward-viewing CE, PE and EQE reached 54.8 cd/A, 60.7 lm/W, and 25.3%, and remained 48.7 cd/A, 43.5 lm/W and 21.3% at the luminance of 1000 cd/m^2, respectively, showing a low efficiency roll-off. These should be the best results of hybrid WOLEDs, simultaneously considering high efficiency, low efficiency roll-off and high CRI above 90, and also indicates that AIEgens are promising material systems to fabricate high performance hybrid WOLEDs, which will greatly reduce the cost in future applications.

Ma and coworkers designed an even simpler structure of three-color hybrid WOLEDs by combining a blue AIEgen TPB-AC and a green AIEgen CP-BP-PXZ as non-doped blue and green fluorescent emitters, respectively, with a red phosphor Ir(dmdppr-mp)$_2$(divm) doped TCTA as red phosphorescent emitter [31]. Among the emitters used, CP-BP-PXZ possesses the properties of AIEgens with delayed fluorescence emission, called aggregation-induced delayed fluorescence (AIDF), which can realize 100% exciton utilization [32]. Therefore, it is expected to obtain high efficiency hybrid WOLEDs by a simple device structure due to the use of non-doped AIEgen emitters. The device structure and the molecular structures of the organic emitters used are shown in Fig. 25. Encouragingly, the resulting hybrid WOLEDs did show EL characteristics with good comprehensive performances. As shown in Fig. 26, the devices had a turn-on voltage as low as 2.4 V and exhibited maximum forward-viewing CE, PE and EQE of 40.6 cd/A, 50.5 lm/W and 20.5%, respectively. And at the luminance of 1000 cd/m^2, the efficiencies still remained 38.7 cd/A, 32.9 lm/W and 18.9%. Especially the spectra were relatively stable, with a small CIE coordinate variation of (0.02, 0.02) from the luminance of 1000 cd/m^2 (0.52, 0.45) to 10,000 cd/m^2 (0.50, 0.43). Importantly, the CRI of the devices reached over 90 at the whole luminance range and the

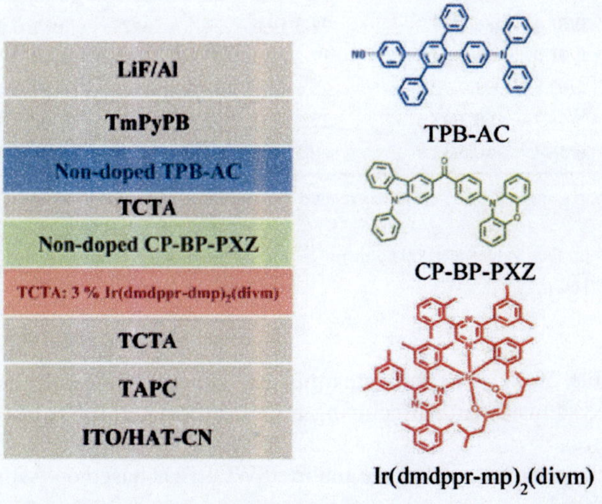

Fig. 25 Schematic diagram of the resulting three-color hybrid WOLEDs based on TPB-AC and CP-BP-PXZ, and molecular structures of TPB-AC, CP-BP-PXZ and Ir(dmdppr-mp)$_2$(divm). Reproduced from [31]

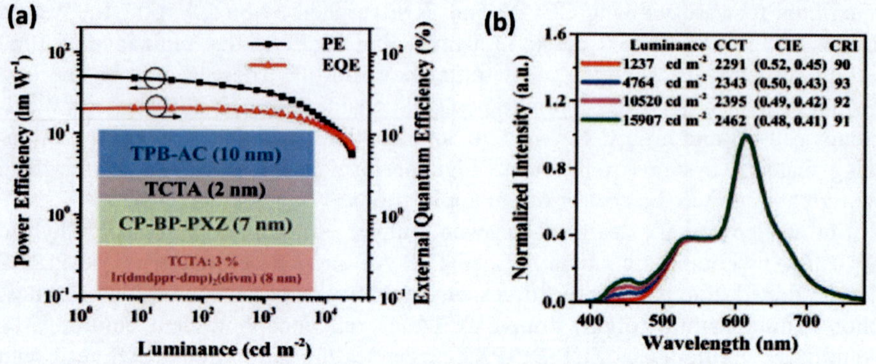

Fig. 26 **a** PE–EQE-luminance characteristics of the resulting three-color hybrid WOLEDs based on TPB-AC and CP-BP-PXZ. *Inset* Structure of emission layers. **b** EL spectra of the resulting three-color hybrid WOLEDs based on TPB-AC and CP-BP-PXZ at different luminance. Reproduced from [31]

Table 3 Summary of EL performances of the reported white organic light-emitting diodes (WOLEDs) based on AIEgens

AIEgens	CE(cd/A)[a]	PE(lm/W)[a]	EQE(%)[a]	CRI[b]	CCT[b]	CIE (x, y)[b]	References
BTPEAn	19.0/–	18.8 (32.0)/19.4	–/–	64	–	(0.34, 0.36)	[28]
	24.8/–	22.9 (38.9)/30.4	–/–	56	–	(0.37, 0.41)	
	38.0/–	75.3/50.1	–/–	46	–	(0.46, 0.51)	
BTPEAn	18.4/–	20.6 (35.0)/–	–/–	82	–	(0.25, 0.27)	[29]
	–/–	8.35 (14.2)/–	–/–	90	–	(0.46, 0.37)	
TPE-TAPBI	48.2/48.17	47.9/41.5	16.5/16.5	–	3170	(0.46, 0.47)	[30]
TPE-TADC	51.8/46.7	56.2/40.9	24.7/20.8	93	2703	(0.48, 0.45)	[24]
3TPA-CN	72.0/61.6	86.7/59.2	22.3/19.6	39	3427	(0.44, 0.48)	[25]
	51.8/51.0	56.6/44.5	17.2/16.8	43	3551	(0.44, 0.46)	
TPB-AC	83.5/69.8	99.9/72.1	25.6/22.1	40	3462	(0.46, 0.48)	[11]
	54.8/48.7	60.7/43.5	25.3/21.3	92	2727	(0.48, 0.45)	
TPB-AC	40.6/38.7	50.5/32.9	20.5/18.9	90	2290	(0.52, 0.45)	[31]

CIR Color rendering index, *CCT* color correlated temperature, *CIE* Commission Internationale de I'Eclairage

[a]Order of maximum, then values at 1000 cd/m^2

[b]Measured at 1000 cd/m^2

maximum value of 93. This indicates that AIEgen molecules are indeed promising material systems for manufacturing high-performance WOLEDs with simple device structures.

The main EL performances of the reported WOLEDs based on AIEgens are summarized in Table 3.

3.3 AIEgens for OLs

OLs have attracted considerable attention because of their unique features of high mechanical flexibility, facile doping of gain materials, high optical quality, simplicity and low-cost fabrication, large area flexibility, etc. However, traditional organic materials as gain medium usually suffer from ACQ, preventing their application in OLs. In contrast, the high efficiency property of AIEgens in solid state can solve the above problems, and it is indicated that they should be good laser medium materials.

Recently, the research groups Yang and Tang worked together to successfully demonstrate a microlaser from an AIEgen material, i.e., tetraphenylethene-containing BODIPY derivative (TPE-BODIPY) [14]. The resulting microlaser was obtained instantly via a surface tension-induced self-assembly approach, i.e., co-aggregation of both AIEgen material TPE-BODIPY and epoxy resin to form microspheres, as shown in Fig. 27. Thus, when a nanosecond pulsed laser (480 nm, 5 ns, 20 Hz) was used to pump the microsphere, an obvious lasing emission was achieved, which is attributed to whispering gallery mode (WGM).

Figure 28a shows the typical measured power dependent spectra of the microlaser with a 59.5 μm diameter. It can be seen clearly that, with increasing pump power, the multiple lasing peaks grew rapidly beyond the lasing threshold. As shown in Fig. 28b, correlating the power dependences of fluorescence intensity and full width at half maximum (FWHM) of fluorescence peaks as a function of excitation power, a clear super-linear curve and a sharp decrease of FWHM were obtained, demonstrating the transition from PL to lasing. The threshold was as low as 1.33 mJ/cm^{-2}, and the quality factor was estimated to be ≈3090 while the lasing peak width was narrow to 0.176 nm at 544.1 nm, which reveals the high quality of the fabricated microlaser. Besides, the lasing modes of the microlaser were also dependent on the size of microsphere, as shown in Fig. 28c, with increasing microsphere size, a clear red-shift of lasing spectra and the increase of free spectral range (FSR) can be readily obtained, which was reasonable in recalling the re-absorption effects of organic materials. Furthermore, by decreasing microsphere diameters, one could observe a single-mode lasing as long as

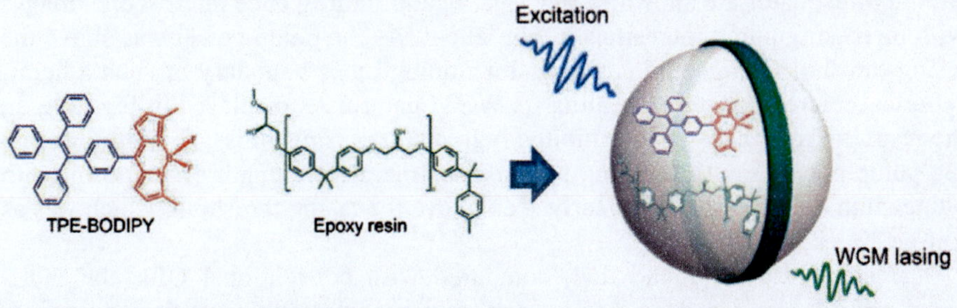

Fig. 27 Molecular structures of TPE-BODIPY and epoxy resin, and schematic of the formed microsphere. Reproduced from [14]

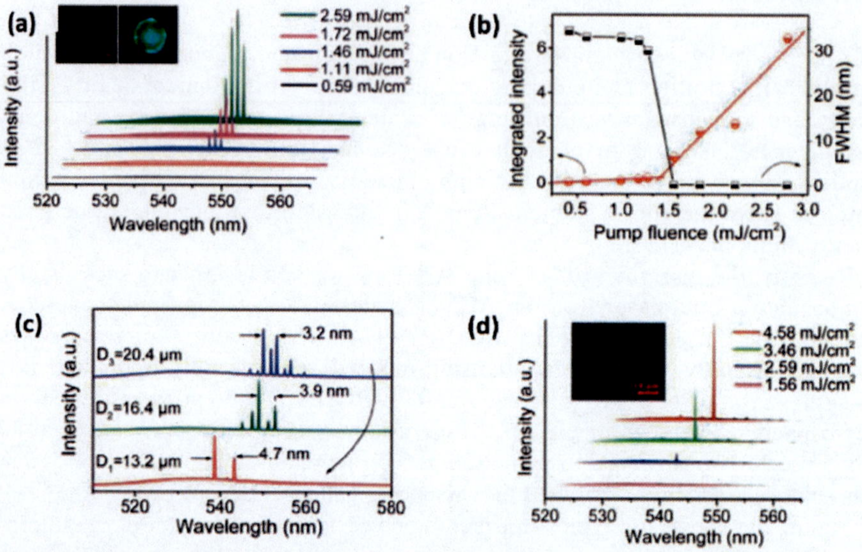

Fig. 28 **a–d** Whispering gallery mode (WGM) lasing characteristics of the resulting microlaser based on TPE-BODIPY. **a** PL spectra of the microlaser with a 59.5 μm diameter as a function of pump fluence. *Inset* Fuorescent image of the lasing microsphere. **b** Full width at half maximum (FWHM) and integrated output intensity as a function of pump fluence. **c** Typical lasing spectra of the microlaser with different diameters. **d** Single-mode lasing spectra of the microlaser with a 8.9 μm diameter. *Inset* Corresponding fluorescent image. Reproduced from [14]

FSR exceeded the gain spectrum range. As shown in Fig. 28d, the single-mode lasing at 538.2 nm for a 8.9 μm diameter microlaser was indeed observed experimentally. The resulting single-mode microlaser exhibited a threshold of 2.24 mJ/cm^2, with FWHM ≈ 0.2 nm.

Zhang et al. [33] also fabricated the same microlaser by using a triphenylamine (TPA)-modified TPAN derivative (BPMT) AIEgen molecule doped in epoxy resin to self-assemble into microsphere on a distributed Bragg reflector (DBR) mirror surface. The pump light source was still a nanosecond laser (351 nm, 10 ns duration, 100 Hz). Bright-field optical microscopy images of the typical hemispherical microresonator are shown in Fig. 29a, b, and fluorescence microscope images with increasing pump intensities in Fig. 29c–f. As the pump power was above the lasing threshold, one could easily find a shining outer boundary of such a hemispherical microresonator, revealing its WGM nature. As displayed in Fig. 29g, h, the resulting microlaser also exhibited high quality lasing emission. With increasing pump power, multiple lasing peaks at the main wavelength of 735.2 nm with a linewidth of 0.23 nm were clearly seen above the lasing threshold, which was as low as 22.3 kW/cm^2.

The above studies found that, compared with conventional OLs, the fabricated microlasers based on AIEgens as gain medium exhibited the advantages of not only low threshold, but also favorable lasing stability, which contributed to the highly efficient emission property of AIEgens in solid state film, thus

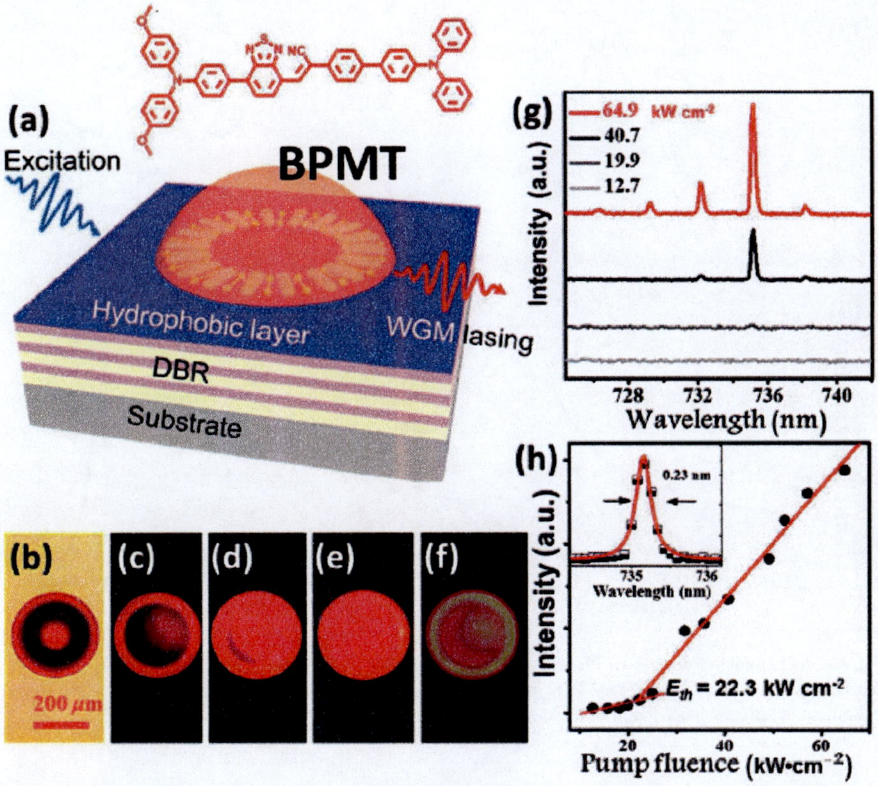

Fig. 29 **a–f** WGM lasing characteristics from the BPMT-doped hemispherical microresonators. **a** Schematic of the optically pumped hemispherical microresonators on a distributed Bragg reflector (DBR) substrate and molecular structure of BPMT. **b** Bright-field and (**c–f**) fluorescence images of the hemisphere with a 36 mm diameter with increasing pump intensities. *Bar* 200 mm. **g** Collected spectra with pump intensities at 12.7, 19.9, 40.7, and 64.9 kW/cm^2. **h** Integrated output intensity as a function of the pump intensity. *Inset* Lorentz fitting (*red line*) of a measured lasing spectral peak mode at 735.2 nm. Reproduced from [33]

making higher concentration doping possible. This provides an alternative way to improve lasing performance of OLs. It is believed that low threshold and high stability OLs will be further developed from the abundant AIEgens.

3.4 AIEgens for OSCs

The excellent luminescent properties of AIEgens also indicate their good photovoltage effect, and it is predicted that high efficiency OSCs should be able to be fabricated by using them in OSCs. In this regard, some OSCs based on AIEgens, not only as active layer, but also as interface layer, have been fabricated successfully and indeed exhibited better efficiency [15, 34–36].

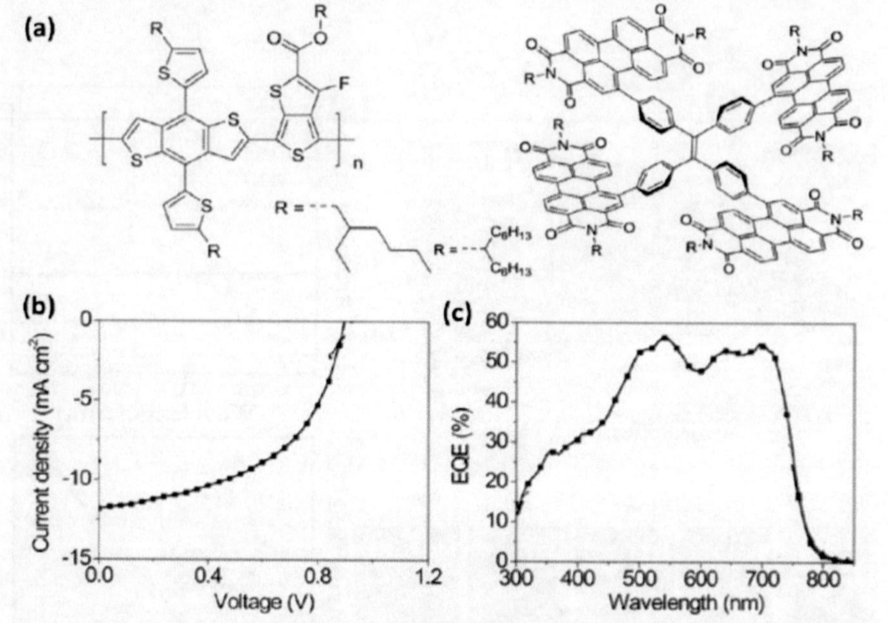

Fig. 30 **a** Chemical structures of PBDTT-F-TT and TPE-PDI_4. **b** *J-V* characteristic of the resulting OSC fabricated from PBDTT-F-TT:TPE-PDI_4 under air mass (AM) 1.5G irradiation at 100 mW/cm^2. **c** External quantum efficiency spectrum of the device. Reproduced from [34]

Yan et al. [34] synthesized a TPE core-based 3D structure small molecular acceptor (TPE-PDI_4) with AIEgen properties, and blended it with PBDTT-F-TT to fabricate OSCs. The power conversion efficiency of the resulting device reached 5.53% with V_{OC} of 0.91 V, J_{SC} of 11.7 mA/cm^2 and FF of 0.52. The molecular structures of TPE-PDI_4 and PBDTT-F-TT, and the photovoltaic characteristics of the resulting OSC are shown in Fig. 30. It can be seen that, due to the highly twisted nature of the TPE core, TPE-PDI_4 exhibited weak molecular aggregation, and formed smooth and amorphous neat films featuring a small domain size, while the 3D structure of TPE-PDI_4 also facilitated the formation of a 3D charge-transport network and thus enabled a reasonably good electron-transport ability (electron mobility of TPE-PDI_4 was as high as 1×10^{-3} cm^2/Vs). All these superior features combined are especially conducive to the achievement of such high efficiency, which also offers a new design approach to develop small molecule materials with 3D structures for high efficiency non-fullerene OSCs.

Furthermore, these authors systematically studied the impact of intramolecular twisting of 3D molecular acceptors on the performance of non-fullerene OSCs [35]. A comparison of the three AIEgens of TPC-PDI_4, TPE-PDI_4 to TPPz-PDI_4, and the corresponding *J–V* and EQE spectral characteristics of the optimized devices are shown in Fig. 31. The experimental results demonstrated that TPPz-PDI_4 showed a lower solubility, stronger aggregation, and thus higher electron mobility than the other two materials. As a result, OSCs based on TPPz-PDI_4 realized higher fill

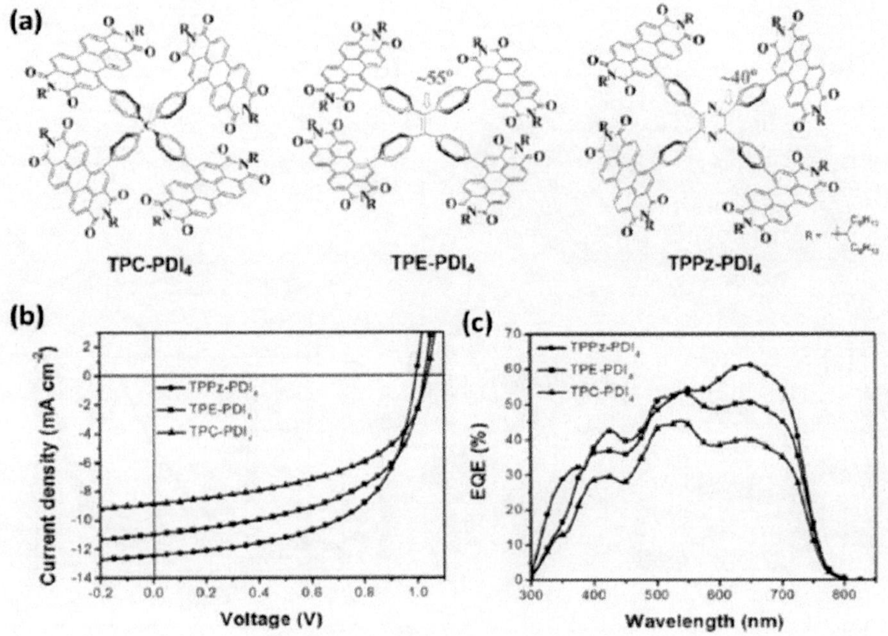

Fig. 31 **a** Chemical structures of TPC-PDI$_4$, TPE-PDI$_4$ and TPPz-PDI$_4$. **b** *J*–*V* characteristic of the resulting three OSCs under AM 1.5G irradiation at 100 mW/cm^2. **c** EQE spectra of the three devices. Reproduced from [35]

factor (0.56) and efficiency (7.1%), which should be attributed to its reduced extent of intramolecular twisting. The study revealed an important structure–property relationship of this class of materials and provided an important guidance in the design and development of high-performance small molecule materials for non-fullerene OSCs.

AIEgen units can also be used to construct effective interface materials for OSCs. As we know, the interface plays a very important role in the performance of OSCs, in which water/alcohol soluble organic cathode interfacial materials are among the most effective material systems [36]. However, possible π–π stacking exists in these materials as well as the derived effect of ACQ in the aggregation state, thus affecting device performance. Li et al. [37] synthesized a type of AIEgen-based cathode interface material by introducing 3-(trimethylammonio)propane-1-sulfonate chain with immobile counter ions to a TPE unit—the famous AIEgens—and used them as an interface layer between aluminum cathode and active layer to PC$_{71}$BM:PTB7-based conventional OSCs. It can be seen that the typical characteristic of the much twisted conformation of AIEgens effectively avoided the possible π–π stacking and the derived effect of ACQ in the aggregation state. As a result, J_{SC}, V_{OC} and FF of the resulting OSCs were improved dramatically, leading to much enlarged power conversion efficiency. For example, by introducing TPE-2 as the interface layer, the power conversion efficiency was enhanced to 8.94%, which was much higher than 3.89% (bare Al cathode), 7.31% (Ca/Al cathode), 7.93% (MeOH/Al), and 8.33%

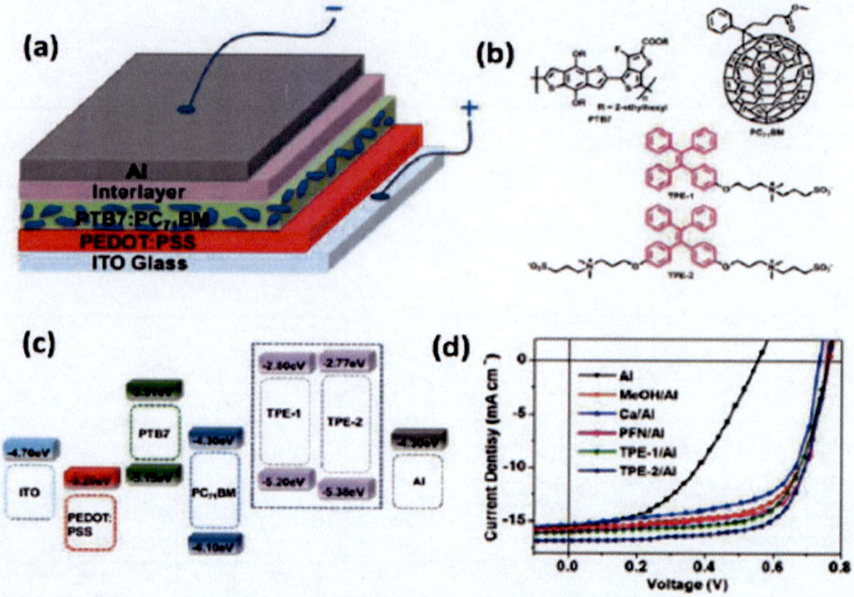

Fig. 32 **a** Structure diagram of the resulting OSCs based on AIEgens as interface layer. **b** Molecular structures of TPE derivative interface layers, $PC_{71}BM$ and PTB7. **c** Energy level diagram of the devices. **d** *J-V* curves of the devices through different treatments under illumination of 100 mW/cm^2 and AM 1.5 G. Reproduced from [37]

(PFN/Al). This should be a relative high efficiency for $PC_{71}BM$:PTB7-based conventional OSCs and confirmed that TPE derivatives could be a new kind of high-efficiency interface material. The device structure used, the molecular structures of the materials used, the relative energy levels, and the *J–V* characteristics of the devices through different treatments are shown in Fig. 32.

Recently, Wei and co-workers [15] found that incorporating a 3D AIEgen, TPE in a J71:ITIC blend can effectively modify the phase separation between J71 (thiophenediyl[4,8-bis[5-(tripropylsilyl)-2-thienyl]benzo[1,2-b:4,5-b′] dithiophene-2,6-diyl]-2,5-thiophenediyl]) and ITIC (3,9-bis(2-methylene-(3-(1,1-dicyanomethylene)-indanone))-5,5,11,11-tetrakis(4-hexylphenyl)-dithieno [2,3-d:20,30 -d′]-s-indaceno[1,2-b:5,6-b′]dithiophene), thus greatly enhancing the power conversion efficiency of the resulting ternary OSCs. By optimizing the doping concentration of TPE, the power conversion efficiency of the resulting ternary OSCs reached 12.16%, which enhanced 21.23% over that of the binary device. The molecular structures of the materials used, and the *J–V* and EQE spectra characteristics of both binary and ternary devices are shown in Fig. 33. Further experimental investigations indicated that the addition of TPE in the binary blends not only increased the electron and hole mobility of the ternary system, but also led to a system with better overall crystallinity compared to binary devices. More importantly, in the ternary blending films, the majority of TPE arranged itself within J71 domains due to high miscibility with J71. Therefore, a

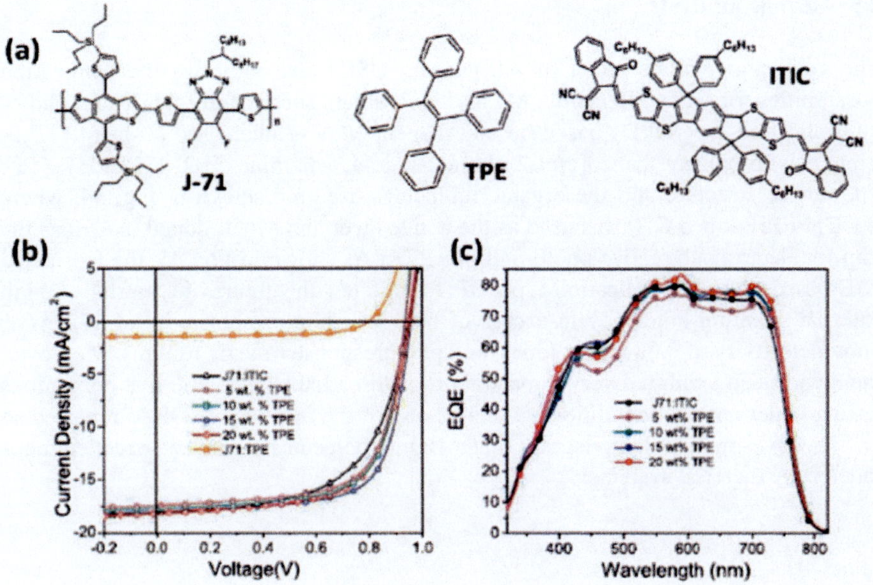

Fig. 33 **a** Chemical structures of J-71, TPE and ITIC. **b** *J–V* characteristics of the resulting both binary and ternary OSCs. **c** EQE spectra of the resulting both binary and ternary devices. Reproduced from [15]

re-organization occurred at the BHJ interface such that the TPE inclusion pushed the ITIC away from the mixed amorphous phase in the ternary blend, ultimately leading to a relatively purer acceptor domain. This rearrangement of the active layer components was the main reason for the enhanced device performance in ternary cells, also demonstrating the advantage of AIEgens with 3D structure as an interface layer in adjusting the film morphology and performance in OSCs.

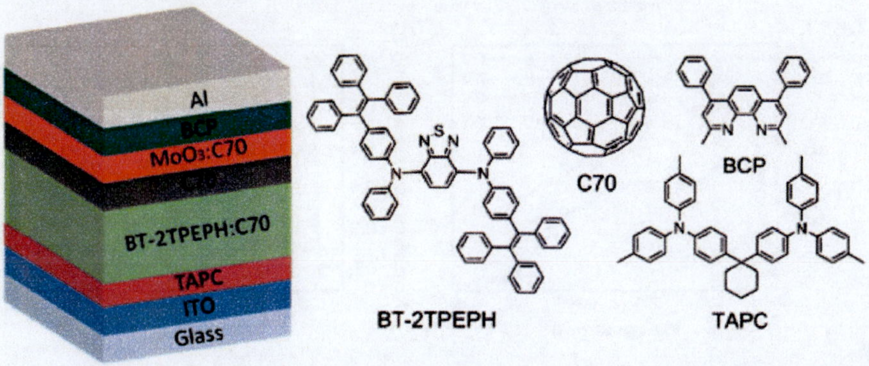

Fig. 34 Schematic of device structure and chemical structures of BT-2TPEPH, C70, BCP and TAPC. Reproduced from [16]

3.5 AIEgens for OPDs

The good photovoltaic effect of AIEgens in OSCs also indicates the application possibilities for OPDs. Recently, Ma and co-workers successfully fabricated photomultiplication-type OPDs based on an AIEgen, *N*[4],*N*[7]-diphenyl-*N*[4],*N*[7]-bis(4-(1,2,2-triphenylvinyl)phenyl)benzo[c][1,2,5]thiadiazole-4,7-diamine (BT-2TPEPH) [16]. The device structure and the organic molecules used are shown in Fig. 34, where BT-2TPEPH doped C70 was used as the active layer and MoO_3 doped in C70 as the trap for electrons to realize photomultiplication. As shown in Fig. 35, the fabricated AIE-based photomultiplication-type OPD exhibited the figures of merits of high external quantum efficiency in excess of 60,000%, photoresponsivity of 172 A/W, photodetectivity of 3.08×10^{12} Jones, and photoresponse time of 1.69 ms. Moreover, the device also exhibited very good stability, with a half-life of 700 h at continuous testing under ambient conditions, which is one of the most stable OPDs reported so far. The investigation demonstrates that AIEgen molecules are indeed excellent photodetective material systems.

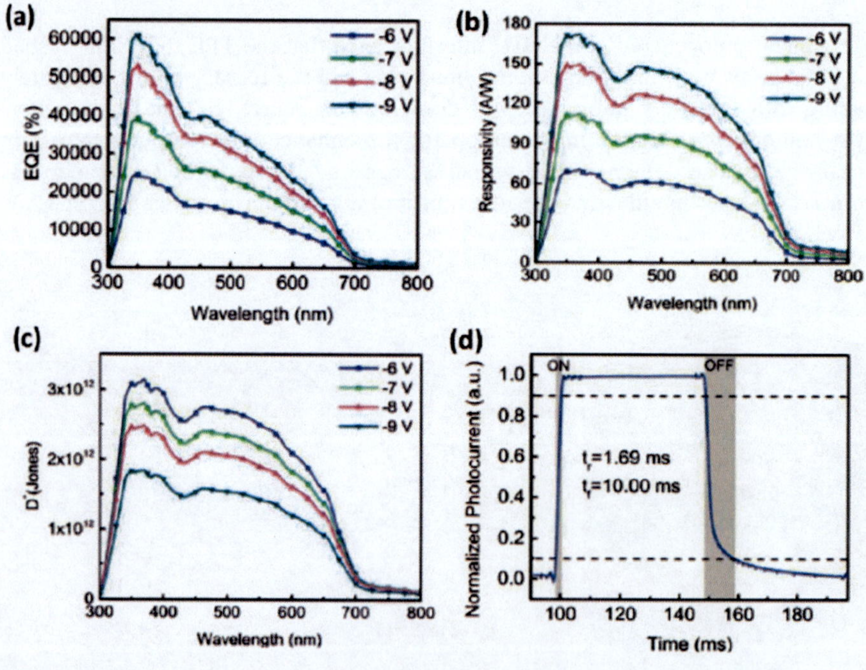

Fig. 35 **a** EQE, **b** responsivity, and **c** specific detectivity characteristics of the resulting AIE-based OPDs at different voltages. **d** Transient photocurrent response (*J–T* curve) at a pulse frequency of 10 Hz and a –6 V bias with a device area of 0.16 cm^2. Reproduced from [16]

4 Conclusion and Outlook

AIEgens have become an important research topic, with optoelectronic devices based on AIEgens becoming an important application field due to their high efficiency emission properties in film. AIEgens clearly exhibit great potential in the development of high performance organic optoelectronic devices, including OLEDs, OLs, OSCs and OPDs due to their excellent multi-functionality. As emitters and hosts, not only high-efficiency blue non-doped fluorescent OLEDs and high-efficiency green/orange/red phosphorescent OLEDs have been developed successfully, but also high efficiency hybrid WOLEDs have been well realized. And, more importantly, the resulting OLEDs echibit extremely low efficiency roll-off at high luminance, which is completely different from the organic emitters with an ACQ feature. As gain medium in OLs, the lasing threshold and stability were greatly improved, which establishes a material foundation for the development of electrically pumped OLs. As we can see, highly efficient OSCs and OPDs based on AIEgens as active and interface layers have also well been implemented due to their good photovoltaic properties, and with the further study of the relative materials and devices, their performance will be further improved. Therefore, it is predicted that AIEgens will exhibit unlimited potential in next-generation organic optoelectronic devices, which will greatly reduce the cost in commercial applications and open new directions for the research of organic optoelectronic materials and devices.

At present, AIEgens with high efficiency and good stability are in urgent need of further development. It is believed that the performances of AIEgens and corresponding organic optoelectronic devices must have more breakthroughs with the deep study and understanding on the effects of packing structures, intermolecular interactions, and aggregate morphologies on the performance of AIEgen macroscopic materials, which will bring about a new generation of commercial material systems in organic optoelectronic industry in the near future.

Acknowledgments The authors gratefully acknowledge the support of the National Natural Science Foundation of China (Grant no. 21788102), and the Guangdong Provincial Key Laboratory of Luminescence from Molecular Aggregates, China (2019B030301003).

References

1. Forrest SR (2020) Organic electronics: foundations to applications. Oxford University Press, Oxford
2. Meng LX, Zhang YM, Wan XJ, Li CX, Zhang X, Wang YB, Ke X, Xiao Z, Ding LM, Xia RX, Yip H-L, Cao Y, Chen YS (2018) Organic and solution-processed tandem solar cells with 17.3% efficiency. Science 361:1094
3. Tang CW, VanSlyke SA (1987) Organic electroluminescent devices. Appl Phys Lett 51:913
4. Baldo MA, OBrien DF, You Y, Shoustikov A, Thompson ME, Forrest SR (1998) High efficiency phosphorescent emission from organic electroluminescent devices. Nature 395:151
5. Uoyama H, Goushi K, Shizu K, Nomura H, Adachi C (2012) Highly efficient organic light-emitting diodes from delayed fluorescence. Nature 492:234

6. Kozlov V, Parthasarathy G, Burrows PE, Khalfin V, Wang J, Chou S, Forrest SR (2000) Structures for organic diodes lasers and optical properties of organic semiconductors under intense optical and electrical excitations. IEEE J Quantum Electron 36:18
7. Wang GD, Adil MA, Zhang JQ, Wei ZX (2019) Large-area organic solar cells: material requirements, modular designs, and printing methods. Adv Mater 31:1805089
8. Yang DZ, Ma DG (2019) Development of organic semiconductor photodetectors: from mechanism to applications. Adv Opt Mater 7:1800522
9. Mei J, Hong YN, Lam JWY, Qin AJ, Tang YH, Tang BZ (2014) Aggregation-induced emission: the whole is more brilliant than the parts. Adv Mater 26:5429
10. Luo JD, Xie ZL, Lam JWY, Cheng L, Tang BZ, Chen HY, Qiu CF, Kwok HS, Zhan XW, Liu YQ, Zhu DB (2001) Aggregation-induced emission of 1-methyl-1,2,3,4,5-pentaphenylsilole. Chem Commun 18:1740
11. Xu Z, Gu JB, Qiao XF, Qin AJ, Tang BZ, Ma DG (2019) Highly efficient deep blue aggregation-induced emission organic molecule: A promising multifunctional electroluminescence material for blue/green/orange/red/white OLEDs with superior efficiency and low roll-off. ACS Photon 6:767
12. Xu Z, Tang BZ, Wang Y, Ma DG (2020) Recent advances in high performance blue organic light-emitting diodes based on fluorescence emitters. J Mater Chem C 8:2614
13. Liu H, Guo JJ, Zhao ZJ, Tang BZ (2019) Aggregation-induced delayed fluorescence. ChemPhotoChem 3:993
14. Liu WW, Yu HK, Hu RR, Xu T, Lun YP, Gan JL, Xu SH, Yang ZM, Tang BZ (2019) Microlasers from AIE-active BODIPY derivative. Small 16:1907074
15. Adil MA, Zhang JQ, Wang YH, Yu JD, Yang C, Lu GH, Wei ZX (2020) Regulating the phase separation of ternary organic solar cells via 3D architectured AIE molecules. Nano Energy 68:104271
16. Guo DC, Xu Z, Yang DZ, Ma DG, Tang BZ, Vadim A (2020) Structure design and performance of photomultiplication-type organic photodetectors based on an aggregation-induced emission material. Nanoscale 12:2648
17. Huang YJ, Wang ZR, Chen Z, Zhang QC (2019) Organic cocrystals: beyond electrical conductivities and field-effect transistors (FETs). Angew Chem Inter Ed 58:9696
18. Brütting W, Frischeisen J (2012) Device efficiency of organic light-emitting diodes. In: Brütting W, Adachi C (eds) Physics of organic semiconductors. Wiley, Weinheim
19. Shi LL, Liang QB, Wang WY, Zhang Y, Li GH, Ji T, Hao YY, Cui YX (2018) Research progress in organic photomultiplication photodetectors. Nanomaterials 8:713
20. Qin W, Yang ZY, Jiang YB, Lam JWY, Liang GD, Kwok HS, Tang BZ (2015) Construction of efficient deep blue aggregation-induced emission luminogen from triphenylethene for nondoped organic light-emitting diodes. Chem Mater 27:3892
21. Zhan XJ, Wu ZB, Lin YX, Xie YJ, Peng Q, Li QQ, Ma DG, Li Z (2016) Benzene-cored AIEgens for deep-blue OLEDs: high performance without hole-transporting layers, and unexpected excellent host for orange emission as a side-effect. Chem Sci 7:4355
22. Lin GW, Chen L, Peng HR, Chen SM, Zhuang ZY, Li YH, Wang BH, Zhao ZJ, Tang BZ (2017) 3,4-Donor- and 2,5-acceptor-functionalized dipolar siloles: synthesis, structure, photoluminescence and electroluminescence. J Mater Chem C 5:4867
23. Liu TX, Zhu LP, Zhong C, Xie GH, Gong SL, Fang JF, Ma DG, Yang CL (2017) Naphthothiadiazole-based near-infrared emitter with a photoluminescence quantum yield of 60% in neat film and external quantum efficiencies of up to 3.9% in nondoped OLEDs. Adv Funct Mater 27:1606384
24. Li YH, Xu Z, Zhu XY, Chen B, Wang ZM, Xiao B, Lam JWY, Zhao ZJ, Ma DG, Tang BZ (2019) Creation of efficient blue aggregation-induced emission luminogens for high-performance nondoped blue OLEDs and hybrid white OLEDs. ACS Appl Mater Interfaces 11:17592
25. Xu Z, Gong YB, Dai YF, Sun Q, Yang DZ, Zhan XJ, Li Z, Tang BZ, Ma DG (2019) High efficiency and low roll-Off hybrid WOLEDs by using a deep blue aggregation-induced emission material simultaneously as blue emitter and phosphor host. Adv Opt Mater 7:1801539
26. Guo XM, Yuan PS, Qiao XF, Yang DZ, Dai YF, Sun Q, Qin AJ, Tang BZ, Ma DG (2020) Mechanistic study on high efficiency deep blue AIE-based organic light-emitting diodes by magneto-electroluminescence. Adv Funct Mater 30:1908704
27. Han PB, Xu Z, Lin CW, Ma DG, Qin AJ, Tang BZ (2020) Tetraphenylbenzene-based AIEgens: horizontally oriented emitters for highly efficient non-doped deep blue OLEDs and hosts for high-performance hybrid WOLEDs. J Mater Chem C 8:7012
28. Liu BQ, Nie H, Zhou XB, Hu SB, Luo DX, Gao DY, Zou JH, Xu M, Wang L, Zhao ZJ, Qin AJ, Peng JP, Ning HL, Cao Y, Tang BZ (2016) Manipulation of charge and exciton distribution based

on blue aggregation-induced emission fluorophors: A novel concept to achieve high-performance hybrid white organic light-emitting diodes. Adv Funct Mater 26:776
29. Liu BQ, Nie H, Lin GW, Hu SB, Gao DY, Zou JH, Xu M, Wang L, Zhao ZJ, Ning HL, Peng JP, Cao Y, Tang BZ (2017) High-performance doping-free hybrid white OLEDs based on blue aggregation-induced emission luminogens. ACS Appl Mater Interfaces 9:34162
30. Chen B, Liu BQ, Zeng JJ, Nie H, Xiong Y, Zou JH, Ning HL, Wang ZM, Zhao ZJ, Tang BZ (2018) Efficient bipolar blue AIEgens for high-performance nondoped blue OLEDs and hybrid white OLEDs. Adv Funct Mater 28:1803369
31. Xu Z, Gu JB, Huang J, Lin CW, Li YZ, Yang DZ, Qiao XF, Qin AJ, Zhao ZJ, Tang BZ, Ma DG (2019) Design and performance study of high efficiency/low efficiency roll-off/high CRI hybrid WOLEDs based on aggregation-induced emission materials as fluorescent emitters. Mater Chem Front 3:2652
32. Huang J, Nie H, Zeng JJ, Zhuang ZY, Gan SF, Cai YJ, Guo JJ, Su SJ, Zhao ZJ, Tang BZ (2017) Highly efficient nondoped OLEDs with negligible efficiency roll-off fabricated from aggregation-induced delayed fluorescence luminogens. Angew Chem Inter Ed 56:12971
33. Lv CY, Liu WW, Luo Q, Yi HY, Yu HK, Yang ZM, Zou B, Zhang YJ (2020) A highly emissive AIE-active luminophore exhibiting deep-red to near-infrared piezochromism and high-quality lasing. Chem Sci 11:4007
34. Liu YH, Mu C, Jiang K, Zhao JB, Li YK, Zhang L, Li ZK, Lai JYL, Hu HW, Ma TX, Hu RR, Yu DM, Huang XH, Tang BZ, Yan H (2015) A tetraphenylethylene core-based 3D structure small molecular acceptor enabling efficient non-fullerene organic solar cells. Adv Mater 27:1015
35. Lin HR, Chen SS, Hu HW, Zhang L, Ma TX, Lai JYL, Li ZK, Qin AJ, Huang XH, Tang BZ, Yan H (2016) Reduced intramolecular twisting improves the performance of 3d molecular acceptors in non-fullerene organic solar cells. Adv Mater 28:8546
36. Duan CH, Zhang K, Zhong CM, Huang F, Cao Y (2013) Recent advances in water/alcohol-soluble π-conjugated materials: New materials and growing applications in solar cells. Chem Soc Rev 42:9071
37. Wang C, Liu ZY, Li MS, Xie YJ, Li BS, Wang S, Xue S, Peng Q, Chen B, Zhao ZJ, Li QQ, Ge ZY, Li Z (2017) The marriage of AIE and interface engineering: convenient synthesis and enhanced photovoltaic performance. Chem Sci 8:3750

REVIEW

Aggregation-Induced Emission: From Small Molecules to Polymers—Historical Background, Mechanisms and Photophysics

Ana Clara B. Rodrigues[1] · J. Sérgio Seixas de Melo[1]

Received: 2 January 2021 / Accepted: 12 February 2021 / Published online: 16 March 2021

Abstract
The enhancement of photoluminescence through formation of molecular aggregates in organic oligomers and conjugated organic polymers is reviewed. A historical contextualization of aggregation-induced emission (AIE) phenomena is presented. This includes the loose bolt or free rotor effect and J-aggregation phenomena, and discusses their characteristic features, including structures and mechanisms. The basis of both effects is examined in key molecules, with a particular emphasis on the AIE effect occurring in conjugated organic polymers with a polythiophene (PT) skeleton with triphenylethylene (TPE) units. Rigidification of the excited state structure is one of the defining conditions required to obtain AIE, and thus, by changing from a flexible ground state to rigid (quinoidal-like) structures, oligo and PTs are among the most promising emerging molecules alongside with the more extensively used TPE derivatives. Molecular structures moving away from the domination of aggregation-caused quenching to AIE are presented. Future perspectives for the rational design of AIEgen structures are discussed.

Keywords AIE · Organic conjugated polymers · Polythiophenes · J-aggregates · Fluorescence

---

Chapter 7 was originally published as Rodrigues, A. C. B. & Seixas de Melo, J. S. Topics in Current Chemistry (2021) 379: 15. https://doi.org/10.1007/s41061-021-00327-9.

✉ J. Sérgio Seixas de Melo
ana.beltran@uc.pt; sseixas@ci.uc.pt

[1] Department of Chemistry, Coimbra Chemistry Centre, University of Coimbra, 3004-535 Coimbra, Portugal

1 Background to the AIE Phenomena: An Historical Perspective

Already 20 years have passed since Tang et al. [1] reported a system that, in 2001, introduced the concept of aggregation-induced emission (AIE) in contrast to the most common opposing phenomenon of quenching of fluorescence occurring in the great majority of fluorophores in solid state as a result of aggregation-caused quenching (ACQ) [2]. Tang et al. pioneered these studies, reporting on the efficient blue emission of 1-methyl-1,2,3,4,5-pentaphenylsilole in thin films, which, upon addition of water (a poor solvent), were shown to induce aggregate formation on these silole derivatives in ethanolic solution [1, 3].

Before the findings Tang et al., there had been only a few reports on enhanced emission in the solid state [4]; these included the enhancement of phosphorescence emission in xanthene dyes (eosin and erythrosine) in mixture of glycerol:water frozen glass [5], excimer emission of hydrocarbons in restricted molecular arrangements [6, 7] and in poor solvent medium [8], and emission of J-aggregates formed in pyridinium-substituted indolizines [9] and in cyanine and merocyanine dyes (Scheme 1) [10, 11]. However, emission enhancement in aggregates, with a new fluorescence band associated with intermolecular interactions, had been reported in the first half of the twentieth century.

1.1 Enhancement of Emission from J-Aggregate Formation

In 1936, Scheibe 12] and Jelley [13, 14] independently observed an unusual behaviour of pseudoisocyanine chloride [also known as 1,1′-diethyl-2,2′-cyanine chloride, PIC chloride (Scheme 1)]: in aqueous solution, the absorption maximum of the spectrum shifted to lower energies (red-shift) when compared with the spectrum of the same dye in ethanol; upon increase of the dye concentration in water this band became more intense and sharp [15]. Dye aggregates with a narrow absorption band that is shifted to longer wavelengths (bathochromically shifted), with very small Stokes shift and increased fluorescence intensity relative to the monomer, are generally designated as Scheibe aggregates or J-aggregates (J denotes Jelley after its discoverer) [11, 16]. In contrast, aggregates with absorption bands shifted to shorter wavelengths (hypsochromically shifted) with respect to the monomer band are designated H-aggregates (H denotes hypsochromic) and exhibit, in most cases, low or almost total absence of fluorescence [17]. The mechanisms associated with these interactions are better understood with the Exciton Model of Kasha, but are out of the scope of this review [16, 18, 19]. The AIE effect partially contrasts with the formation of emissive H- (non-emissive) or J- (emissive) aggregates because AIE interaction of two molecules leads to a rigidification of the molecular structure—not mandatory in J-aggregates—thus decreasing the efficiency of the radiationless internal conversion channel and increasing the efficiency of the radiative de-excitation channel.

Enhancement of emission caused by J-aggregates has been described in several systems (Scheme 1). Oelkrug et al. [20] showed the role of J-aggregation in

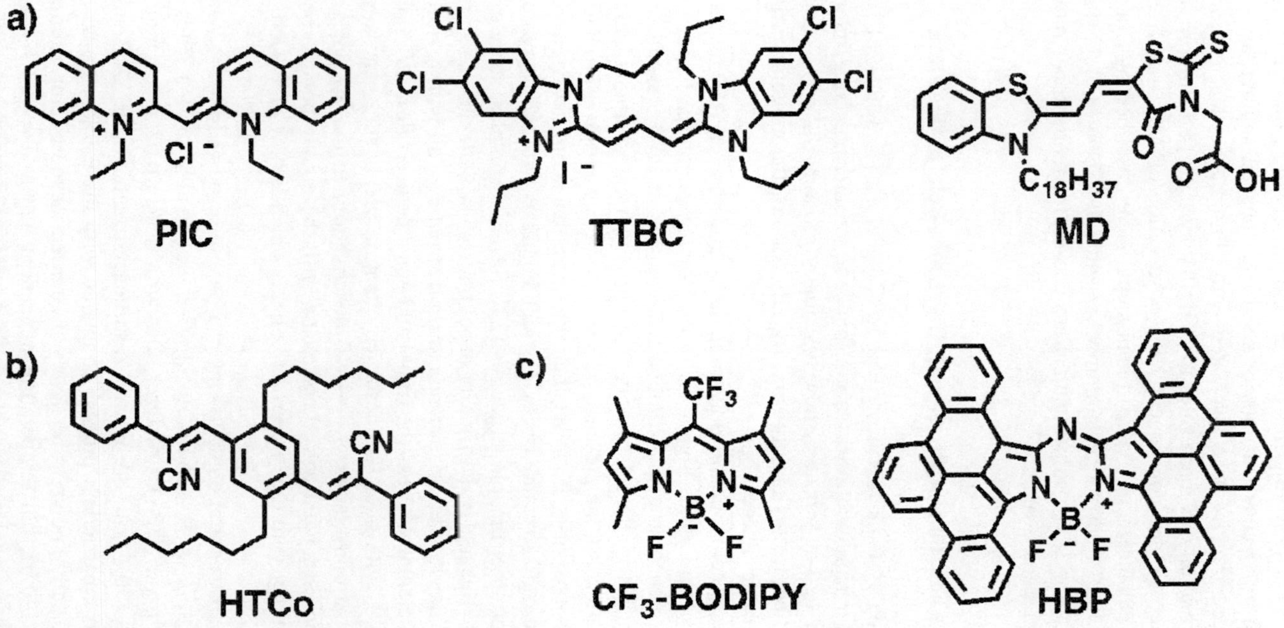

Scheme 1 **a–c** Structures of compounds reported to form emissive J-aggregates. **a** Cyanine and merocyanine derivatives such as 1,1′-diethyl-2,2′-cyanine chloride (PIC chloride); 1,1′,3,3′-tetraethyl-5,5′,6,6′-tetrachlorobenzimidazolocarbocyanine iodide (TTBC iodide) and (3-carboxymethyl-5-[2-(3-octadecyl-2-benzothiazolinylidene) ethylidene]-2-thioxo-4-thiazolidinone (MD). **b** Oligophenylenevinylenes, such as 1,4-bis(α-cyanostyryl)-2,5-di-*n*-hexylbenzene (HTCo). **c** 4,4-Difluoro-4-bora-3a,4a-diaza-s-indacene (BODIPY) systems: CF_3-BODIPY and hexaphenylazaBODIPY (HPB), which are recently reported aggregation-induced emission (AIE)-luminogens

enhancement of emission in solid state oligophenylenevinylenes, with a change in fluorescence quantum yield from practically zero in solution to 60% in nanoparticles or in films, further justifying this phenomenon as arising from the rigid environments provided by viscous solvents, or solid state phases, suppressing the torsion-induced molecular movement associated with nonradiative deactivation [21]. Reports on the formation of J-aggregates in BODIPY systems (4,4-difluoro-4-bora-3a,4a-diaza-s-indacene), and consequently the observation of AIE or aggregation-induced emission enhancement (AIEE) phenomena, were described by Choi et al. [22] with meso-BODIPY derivatives and, more recently, by Sheng et al. [23] in aza-BODIPY. In these works, the substituents were chosen for their proper molecular packing into a J-aggregate structure.

1.2 Monomer-Like Emission Enhancement: The "Loose Bolt" or "Free Rotor" Effect

The AIE-luminogen systems first described by Tang et al. [1, 3, 24–26] resulted from intramolecular rigidification as a consequence of intermolecular interaction [27], but, contrary to J-aggregates, show essentially monomer-like emission [28]. Indeed, the AIE effect results mainly from rigidification of the monomer unit, as a result of this (aggregate) interaction.

In general, rigid systems such as polyaromatic hydrocarbons (PAH) display photoluminescence quantum yields higher than their polyene counterparts (identical or similar number of carbon atoms). Although the emission of PAH depends on symmetry considerations (with the relative position of the two lowest singlet lying states having a determinant role), the presence of heteroatom(s), and solvent (polarity and viscosity), they are most often fluorescent compounds. In a simplified view, this can be attributed to the rigid structure of PAH, which contrasts with the degree of freedom of polyenes or other flexible structures [29]. Several examples, some given in Scheme 2, illustrate this molecular rigidity in comparison with flexible analogues. In Scheme 2, the left hand panel corresponds to rigid structures, all with high fluorescence emission efficiency, whereas the right hand panel depicts "flexible" molecular structures ("rotors"), with poor luminescence quantum yields and, generally, high radiationless internal conversion ($S_1 \rightsquigarrow S_0$) deactivation. In the examples given, the twisting motion associated with the skeleton with the double bond(s) becomes locked, and an analogy with the "free rotor" effect, which is now precluded, seems adequate.

In Scheme 2, the structural comparisons can be envisaged as follows. With azabenzene (AB), with $\phi_F \approx 0$, simple blocking of the twisting motion leads to the 9,10-diazaphenanthrene (DPA) structure, now with a $\phi_F = 0.02$ in ethanol and a lifetime of 4.27 ns [30]. With the E- and Z-stilbene structures (the Z-cis structure photoisomerizes in a barrierless process compared with the E-trans structure) [31, 32], which are basically known to be poorly fluorescent at room temperature (with values of $\phi_F = 0.04$–0.05, but temperature dependent and at T = 83.2 K—another way to rigidify the molecular structure—with $\phi_F \sim 0.95$ [33]) can, again by blocking the twisting motion associated with the central double bond, lead to the structures Ind-Ind (indeno[2,l-a]indene) and DPCB (1,2-diphenylcyclobutene), which are

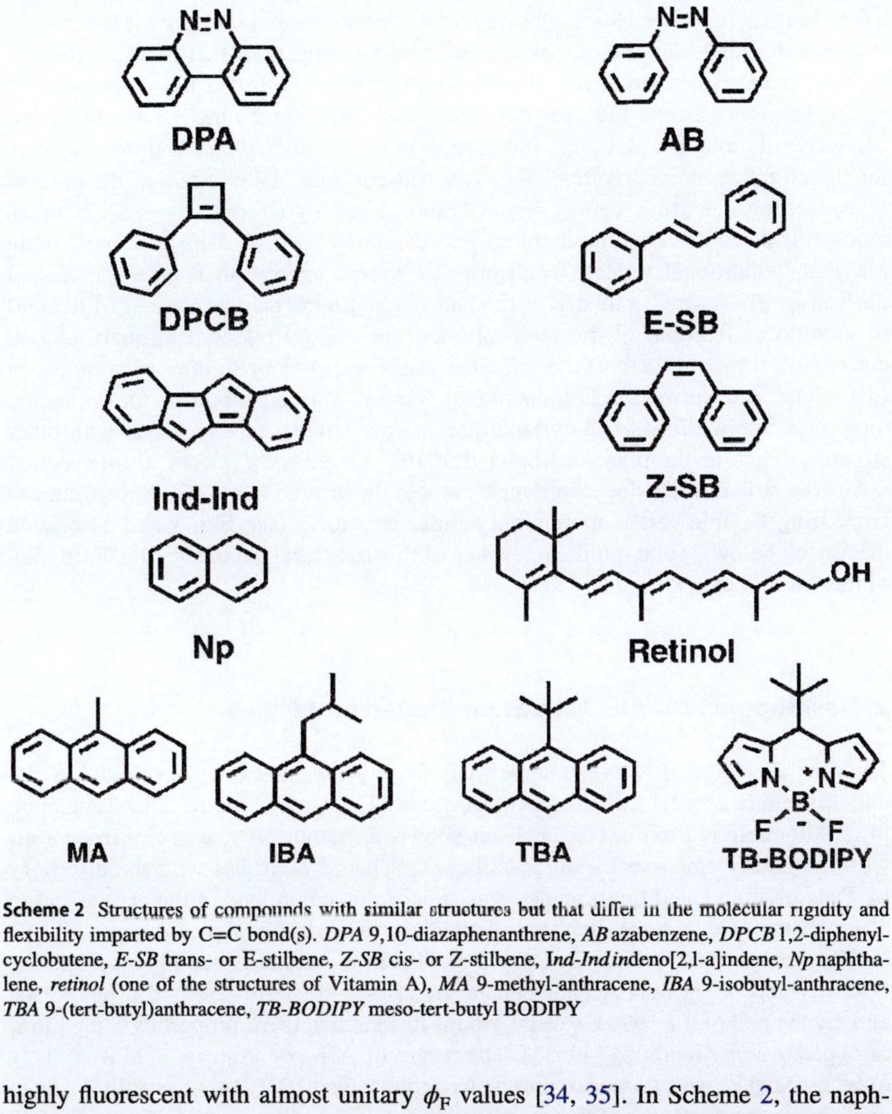

Scheme 2 Structures of compounds with similar structures but that differ in the molecular rigidity and flexibility imparted by C=C bond(s). *DPA* 9,10-diazaphenanthrene, *AB* azabenzene, *DPCB* 1,2-diphenylcyclobutene, *E-SB* trans- or E-stilbene, *Z-SB* cis- or Z-stilbene, *Ind-Ind* indeno[2,1-a]indene, *Np* naphthalene, *retinol* (one of the structures of Vitamin A), *MA* 9-methyl-anthracene, *IBA* 9-isobutyl-anthracene, *TBA* 9-(tert-butyl)anthracene, *TB-BODIPY* meso-tert-butyl BODIPY

highly fluorescent with almost unitary ϕ_F values [34, 35]. In Scheme 2, the naphthalene and retinol pair constitutes another classical example where molecules with the same number of carbon–carbon double bonds have differences of almost one order of magnitude in ϕ_F (Np, $\phi_F=0.21$ [36] in ethanol and retinol with $\phi_F=0.0298$ in hexane [37], the two obtained at $T=293$ K), again due to rigidity associated to Np and not present in retinol. There is an additional type of molecule, the 9-alkyl-substituted anthracenes, where this rotor or loose bolt effect is particular notorious and imparts a gain in rigidity of the molecules, in particular, when a comparison of the emission efficiency is made at room temperature between 9-methyl-anthracene (MA, $\phi_F=0.29$), 9-isobutyl-anthracene (IBA, $\phi_F=0.39$) and 9-(*tert*-butyl) anthracene (TBA, $\phi_F=0.011$ in *n*-hexane) [38, 39]. The two structures (MA and

TBA) possess the same chromophoric core (anthracene) and therefore it is not fluorescence deactivation (and associated radiative rate constant) that explains the difference in behavior. Moreover, since triplet formation is found non-populated for 9-(*tert*-butyl)anthracene [39], the large difference between the methyl and tert-butyl derivatives is found in the large increase in the internal conversion deactivation in this latter anthracene derivative. This is a consequence of the fact that the process of converting electronic energy into vibrational energy (from $S_1 \sim\sim\rightarrow S_0$) is much more efficient with the *t*-butyl-anthracene derivative, where there are many more low-lying vibrational modes, to absorb this excess energy, than with the methyl derivative. This increase in size of the substituent augments the number of degrees of vibrational freedom of the molecule, and the energy released through internal conversion turns out to be more efficient at the expense of fluorescence emission (the triplet state formation is inoperative). That is, the loose bolt or rotor effect is more effective with 9-(*tert*-butyl)anthracene. An identical effect is seen with other structures such as the meso-tert-butyl BODIPY, where a S_1/S_0 conical intersection favors the radiationless deexcitation pathway to the ground state [40]. Recent studies comparing flexible versus more rigid similar structures (see Fig. 1 and associated discussion below), point out the relevance of this structural balance in the discussion of the AIE effect [41].

2 Development of AIE-Active Luminogen Molecules

After the account of AIE in pentaphenylsilole [1, 3], another fluorogen found to display this effect was 1,1,2,2-tetraphenylethylene (TPE), again reported by Tang et al. [42]. Although TPE emission has been known for about 50 years [43], in contrast with the poorly emissive E- and Z-stilbenes (Scheme 2), it was with the discovery by Tang of augmented fluorescence emission of this compound in the crystal and in water:acetonitrile mixtures that this effect began to be deeply explored.

The π twist of the central C=C bond of TPE in the excited state [44–46], and the viscosity-dependent fluorescence enhancement [47, 48] were reported in the 1980s and, by the end of the 1990s, several groups had studied these properties using ultrafast spectroscopy methods [49–51]. The origin of AIE phenomena in TPE is likely to be associated with the loose bolt or free rotor effect [52]. This basically leads to an increase (or decrease) of the coupling modes of the vibronic transitions [52, 53]. In systems where there is the possibility of changing from a C=C double bond to a carbon–carbon bond with partial single bond character (upon excitation), this "loose bolt effect" may therefore be considered as the genesis of the AIE effect. Indeed, most initial compounds showing AIE were based on a TPE central core, i.e., there was, in the excited state, a rigidification of these TPE-like molecules, resulting from aggregate interaction, with a decrease in the radiationless internal conversion decay pathway. Actually, this intermolecular interaction involving TPE molecules would result not in a particular emissive aggregate, as with J-aggregates, but in a structural rigidification of the molecule (in the aggregate state) that would preclude the "loose

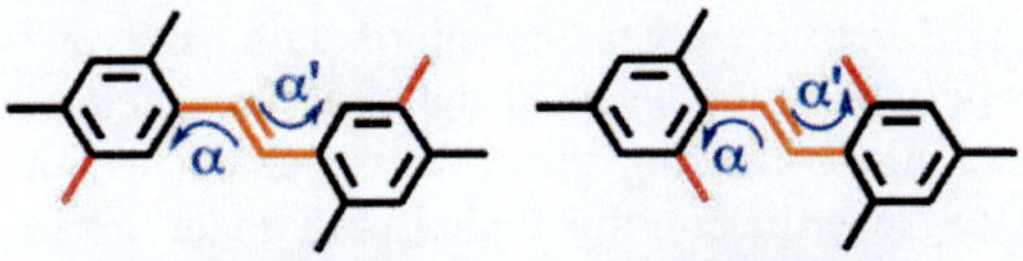

	2,4,5-TMe-DPE	2,4,6-TMe-DPE
$\|\alpha\| = \|\alpha'\|$	21.6°	41.7°
ϕ_F^{THF}	0.134	0.006
τ_F^{THF}	45.2 ps/363 ps	2.1 ps
k_R^{THF}	5.1×10^8 s^{-1}	2.9×10^9 s^{-1}
k_{NR}^{THF}	3.3×10^9 s^{-1}	4.7×10^{11} s^{-1}

Fig. 1 Structures of 2,4,5-TMe-DPE and 2,4,6-TMe-DPE along with their photophysical data measured in tetrahydrofuran (THF) solution. ϕ_F Fluorescence quantum yield of solution, τ_F fluorescence decay times, k_R radiative rate constant, k_{NR} nonradiative rate constant [41]

bolt effect" and therefore the "movement" otherwise associated with this highly efficient internal conversion radiationless deactivation.

The pioneering studies of Tang et al. led to a rapid increase in the number of known luminogens with AIE characteristics (AIEgens) [54, 55] and innumerable AIE materials have been developed [56–58]. Other systems with AIE properties have been rediscovered [e.g., AIE building blocks of quinoline-malonitrile (QM) and polyaryl-substituted pyrrole (PAP) in Scheme 3] [54, 59–62]. To design and rediscover new AIEgen molecules it is therefore of utmost relevance to equate the mechanisms leading to the occurrence of AIE and AIEE. The free rotor or loose bolt effect associated with the introduction of the tert-butyl group (or similar groups with a high number of low-lying vibrational modes) will be of relevance in the design of AIE active polymers described in the section on Design and photophysical studies of AIE-active polymers, and is particularly relevant in the section on TPE-based conjugated polymers where a tert-butyl derivative increases its photoluminescence yield in the solid state and in a solvent mixture where a bad solvent (water) is dominant.

2.1 AIE Working Mechanisms

The working mechanisms of AIE have been investigated extensively [63–65]. One such proposed mechanism is restriction of the intramolecular rotation (RIR) [66], which can be measured through temperature-, viscosity-, and intramolecular-steric-hindrance-dependent luminescence quantum efficiency. Restriction of intramolecular vibration (RIV) [67] was also proposed to account for the strong luminescence in the solid state. A more generalized mechanistic model for AIE was proposed by combining RIR and RIV into a global principle of restriction of intramolecular motions (RIM) [68]. Photocyclization [69–71], suppression of the formation of twisted intramolecular charge transfer state [63, 72–74] and Z/E isomerization [75–77] were other hypotheses proposed to explain the occurrence of AIE. RIM was shown to be the main mechanistic cause for AIE effects observed in many luminogen systems based on pure aromatic molecular rotors such as hexaphenylsilole

(HPS). With TPE, recent studies and interpretations favor both RIM [68] and Z/E isomerization [77] being involved in the emission quenching of TPE in solution. Therefore, blocking or limiting these two mechanisms, by aggregate interaction, will lead to AIE. Nevertheless, this topic is still under debate, particularly regarding which molecular mechanism (RIM or isomerization) plays the major role in the AIE effect [78].

Analysis of the potential energy surfaces (PES) of model propeller-shaped structure also showed that the enhancement of emission in the solid state can be associated with the decrease of the excited state decay caused by nonradiative mechanisms [79]. Moreover, computational theoretical studies indicate that the increase of steric hindrance improves the fluorescence quantum efficiency and that rotation of aromatic substituents and π–π stacking interactions are the dominant pathways for the nonradiative decay of AIEgens [80]. Furthermore, the combination of ultrafast photophysical techniques, quantum chemistry calculations and strategic design of new molecules led to three additional proposed AIE mechanisms: restriction of the excited-state "double-bond" torsion (ESDBT) [41], restriction access to conical intersection (RACI) [81, 82] and dark state in heteroatoms systems (RADS) [83].

2.1.1 Stilbene-Like Molecules

Aiming to shed light on the potential reasons why some stilbene-like molecules are emissive in solution, whereas other are not, photophysical studies of two model stilbene derivatives, di-*o*-methyl substituted 2,4,6-TMe-DPE and mono-*o*-methyl substituted 2,4,5-TMe-DPE, were recently conducted [41]. While the twisted structure of 2,4,6-TMe-DPE, which exhibits an AIE effect, was poorly fluorescent in tetrahydrofuran (THF), with a fluorescence quantum yield, $\phi_F=0.006$, 2,4,5-TMe-DPE, a structurally planar molecule, displayed $\phi_F=0.134$ in the same solvent (Fig. 1) [41]. Femtosecond transient absorption (fs-TA) measurements for these two molecules in THF showed that two processes are involved in the excited-state deactivation of 2,4,5-TMe-DPE, highlighted by the two time constants ($\tau_1=45.2$ ps and $\tau_2=363$ ps), while a much faster process occurs with 2,4,6-TMe-DPE ($\tau=2.1$ ps). The data also indicated that the twisted 2,4,6-TMe-DPE displays a faster excited-state molecular motion than the structurally planar 2,4,5-TMe-DPE ($k_{NR}=4.7\times10^{11}$ s^{-1} vs. 3.3×10^9 s^{-1}), thus showing the clear dominance of the radiationless decay pathways with 2,4,6-TMe-DPE. The restricted molecular motion in the planar 2,4,5-TMe-DPE made the radiative decay comparable with the nonradiative decay, thus clarifying the $\phi_F=0.134$ obtained in solution.

2.1.2 Quantum Mechanical Calculations

Time-dependent density functional theory (TDDFT) quantum mechanical theory was used to perform calculations in the excited state of 2,4,6-TMe-DPE and 2,4,5-TMe-DPE [41]. The authors concluded that most of the excited state energy of 2,4,6-TMe-DPE was released nonradiatively via the excited-state "double-bond" torsion (ESDBT) mechanism; however, ESDBT was suppressed in 2,4,5-TMe-DPE, and this justified its higher radiative rate constant value, k_R (see Fig. 1) [41].

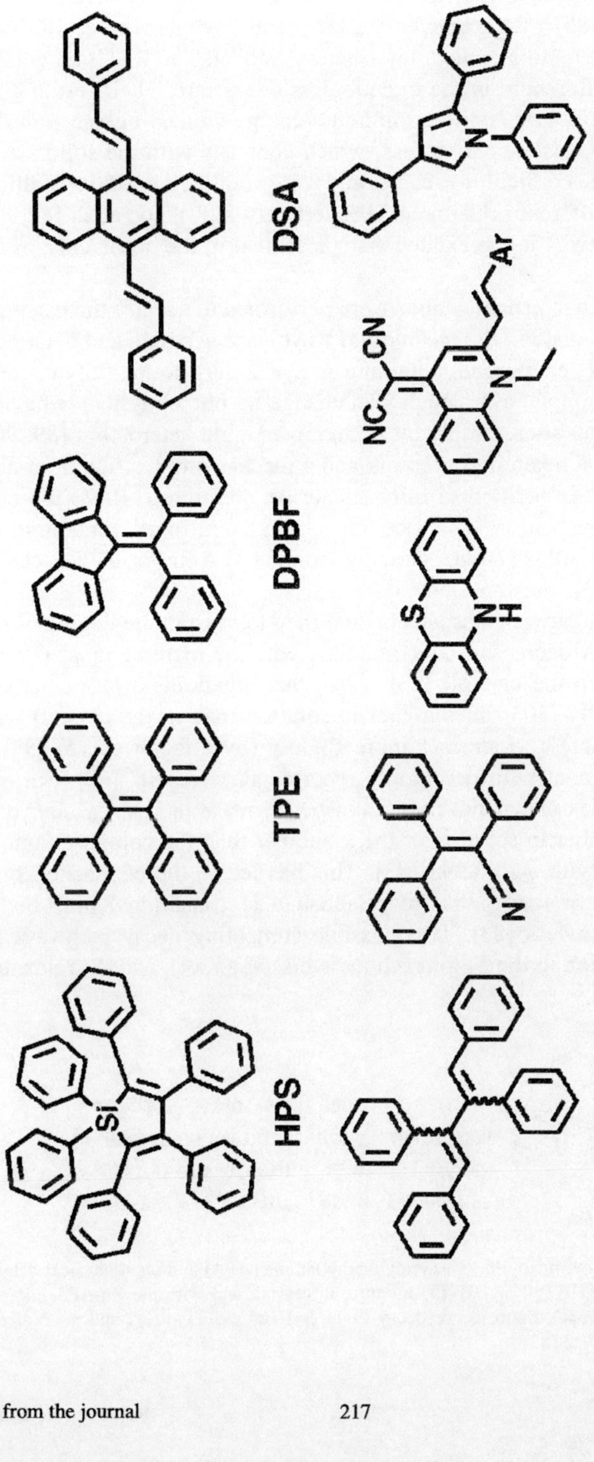

Scheme 3 Molecular structure of some AIE-active luminogen cores. *HPS* Hexaphenylsilole; *TPE* tetraphenylethene; *DPBF* diphenylbenzofulvene; *DSA* 9,10-distyrylanthracene; *TPBD* tetraphenyl-1,4-butadiene; *TPAN* triphenylacrylonitrile; *PTZ* phenothiazine; *QM* quinoline-malonitrile; *PAP* polyaryl-substituted pyrrole

In order to further understand the relationship between intramolecular motions and AIE properties, Blancafort and Morokuma conducted independent quantum chemical investigations [81, 84–86]. They analyze the deactivation processes of AIEgens DPBF [85] , tetraphenylsilole [86], and phenyleneimide [87] with the RACI (restricted access to a conical intersection) model. The RACI model explains the AIE effect by differences in the radiationless decay rate. Their results predicted a largely distorted structure near the minimum energy conical intersection (MECI). In solution, the MECI is easy to access, which contrasts with the solid state situation, where it becomes difficult to access, and it is prohibited in the crystalline state. Later, by studying different substituted TPE derivatives, Kokado et al. [82, 88] indicated that the C=C twist in the excited state is critical in the occurrence of the AIE phenomenon.

Many of these mechanistical studies were performed to specify the exact motions and decay pathways, making the meaning of RIM more accurate and detailed. However, the behavior of heteroatom-containing AIEgens may not be fully explained by RIM. Indeed, for example, a molecule can change its photophysical behavior from AIE to ACQ characteristics, simply by changing a single heteroatom [89, 90]. Two excited states must be taken into account, and a mechanism, restriction of access to dark state (RADS), is specified to fully equate (in addition to RIM) the complete the picture of AIE mechanism. To probe this concept, a nitrogen-containing AIEgen named 9-anthyl-methyl)bis (2-pyridylmethyl)amine (APA in Fig. 2) was chosen as a model compound [83].

From solution to aggregate suspension and then to crystal, the degree of molecular freedom gradually decreases, concomitantly with the increase in ϕ_F (from 0.006 in solution to 0.39 in the crystal) [83]. Also, the radiationless rate constant, k_{NR}, decreases dramatically (103 times higher in solution than in the crystal) while the radiative rate constant, k_R, increases more slightly (by a factor of 15) [83]. These results are in agreement with commonly reported RIM results [68]. Surprisingly, the zinc-APA complex shows not only a drastic decrease in k_{NR} (ca. four orders of magnitude, 5247 higher in solution of the monomer than the complex), but also an enhancement of k_R with ϕ_F=0.998 [83]. This has led to the conclusion that RIM and the availability of lone pairs (to establish n,π* transitions) may both influence the emission process [83]. Two possible competing decay pathways may be involved: (1) relaxation to the S_1 to reach the emissive (π,π*); and (2) relaxation to a

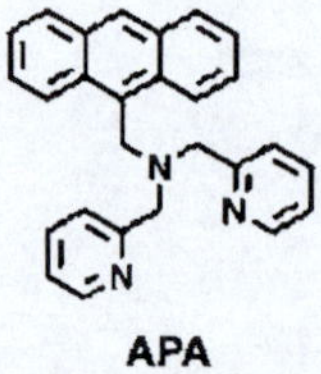

APA

	ϕ_F	k_R	k_{NR}
THF	0.006	$6.4\times10^6\ s^{-1}$	$1060\times10^6\ s^{-1}$
90W:10THF	0.028	$9.1\times10^6\ s^{-1}$	$316\times10^6\ s^{-1}$
Crystal	0.39	$10.3\times10^6\ s^{-1}$	$16.1\times10^6\ s^{-1}$
Zn-Complex	0.998	$101\times10^6\ s^{-1}$	$0.2\times10^6\ s^{-1}$

Fig. 2 Structures of 9-anthyl-methyl)bis (2-pyridylmethyl)amine (APA) and photophysical data obtained in THF, 90%water:10% THF (90 W:10THF) mixture, in crystal, and with the Zn-APA complex. Fluorescence quantum yield (ϕ_F), fluorescence decay times (τ_F) and radiative (k_R) and nonradiative (k_{NR}) rate constants (k_{NR}) are given [83]

dark state (n, π*), leading to fast non-radiative decay. Since the (n, π*) state is lower in energy than (π, π*), it results in weak fluorescence and short lifetime value [83]. This shows that, even in molecules where the AIE effect is potentially present, the balance between the increase or decrease of photoluminescence may also depend on the nature of the two lowest lying singlet excited states, and not only due to a restriction of the molecular motion process.

3 Photophysical Studies of AIE-Luminogens

3.1 TPE Derivatives

A good–bad solvent mixture constitutes a consensual strategy to study AIE behavior in dilute solutions of potential AIEgens [91]. Since TPE derivatives are typically aromatic-like (hydrophobic) molecules, water is a poor solvent, and is therefore used as the most common bad solvent in combination with water-miscible good solvents such as THF, acetonitrile, dimethylsulfoxide and 1,4-dioxane. Formation of nanoaggregates in suspension, in solvent–nonsolvent mixtures, may also lead to the Mie scattering effect [92], detected by the appearance of a long wavelength tail in the absorption spectrum [93].

TPE is, as described, one of the most flowing sources of AIEgens. This is in part due to its structural modification (rigidification) in the excited state, but also to the development of simplified strategies of synthesis and derivatization of this molecule [43]. Scheme 4 depicts some illustrative **TPE** derivatives synthesized by several research groups [94–98].

Increases in the photoluminescence efficiency in the solid state, or from a bad-solvent to a good-solvent fraction, may be significantly different depending on the system; this happens even among several TPE derivatives (Table 1). The AIE factor (α_{AIE}), defined as the ratio between the ϕ_F values of the aggregate and of the monomer (isolated molecule in a good solvent), constitutes a parameter that establishes a qualitative manner of measuring the emission increase due to the AIE. Table 1 summarizes some of the relevant literature data on TPE derivatives with data obtained from photophysical studies [94–98]. As noticed in Table 1, the value of α_{AIE} is structure dependent, and so is the dominant intramolecular motion, restricted when aggregation occurs, acting as an active dissipation factor for the radiationless deactivation [99]. In Table 1 and Scheme 4, it is worth noting the highest α_{AIE} values for compounds **15** and **16**, which are ca. 9.5 times higher than those of TPE.

AIEgens oligomers generally possess twisted conformational structures in the ground state, and hence afford a more moderate packing pattern than traditional luminogens with a planar structure. This facilitates the interplay between different morphologies by heating, solvent fuming, mechanical stimuli such as pressing, grinding and so forth, in particular upon structural change promoted by light excitation [100]. Emission of many AIEgens can be switched through modulating morphology of the luminogen between amorphous and crystalline states [101]. Consequently, several AIEgens present mechanoluminescence (ML) properties [26, 101–106], which have been the subject of extensive reviews in the past 5 years

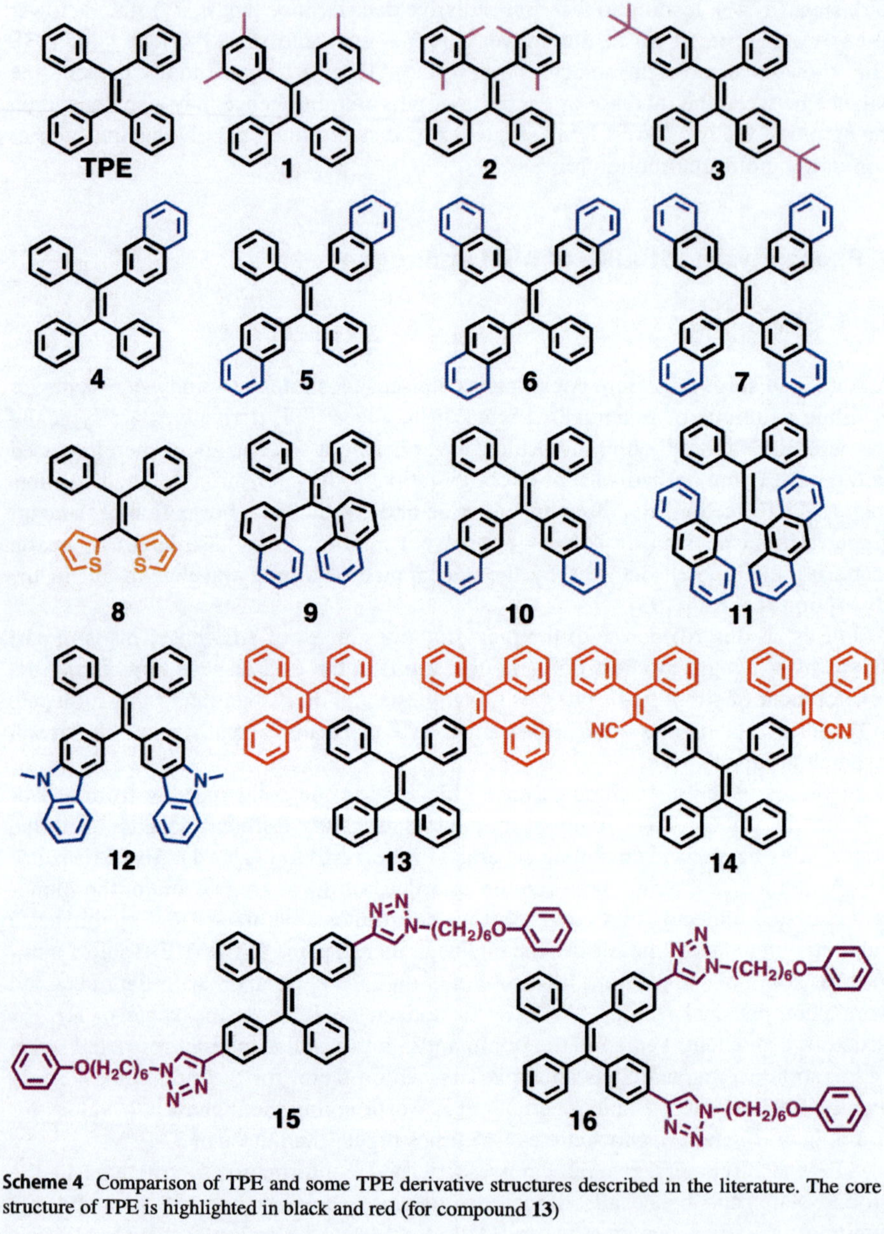

Scheme 4 Comparison of TPE and some TPE derivative structures described in the literature. The core structure of TPE is highlighted in black and red (for compound **13**)

[100, 107–110]. Among those AIE-active and ML responsive compounds, TPE [108, 111], distyrylanthracene (DSA) [110], diphenylbenzonfulvene (DPBF) [26, 101, 106] and phenothiazine (PTZ) [112–116] derivatives are likely to be within the more extensively investigated (Scheme 3).

Table 1 Comparison of fluorescence quantum yields for tetraphenylethylene (TPE) derivatives in solution, mixture of good solvent/bad solvent and in solid state, both powder and film, where data are available, along with the aggregation-induced emission (AIE) factor (α_{AIE}), which discriminates the enhancement of the fluorescence upon aggregation

Compound	Solution	Good solvent/ bad solvent	Powder	Film	α_{AIE} [a]
TPE	0.003	0.25	0.23	0.26	87
1	0.006	0.075	0.30	n.a.[b]	50
2	0.60	0.75	0.98	n.a	1.6
3	0.024	0.20	0.14	0.46	19
4	0.003	n.a	n.a	0.45	56
5	0.003	0.043	0.14	0.12	47
6	0.016	n.a	n.a	0.37	23
7	0.017	n.a	n.a	0.22	13
8	0.002	0.007	n.a	n.a	3.3
9	0.017	0.13	n.a	n.a	7.3
10	0.004	0.04	n.a	n.a	10
11	0.097	0.18	n.a	n.a	1.9
12	0.004	0.13	n.a	n.a	33
13	0.004	0.46	n.a	n.a	115
14	0.003	0.13	n.a	n.a	43
15	0.001	0.39	1.0	n.a	833
16	0.001	0.40	1.0	n.a	833

Compound numbers are related to structures in Scheme 4. Data obtained from [94–98]

[a] $\alpha_{AIE} = \Phi_F^{agg}/\Phi_F^{sol}$, where ϕ_F^{agg} = highest fluorescence quantum yield of the aggregate form and ϕ_F^{sol} = fluorescence quantum yield diluted in solution

[b] Data not available

3.2 Phenothiazine Derivatives

PTZ is a well-known heterocyclic compound [117], with a twisted butterfly-like structure, with electron-rich nitrogen and sulfur heteroatoms, and therefore with a strong electron donor character. This favors the presence of a lowest lying n,π^* singlet excited state and favors intersystem crossing between the excited singlet and triplet states (with different state configurations) according to El-Sayed rules [118], opening a potential new emissive channel and class of molecules: PTZ-based luminogens with thermally activated delayed fluorescence (TADF) [119, 120]. Due to its non-planar bending "butterfly" shaped conformation [121], PTZ's ring may avoid intermolecular interactions, such as π–π stacking, which are typically responsible for ACQ. This has led to the use of the PTZ group as a basic chromophoric structure platform for the design of new AIE luminogens [122, 123].

Tang et al. [124] prepared a series of phenothiazine-based red/near infrared (NIR) fluorophores, found AIE-active, using 10-butyl-phenothiazine-3-carboxaldehyde as starting material. Among them, PTZ1 and PTZ6 presented higher ϕ_F, of 0.234 and 0.207, respectively, with $\lambda^{em} > 610$ nm in the solid state (Fig. 3). In these PTZ compounds, the mechanism of the AIE behavior was further explored

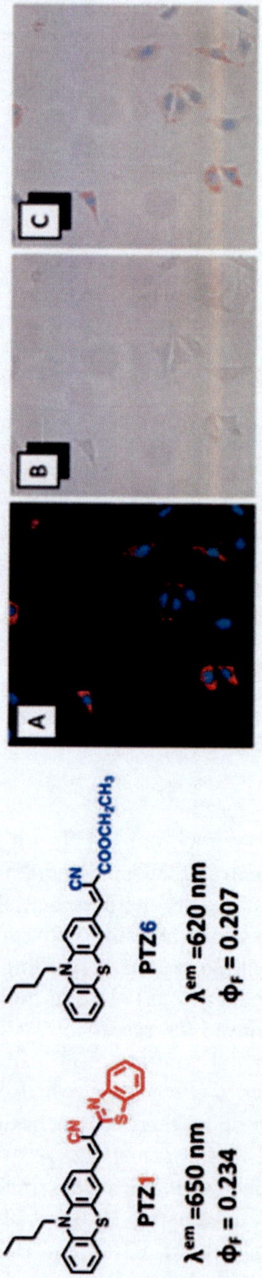

Fig. 3 Left panel Chemical structures of PTZ1 and PTZ6 along with their respective emission maxima and fluorescence quantum yield in solid state. Images on the right display confocal microscopy images of HeLa cells stained with PTZ6: **a** fluorescence, **b** bright-field and **c** merged (nuclei stained by commercially available nuclei imaging agent Hoechst 33342). Adapted from reference [124], with permission from RSC, Copyright 2019

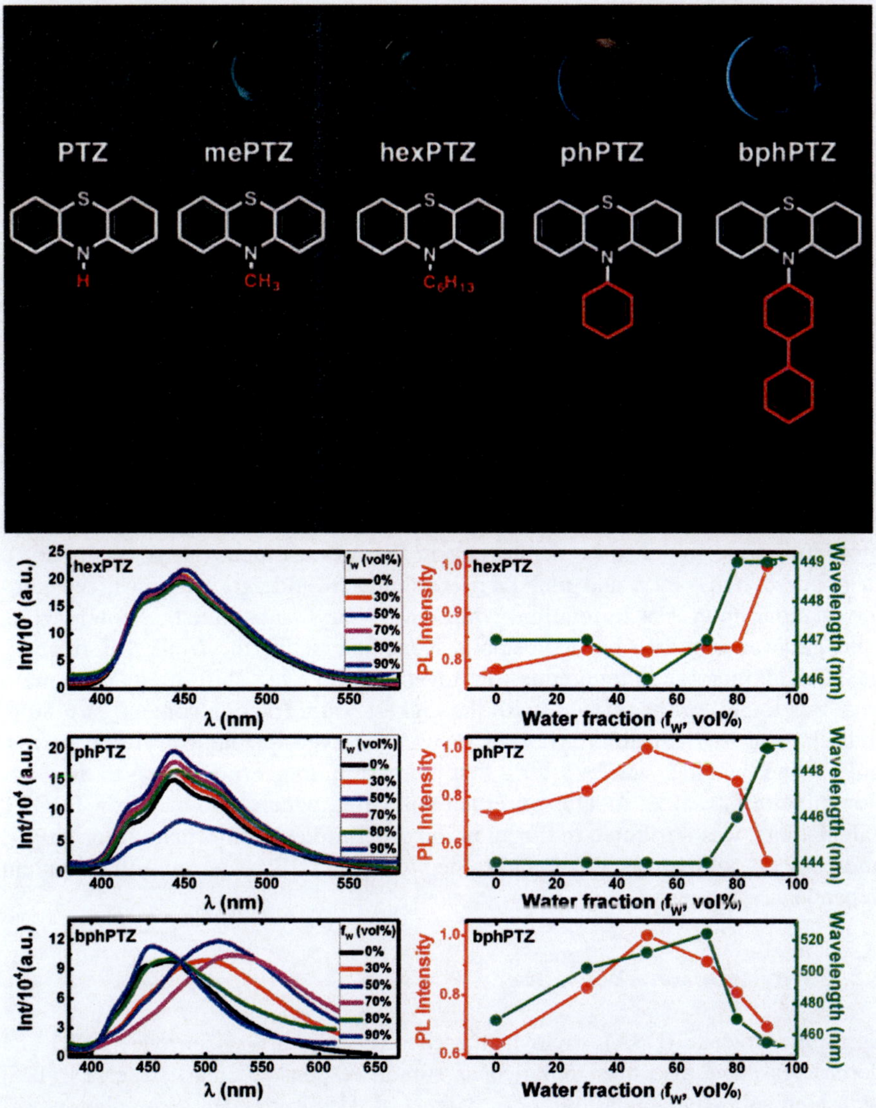

Fig. 4 Photophysical studies of *N*-alkyl and *N*-aryl substituted phenothiazines (PTZs). Structures, acronyms and solid state emission (top images thin films, bottom images powder) under UV-illumination. Fluorescence emission spectra for the *N*-substituted PTZs in THF:water mixtures and their respective correlation of photoluminescence intensity and fluorescence emission maxima (λ^{em}) with the mixture water fraction (f_W=0–90%). Adapted from reference [125], with permission from Elsevier, Copyright 2020

[124]. Single-crystal analysis of PTZ1 suggested that the butterfly-shaped geometry was partially responsible for the higher emission efficiency in solid or aggregated states than in solution, with the *n*-butyl group acting as a spacer to

separate neighboring molecules. The high ϕ_F and red/NIR emission made those dyes favorable for biological imaging, with PTZ6 being used to stain HeLa cells. Indeed, PTZ6 was found to accumulate mainly in lipid droplets (LDs), with better a performance when compared to traditional LD tracking agents such as Nile red. This selectively of PTZ6 was suggest to result from the hydrophobic *n*-butyl group.

The design of new PTZ derivatives with AIE properties require a deeper understanding of the photophysical properties and photostability of this class of compounds. A comprehensive study of the photophysical properties of five *N*-alkyl and *N*-aryl substituted PTZs was reported recently (Fig. 4) [125].

The electronic spectral properties (of both singlet and triplet states) and fluorescence quantum yields (ϕ_F) in solution were found to be poorly affected by substitution. The *N*-methyl and *N*-phenyl substituted phenothiazines, mePTZ and phPTZ, respectively, showed higher ϕ_F values in the solid state, either in thin films or powder, than in solution. It was found that the solid state (powders) of mePTZ and bphPTZ derivatives exhibit room temperature phosphorescence (RTP, Fig. 4). Moreover, and more interestingly, the photostability and AIE properties were found to be dependent on the *N*-substituent (alkyl versus aryl). In polar solvents, PTZ and mePTZ were found to undergo self-sensitized photooxidation upon UV-irradiation while the *N*-aryl substituted phenothiazine, bphPTZ, was found to be photostable. Nonetheless, for the *N*-alkyl derivative, hexPTZ, an increase in total emission, together with a red-shift of emission maxima, was found for the mixture with the highest water fraction content, $f_W > 80\%$, thus showing AIE. A different behavior was observed for the *N*-aryl derivatives, phPTZ and bphPTZ, where, after a first increase in total emission up to the fraction mixture $f_W = 50\%$, ACQ is observed thereafter, which, also based on TDDFT calculations, was attributed to the adoption of a different conformation for phPTZ and bphPTZ, which was also found to be water (in THF: water mixtures) content dependent.

3.3 Distyrylanthracene Derivatives

Distyrylanthracene (DSA), divinylanthacenes (DVA), distyrylbenzene (DSB) and derivatives, have also been reported as stimuli responsive smart materials [126] with high solid-state emission [127]. Tian et al. [128] provided some photophysical insights on the origin of the AIE phenomenon in a series of DSA derivatives (Fig. 5), using steady-state and time-resolved fluorescence.

In dilute solutions, DSA derivatives display a twisted structure in the ground state that, upon excitation, ultimately relaxes, within several picoseconds, to a planar structure [128]. With the DSA derivatives, a more extended conjugation is observed in the excited state (electron cloud more delocalized) leading to a more stabilized and rigid structure. Comparing the derivatives, those with electron-donating groups (BMOSA and BTVA) have higher radiationless rate constant values, k_{NR}, whereas the DSA-derivative with electron-withdrawing group (BP2VA) display the smallest k_{NR}. In good solvent solutions, the radiationless decay channel dominates the

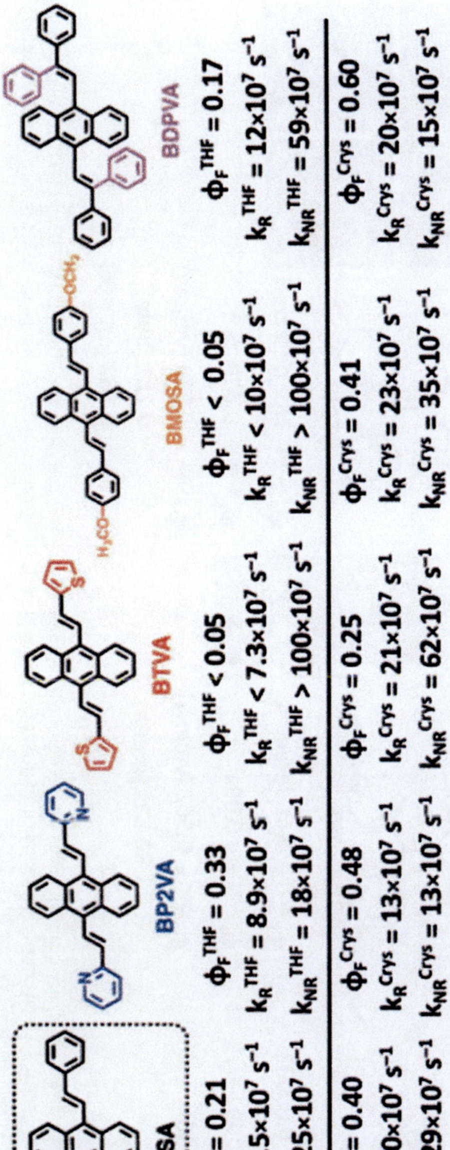

Fig. 5 Structures and acronyms of distyrylanthracene (DSA) derivatives, along with their respective fluorescence quantum yields (ϕ_F) and rate constants in THF and crystal (Crys) [128]

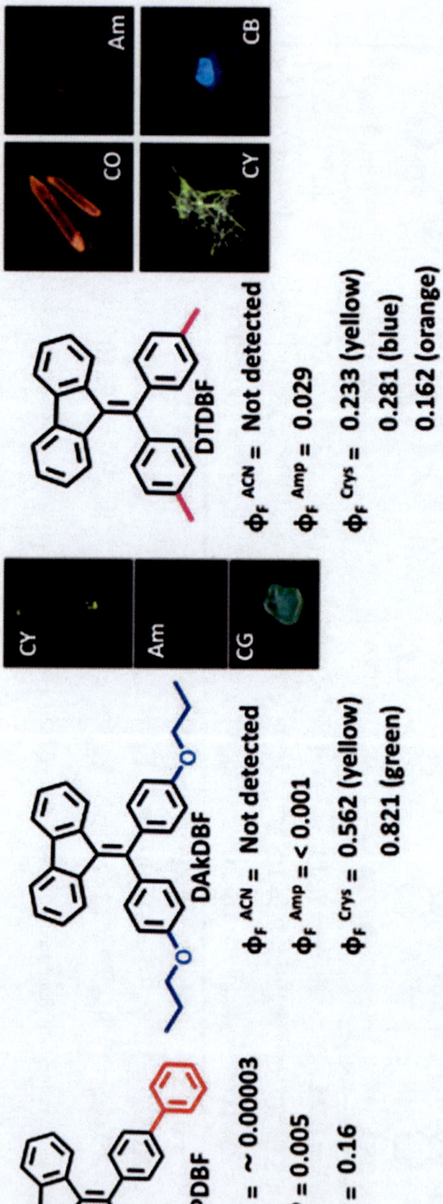

Fig. 6 Structures and respective fluorescence quantum yields (ϕ_F) of diphenylbenzonfulvene (DPBF) derivatives BpPDBF, DAkBDF and DTDBF solids. Photos of DAkBDF and DTDBF solids under UV illumination: *Am* amorphous powder, *CY* yellow crystal, *CG* green crystal, *CO* orange crystal, *CB* blue crystal. Adapted from Refs. [101] and [106], with permission from Wiley, Copyright 2020

deactivation of the excited state (k_{NR} is at least three times higher than k_R) of DSA and derivatives. In contrast, the nonradiative deactivation is greatly suppressed by molecular stacking in the aggregates, here studied either in single crystal structures or supported in polymethylmethacrylate (PMMA) films, and consequently the fluorescence is significantly enhanced (e.g., for BMOSA, ϕ_F in crystal is over eight times higher than in THF solution).

4 AIE and CIE in Small Molecules

Diphenylbenzonfulvene (DPBF) derivatives are known to display another AIE-related property: crystallization-induced emission (CIE). DPBF derivatives are almost non-emissive in solution and amorphous state, but highly emissive in the crystalline state [111]. BpPDBF (Fig. 6) crystals emit an intense blue light upon UV-excitation, while its amorphous powders emit a faint yellow light; in the good solvent acetonitrile, BpPDBF is poorly emissive [26]. Notably, the ϕ_F of BpPDBF crystal is 32 times higher than that of its amorphous powder. The CIE effect is noticed even more dramatically for DAkBDF (Fig. 6); indeed, while its amorphous powder is practically non-emissive (<0.001), in its crystal structures it shows high ϕ_F values (0.56–0.82) [106].

Dong et al. [101] replaced alkoxy phenyl groups with tolyl groups with ditolyldibenzofulvene (DTDBF, Fig. 6) leading to a twisted conformation, bulky conjugation core and peripheral tolyl groups. As described with DPBF [24] and other derivatives [25, 129], DTDBF preserved the AIE-active core: while it is nearly non-emissive in pure acetonitrile, its emission intensity was gradually augmented upon gradual addition of water in acetonitrile:water mixtures. For DTDBF, more interesting was to observe three single crystals of this compound with varied emission colors (Fig. 6). The molecular conformations, packing patterns and interactions in the three crystal structures of DTDBF were investigated. Specific strong intermolecular interactions (such as π–π stacking or H-/J-aggregates) were

Table 2 Comparison of the optical and photophysical properties of ditolyldibenzofulvene (DTDBF) solids in crystalline and amorphous states [101]. See sample structures in Fig. 6

Sample	λ^{em} (nm)[a]	ϕ_F[b]	A1/A2[c]	τ_1 (ns)	τ_2 (ns)	$k_F \times 10^{-7}$ (s^{-1})[d]	$k_{NR} \times 10^{-8}$ (s^{-1})[e]
CO	586	0.162	100/0	1.7		9.53	4.99
CY	545	0.233	100/0	2.9		8.03	2.64
CB	461	0.281	100/0	39.2		0.72	0.18
Am	557	0.029	80/20	0.9	1.9	2.64[f]	8.83[f]

[a]Emission maximum

[b]Fluorescence quantum yield determined with an integrating sphere with $\lambda^{ex} = 340$ nm

[c]Fraction (A, %) and lifetime (τ, ns) of shorter (1) or longer lived species (2)

[d]$k_F = \frac{\phi_F}{\tau_F}$;

[e]$k_{NR} = \frac{(1-\phi_F)}{\tau_F}$;

[f]Determined with an average lifetime, $<\tau> = 1.1$ ns

ruled out due to the twisted molecular conformations in all the crystal structures of DTDBF. Thus, the emissions of those crystals should occur from a single molecule which depends mainly on its molecular conformation. The asymmetric units of CB, CY and CO contain one, three, and six molecules of DTDBF respectively. The more C–H···π interactions presented in the crystal structure, the higher the photoluminescence efficiency (Table 2). The existence of these weak interactions between molecules helps lock the motion of the aryl rings, and, hence, to rigidify the molecular conformation [100], with consequent increase in photoluminescence efficiency due to CIE. While the three crystal structures display single exponential fluorescence decays, the amorphous solid of DTDBF displays a bi-exponential decay, with 80% contribution of the short-lived component and 20% of the long-lived component, suggesting the availability of fast decay processes due to the loose packing pattern (Table 2). The lifetime increases promptly with the number of C–H···π interactions when comparing the three crystal structures, which also agrees with the increase in photoluminescence efficiency. Nonetheless, the fluorescence quantum yield in crystalline form is at least five times higher than in amorphous powder.

Another luminogen core molecule that matches both ML properties with high emission efficiency in the solid state (CIE) is triphenylacrylonitrile (TPAN), see structure in Fig. 7 [111]. Combination of TPAN with diaryl [130] and tryphenylamine [131, 132] groups introduces donor–acceptor (D–A) characteristics into these systems, making these AIE-active luminogens into molecules with intramolecular charge transfer (ICT) character and ϕ_F values close to 1.0 in solid powder. The TPAN moiety was also attached to the backbone of electron-rich *N*-heterocycles to design D–A polymers that feature AIE properties (see next section) [133–136].

5 Design and Photophysical Studies of AIE-Active Polymers

Until 2 years ago the literature regarding the AIE effect in polymers was relatively limited, particularly when compared with oligomers showing this effect. This has generated increased interest, which led to a growth in publications of AIE effect in polymers from ~ 100 papers in 2016 to 190 in 2019. Although less attention has been devoted to polymers with AIE, several reviews have been published concerning synthetic strategies and smart material applications [56, 57, 137, 138] of AIE polymers, their mechanochromic fluorescent properties [139] and some specific applications of these polymers in biological systems [140] and as detector sensors for explosives [141].

5.1 Building AIE-Active Polymers from Small Molecules with AIE Properties

AIE-active polymers built from a combination of polycarbazoles and polytriphenylamines backbones (donor) and TPAN (acceptor) moieties were studied for the detection of 1,3,5-trinitrobenzene (TNB), used as prototypical nitroaromatic explosive, in THF:water mixtures [133]. Detection of TNB by fluorescence spectroscopy

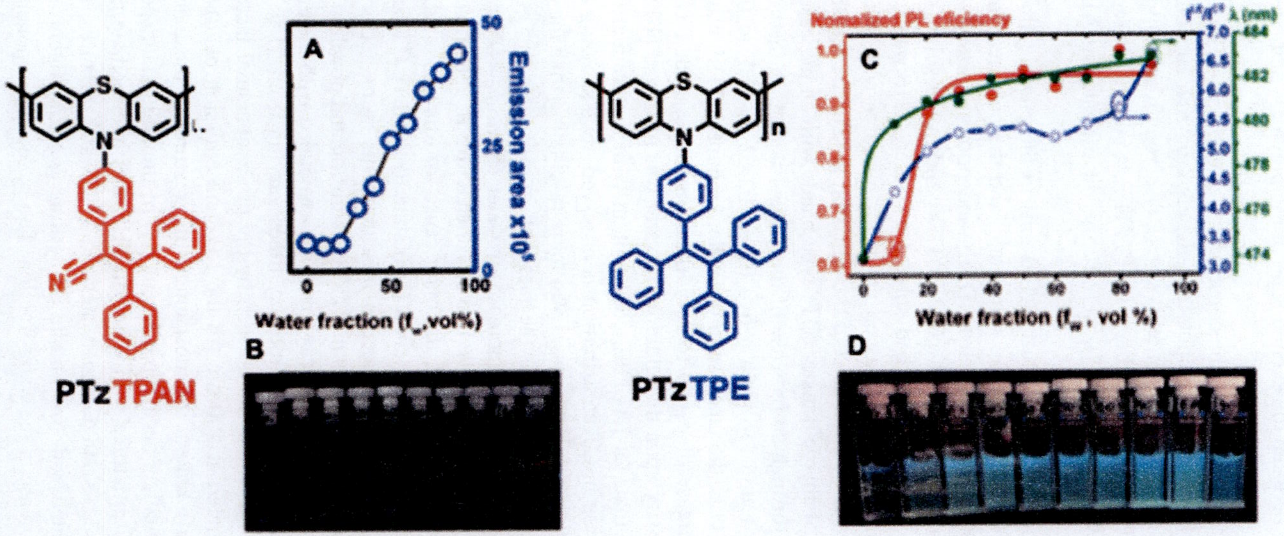

Fig. 7 Photophysical studies of PTZ-based polymers. PTZ-triphenylacrylonitrile (PTzTPAN): **a** Correlation of emission area with increasing water fraction in THF:water mixtures and **b** photos under UV irradiation (with $\lambda^{ex} = 365$ nm) of PTzTPAN in THF:water mixtures (from left 0% to right 90% with increments of water fraction (v/v) of 10%). PTZ-tetraphenylethene (PTzTPE): **c** correlation of emission area, maximum emission wavelength and ratio of intensity of locally excited (LE) versus charge transfer (CT) emission band with increasing water fraction for in THF:water mixtures for PTzTPE. **d** Photos under UV irradiation green fluorescence emission of PTzTPE in THF:water mixtures (from left 0% to right 90% with increments of water fraction (v/v) of 10%). Adapted from reference [142] with permission from ACS, Copyright 2018

at low polymer concentrations by the quenching of the emission of the polymers upon addition of TNB was investigated. The quenching mechanism was found to be based on electron transfer between the excited state of the polymers and the nitroaromatic analyte [133, 141].

The TPAN structure was also used for the construction of another AIE-active polymer, PTzTPAN (Fig. 7) [142]. With this polymer, the electron-rich backbone used was from another AIE building block: phenothiazine (PTz acronym in the polymer). This PTz-based polymer PTzTPAN, together with PTzTPE, a polymer based on the PTz backbone and the TPE moiety (Fig. 7), were further investigated as AIE luminogens [142].

PTzTPAN displays a significant red-shift (~100 nm) in the photoluminescence spectra on going from THF solution to the solid state, and the polymer was found to be non-emissive in polar solvents, such as DMF, and poorly emissive in non-polar solvents (ϕ_F=0.006 in dioxane and 0.002 in THF) and in the solid (powder) with ϕ_F=0.002 [142]. In classical good:bad solvent mixtures, THF:water, a clear enhancement of fluorescence emission was observed for mixtures with water fraction (f_w)≥40%, also followed by a red-shift of the emission [142]. The red-shift was attributed to the behavior of charge transfer (CT)-type bands with an increase in solvent polarity, in this case through an increase of the water fraction. In the good:bad solvent mixtures, ϕ_F was found to increase five times: 0.002 in THF to 0.010 in f_w=90% [142]. With PTzTPE, the increase in the water amount in THF: water mixtures led to a gradual increase of ϕ_F up to f_w=40%. Nonetheless, for PTzTPE, a decrease in CT band intensity upon the addition of water was observed [142]. Very interestingly, this showed that, for PTzTPE, concomitant with the gradual increase in the AIE effect, a decrease in CT contribution is observed, thus showing that the latter is in competition with the AIE process. Even more fascinating than the competition between the CT and AIE effects was the fact that, in dioxane:water mixtures, PTzTPE showed ACQ with the gradual increase of water (from ϕ_F=0.14 in pure dioxane to 0.042 in f_w=90%). Indeed, selective AIE behaviour in specific solvents was previously reported for D–A systems; the main cause for this was attributed to the transition from the emissive locally excited (LE) state to the dark CT state [72, 143]. Furthermore, time-resolved fluorescence measurements were performed to characterize the coexistence of these states in PTzTPE solution and powder (Fig. 8). In general, the shortest and the longest decay times are associated with the emission of the LE and CT species (τ_1 and τ_3) with decay times varying from 51 to 103 ps (LE) and 4.7 to 5.6 ns (CT), which can be seen by the pre-exponential factors, a_{i1}, when the decays are collected at 500 nm (in the LE state emission band) and a_{i3} for λ^{em}>600 nm (CT band). One of the most interesting features of these decays is the presence of the intermediate decay component (τ_2) in the solid state (powder), dioxane, and in the solvent mixtures with water; in THF, this intermediate lifetime, τ_2, is found to be absent (Fig. 8). In dioxane:water and THF:water mixtures, the pre-exponential values associated with this middle decay time (τ_2) component increase concomitantly with the increase in the water fraction. The τ_2 decay component could therefore be assigned to the emission of the polymer emissive aggregate, which was found to be present even in dioxane. Furthermore, fs-transient absorption (fs-TA) data obtained in THF shed light on the observed behavior, showing that aggregation

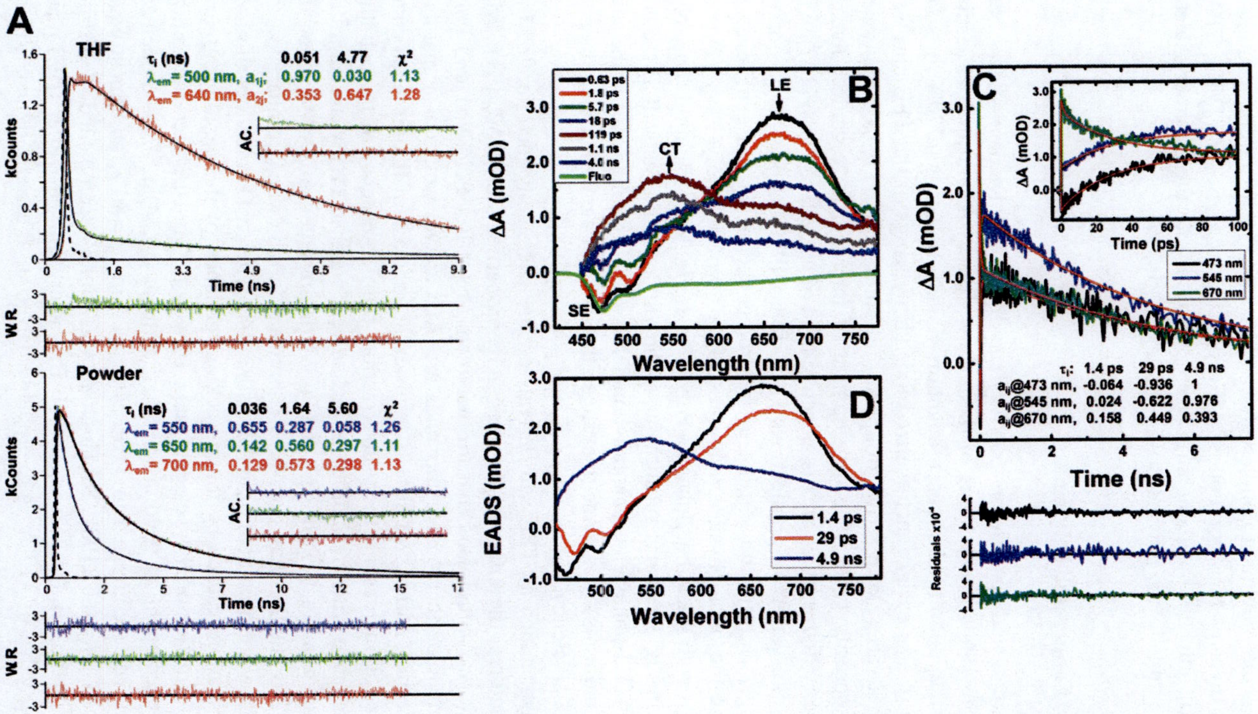

Fig. 8 **a** Room temperature fluorescence decays for PTzTPE collected with excitation at 404 and 451 nm in THF solution and in the solid state (powder), respectively. **b** Room temperature femtosecond time-resolved transient absorption (fs-TA) data for PTzTPE in THF solution collected with excitation at 430 nm. **c** Representative kinetic traces. **d** Evolution-associated difference spectra (EADS) obtained from global analysis of the TA data. *Insets* in **c** are the decays at shorter times. Adapted from reference [142] with permission from ACS, Copyright 2018

of the PTzTPE polymer indeed does not occur in THF. The fs-TA studies also showed that the CT state was formed both from direct excitation in the ground-state or at the expense of the decay of the LE excited state.

As shown in the previous examples, the general strategy for the preparation of AIE polymers involves incorporation of AIEgens, such as HPS, TPE [136], TPAN and DSA [140], into the backbone structure or grafted onto the polymer skeleton with different synthetic and polymerization strategies [56, 57].

5.2 TPE-Based Conjugated Polymers

TPE-based conjugated polymers—poly(arylene-diphenyl-vinylenes)—have been known since the 1960s [144]. Some of them show distinct AIE properties leading to high solid-state photoluminescence quantum yields—higher than 70% for some poly-TPEs [145]. The easily functionalized four phenyl rings of the TPE unit were combined with a variety of reactive groups in the synthesis of TPE-based polymers with inherent porosity [146]. The development of microporous organic polymers (MOPs) with very high surface area is an active field of current research [147], with high potential in a variety of applications [145, 148, 149], such as energy storage, light harvesting, catalysis and sensing of hazardous chemicals and explosives. A recent study on the photophysical properties of two types of poly(1,4-phenylene-diarylvinylene)s with 4-tert-butylphenyl (Poly-t-Bu in Fig. 9) or naphthyl (Poly-Np1 and Poly-Np2 in Fig. 9) as aryl substituents at the vinylene units as well as the corresponding poly(1,3-phenylene-dinaphthylvinylene), either in solution or in the solid state (powder and films), was made to evaluate the effect of aggregation on the photoluminescence efficiency, but also the polymer porosity [98].

When AIEgens, such as TPE and derivatives, are chemically incorporated into polymeric structures, they may import their AIE properties [138, 150, 151]. With this in mind, it was anticipated that the polymers Poly-t-Bu, Poly-Np1 and Poly-Np2, whose monomeric model compounds are AIE active, would also maintain their AIE properties. Although Poly-t-Bu showed an increase of fluorescence quantum yield upon the increase of the water fraction, the fluorescence of the naphthyl polymers is quenched with the addition of bad-solvent water in THF:water mixtures. Actually, as shown in Fig. 9, Poly-Np1 and Poly-Np2 presented ACQ. Moreover, the ϕ_F of Poly-Np2 is smaller than Poly-Np1 either in solution (when compared at the same f_W in THF:water mixtures) or in solid state [98]. The ACQ behaviour of naphthyl polymers may be caused by the π-stacking tendency of the naphthyl groups, leading to the formation of H-type aggregates. However, Poly-t-Bu presented a high photoluminescence quantum yield in solid state with $\phi_F = 0.64$ in film—an increase of ca. 14 times when compared to THF. Moreover, in $f_W = 90\%$, an almost equal value is obtained, $\phi_F = 0.62$ [98]. This high ϕ_F was found accompanied by distinct microporosity in the solid state with a high Brunauer–Emmett–Teller (BET) surface area, S_{BET}, value of 417 $m^2\ g^{-1}$, probably driven by the steric demand of the tert-butyl groups. Conversely, the occurrence of π–π interactions may also lead to tight packing of Poly-Np1 and Poly-Np2, since these polymers do not show intrinsic microporosity ($S_{BET} < 17\ m^2\ g^{-1}$) [98].

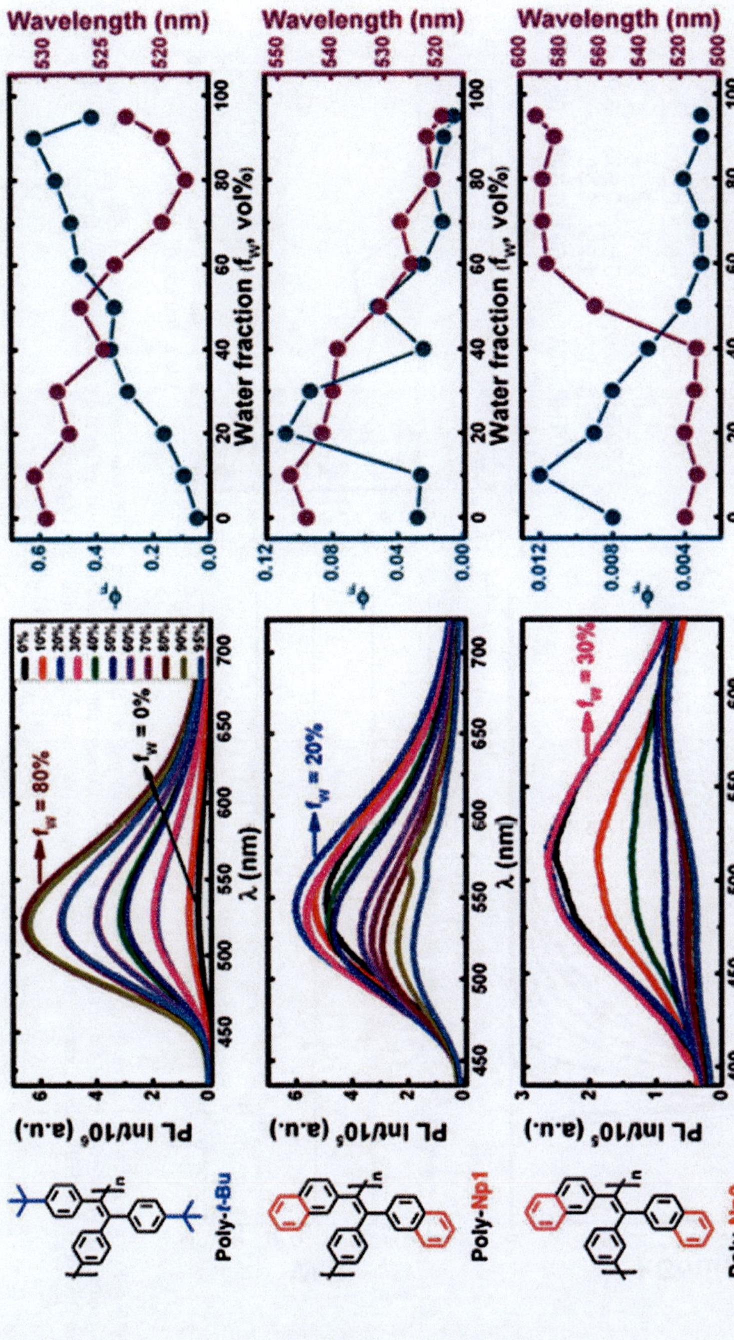

Fig. 9 Room temperature fluorescence emission [photoluminescence (PL)] spectra of the polymers Poly-t-Bu, Poly-Np1 and Poly-Np2 in THF:water mixtures and their respective correlation of the fluorescence quantum yield and emission maxima with increasing water fraction, f_W=0–95%. Adapted from reference [98] with permission from RSC, Copyright 2020

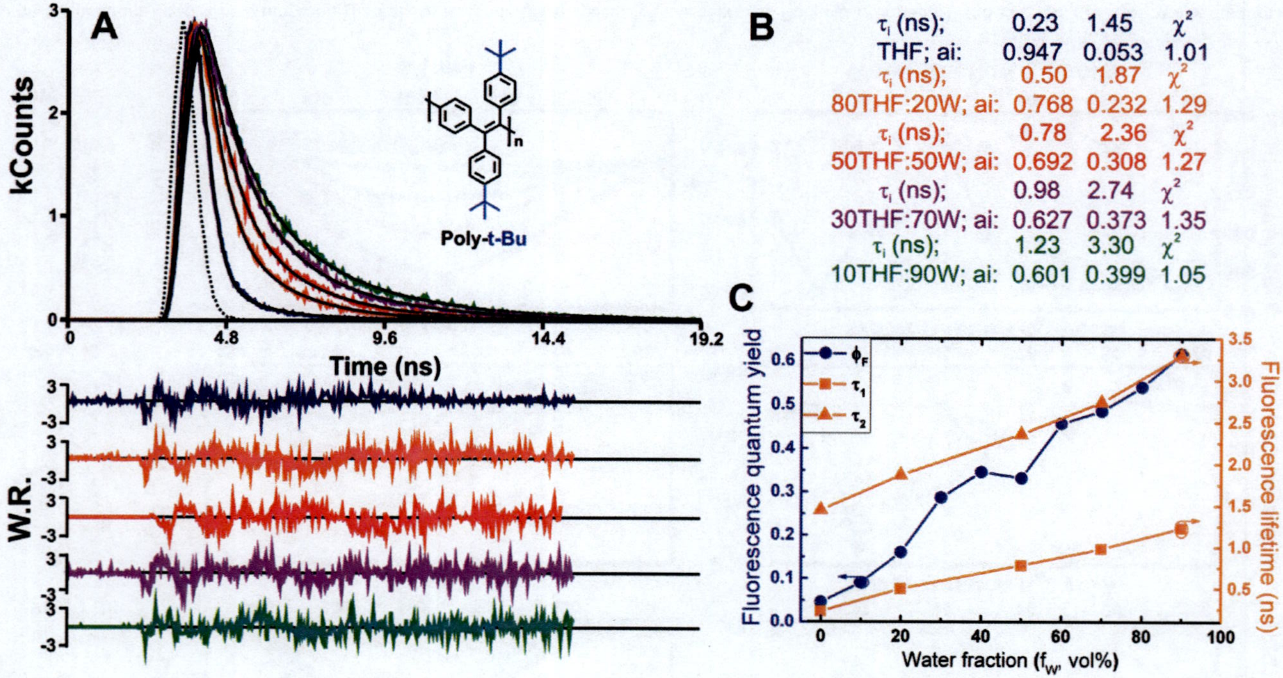

Fig. 10 **a** Room temperature fluorescence decays for Poly-t-Bu collected in THF solution and in THF:water mixtures. **b** Fluorescence decay times and χ^2 values of the analyzed decays of Poly-t-Bu in the different THF:water mixture. **c** Correlation of fluorescence quantum yields and fluorescence decay times for Poly-t-Bu with increasing water fraction (f_W) in THF:water mixtures. Adapted from reference [98] with permission from RSC, Copyright 2020

Time-resolved fluorescence decays were also obtained in order to gain further insight into the precise nature of the AIE phenomenon in Poly-t-Bu. The fluorescence decays of Poly-t-Bu in THF and THF:water mixtures were found to be well fitted with a bi-exponential decay law, thus indicating the presence of two emitting species in the aggregate region ($\lambda^{em}=525$ nm) [98]. This possibly indicates that the emissive aggregate results from two main conformers with different contributions to the photoluminescence decay. For Poly-t-Bu in THF, a fast decay component, τ_1, in the 0.23–1.2 ns range together with a longer decay time component, τ_2, varying from 1.5 to 3.3 ns, were found on going from 0 to 90% water content in THF:water mixtures [98].

As seen in Fig. 10, the fluorescence quantum yields and the fluorescence lifetimes of Poly-t-Bu follow a similar trend, i.e., a concomitant increase of these parameters with the increase in water content. In addition, the determination of the radiative (k_F) and radiationless (k_{NR}) rate constants (avoiding average lifetime contributions and instead considering the longer decay time, τ_2, usually associated to a higher fractional contribution, C_i, to the decays) clearly showed a decrease of k_{NR} and an increase of k_F with f_w. The radiationless decay is therefore dominant up to $f_w=70\%$, and, from there onwards, the radiative deactivation begins to dominate, mirroring a high contribution of this deactivation channel due to the AIE effect [98].

Conjugated polymers (CPs) have great potential in various applications ranging from electronics, aerospace, sensing, catalysis and energy; hence there is intensive investigation of these materials [152]. Their electronic properties are, however, highly dependent on the physical conformation of the polymeric chains and the way these chains pack together in films [153]. An obstacle to their development is the significant fluorescence quenching in the solid state that generally occurs due to strong interchain interaction, as in the example of the polythiophenes [154]. On the other hand, oligo- and poly-thiophenes embrace one of the characteristics associated with the AIE effect: structural rigidification in the excited state. Whereas in molecules where this AIE effect occurs, the generic mechanism involves rigidification associated to close proximity of another molecule (aggregation-induced), in oligo- and poly-thiophenes this essentially results from the quinoidal-like structure (extended π-conjugation), in contrast with the possibility of a more twisted conformer(s) (and this depends on the substitution) in the ground state [155–158]. This obviously paves the way for the study of oligo-and poly-thiophenes as potential AIEgens.

To try to overcome the ACQ effect, two TPE-polythiophenes, homoPT and coPT (structures in Fig. 11), with different degrees of TPE substitution were recently investigated [153]. The introduction of the propeller-shaped TPE units prevent the molecules from packing in a π–π stacking process in the polymeric aggregate, therefore precluding emission quenching of these polythiophenes in the solid state. To understand the role of the sterically bulky TPE pendant groups on the inhibition of intra- and/or interchain packing (aggregation) of these conjugated polymers, photohysical studies were performed with poly(3-hexylthiophene), P3HT, with similar average molecular weight, used as a "pure" model polythiophene, and a TPE-thiophene (T-TPE) model oligomeric compound. Unlike what was predicted (TPE

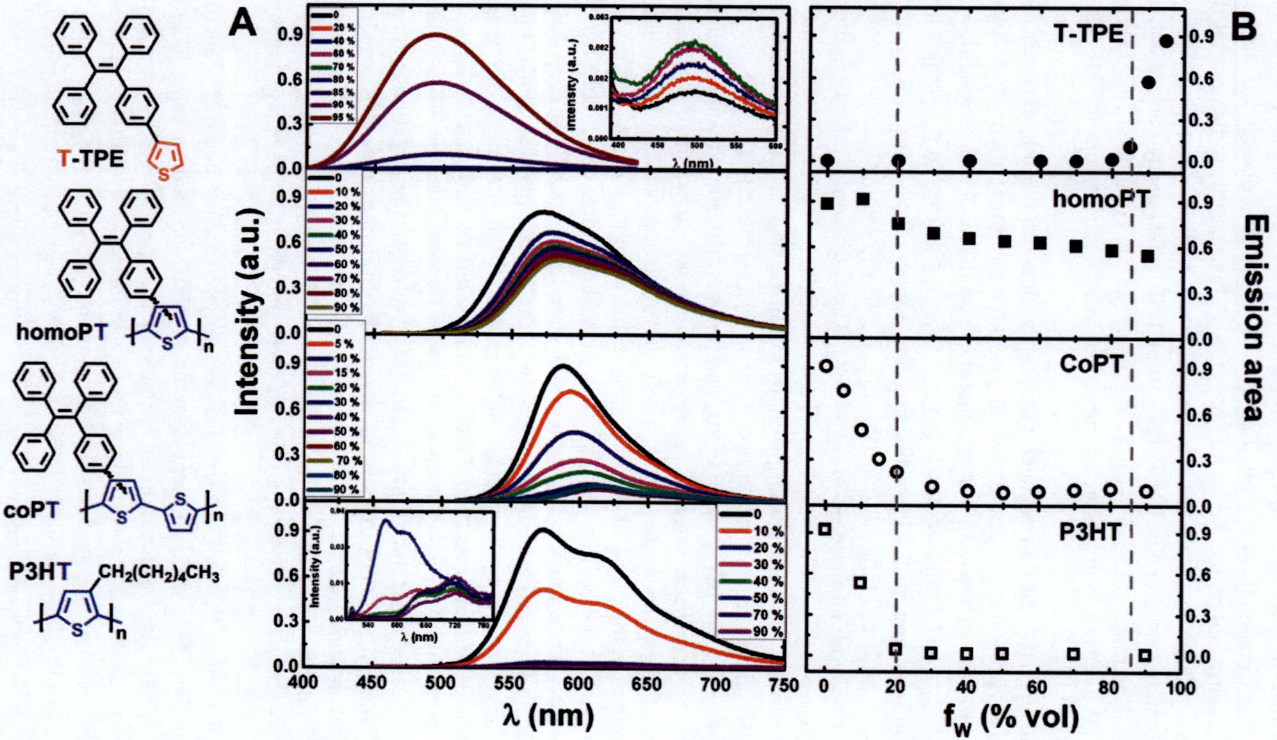

Fig. 11 Structures and acronyms of model compound and poly-thiophenes. **a** Fluorescence emission spectra and **b** correlation of emission area with increasing water fractions (f_w) in THF/water mixtures for the polythiophenes, homoPT, coPT, and P3HT, together with the model compound T-TPE. Adapted from reference [153], with permission from ACS, Copyright 2020

preventing π–π stacking and consequent quenching of emission), homoPT and coPT, as well as P3HT, showed lower ϕ_F values in thin films than in solution of good solvents. Among the three polymers, photoluminescence quenching was found higher for P3HT, with a ϕ_F decreasing from 0.19 in toluene to 0.009 in thin films [153]. It is interesting to realize that emission quenching, followed by the red-shifted emission spectra in the solid state, was previously attributed to emission from interchain (aggregate) excited states in P3HT films, where symmetry considerations reduce the intensity of the high energy emission [159]. Amongst the poly(TPE-thiophene) polymers, the quenching was found more pronounced for coPT ($\phi_F = 0.22$ in toluene versus $\phi_F = 0.03$ in the solid film—a ratio of ~ 7) than for homoPT (0.13 versus 0.08, a ratio of ~ 1.6). These results suggest that for the PT with a TPE substituent in each thiophene unit, homoPT, the higher steric hindrance reduces planarization of the main backbone polymer chain in the film (although imparting rigidification of the skeketon) and thus the ACQ process is less efficient than for the other polythiophene counterparts. To further study these polymers, lead ACQ compounds in solvent-nonsolvent mixtures were investigated in THF:water mixtures and correlated with the monomer T-TPE (Fig. 11).

For T-TPE, a clear enhancement of the emission band was observed for mixtures with a $f_W \geq 90\%$, thus showing the presence of the AIE effect. For the polymers, a red-shift of the emission was found on going from THF to $f_W = 90\%$ (where the highest degree of aggregation is expected), see Fig. 11. For the coPT and P3HT polymers, ACQ is clearly dominant, starting immediately from very low water fractions, whereas for homoPT, in THF:water mixtures, suppression of fluorescence emission was less pronounced than for coPT; indeed, only a minor gradual decrease of fluorescence was observed up to the mixture with the highest water content. Thus, contrary to what is observed for the coPT polymer, the characteristic thiophene-TPE monomeric unit of the homoPT polymer restricts (although not completely) formation of non-emissive aggregates. This shows that a higher degree of TPE substituents (as in homoPT) restricts formation of non-emissive aggregates, basically balancing the AIE and ACQ effects (Fig. 11).

Another interesting feature coming from the study of these polymers is the observation from time-resolved fluorescence studies, as a function of temperature, that for homoPT and coPT, the deactivation of the excited state occurs mainly through energy migration along the polymer backbone. Moreover, for P3HT, the excited state decay involves competition between conformational relaxation (associated with decay times in the 12.8–15.4 ps range) and energy migration (~ 80 ps) [153]. Again, absence of conformational relaxation for TPE-substituted polythiophenes can be attributed to the restriction of torsional movements promoted by the bulky TPE substituents in these polythiophenes. The study showed that, even though the strategy of incorporation AIE-luminogen into the polymer backbone could not circumvent the ACQ effect in all polythiophenes, restriction in the formation of nonemissive aggregates was observed. Indeed, T-TPE showed AIE, and with homoPT the ACQ and AIE effects apparently cancel each other out [153].

6 Concluding Remarks

In the past 20 years, and associated with the discovery of the AIE effect, a plethora of AIEgen molecules has been developed. The challenges are now in the development of applications to take real advantage of this effect. Despite the enormous number of structures, the rational design of new structures remains a challenge. The future design of innovative AIE systems, small molecules (oligomers) and polymers, will take advantage of known photophysical mechanisms to prevent ACQ phenomena in solid, aggregate or confined media. Despite the enormous number of studies in the past two decades on this effect, of making emissive molecules that, otherwise isolated, are non-emissive, this seems just the beginning of a chapter of exciting new research. The rediscovery of molecules that may suffer structural changes in the excited state (e.g., molecular rigidity or the decrease of the number of low lying vibrational levels by minimizing the contribution via the loose bolt effect) opens a new route for an enormous variety of AIEgens. Additional strategies to increase the photoluminescence emission in the solid, or aggregate state, will surely include the design of J-aggregate forming compounds, molecules with room temperature phosphorescence and thermally activated delayed fluorescence.

Acknowledgements The authors acknowledge financial support from Project "Hylight" (no. 031625) 02/SAICT/2017, PTDC/QUI-QFI/31625/2017, which is funded by the Portuguese Science Foundation (Fundação para Ciência e Tecnologia, FCT) and COMPETE Centro 2020. We also acknowledge funding by Fundo Europeu de Desenvolvimento Regional (FEDER) through COMPETE and project ROTEIRO/0152/2013. The Coimbra Chemistry Centre (CQC) is supported through projects UIDB/00313/2020 and UIDP/00313/2020, co-funded by FCT and COMPETE 2020.

Compliance with Ethical Standards

Conflict of interest The authors declare that there is no conflict of interests.

References

1. Luo J, Xie Z, Lam JW, Cheng L, Chen H, Qiu C, Kwok HS, Zhan X, Liu Y, Zhu D, Tang BZ (2001) Aggregation-induced emission of 1-methyl-1,2,3,4,5-pentaphenylsilole. Chem Commun 18:1740–1741. https://doi.org/10.1039/b105159h
2. Birks JB (1970) Photophysics of aromatic molecules. Wiley, London
3. Tang BZ, Zhan X, Yu G, Sze Lee PP, Liu Y, Zhu D (2001) Efficient blue emission from siloles. J Mat Chem 11(12):2974–2978. https://doi.org/10.1039/b102221k
4. Jenekhe SA, Chen XL (2000) Supramolecular photophysics of self-assembled block copolymers containing luminescent conjugated polymers. J Phys Chem B 104(27):6332–6335. https://doi.org/10.1021/jp000896u
5. Pant DD, Bhagchandani CL, Pant KC, Verma SP (1971) Aggregation in xanthene dyes, exciton emission and phosphorescence enhancement. Chem Phys Lett 9(6):546–547. https://doi.org/10.1016/0009-2614(71)85121-7
6. Winnik FM (1993) Photophysics of preassociated pyrenes in aqueous polymer solutions and in other organized media. Chem Rev 93(2):587–614. https://doi.org/10.1021/cr00018a001
7. Kitamura T, Takahashi Y, Yamanaka T, Uchida K (1991) Luminescence associated with the molecular aggregation of hydrocarbons doped in amorphous silica glasses. J Lumin 48–49:373–376. https://doi.org/10.1016/0022-2313(91)90141-h

8. Rudalevige T, Francis AH, Zand R (1998) Spectroscopic studies of fullerene aggregates. J Phys Chem A 102(48):9797–9802. https://doi.org/10.1021/jp9832591
9. Carter PW, DiMagno SG, Porter JD, Streitwieser A (1993) pi-Stacking and aggregation of pyridinium-substituted indolizin. J Phys Chem 97(5):1085–1096. https://doi.org/10.1021/j100107a017
10. Xu W, Guo H, Akins DL (2001) Aggregation and exciton emission of a cyanine dye encapsulated within mesoporous MCM-41. J Phys Chem B 105(32):7686–7689. https://doi.org/10.1021/jp004154c
11. Möbius D (1995) Scheibe aggregates. Adv Mater 7(5):437–444. https://doi.org/10.1002/adma.19950070503
12. Scheibe G (1937) Über die Veränderlichkeit der Absorptionsspektren in Lösungen und die Nebenvalenzen als ihre Ursache. Angew Chem-Ger Edit 50(11):212–219. https://doi.org/10.1002/ange.19370501103
13. Jelley EE (1936) Spectral absorption and fluorescence of dyes in the molecular state. Nature 138(3502):1009–1010. https://doi.org/10.1038/1381009a0
14. Jelley EE (1937) Molecular, nematic and crystal states of I: I-diethyl–cyanine chloride. Nature 139(3519):631–631. https://doi.org/10.1038/139631b0
15. Wurthner F, Kaiser TE, Saha-Moller CR (2011) J-aggregates: from serendipitous discovery to supramolecular engineering of functional dye materials. Angew Chem Int Ed Engl 50(15):3376–3410. https://doi.org/10.1002/anie.201002307
16. Kasha M, Rawls HR, El-Bayoumi MA (1965) The exciton model in molecular spectroscopy. Pure Appl Chem 11(3–4):371–392. https://doi.org/10.1351/pac196511030371
17. Brixner T, Hildner R, Köhler J, Lambert C, Würthner F (2017) Exciton transport in molecular aggregates—from natural antennas to synthetic chromophore systems. Adv Energy Mater 7:16. https://doi.org/10.1002/aenm.201700236
18. Kasha M (1976) Molecular Excitons in Small Aggregates. In: Di Bartolo B, Pacheco D, Goldberg V (eds) Spectroscopy of the excited state, vol 12, vol NATO Advanced Study Institutes Series (Series B: Physics). Springer, Boston. https://doi.org/10.1007/978-1-4684-2793-6_12
19. Deshmukh AP, Koppel D, Chuang C, Cadena DM, Cao J, Caram JR (2019) Design principles for two-dimensional molecular aggregates using Kasha's Model: tunable photophysics in near and short-wave infrared. J Phys Chem C 123(30):18702–18710. https://doi.org/10.1021/acs.jpcc.9b05060
20. Oelkrug D, Tompert A, Egelhaaf H-J, Hanack M, Steinhuber E, Hohloch M, Meier H, Stalmach U (1996) Towards highly luminescent phenylene vinylene films. Synth Methods 83(3):231–237. https://doi.org/10.1016/s0379-6779(96)04484-0
21. Oelkrug D, Tompert A, Gierschner J, Egelhaaf H-J, Hanack M, Hohloch M, Steinhuber E (1998) Tuning of fluorescence in films and nanoparticles of oligophenylenevinylenes. J Phys Chem B 102(11):1902–1907. https://doi.org/10.1021/jp973225d
22. Choi S, Bouffard J, Kim Y (2014) Aggregation-induced emission enhancement of a meso-trifluoromethyl BODIPY via J-aggregation. Chem Sci 5(2):751–755. https://doi.org/10.1039/c3sc52495g
23. Sheng W, Zheng Y-Q, Wu Q, Chen K, Li M, Jiao L, Hao E, Wang J-Y, Pei J (2020) Synthesis, characterization, and tunable semiconducting properties of aza-BODIPY derived polycyclic aromatic dyes. Sci China Chem 63(9):1240–1245. https://doi.org/10.1007/s11426-020-9807-7
24. Dong Y, Lam JWY, Qin A, Li Z, Sun J, Kwok HS, Tang BZ (2006) Aggregation-induced emission, vol 6333. SPIE optics + photonics. SPIE, Bellingham WA. https://doi.org/10.1117/12.679373
25. Tong H, Dong Y, Haussler M, Lam JW, Sung HH, Williams ID, Sun J, Tang BZ (2006) Tunable aggregation-induced emission of diphenyldibenzofulvenes. Chem Commun 10:1133–1135. https://doi.org/10.1039/b515798f
26. Dong Y, Lam JW, Qin A, Li Z, Sun J, Sung HH, Williams ID, Tang BZ (2007) Switching the light emission of (4-biphenylyl)phenyldibenzofulvene by morphological modulation: crystallization-induced emission enhancement. Chem Commun 1:40–42. https://doi.org/10.1039/b613157c
27. Ren Y, Lam JW, Dong Y, Tang BZ, Wong KS (2005) Enhanced emission efficiency and excited state lifetime due to restricted intramolecular motion in silole aggregates. J Phys Chem B 109(3):1135–1140. https://doi.org/10.1021/jp046659z
28. Wurthner F (2020) Aggregation-induced emission (AIE): a historical perspective. Angew Chem Int Ed Engl 59(34):14192–14196. https://doi.org/10.1002/anie.202007525
29. Valeur B, Berberan-Santos MN (2012) Structural effects on fluorescence emission. In: Molecular fluorescence. Wiley, New York, pp 75–107. https://doi.org/10.1002/9783527650002.ch4

30. Takezaki M, Hirota N, Terazima M (1996) Excited state dynamics of 9,10-diazaphenanthrene studied by the time-resolved transient grating method. J Phys Chem 100(24):10015–10020. https://doi.org/10.1021/jp9602540
31. Tabuchi M, Momotake A, Kanna Y, Nishimura Y, Arai T (2011) Extremely efficient and long lifetime fluorescence of cis-stilbene contained in a rigid dendrimer. Photochem Photobiol Sci 10(10):1521–1523. https://doi.org/10.1039/C1PP05126A
32. Saltiel J, Megarity ED, Kneipp KG (1966) The mechanism of direct CIS-trans photoisomerization of the stilbenes. J Am Chem Soc 88(10):2336–2338. https://doi.org/10.1021/ja00962a057
33. Charlton JL, Saltiel J (1977) An analysis of trans-stilbene fluorescence quantum yields and lifetimes. J Phys Chem 81(20):1940–1944. https://doi.org/10.1021/j100535a011
34. Saltiel J, Zafiriou OC, Megarity ED, Lamola AA (1968) Tests of the singlet mechanism for cis-trans photoisomerization of the stilbenes. J Am Chem Soc 90(17):4759. https://doi.org/10.1021/ja01019a063
35. DeBoer CD, Schlessinger RH (1968) Multiplicity of the photochemically reactive state of 1,2-diphenylcyclobutene. J Am Chem Soc 90(3):803–804. https://doi.org/10.1021/ja01005a052
36. Valeur B (2002) Molecular fluorescence: principles and applications. Wiley, Weinheim
37. Tsin ATC, Pedrozo-Fernandez HA, Gallas JM, Chambers JP (1988) The fluorescence quantum yield of vitamin A2. Life Sci 43(17):1379–1384. https://doi.org/10.1016/0024-3205(88)90304-9
38. Schoof S, Güsten H, Von Sonntag C (1978) Fluoreszenz und Fluoreszenzlöschung von meso-substituierten Anthracenderivaten in Lösung. Berichte Bunsengesellschaft Phys Chem. 82(10):1068–1073. https://doi.org/10.1002/bbpc.19780821009
39. Guesten H, Mintas M, Klasinc L (1980) Deactivation of the fluorescent state of 9-tert-butylanthracene. 9-tert-Butyl-9,10(Dewar anthracene). J Am Chem Soc 102(27):7936–7937. https://doi.org/10.1021/ja00547a022
40. Jiao L, Yu C, Wang J, Briggs EA, Besley NA, Robinson D, Ruedas-Rama MJ, Orte A, Crovetto L, Talavera EM, Alvarez-Pez JM, Van der Auweraer M, Boens N (2015) Unusual spectroscopic and photophysical properties of meso-tert-butylBODIPY in comparison to related alkylated BODIPY dyes. RSC Adv 5(109):89375–89388. https://doi.org/10.1039/C5RA17419H
41. Zhang H, Liu J, Du L, Ma C, Leung NLC, Niu Y, Qin A, Sun J, Peng Q, Sung HHY, Williams ID, Kwok RTK, Lam JWY, Wong KS, Phillips DL, Tang BZ (2019) Drawing a clear mechanistic picture for the aggregation-induced emission process. Mater Chem Front 3(6):1143–1150. https://doi.org/10.1039/C9QM00156E
42. Dong Y, Lam JWY, Qin A, Liu J, Li Z, Tang BZ, Sun J, Kwok HS (2007) Aggregation-induced emissions of tetraphenylethene derivatives and their utilities as chemical vapor sensors and in organic light-emitting diodes. Appl Phys Lett 91:1. https://doi.org/10.1063/1.2753723
43. Tanaka Y, Machida T, Noumi T, Sada K, Kokado K (2020) Emissive tetraphenylethylene (TPE) derivatives in a dissolved state tightly fastened by a short oligo(ethylene glycol) chain. Org Chem Front 7(18):2649–2656. https://doi.org/10.1039/d0qo00839g
44. Stegemeyer H (1968) Luminescence of sterically hindered arylethylenes. Ber Bunsen Phys Chem 72(2):335–340. https://doi.org/10.1002/bbpc.19680720264
45. Fischer G, Fischer E, Stegemeyer H (1973) Fluorescence and absorption spectra of sterically hindered arylethylenes in the neat glassy state. Ber Bunsen Phys Chem 77(9):685–687. https://doi.org/10.1002/bbpc.19730770907
46. Leigh WJ, Arnold DR (1981) Merostabilization in radical ions, triplets, and biradicals. 6. The excited state behaviour of para-substituted tetraphenylethylenes. Can J Chem 59(21):3061–3075. https://doi.org/10.1139/v81-448
47. Sharafy S, Muszkat KA (1971) Viscosity dependence of fluorescence quantum yields. J Am Chem Soc 93(17):4119–4125. https://doi.org/10.1021/ja00746a004
48. Klingenberg HH, Lippert E, Rapp W (1973) The influence of an intramolecular relaxation process on the frequency-time behaviour or the fluorescence of tetraphenylethylene in highly viscous solutions. Chem Phys Lett 18(3):417–419. https://doi.org/10.1016/0009-2614(73)80206-4
49. Barbara PF, Rand SD, Rentzepis PM (1981) Direct measurements of tetraphenylethylene torsional motion by picosecond spectroscopy. J Am Chem Soc 103(9):2156–2162. https://doi.org/10.1021/ja00399a003
50. Greene BI (1981) Observation of a long-lived twisted intermediate following picosecond UV excitation of tetraphenylethylene. Chem Phys Lett 79(1):51–53. https://doi.org/10.1016/0009-2614(81)85286-4

51. Zijlstra RWJ, van Duijnen PT, Feringa BL, Steffen T, Duppen K, Wiersma DA (1997) Excited-state dynamics of tetraphenylethylene: ultrafast stokes shift, isomerization, and charge separation. J Phys Chem A 101(51):9828–9836. https://doi.org/10.1021/jp971763k
52. Turro NJ (1991) Modern molecular photochemistry. University Science Books, Mill Valley
53. Becker RS (1969) Theory and interpretation of fluorescence and phosphorescence. Wiley, New York
54. Xu S, Duan Y, Liu B (2020) Precise molecular design for high-performance luminogens with aggregation-induced emission. Adv Mater 32(1):e1903530. https://doi.org/10.1002/adma.201903530
55. Yang J, Chi Z, Zhu W, Tang BZ, Li Z (2019) Aggregation-induced emission: a coming-of-age ceremony at the age of eighteen. Sci China Chem 62(9):1090–1098. https://doi.org/10.1007/s11426-019-9512-x
56. Hu R, Qin A, Tang BZ (2020) AIE polymers: synthesis and applications. Prog Polym Sci. https://doi.org/10.1016/j.progpolymsci.2019.101176
57. Zhou S-Y, Wan H-B, Zhou F, Gu P-Y, Xu Q-F, Lu J-M (2019) AIEgens-lightened functional polymers: synthesis, properties and applications. Chin J Polym Sci 37(4):302–326. https://doi.org/10.1007/s10118-019-2217-0
58. Li J, Wang J, Li H, Song N, Wang D, Tang BZ (2020) Supramolecular materials based on AIE luminogens (AIEgens): construction and applications. Chem Soc Rev 49(4):1144–1172. https://doi.org/10.1039/c9cs00495e
59. Pina J, Pinheiro D, Nascimento B, Pineiro M, Seixas de Melo JS (2014) The effect of polyaromatic hydrocarbons on the spectral and photophysical properties of diaryl-pyrrole derivatives: an experimental and theoretical study. Phys Chem Chem Phys 16(34):18319–18326. https://doi.org/10.1039/c4cp01797h
60. Chang D, Chen J, Liu Y, Huang H, Qin A, Deng GJ (2021) Metal-free synthesis and photophysical properties of 1,2,4-triarylpyrroles. J Org Chem 86:110–127 https://doi.org/10.1021/acs.joc.0c01788
61. Chen M, Li L, Wu H, Pan L, Li S, He B, Zhang H, Sun JZ, Qin A, Tang BZ (2018) Unveiling the different emission behavior of polytriazoles constructed from pyrazine-based AIE monomers by click polymerization. ACS Appl Mater Interfaces 10(15):12181–12188. https://doi.org/10.1021/acsami.8b03178
62. Guo Z, Yan C, Zhu W-H (2020) High-performance quinoline-malononitrile core as a building block for the diversity-oriented synthesis of AIEgens. Angew Chem Int Ed Engl 59(25):9812–9825. https://doi.org/10.1002/anie.201913249
63. Qian Y, Cai MM, Xie LH, Yang GQ, Wu SK, Huang W (2011) Restriction of photoinduced twisted intramolecular charge transfer. ChemPhysChem 12(2):397–404. https://doi.org/10.1002/cphc.201000457
64. Zhou P, Li P, Zhao Y, Han K (2019) Restriction of flip-flop motion as a mechanism for aggregation-induced emission. J Phys Chem Lett 10(21):6929–6935. https://doi.org/10.1021/acs.jpclett.9b02922
65. Yin PA, Wan Q, Niu Y, Peng Q, Wang Z, Li Y, Qin A, Shuai Z, Tang BZ (2020) Theoretical and experimental investigations on the aggregation-enhanced emission from dark state: vibronic coupling effect. Adv Electron Mater 6:7. https://doi.org/10.1002/aelm.202000255
66. Hong Y, Lam JW, Tang BZ (2009) Aggregation-induced emission: phenomenon, mechanism and applications. Chem Commun 29:4332–4353. https://doi.org/10.1039/b904665h
67. Mei J, Hong Y, Lam JW, Qin A, Tang Y, Tang BZ (2014) Aggregation-induced emission: the whole is more brilliant than the parts. Adv Mater 26(31):5429–5479. https://doi.org/10.1002/adma.201401356
68. Leung NLC, Xie N, Yuan W, Liu Y, Wu Q, Peng Q, Miao Q, Lam JWY, Tang BZ (2014) Restriction of intramolecular motions: the general mechanism behind aggregation-induced emission. Chem Eur J 20(47):15349–15353. https://doi.org/10.1002/chem.201403811
69. Mazzucato U, Spalletti A (2009) Competition between photoisomerization and photocyclization of the cis isomers of n-styrylnaphthalenes and -phenanthrenes. J Phys Chem A 113(52):14521–14529. https://doi.org/10.1021/jp904017e
70. Prlj A, Doslic N, Corminboeuf C (2016) How does tetraphenylethylene relax from its excited states? Phys Chem Chem Phys 18(17):11606–11609. https://doi.org/10.1039/c5cp04546k
71. Gao YJ, Chang XP, Liu XY, Li QS, Cui G, Thiel W (2017) Excited-state decay paths in tetraphenylethene derivatives. J Phys Chem A 121(13):2572–2579. https://doi.org/10.1021/acs.jpca.7b00197

72. Gao BR, Wang HY, Hao YW, Fu LM, Fang HH, Jiang Y, Wang L, Chen QD, Xia H, Pan LY, Ma YG, Sun HB (2010) Time-resolved fluorescence study of aggregation-induced emission enhancement by restriction of intramolecular charge transfer state. J Phys Chem B 114(1):128–134. https://doi.org/10.1021/jp909063d
73. Sasaki S, Drummen GPC, Konishi G-i (2016) Recent advances in twisted intramolecular charge transfer (TICT) fluorescence and related phenomena in materials chemistry. J Mater Chem C 4(14):2731–2743. https://doi.org/10.1039/c5tc03933a
74. Liu J, Zhang C, Dong J, Zhu J, Shen C, Yang G, Zhang X (2017) Endowing a triarylboron compound showing ACQ with AIE characteristics by transforming its emissive TICT state to be dark. RSC Adv 7(24):14511–14515. https://doi.org/10.1039/c7ra00426e
75. Tseng N-W, Liu J, Ng JCY, Lam JWY, Sung HHY, Williams ID, Tang BZ (2012) Deciphering mechanism of aggregation-induced emission (AIE): Is E-Zisomerisation involved in an AIE process? Chem Sci 3(2):493–497. https://doi.org/10.1039/c1sc00690h
76. Garg K, Ganapathi E, Rajakannu P, Ravikanth M (2015) Stereochemical modulation of emission behaviour in E/Z isomers of diphenyldipyrroethene from aggregation induced emission to crystallization induced emission. Phys Chem Chem Phys 17(29):19465–19473. https://doi.org/10.1039/c5cp02400e
77. Yang Z, Qin W, Leung NLC, Arseneault M, Lam JWY, Liang G, Sung HHY, Williams ID, Tang BZ (2016) A mechanistic study of AIE processes of TPE luminogens: intramolecular rotation vs configurational isomerization. J Mater Chem C 4(1):99–107. https://doi.org/10.1039/c5tc02924d
78. Xie Y, Li Z (2019) Recent advances in the Z/E isomers of tetraphenylethene derivatives: stereoselective synthesis, AIE mechanism, photophysical properties, and application as chemical probes. Chem Asian J 14(15):2524–2541. https://doi.org/10.1002/asia.201900282
79. Stojanović L, Crespo-Otero R (2019) Understanding aggregation induced emission in a propeller-shaped blue emitter. ChemPhotoChem 3(9):907–915. https://doi.org/10.1002/cptc.201900075
80. Xu Y, Wang F, Chen X, Liu Y, Zhou Z, Teng B (2019) A new strategy of design and development of aggregation-induced emission materials based on a deep insight into mechanism. J Phys Chem C 123(48):29379–29385. https://doi.org/10.1021/acs.jpcc.9b06198
81. Suzuki S, Sasaki S, Sairi AS, Iwai R, Tang BZ, Konishi G-i (2020) Principles of aggregation-induced emission: design of deactivation pathways for advanced AIEgens and applications. Angew Chem Int Ed Engl 59(25):9856–9867. https://doi.org/10.1002/anie.202000940
82. Kokado K, Sada K (2019) Consideration of molecular structure in the excited state to design new luminogens with aggregation-induced emission. Angew Chem Int Ed Engl 58(26):8632–8639. https://doi.org/10.1002/anie.201814462
83. Tu Y, Liu J, Zhang H, Peng Q, Lam JWY, Tang BZ (2019) Restriction of access to the dark state: a new mechanistic model for heteroatom-containing AIE systems. Angew Chem Int Ed Engl 58(42):14911–14914. https://doi.org/10.1002/anie.201907522
84. Crespo-Otero R, Li Q, Blancafort L (2019) Exploring potential energy surfaces for aggregation-induced emission—from solution to crystal. Chem Asian J 14(6):700–714. https://doi.org/10.1002/asia.201801649
85. Li Q, Blancafort L (2013) A conical intersection model to explain aggregation induced emission in diphenyl dibenzofulvene. Chem Commun 49(53):5966–5968. https://doi.org/10.1039/C3CC41730A
86. Peng X-L, Ruiz-Barragan S, Li Z-S, Li Q-S, Blancafort L (2016) Restricted access to a conical intersection to explain aggregation induced emission in dimethyl tetraphenylsilole. J Mater Chem C 4(14):2802–2810. https://doi.org/10.1039/C5TC03322E
87. Suzuki S, Maeda S, Morokuma K (2015) Exploration of quenching pathways of multiluminescent acenes using the GRRM method with the SF-TDDFT method. J Phys Chem A 119(47):11479–11487. https://doi.org/10.1021/acs.jpca.5b07682
88. Kokado K, Machida T, Iwasa T, Taketsugu T, Sada K (2018) Twist of C=C bond plays a crucial role in the quenching of AIE-active tetraphenylethene derivatives in solution. J Phys Chem C 122(1):245–251. https://doi.org/10.1021/acs.jpcc.7b11248
89. Yoshii R, Nagai A, Tanaka K, Chujo Y (2013) Highly emissive boron ketoiminate derivatives as a new class of aggregation-induced emission fluorophores. Chem Eur J 19(14):4506–4512. https://doi.org/10.1002/chem.201203703
90. Nie H, Hu K, Cai Y, Peng Q, Zhao Z, Hu R, Chen J, Su S-J, Qin A, Tang BZ (2017) Tetraphenylfuran: aggregation-induced emission or aggregation-caused quenching? Mater Chem Front 1(6):1125–1129. https://doi.org/10.1039/C6QM00343E

91. Baig MZK, Prusti B, Roy D, Chakravarty M (2019) Positional variation of monopyridyl-N in unsymmetrical anthracenyl pi-conjugates: difference between solution- and aggregate-state emission behavior. ACS Omega 4(3):5052–5063. https://doi.org/10.1021/acsomega.9b00046
92. Tang Y, Tang BZ (2019) Principles and applications of aggregation-induced emission. Springer, Berlin. https://doi.org/10.1007/978-3-319-99037-8
93. Wang Y, Liu T, Bu L, Li J, Yang C, Li X, Tao Y, Yang W (2012) Aqueous nanoaggregation-enhanced one- and two-photon fluorescence, crystalline J-aggregation-induced red shift, and amplified spontaneous emission of 9,10-bis(p-dimethylaminostyryl)anthracene. J Phys Chem C 116(29):15576–15583. https://doi.org/10.1021/jp3031094
94. Zhou J, Chang Z, Jiang Y, He B, Du M, Lu P, Hong Y, Kwok HS, Qin A, Qiu H, Zhao Z, Tang BZ (2013) From tetraphenylethene to tetranaphthylethene: structural evolution in AIE luminogen continues. Chem Commun 49(25):2491–2493. https://doi.org/10.1039/C3CC00010A
95. Wang J, Mei J, Hu R, Sun JZ, Qin A, Tang BZ (2012) Click synthesis, aggregation-induced emission, E/Z isomerization, self-organization, and multiple chromisms of pure stereoisomers of a tetraphenylethene-cored luminogen. J Am Chem Soc 134(24):9956–9966. https://doi.org/10.1021/ja208883h
96. Zhang G-F, Chen Z-Q, Aldred MP, Hu Z, Chen T, Huang Z, Meng X, Zhu M-Q (2014) Direct validation of the restriction of intramolecular rotation hypothesis via the synthesis of novel ortho-methyl substituted tetraphenylethenes and their application in cell imaging. Chem Commun 50(81):12058–12060. https://doi.org/10.1039/C4CC04241G
97. Zhang G-F, Wang H, Aldred MP, Chen T, Chen Z-Q, Meng X, Zhu M-Q (2014) General synthetic approach toward geminal-substituted tetraarylethene fluorophores with tunable emission properties: X-ray crystallography. Aggreg Induced Emiss Piezofluorochrom Chem Mater 26(15):4433–4446. https://doi.org/10.1021/cm501414b
98. Rodrigues ACB, Geisler IS, Klein P, Pina J, Neuhaus FJH, Dreher E, Lehmann CW, Scherf U, Seixas de Melo JS (2020) Designing highly fluorescent, arylated poly(phenylene vinylene)s of intrinsic microporosity. J Mater Chem C 8(7):2248–2257. https://doi.org/10.1039/c9tc06028f
99. Cai Y, Du L, Samedov K, Gu X, Qi F, Sung HHY, Patrick BO, Yan Z, Jiang X, Zhang H, Lam JWY, Williams ID, Lee Phillips D, Qin A, Tang BZ (2018) Deciphering the working mechanism of aggregation-induced emission of tetraphenylethylene derivatives by ultrafast spectroscopy. Chem Sci 9(20):4662–4670. https://doi.org/10.1039/C8SC01170B
100. Dong YQ, Lam JW, Tang BZ (2015) Mechanochromic luminescence of aggregation-induced emission luminogens. J Phys Chem Lett 6(17):3429–3436. https://doi.org/10.1021/acs.jpclett.5b01090
101. Duan Y, Ma H, Tian H, Liu J, Deng X, Peng Q, Dong YQ (2019) Construction of a luminogen exhibiting high contrast and multicolored emission switching through combination of a bulky conjugation core and tolyl groups. Chem Asian J 14(6):864–870. https://doi.org/10.1002/asia.201801608
102. Dong Y, Lam JWY, Li Z, Qin A, Tong H, Dong Y, Feng X, Tang BZ (2005) Vapochromism of hexaphenylsilole. J Inorg Organomet 15(2):287–291. https://doi.org/10.1007/s10904-005-5546-0
103. Zheng M, Sun M, Li Y, Wang J, Bu L, Xue S, Yang W (2014) Piezofluorochromic properties of AIE-active 9,10-bis(N-alkylpheno-thiazin-3-yl-vinyl-2)anthracenes with different length of alkyl chains. Dyes Pigments 102:29–34. https://doi.org/10.1016/j.dyepig.2013.10.020
104. Zhang G, Sun J, Xue P, Zhang Z, Gong P, Peng J, Lu R (2015) Phenothiazine modified triphenylacrylonitrile derivates: AIE and mechanochromism tuned by molecular conformation. J Mater Chem C 3(12):2925–2932. https://doi.org/10.1039/c4tc02925a
105. Huang B, Chen WC, Li Z, Zhang J, Zhao W, Feng Y, Tang BZ, Lee CS (2018) Manipulation of molecular aggregation states to realize polymorphism, AIE, MCL, and TADF in a single molecule. Angew Chem Int Ed Engl 57(38):12473–12477. https://doi.org/10.1002/anie.201806800
106. Luo X, Li J, Li C, Heng L, Dong YQ, Liu Z, Bo Z, Tang BZ (2011) Reversible switching of the emission of diphenyldibenzofulvenes by thermal and mechanical stimuli. Adv Mater 23(29):3261–3265. https://doi.org/10.1002/adma.201101059
107. Zhao J, Chi Z, Zhang Y, Mao Z, Yang Z, Ubba E, Chi Z (2018) Recent progress in the mechanofluorochromism of cyanoethylene derivatives with aggregation-induced emission. J Mater Chem C 6(24):6327–6353. https://doi.org/10.1039/c8tc01648h
108. Yang Z, Chi Z, Mao Z, Zhang Y, Liu S, Zhao J, Aldred MP, Chi Z (2018) Recent advances in mechano-responsive luminescence of tetraphenylethylene derivatives with aggregation-induced emission properties. Mater Chem Front 2(5):861–890. https://doi.org/10.1039/c8qm00062j

109. Ubba E, Tao Y, Yang Z, Zhao J, Wang L, Chi Z (2018) Organic mechanoluminescence with aggregation-induced emission. Chem Asian J 13(21):3106–3121. https://doi.org/10.1002/asia.201800926
110. Zhao J, Chi Z, Yang Z, Mao Z, Zhang Y, Ubba E, Chi Z (2018) Recent progress in the mechanofluorochromism of distyrylanthracene derivatives with aggregation-induced emission. Mater Chem Front 2(9):1595–1608. https://doi.org/10.1039/c8qm00130h
111. Lin Y, Chen G, Zhao L, Yuan WZ, Zhang Y, Tang BZ (2015) Diethylamino functionalized tetraphenylethenes: structural and electronic modulation of photophysical properties, implication for the CIE mechanism and application to cell imaging. J Mater Chem C 3(1):112–120. https://doi.org/10.1039/c4tc02161d
112. Yang J, Gao X, Xie Z, Gong Y, Fang M, Peng Q, Chi Z, Li Z (2017) Elucidating the excited state of mechanoluminescence in organic luminogens with room-temperature phosphorescence. Angew Chem Int Ed Engl 56(48):15299–15303. https://doi.org/10.1002/anie.201708119
113. Xue P, Ding J, Chen P, Wang P, Yao B, Sun J, Sun J, Lu R (2016) Mechanical force-induced luminescence enhancement and chromism of a nonplanar D–A phenothiazine derivative. J Mater Chem C 4(23):5275–5280. https://doi.org/10.1039/c6tc01193d
114. Ma C, Zhang X, Yang Y, Ma Z, Yang L, Wu Y, Liu H, Jia X, Wei Y (2016) Halogen effect on mechanofluorochromic properties of alkyl phenothiazinyl phenylacrylonitrile derivatives. Dyes Pigments 129:141–148. https://doi.org/10.1016/j.dyepig.2016.02.028
115. Yang J, Qin J, Geng P, Wang J, Fang M, Li Z (2018) Molecular conformation-dependent mechanoluminescence: same mechanical stimulus but different emissive color over time. Angew Chem Int Ed Engl 57(43):14174–14178. https://doi.org/10.1002/anie.201809463
116. Chen Y, Xu C, Xu B, Mao Z, Li J-A, Yang Z, Peethani NR, Liu C, Shi G, Gu FL, Zhang Y, Chi Z (2019) Chirality-activated mechanoluminescence from aggregation-induced emission enantiomers with high contrast mechanochromism and force-induced delayed fluorescence. Mater Chem Front 3(9):1800–1806. https://doi.org/10.1039/c9qm00312f
117. Massie SP (1954) The chemistry of phenothiazine. Chem Rev 54(5):797–833. https://doi.org/10.1021/cr60171a003
118. El- Sayed MA (1968) Triplet state—its radiative and nonradiative properties. Accounts Chem Res 1(1):8–16
119. Okazaki M, Takeda Y, Data P, Pander P, Higginbotham H, Monkman AP, Minakata S (2017) Thermally activated delayed fluorescent phenothiazine–dibenzo[a, j]phenazine–phenothiazine triads exhibiting tricolor-changing mechanochromic luminescence. Chem Sci 8(4):2677–2686. https://doi.org/10.1039/c6sc04863c
120. Nobuyasu RS, Ward JS, Gibson J, Laidlaw BA, Ren Z, Data P, Batsanov AS, Penfold TJ, Bryce MR, Dias FB (2019) The influence of molecular geometry on the efficiency of thermally activated delayed fluorescence. J Mater Chem C 7(22):6672–6684. https://doi.org/10.1039/c9tc00720b
121. Wagner E, Filipek S, Kalinowski MK (1988) Visible absorption spectra of the phenothiazine radical cation and its 10-substituted derivatives. Monatsh Chem 119(8–9):929–932
122. Huang L, Wen X, Liu J, Chen M, Ma Z, Jia X (2019) An AIE molecule featuring changeable triplet emission between phosphorescence and delayed fluorescence by an external force. Mater Chem Front 3(10):2151–2156. https://doi.org/10.1039/c9qm00509a
123. Shi R, Chen H, Qi Y, Huang W, Yin G, Wang R (2019) From aggregation-induced to solution emission: a new strategy for designing ratiometric fluorescent probes and its application for in vivo HClO detection. Analyst 144(5):1696–1703. https://doi.org/10.1039/c8an01950a
124. Gong J, Han J, Liu Q, Ren X, Wei P, Yang L, Zhang Y, Liu J, Dong Y, Wang Y, Song X, Tang BZ (2019) An ideal platform of light-emitting materials from phenothiazine: facile preparation, tunable red/NIR fluorescence, bent geometry-promoted AIE behaviour and selective lipid-droplet (LD) tracking ability. J Mater Chem C 7(14):4185–4190. https://doi.org/10.1039/c9tc00143c
125. Rodrigues ACB, Pina J, Seixas de Melo JS (2020) Structure-relation properties of N-substituted phenothiazines in solution and solid state: Photophysical, photostability and aggregation-induced emission studies. J Mol Liq 317:113966. https://doi.org/10.1016/j.molliq.2020.113966
126. Jia XR, Yu HJ, Chen J, Gao WJ, Fang JK, Qin YS, Hu XK, Shao G (2018) Stimuli-responsive properties of aggregation-induced-emission compounds containing a 9,10-distyrylanthracene moiety. Chem Eur J 24(71):19053–19059. https://doi.org/10.1002/chem.201804315
127. Freudenberg J, Rominger F, Bunz UH (2015) New aggregation-induced emitters: tetraphenyldistyrylbenzenes. Chem Eur J 21(47):16749–16753. https://doi.org/10.1002/chem.201502877

128. Zhang J, Ma S, Fang H, Xu B, Sun H, Chan I, Tian W (2017) Insights into the origin of aggregation enhanced emission of 9,10-distyrylanthracene derivatives. Mater Chem Front 1(7):1422–1429. https://doi.org/10.1039/C7QM00032D
129. Tong H, Dong Y, Hong Y, Häussler M, Lam JWY, Sung HHY, Yu X, Sun J, Williams ID, Kwok HS, Tang BZ (2007) Aggregation-induced emission: effects of molecular structure, solid-state conformation, and morphological packing arrangement on light-emitting behaviors of diphenyldibenzofulvene derivatives. J Phys Chem C 111(5):2287–2294. https://doi.org/10.1021/jp0630828
130. Gong Y, Tan Y, Liu J, Lu P, Feng C, Yuan WZ, Lu Y, Sun JZ, He G, Zhang Y (2013) Twisted D-pi-A solid emitters: efficient emission and high contrast mechanochromism. Chem Commun 49(38):4009–4011. https://doi.org/10.1039/c3cc39243k
131. Yuan WZ, Gong Y, Chen S, Shen XY, Lam JWY, Lu P, Lu Y, Wang Z, Hu R, Xie N, Kwok HS, Zhang Y, Sun JZ, Tang BZ (2012) Efficient solid emitters with aggregation-induced emission and intramolecular charge transfer characteristics: molecular design, synthesis, photophysical behaviors, and OLED application. Chem Mater 24(8):1518–1528. https://doi.org/10.1021/cm300416y
132. Yuan WZ, Tan Y, Gong Y, Lu P, Lam JW, Shen XY, Feng C, Sung HH, Lu Y, Williams ID, Sun JZ, Zhang Y, Tang BZ (2013) Synergy between twisted conformation and effective intermolecular interactions: strategy for efficient mechanochromic luminogens with high contrast. Adv Mater 25(20):2837–2843. https://doi.org/10.1002/adma.201205043
133. Dong W, Pina J, Pan Y, Preis E, Seixas de Melo JS, Scherf U (2015) Polycarbazoles and polytriphenylamines showing aggregation-induced emission (AIE) and intramolecular charge transfer (ICT) behavior for the optical detection of nitroaromatic compounds. Polymer 76:173–181. https://doi.org/10.1016/j.polymer.2015.08.064
134. Su X, Gao Q, Wang D, Han T, Tang BZ (2020) One-step multicomponent polymerizations for the synthesis of multifunctional AIE polymers. Macromol Rapid Commun. https://doi.org/10.1002/marc.202000471
135. Liu J, Lam JWY, Tang BZ (2009) Aggregation-induced emission of silole molecules and polymers: fundamental and applications. J Inorg Organomet 19(3):249. https://doi.org/10.1007/s10904-009-9282-8
136. Qin A, Lam JWY, Tang BZ (2012) Luminogenic polymers with aggregation-induced emission characteristics. Prog Polym Sci 37(1):182–209. https://doi.org/10.1016/j.progpolymsci.2011.08.002
137. Li Y, Liu S, Han T, Zhang H, Chuah C, Kwok RTK, Lam JWY, Tang BZ (2019) Sparks fly when AIE meets with polymers. Mater Chem Front 3(11):2207–2220. https://doi.org/10.1039/c9qm00404a
138. Qiu Z, Liu X, Lam JWY, Tang BZ (2019) The marriage of aggregation-induced emission with polymer science. Macromol Rapid Commun 40(1):e1800568. https://doi.org/10.1002/marc.201800568
139. Pucci A (2019) Mechanochromic fluorescent polymers with aggregation-induced emission features. Sensors 19:22. https://doi.org/10.3390/s19224969
140. Zhan R, Pan Y, Manghnani PN, Liu B (2017) AIE polymers: synthesis, properties, and biological applications. Macromol Biosci 17:5. https://doi.org/10.1002/mabi.201600433
141. Zhou H, Chua MH, Tang BZ, Xu J (2019) Aggregation-induced emission (AIE)-active polymers for explosive detection. Polym Chem 10(28):3822–3840. https://doi.org/10.1039/c9py00322c
142. Rodrigues ACB, Pina J, Dong W, Forster M, Scherf U, Seixas de Melo JS (2018) Aggregation-induced emission in phenothiazine–TPE and –TPAN polymers. Macromolecules 51(21):8501–8512. https://doi.org/10.1021/acs.macromol.8b01758
143. Gao B-R, Wang H-Y, Yang Z-Y, Wang H, Wang L, Jiang Y, Hao Y-W, Chen Q-D, Li Y-P, Ma Y-G, Sun H-B (2011) Comparative time-resolved study of two aggregation-induced emissive molecules. J Phys Chem C 115(32):16150–16154. https://doi.org/10.1021/jp2027753
144. Hoyt JM, Frank CE, Koch K (1967) Polymerization of bis(α-Haloalkyl) aromatic compounds
145. Baysec S, Preis E, Allard S, Scherf U (2016) Very high solid state photoluminescence quantum yields of poly(tetraphenylethylene) derivatives. Macromol Rapid Commun 37(22):1802–1806. https://doi.org/10.1002/marc.201600485
146. Dalapati S, Gu C, Jiang D (2016) Luminescent porous polymers based on aggregation-induced mechanism: design. Synth Funct Small 12(47):6513–6527. https://doi.org/10.1002/smll.201602427
147. Jansen JC, Esposito E, Fuoco A, Carta M (2019) Microporous organic polymers: synthesis, characterization, and applications. Polymers (Basel) 11:5. https://doi.org/10.3390/polym11050844

148. Patra A, Scherf U (2012) Fluorescent microporous organic polymers: potential testbed for optical applications. Chem Eur J 18(33):10074–10080. https://doi.org/10.1002/chem.201200804
149. Palma-Cando A, Woitassek D, Brunklaus G, Scherf U (2017) Luminescent tetraphenylethene-cored, carbazole- and thiophene-based microporous polymer films for the chemosensing of nitroaromatic analytes. Mater Chem Front 1(6):1118–1124. https://doi.org/10.1039/c6qm00281a
150. Pina J, Seixas de Melo J, Burrows HD, Bilge A, Farrell T, Forster M, Scherf U (2006) Spectral and photophysical studies on cruciform oligothiophenes in solution and the solid state. J Phys Chem B 110(31):15100–15106. https://doi.org/10.1021/jp060707t
151. De Nisi F, Francischello R, Battisti A, Panniello A, Fanizza E, Striccoli M, Gu X, Leung NLC, Tang BZ, Pucci A (2017) Red-emitting AIEgen for luminescent solar concentrators. Mater Chem Front 1(7):1406–1412. https://doi.org/10.1039/c7qm00008a
152. Zhang Z, Liao M, Lou H, Hu Y, Sun X, Peng H (2018) Conjugated polymers for flexible energy harvesting and storage. Adv Mater 30(13):e1704261. https://doi.org/10.1002/adma.201704261
153. Pina J, Rodrigues ACB, Alnady M, Dong W, Scherf U, Seixas de Melo JS (2020) Restricted aggregate formation on tetraphenylethene-substituted polythiophenes. J Phys Chem C 124(25):13956–13965. https://doi.org/10.1021/acs.jpcc.9b10908
154. Raithel D, Baderschneider S, de Queiroz TB, Lohwasser R, Köhler J, Thelakkat M, Kümmel S, Hildner R (2016) Emitting species of poly(3-hexylthiophene): from single. Isolated Chains Bulk Macromol 49(24):9553–9560. https://doi.org/10.1021/acs.macromol.6b02077
155. Becker RS, Seixas de Melo JS, Maçanita AL, Elisei F (1996) Comprehensive evaluation of the absorption, photophysical, energy transfer, structural, and theoretical properties of α-oligothiophenes with one to seven rings. J Phys Chem 100(48):18683–18695. https://doi.org/10.1021/jp960852e
156. Pina J, Burrows HD, Becker RS, Dias FB, Maçanita AL, Seixas de Melo JS (2006) Photophysical Studies of α, ω-dicyano-oligothiophenes NC($_C$4$_H$2S$_)$nCN (n = 1–6). J Phys Chem B 110(13):6499–6505. https://doi.org/10.1021/jp055455v
157. Pina J, Seixas de Melo JS, Burrows HD, Maçanita AL, Galbrecht F, Bünnagel T, Scherf U (2009) Alternating binaphthyl–thiophene copolymers: synthesis, spectroscopy, and photophysics and their relevance to the question of energy migration versus conformational relaxation. Macromolecules 42(5):1710–1719. https://doi.org/10.1021/ma802395c
158. Pina J, Eckert A, Scherf U, Galvão AM, Seixas de Melo JS (2018) Cyclopentadithiophene derivatives: a step towards an understanding of thiophene copolymer excited state deactivation pathways. Mater Chem Front 2(1):149–156. https://doi.org/10.1039/C7QM00440K
159. Clark J, Silva C, Friend RH, Spano FC (2007) Role of intermolecular coupling in the photophysics of disordered organic semiconductors: aggregate emission in regioregular polythiophene. Phys Rev Lett 98(20):206406. https://doi.org/10.1103/PhysRevLett.98.206406

REVIEW

Aggregation-Induced Emission Fluorescent Gels: Current Trends and Future Perspectives

Javad Tavakoli[1] · Amin Jamshidi Ghahfarokhi[2] · Youhong Tang[2]

Received: 8 November 2020 / Accepted: 22 December 2020 / Published online: 5 February 2021

Abstract
The development of fluorescent gels, if not the current focus, is at the center of recent efforts devoted to the invention of a new generation of gels. Fluorescent gels have numerous properties that are intrinsic to the gel structure, with additional light-emitting properties making them attractive for different applications. This review focuses on current studies associated with the development of fluorescent gels using aggregation-induced emission fluorophores (AIEgens) to ultimately suggest new directions for future research. Here, we discuss major drawbacks of the methodologies used frequently for the fabrication of fluorescent gels using traditional fluorophores compared to those using AIEgens. The fabrication strategies to develop AIE-based fluorescent gels, including physical mixing, soaking, self-assembly, non-covalent interactions, and permanent chemical reactions, are discussed thoroughly. New and recent findings on developing AIE-active gels are explained. Specifically, physically prepared AIE-based gels including supramolecular, ionic, and chemically prepared AIE-based gels are discussed. In addition, the intrinsic fluorescent properties of natural gels, known as clustering-triggered fluorescent gel, and new and recent relevant findings published in peer-reviewed journals are explained. This review also revealed the biomedical applications of AIE-based fluorescent hydrogels including drug delivery, biosensors, bioimaging, and tissue engineering. In conclusion, the current research situation and future directions are identified.

Keywords Fluorescent gels · Hydrogels · Aggregation-induced emission · AIE polymers · Biomedical applications

Chapter 8 was originally published as Tavakoli, J., Ghahfarokhi, A. J. & Tang, Y. Topics in Current Chemistry (2021) 379: 9. https://doi.org/10.1007/s41061-020-00322-6.

✉ Youhong Tang
youhong.tang@flinders.edu.au

[1] Centre for Health Technologies, School of Biomedical Engineering, Faculty of Engineering and Information Technology, University of Technology Sydney, Ultimo, NSW 2007, Australia

[2] Institute for NanoScale Science and Technology, College of Science and Engineering, Flinders University, Adelaide, SA 5042, Australia

1 Introduction

Gels have substantial implications from basic science to practical applications, playing a pivotal role in the development of templates for biomedical devices at the micro and nanoscale, the creation of novel microactuators and nanosensors, and the construction of tissue-engineered fabricates [1–3]. Researchers classify polymeric gels in different ways based on polymer type and structure as well as the nature of crosslinking agents often leading to the formation of physically or chemically crosslinked gels. When physical interactions are incorporated into the formation of gels, physical entanglement of polymer molecules, the formation of non-covalent bonds, and the creation of crystalline regions are responsible for the development of a durable network with physical stability upon swelling. As an example, supramolecular gels are often formed as a consequence of the molecular self-assembly process to shape an aqueous mixture containing a three-dimensional (3D) supramolecular network [4, 5]. Noncovalent interactions are central to the formation of the network, holding supramolecules encapsulating a large amount of water and exhibiting viscoelastic characteristics. Supramolecular gels usually flow under low magnitude shear forces, resemble the structural complexity of the extracellular matrix in living tissues, and are highly hydrated. Hence, they are used widely for the design of new biomaterials [6–8].

In contrast, chemically crosslinked gels are produced based on chemical reactions to create crosslinks between polymeric chains forming a relatively strong 3D network [9]. Chemically crosslinked gels—novel polymeric materials with controllable viscoelastic properties—are dimensionally stable polymeric networks that can take up large amounts of water or biological fluids, with their distinct 3D structures being maintained during the swelling process [10]. The unique properties of chemically crosslinked gels include high water content, biocompatibility, and tunable viscoelasticity.

These properties, along with their flexibility to form various constructs via different fabrication methods, have made them attractive for diverse biomedical applications [11–13]. Resembling soft tissue's structure and function and providing an extremely rare environment for cell growth, tissue-engineered scaffolds made of gels have been widely used in tissue engineering and regenerative medicine [14–17]. Besides, the ability of gels to preserve pharmaceutical agents within their structure and control their release over a sustained period has made them highly attractive in the development of novel pathways for drug delivery [18, 19]. Over the past decade, extensive efforts have been made to improve the mechanical, physical, and physicochemical properties of gels, which has led to expansion of their application beyond biomedical areas [20]. For example, highly porous and resilient gels with improved physicochemical properties have been developed for the subtraction of heavy metals, elimination of toxic components, and dye retrieval from various industrial wastes [21–23]. Enhancement of the mechanical properties of chemically crosslinked gels has resulted in the development of more appropriate transducers, based on the gels' mechanical work [24, 25], and more efficient gel-based sensors have been created with mechanically resilient hydrogels onboard [26, 27].

Current research aimed at developing a new generation of gels (chemically and physically crosslinked gels) is dedicated largely to the invention of fluorescent gels [28–30]. Fluorescent gels possess numerous properties that are intrinsic to the gel structure, with the additional property of being light-emitting. A crucial role in this light-emitting property is played by the successful crosslinking of fluorescent molecules incorporated into the gel backbone, or the assembly of fluorophores and molecular gelators, via either permanent (covalent) or temporary (noncovalent) bonds [31]. The ability of gels to emit light makes them suitable for advanced applications in chemical and environmental sensing, biomaterials, bioimaging, and tissue engineering [32]. Unfortunately, the current problem limiting the large-scale production of fluorescent gels is their synthesis, which involves multi-part sequences and complex chemistry [33]. Other main drawbacks are the aggregation of fluorophores quenching the light intensity and leading to significant reduction of fluorescence and photobleaching of fluorophores [34–36]. Moreover, the chemical and mechanical stability of the gels may be affected by the presence of bulky and hydrophobic fluorophores, which can destabilize gels construction during frequent swelling–deswelling processes [37]. Overall, developing bright and stable fluorescent gels for practical applications remains an arduous challenge. The main obstacle here is the irrepressible quenching of traditional fluorophores in the aggregated state, known as aggregation-caused quenching [38]. This characteristic results in negligible brightness, which is unlikely to be sensitive enough to fulfill the basic conditions required for the successful execution of experiments with reliable results to form advanced strategies.

Aggregation-induced emission (AIE), a new concept proposed in 2001 by Ben Zhong Tang from the Hong Kong University of Science and Technology, has motivated researchers across the globe to initiate different novel studies with AIE fluorophores (AIEgens)—an innovative category of advanced material with exceptional performance in numerous fields of research. In contrast to conventional fluorescent dyes, where aggregation caused quenching (ACQ), AIEgens exhibit highly fluorescent properties in the aggregated or solid state. As such, AIEgens provide a powerful platform enabling breakthrough studies in science, engineering, and medicine in multiscale landscapes [39–42]. Recent studies using AIEgens have demonstrated new opportunities to overcome the major drawbacks associated with developing fluorescent gels [43]. AIEgens enjoy high emission efficiency, exceptional photostability, and incomparable sensitivity, and exhibit enhanced fluorescent properties in an aggregated state [44, 45]. Therefore, compared to traditional fluorophores, confining AIEgens within a gel network or a sol–gel medium, even at high concentrations, is more likely to result in the formation of a stable fluorescent gel with enhanced light intensity under irradiation. The restriction of the intramolecular motion of AIEgens embedded, physically or chemically, within the gel network causes increased photon emissions from gels made with AIEgens [46]. Compared to traditional fluorophores, the great diversity of molecular structure, ranging from pure hydrocarbons to heteroatom-containing compounds and organometallic complexes, is another advantage of AIEgens, making them predominant in the creation of more diverse fluorescent gels and thus expanding their applications [47, 48]. Meanwhile, another aspect that makes AIEgens attractive is their ability to successfully characterize the structure–function relationship of gels, i.e., the visualization of crystalline regions and accurate measurement of both swelling and degradation processes, which was not attainable before [49, 50]. This review will describe new findings relevant to the incorporation of AIEgens in the development of fluorescent gels to introduce initiatives for the development of new hydrophilic platforms for biomedical applications.

2 Fabrication Strategy

A simple method to prepare fluorescent gels is the diffusion of an AIEgen into the as-prepared gel network by soaking [51, 52]. The swelling of a gel in the AIEgen solution results in penetration of the AIEgen, creating a fluorescent gel. However, this method is not efficient since, firstly, the majority of AIEgens are hydrophobic, making their diffusion process abortive, especially in water. Secondly, the diffusion of the AIEgen is controlled mainly by the swelling process; hence, the fluorescent property of the final gel depends strongly on its structural parameters (i.e., crosslinking density) with highly swellable gels becoming more fluorescent than those with low swelling capacity. On the other hand, a lack of appropriate control over the diffusion process may lead to non-uniform distribution of the AIEgen inside the gel structure, resulting in non-homogenous light emission. The main drawback is the loss of luminescence due to photo-leaching occurring over time during the re-swelling process. Another strategy to develop fluorescent gels is the physical mixing of AIEgen with gelators [53]. Hydrogelators are small molecules that are self-assembled in water to create a 3D network exhibiting viscoelastic behavior [54]. During the self-assembly process, water is encapsulated within the structure of the gel, developing an aqueous mixture that is unable to flow in the absence of external forces. When an organic solvent is responsible for the initiation of the self-assembly process, the final gel is known as an organogelator [55]. This strategy also suffers from leaching of the AIEgen entities, since the gel network is shaped due to noncovalent interactions between the hydrogelator and AIEgen, leading to a reduction in fluorescent properties [56].

Another disadvantage of this method is the effect of AIEgens on the stability of hydrogelators. However, since molecular self-assembly is a predominant route in the progression of life, this strategy is most likely one of the best approaches to developing fluorescent gels involving biomolecules and AIEgen, mimicking the physiological behavior of their native counterparts [57]. Some studies have modified this strategy by loading AIEgen-labelled nanoparticles to further secure their availability within the gel network [58–60]. This approach relies on the functionalization of nanoparticles and subsequent labelling with AIEgens. The physical mixing of AIEgen-labelled nanoparticles within the polymeric gels, known as loading, then takes place before crosslinking. The proportion of AIEgen-labelled nanoparticles relative to polymer molecules tunes the fluorescent properties of the polymeric gels. Alternatively, temporary crosslinking of polymeric gels using noncovalent interactions, including electrostatic, hydrogen, and ionic bonds as well as host–guest complexation reactions can be considered another strategy to create fluorescent polymeric gels. The gilt-edged structural hierarchy, along with the consequent dynamic host–guest interaction, will further enhance the stability of the gel network, making it capable of being self-healed or enviro-sensitive, which are important for tuning both the associated stiffness and fluorescent properties [61–63].

Another effective technique to create fluorescent gels is to crosslink fluorescent polymers via covalent bonds, where the final gel network contains fluorescent polymers as the backbone, connecting via crosslinking agents. Several approaches have been proposed for the construction of AIE polymers, which is the first step towards creating fluorescent gels. Synthesis methods such as post-polymerization, polymerization initiates by AIEgens, self-assembly, and direct polymerization of AIEgens have been identified [64–67]. Reversible addition-fragmentation chain transfer (RAFT) and atom transfer radical (ATR) polymerizations have been used to prepare AIE polymers when AIEgen was selected as the initiator. Tetraphenylthiophene and tetraphenylethylene (TPE)-based AIEgens have been used as initiators to develop AIE polymers. Remarkably, introducing just one small AIE molecule into a polymer chain structure results in the development of an AIE-polymer with a fluorescent property. Studies have shown that metal-catalyzed polymerizations (i.e., Sonogashira, Suzuki, McMurry) can effectively incorporate AIEgen monomers to synthesize AIE polymers [68]. The modification of polymers using click reactions to ultimately attach the AIEgen pendant group to the polymer backbone is known as post-functionalization—another method for developing fluorescent polymers. On formation of amphiphilic polymers containing AIE-pendant groups, these can then self-assemble into nanoparticles or micelles in aqueous solutions, developing fluorescent properties.

Recently, a few researchers have reported polymers comprising flexible chains and tunable structures that contain nonconventional fluorophores and can emit strong light under certain circumstances. These non-aromatic polymeric gels exhibit room temperature fluorescence with clustering-triggered emission characteristics. It is believed that the aggregation (clustering) of pendant groups, changing the molecular structure, and the involvement of π electrons are likely to restrict inter- and intramolecular interactions and modulate fluorescent emission. Poly(amino acid) of ε-poly-L-lysine, sodium alginate, polyacrylic acid, polyacrylamide, and poly(*N*-isopropyl acrylamide) are among a few polymeric gels that can emit intense blue light in concentrated solutions and solid state [69–71]. A summary of fabrication strategies to develop fluorescent polymeric gels is presented in Table 1.

3 Developing AIE-Active Gels

3.1 Self-Assembled Fluorescent Supramolecular Gels

The AIE phenomenon in supramolecular gels has also been reported by many researchers. Lopez et al. [72] developed a fluorescent gel based on 5-(4-nonyl phenyl)-7-azaindole, which exhibited a red shift compared to the emission band observed in its dilute state and emitted blue in the solid-state. They also observed a reversible emission switching at 306 nm during the frequent sol–gel transition (Fig. 1).

Table 1 Summary of fabrication strategies to develop fluorescent gels

Strategy	Pros and cons	References
Physically crosslinked	Advantages Easy to perform Mimics physiological behavior (molecular self-assembly is a predominant route in life) Disadvantages AIEgen leaching (noncovalent interactions between the hydrogelator and AIEgen) Structural instability (hydrophobicity of AIEgens) Weak fluorescent property (AIEgen emission is obscured by both nanoparticle and gel) Limited loading (affected by the size and concentration of nanoparticles) Limited AIEgens can be incorporated	[51–63]
Chemically crosslinked	Advantages Enhanced fluorescent property Excellent structural stability with secured AIEgen Disadvantages Complex chemistry	[64–68]

AIEgen Aggregation-induced emission fluorophore

Bhattacharya and Samanta [73] synthesized a hydrogelator based on phenylenedivinylene bis-*N*-octyl pyridinium salt with an AIE feature, high melting temperature, and superior mechanical properties even at a low concentration of the gelator. They reported a morphology transition from fibre-to-coil-to-tube depending on gelator concentration, and observed that the fluorescent property was controlled by temperature and salt concentration. The hydrogel emitted different colors varying between blue, white, and orange. A review by Tang et al. [74] described the contribution of organic AIEgens in the development of self-assembled organogels with gelation-enhanced emission characteristics. They revealed that AIEgens including TPE, silole, and cyanostilbene derivatives can self-assemble into organogels, generating high fluorescent properties in the gel state. Yang et al. [75] developed an organogelator containing 3,5-di(pyridine-2-yl)phenyl, fluorenyl, and 3,4,5-tris(dodecyloxy)phenyl groups that can self-assemble into one-dimensional (1D) 300 nm nanofibers. The self-assembled nanofibers emitted a strong cyan fluorescence that can be used as a chemosensor to detect acid vapors (Fig. 2a). By bonding a stigmasterol group to TPE via click chemistry, Tang et al. [76] created a gelator with a flexible structure that self-assembled through cooling-heating cycles into a gel demonstrating a strong AIE characteristic compared to its solution counterpart. The AIE-active organogel self-assembled into a 3D network of fibrous aggregates emitting blue light (Fig. 2b). The researchers observed that tuning the flexibility of the AIE-active gelator, by changing the length of the alkyl chain between two TPE units, affected the final fluorescent property of the gelator, with lower flexibility resulting in less emission. Zhang et al. [77] developed a hyperbranched AIE gelator based on TPE and benzyl ether relying on the formation of multivalent π–π interactions with

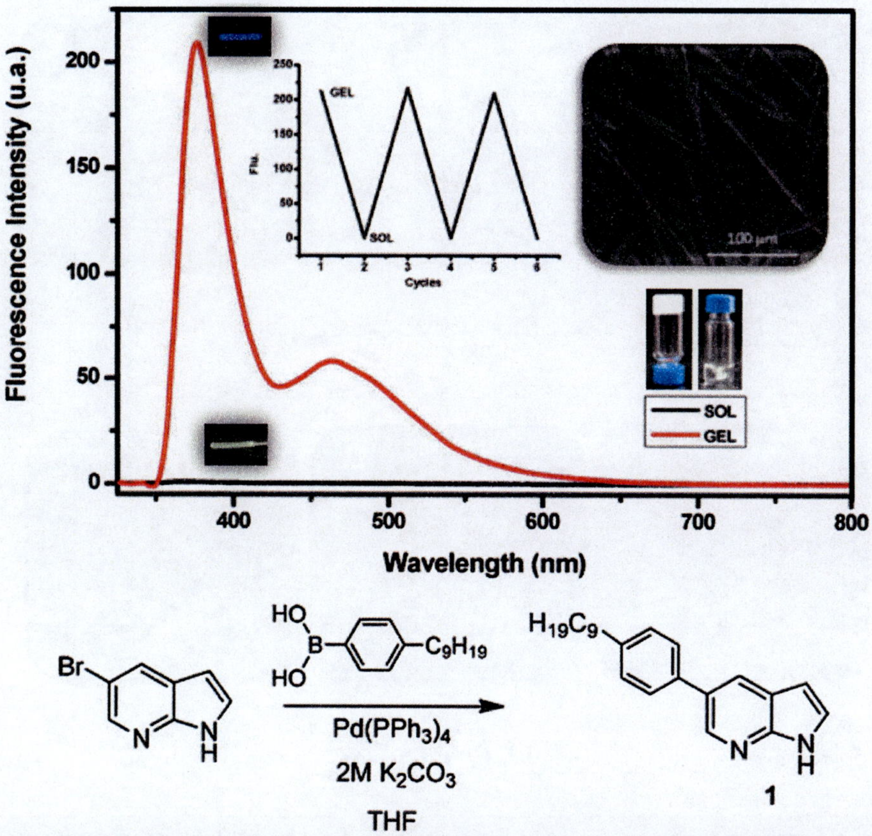

Fig. 1 Properties of a fluorescent gel based on 5-(4-nonyl phenyl)-7-azaindole with fibrillar structure in gel state emitting blue light under ultraviolet (UV) irradiation. The aggregation-induced emission (AIE) compound (synthesis scheme presented) exhibits a frequent on–off fluorescent property under sol–gel cyclic conditions (modified from [72] with permission). *THF* Tetrahydrofuran

several solvents to form a fluorescent gel with its emission color modulated by multiple stimuli (Fig. 2c). The researchers synthesized silole–cholesterol and silole–urea compounds with strong intermolecular interaction occurring between silole and the moieties, leading to the formation of fluorescent organogels in polar solvents. They found a worm-like structure for both xerogels with a reversible emission property transiting between sol and gel states as well as between heating and cooling cycles (Fig. 2d) [78].

Tetraphenylsiloles containing long alkoxyl chains were prepared by Wan et al., and these self-assembled in hydrocarbon solvents, forming a stable gel with a twisted nanofibrous structure. A reversible sol–gel process was observed for the AIE-active gel occurring at room and high temperatures, respectively, with the fluorescent intensity approximately 100 times higher for the gel compared to the solution (Fig. 2e). They also observed that the self-assembly property and morphology

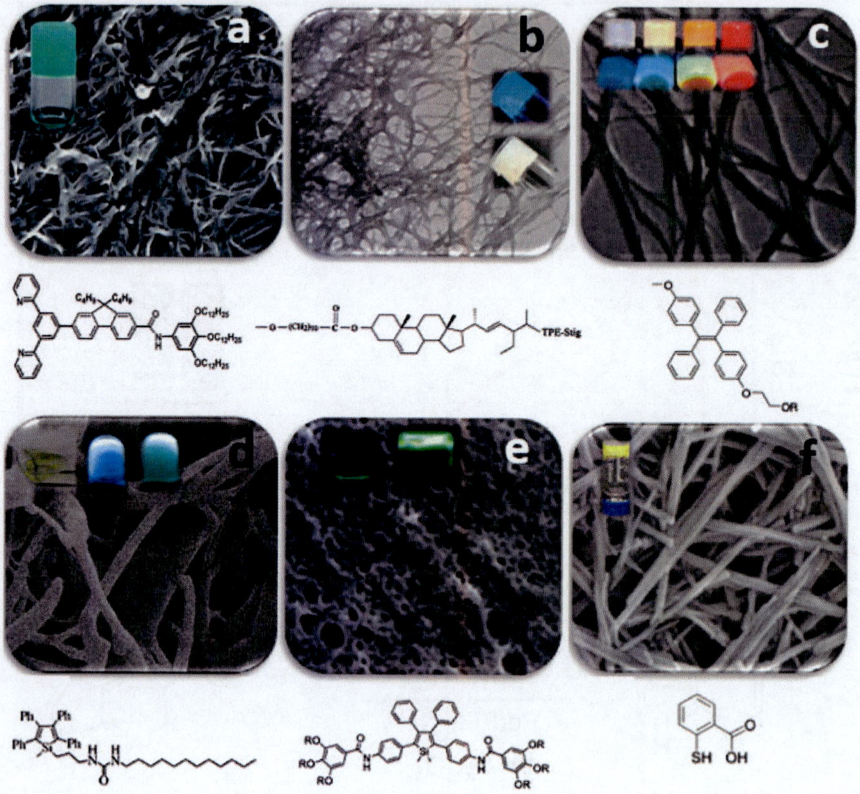

Fig. 2 Scanning and transmission electron micrographs, photographs (daylight and under UV irradiation) and chemical structures of **a** an organo-gelator containing 3,5-di(pyridine-2-yl)phenyl, fluorenyl, and 3,4,5-tris(dodecyloxy)phenyl groups (modified from [75] with permission), **b** a gelator developed through bonding a stigmasterol group to tetraphenylethylene (TPE) via click chemistry (modified from [76] with permission), **c** a hyperbranched AIE gelator based on TPE and benzyl ether (modified from [77] with permission), **d** a gelator containing silole-cholesterol (modified from [78] with permission), **e** a thermoresponsive AIE-active silole derivatives gel (modified from [79] with permission), and **f** a self-assembled silver nanocluster with AIE effect (modified from [80] with permission)

of the gels were extremely affected by the length of the alkoxy chains, reflecting the role of hydrogen bonds and π stacking interactions in the formation of organogels. The combination of these interactions was considered as the route to restrict intermolecular interactions, tuning the fluorescent property of the organogels [79]. Sun et al. [80] constructed a self-assembled silver nanocluster with AIE effect that was able to form nanofibers when interacting with succinic acid. They found the compact structure along with the argentophilic interactions to be responsible for development of the fluorescent hydrogel (Fig. 2f).

Several thermoreversible organogelators, based on cyanostilbene, have been developed by Park and co-workers [81, 82], forming a 3D fibrous mat consisting of crystalline nanofibers or bundles of fibrous aggregates. The organogelators consisted of phenyl rings and cyano groups with π–π and dipole–dipole interactions between them developing fluorescent properties. The quantum yield of the gels was more than 50%, which was far higher (approximately 31-fold) than that of the in-solution state. These results revealed green and blue emissions for β-cyano- and trifluoromethyl-substituted distyrylbenzene organogels excited under 365 nm UV irradiation (Fig. 3a) [81, 82]. The development of a transparent organogel with a high fluorescent property, based on the substitution of 3,5-bis(trifluoromethyl)phenyl group with pyridine ring in cyanostilbene, was also reported by Park's group [81, 82]. The formation of hydrogen bonds between tartaric acid or fluoromethylbenzoic acid and pyridine ring was responsible for the self-assembly process, providing the organogel with transparency and high fluorescent properties. It was suggested that, based on a red shift in emission spectra, the self-assembled organogel adopted a planar conformation in J-aggregation (Fig. 3b). Besides, the chiral characteristic of tartaric acid directed the robust circular dichroism signal perceived in the gel [83]. In another study, Park's group [84] synthesized a functional fluorescent gelator that was self-assembled even at a low concentration (0.2 wt%) using both polar and nonpolar solvents. They found that the fluorescent property of the gel was reversibly modulated by changing temperature and proton induction. The researchers modified the aromatic structure of 1-cyano-trans-1,2-bis-(3′,5′-bis-trifluoromethyl-biphenyl) ethylene by including a photoacid-active pyridine unit. This resulted in the development of a stimuli-responsive gelator with a high-performance characteristic to turn on fluorescence, with a thermal- or proton-induced gel-to-sol transition tuning the final fluorescent property (Fig. 3c) [84].

Zhang et al. [85] designed a salt-responsive hydrogelator where TPE was grafted to a salt-responsive peptide, creating a fluorescent hydrogel with the emission enhanced upon gelation. Their experiment revealed that the fluorescent property of the self-assembled gel was not affected by other stimuli i.e., pH, temperature, and solvent, but rather by salt concentration. The TPE–peptide hydrogelator was highly fluorescent when exposed to a high concentration NaCl solution (1.5 M) emitting a high-intensity blue light (466 nm) with a decreasing intensity in lower salt concentrations. They also observed that the self-assembly process of the TPE-peptide was initiated at 500 mM salt. The relative intensity of the gel in the presence of 1 M NaCl was seven times higher than that when exposed to 10 mM salt [85]. Zhang et al. [86] designed a supramolecular AIE gel, based on Pillar[5]arene, which could detect low concentrations of ions Fe^{3+}, Hg^{2+}, Ag^{+}, and Br^{-}, exhibiting fluorescent properties with bright blue-white emission. Flip-Leon et al. [87] developed a series of 1,8-naphthalimide derivatives with AIE effect, leading to the creation of stimuli-responsive supramolecular hydrogels. They reported a structural transition between the nanoparticle and fibrillar morphologies for gelator and non-gelator molecules, respectively. The gelator exhibited gelation-enhanced emission with the fluorescent property and quantum yield increase of 4700% and 39%, respectively [87]. Hong et al. [88] fabricated an AIE-active hydrogel incorporating benzimidazolium assembled with α-cyclodextrin, which showed a thermoreversible sol–gel transition with a white fluorescent property. They used the gel successfully to detect organic dye molecules.

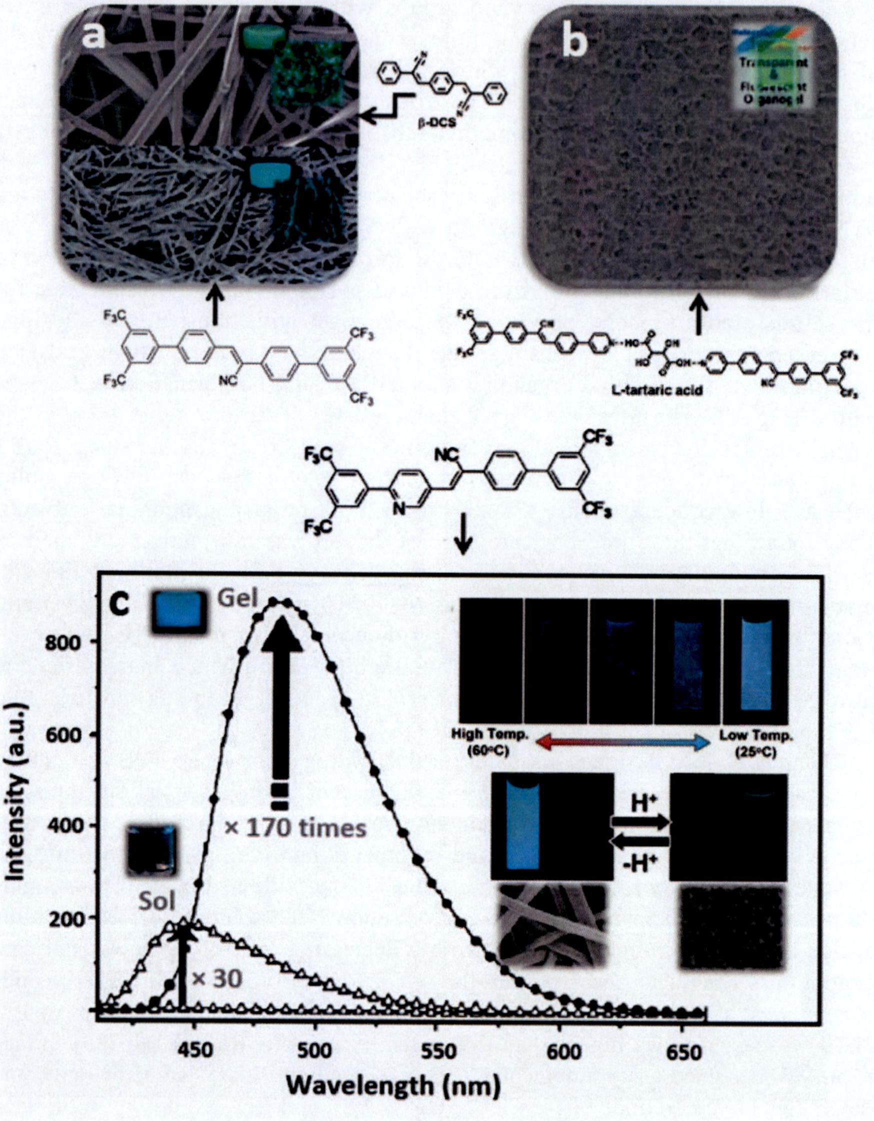

Fig. 3 **a** Scanning electron and fluorescent microscopic images and photographs of green and blue emissions for β-cyano- and trifluoromethyl-substituted distyrylbenzene organogels excited under 365 nm UV irradiation (modified from [81, 82] with permission). **b** Scanning electron microscopy and photograph of a transparent organogel with a high fluorescent property, based on substitution of 3,5-bis(trifluoromethyl) phenyl group with pyridine ring in cyanostilbene (modified from [83] with permission). **c** A functional fluorescent gelator prepared through the modification of the aromatic structure of 1-cyano-trans-1,2-bis-(3′,5′-bis-trifluoromethyl-biphenyl) ethylene by including a photoacid-active pyridine unit. The gel was reversibly modulated by changing temperature and proton induction (modified from [84] with permission)

Based on the electrostatic interaction formed between poly(sodium p-styrene sulfonate) and TPE derivatives, Wang et al. [89] developed a fluorescent hydrogel that was sensitive to biomolecules with the addition of adenosine triphosphate reducing the fluorescent intensity. Their results showed the diffusion of the gel over the cell membrane entering the cytoplasm, indicating the gel's capability for use in cell imaging [89]. By co-assembly of aromatic amino acids and bipyridine, Xing et al. [90] developed light irradiation-triggered luminescence gels without phase separation and photobleaching problems. They suggested employing amino acid residues of short peptides as a suitable method for fabricating self-assembled polymeric gels with the AIE effect [90].

Cheng et al. [91] used a photoisomeric cyanostilbene conjugated cationic surfactant and developed a fluorescent hydrogel that can be turned off and on in response to UV light through organic–inorganic supramolecular co-assembly. A fluorescent quenching was found due to the trans–cis photoisomerization of the cyanostilbene unit occurring during UV exposure, but this can be recovered when subjected to an annealing process [91]. A summary of self-assembled fluorescent supramolecular gels with their key features is presented in Table 2.

3.2 Chemically Prepared Fluorescent Gels

In recent years, numerous AIE-based hydrogels have been developed with AIEgens either covalently conjugated or physically bonded within the hydrogel structure. Kokado et al. [92] developed a poly(γ-glutamic acid) hydrogel using an o-carborane AIEgen, as a crosslinking reagent in combination with ethylene glycol diglycidyl ether. Water-sensitive luminescence with a lower fluorescent intensity was observed in the swollen state [92]. They observed a reversible emission property with stronger emission associated with higher crosslinking density, resulting in an absolute quantum yield reaching 0.4 in the dry state (Fig. 4a). Jiang et al. [93] prepared fluorescent hydrogel nanoparticles (ca. 300 nm) with their fluorescent intensities being reliant on temperature changing over a wide temperature range (2–80 °C). The nanoparticles were core–shell in structure with nanoparticles consisting of poly[styrene-co-(*N*-isopropyl acrylamide)] as the core and poly[(*N*-isopropyl acrylamide)-co-polyacrylic acid)] as the shell. A water-soluble TPE derivative containing two positive charge sites per molecule was immobilized on the surface of the shell via electrostatic interaction occurring between the negative charges of carboxyl groups (polyacrylic acid) and positive charges from the AIEgen. The fluorescent characteristic of the hydrogel was reported as reversible upon exposure to frequent cooling–heating cycles (Fig. 4b) [93].

Zhou et al. [94] also developed AIE-based fluorescent hydrogels through the copolymerization of *N*-isopropyl acrylamide and TPE-acrylate monomers. The interpenetrating polymer network was pH- and thermoresponsive, with decreasing pH enhancing the fluorescent property [94]. Tang et al. also reported the development of a fluorescent hydrogel thermometer based on poly(*N*-isopropyl acrylamide)-labelled-AIEgen via direct polymerization. Reasonable details in the thermal transitions were detectable using the aqueous polymer solution with

Table 2 Summary of self-assembled AIEgen gels with fluorescent properties

AIEgen	Key features	References
5-(4-Nonyl phenyl)-7-azaindole	Blue (solid) and red (dilute) emission with a reversible emission switching at 306 nm during the frequent sol–gel transition	[72]
Phenylenedivinylene bis-N-octyl pyridinium salt	Morphology transition from fibre-to-coil-to-tube (based on the gelator concentration) with the fluorescent property controlled by temperature and salt concentration	[73]
TPE, silole, and cyanostilbene derivatives	Gelation-enhanced emission	[74]
3,5-Di(pyridine-2-yl)phenyl, fluorenyl, 3,4,5-tris(dodecyloxy)phenyl	Self-assembly into one-dimensional 300 nm nanofibers, emitting a strong cyan fluorescence when exposed to acid vapours	[75]
TPE derivative (alkyl chain between two TPE units)	Self-assembly through cooling-heating cycles demonstrating a strong AIE characteristic affected by the length of alkyl chain	[76]
Hyperbranched AIE gelator based on TPE and benzyl ether	The fluorescent property was modulated by multiple stimuli	[77]
Silole-cholesterol Silole-urea	Fluorescent organogels in polar solvents with worm-like structure	[78]
Tetraphenylsiloles containing alkoxyl chains	Creation of a twisted nanofibrous structure with enhanced fluorescent property (> 100 times) in gel state compared to solution	[79]
Silver nanocluster	Fluorescent nanofibers formation when interacting with succinic acid	[80]
Cyanostilbene	Forming a 3D fibrous mat consisting of representing chiral characteristics	[81–84]
TPE-peptide	The fluorescent property of the self-assembled gel was affected by salt concentration	[85]
Pillar[5]arene	Low concentrations of ions Fe^{3+}, Hg^{2+}, Ag^{+}, and Br^{-}, were detected based on the fluorescent property	[86]
1,8-Naphthalimide derivatives	The gelator exhibited elation-enhanced emission with high quantum yield (4700%)	[87]
Benzimidazolium-α cyclodextrin	Thermoreversible sol–gel transition to detect organic dye molecules	[88]
TPE-poly(sodium p-styrene sulfonate)	Adenosine triphosphate sensitive fluorescent gel applicable for cell imaging	[89]
Aromatic amino acids and bipyridine	Creation of light irradiation-triggered luminescence gels	[90]

TPE Tetraphenylethylene

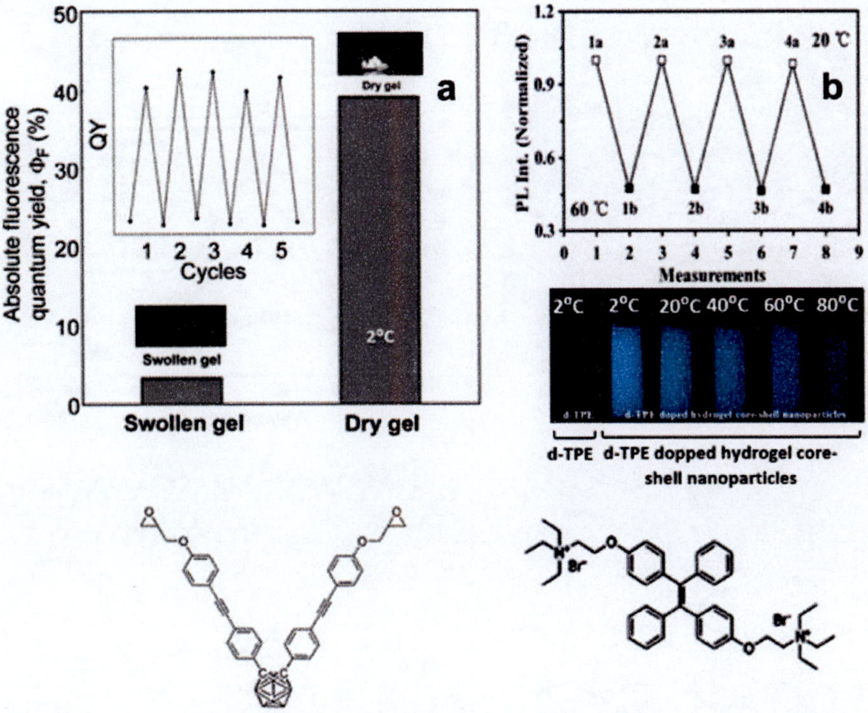

Fig. 4 **a** Fluorescence quantum yields of poly(γ-glutamic acid) hydrogels (chemical structure presented) in the swollen and dry state including on–off quantum yields during five frequent swelling and drying cycles and camera images of the hydrogel captured in dry and swelling states under UV lamp radiation (modified from [92] with permission). **b** Change in fluorescent intensity, corresponding photographs, chemical structure and scanning electron micrograph of thermally responsive nano hydrogels exhibiting different brightness at different temperatures (modified from [93] with permission)

working mode adjustable via changing the labelling degree of the polymer [95]. The development of a fluorescent self-healing hydrogel based on TPE with thermoresponsive properties, emitting blue light at high temperature (70 °C), was reported by Wang et al. [96]. The hydrogel was biocompatible and exhibited a pH-sensitive drug release capability, which made it attractive for biomedical applications [96]. Based on phenyldiaminotriazine, Galindo et al. [97] synthesized a physically crosslinked hydrogel, exhibiting fluorescent property with blue light emission, by radical polymerization of vinyl and acrylic monomers. Both the formation of the hydrogel as well as its AIE behavior were based on the creation of hydrogen bonds between the triazine moieties and the polymer chain. Their results revealed the impact of swelling ratio on the fluorescent intensity of hydrogel, with increasing swelling ratio leading to a 50% decrease in intensity (Fig. 5a) [97]. A pH-sensitive fluorescent hydrogel made from a pH-sensitive peptide capped with TPE having a higher emission at higher pH was developed

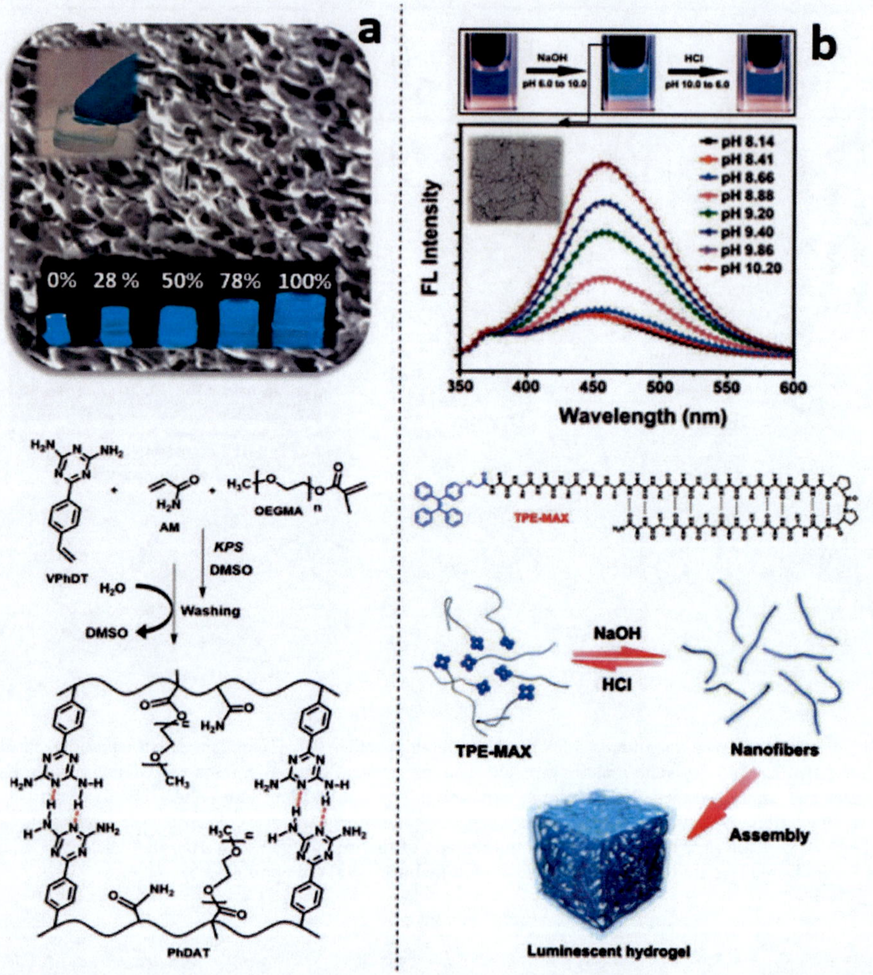

Fig. 5 **a** Scanning electron microscopy and photographs of a phenyldiaminotriazine-based fluorescent hydrogel prepared through physical interaction. Photographs were captured in daylight and under UV irradiation at different swelling ratios (modified from [97] with permission). The scheme presented shows synthesis route of the phenyldiaminotriazine-based fluorescent hydrogel. **b** Sensitivity to pH, transmission electron microscopy image, and change in fluorescent intensities of a peptide-based fluorescent hydrogel, including the chemical structural of the compound and a schematic drawing representing the associated assembly process (modified from [98] with permission).

by Zhang et al.; their results showed a reversible response to pH change with the hydrogel being emissive for pH values adjusted between 6 and 10 (Fig. 5b) [98].

Several studies have reported the AIE effect of naphthalimide with versatile applications as smart materials. Recently, Liu et al. [99] prepared a naphthalimide-based AIE active polymeric hydrogel through the polymerization of *N*-isopropyl acrylamide, hydroxyethyl methacrylate, and 4-phenoxy-*N*-allyl-1,8-naphthalimide monomers. The AIE monomer was modified with the phenoxy group to change its molecular conformation to a propeller shape when compared to the planar naphthalimide molecule. The semi-interpenetrating hydrogel network exhibited a fluorescence pattern switch, emitting a blue light with a thermoresponsive properties affecting both the volume phase transition and emission properties. The AIE hydrogel showed two simultaneous optical properties regulated by just a single stimulus (Fig. 6) [99].

Hou et al. developed thermoresponsive self-healing hydrogels exhibiting the AIE effect based on TPE-acrylamide copolymer crosslinked by diacylhydrazide; the hydrogels revealed an enhanced fluorescent property above the lower critical temperature that was regulated to body temperature [100]. Zhao et al. developed a chemically prepared fluorescent hydrogel based on tetra-(4-pyridylphenyl)ethylene that exhibited concurrent changes in fluorescence color, brightness, and shape in response to a single stimulus. The bi-layered hydrogel consisted of poly(acrylamide-r-sodium 4-styrene sulfonate) as a matrix and AIEgen as the core element and was pH-sensitive. At an acidic pH, the hydrogel was brighter under UV exposure while

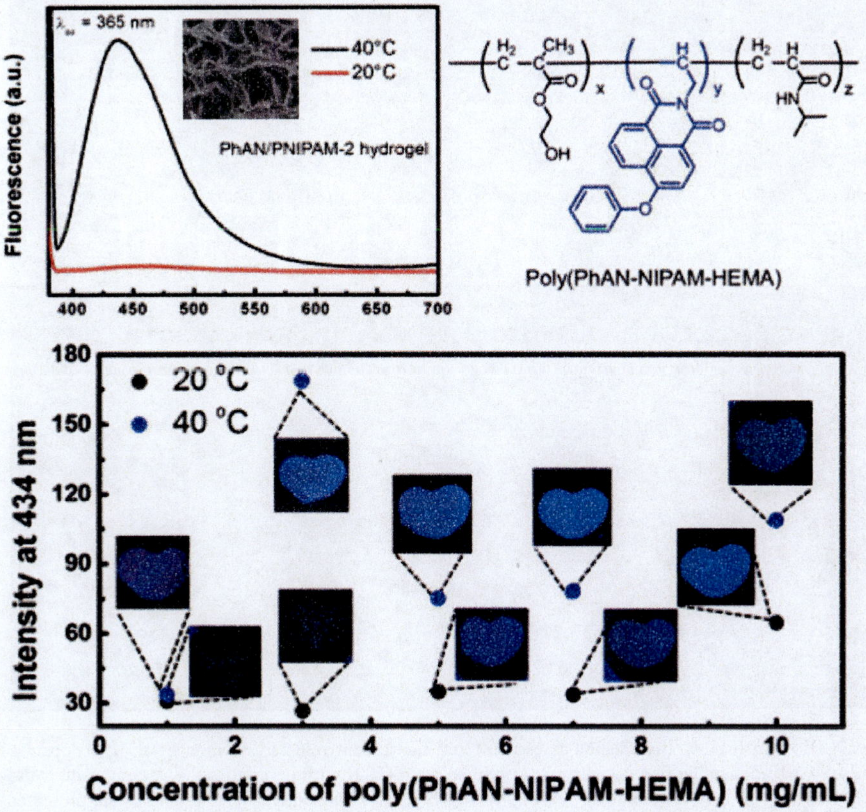

Fig. 6 The fluorescent intensity (*inset* scanning electron microscopic image) and chemical structure of the chemically cross-linked naphthalimide-based AIE-active polymeric hydrogel with higher emission at a higher temperature (modified from [99] with permission)

its shape changed, representing a multi-responsive actuator with a variety of applications such as 3D printing, artificial muscles, and wearable devices [101]. Zhang et al. [102] covalently grafted an AIEgen into the surface of mesoporous dopamine and developed an injectable fluorescent hydrogel by loading the AIE-grafted-dopamine microspheres into the carboxymethyl nanocrystalline cellulose gel structure via photo-crosslinking strategy. Their results revealed a strong anti-photobleaching property for the hydrogel with enhanced biocompatibility and photoacoustic imaging characteristic (Fig. 7a–d) [102].

An interesting study was performed by Ji et al. [103], who employed the classic bilayer approach to construct a composite hydrogel consisting of three different self-healing hydrogels. These three layers were covalently crosslinked and exhibited different fluorescent colors with a robust self-healing property. The tensile tests proved a stable connection between layers, while the self-healing property of the three-layered hydrogel was controlled by dynamic covalent bonds, ionic interactions, and

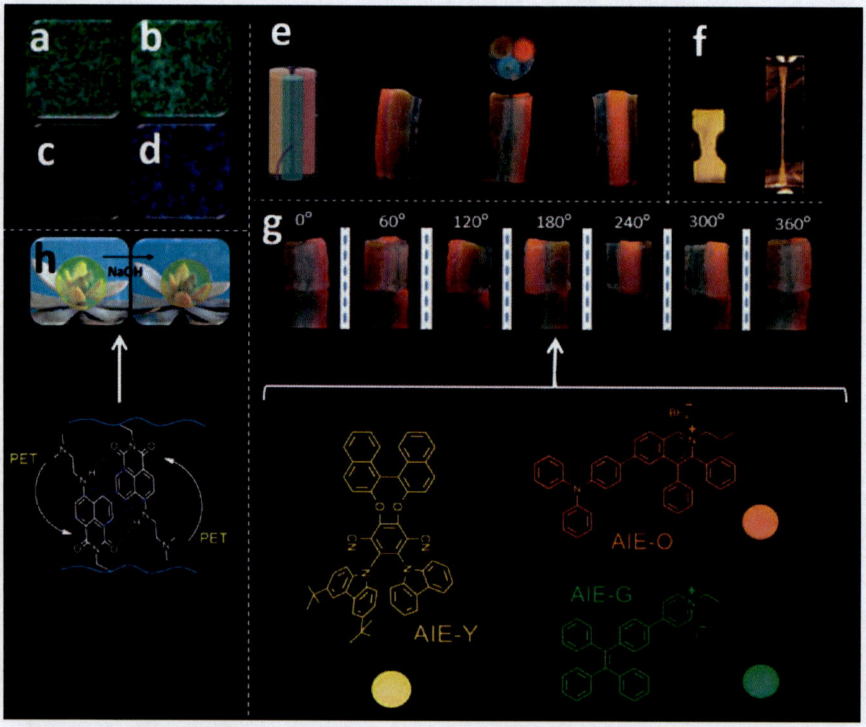

Fig. 7 Biocompatibility (live/death) assessment and fluorescent microscopic images of **a**, **c** dopamine and **b**, **d** AIE-dopamine-based hydrogels (modified from [102] with permission). **e** A composite hydrogel consisting of three different self-healing hydrogels (the associated chemical structures are presented) exhibiting different fluorescent colors under UV irradiation **f** with high extensibility and **g** self-healing property associated with different relative orientations (modified from [103] with permission). **h** Aggregation of the covalently immobilized naphthalimide groups within the acrylamide hydrogel structure leading to an intense yellow-green emission at low pH (modified from [104] with permission)

hydrogen bonding. They also found a strong relationship between the relative orientation of layers and the self-healing property of the hydrogel, with 0° and 360° orientations exhibiting the highest self-healing property and showing the biggest breaking values under strain (Fig. 7e–g) [103]. Li et al. used radical copolymerization to synthesize a highly fluorescent hydrogel using naphthalimide-containing monomer and acrylamide in the presence of chitosan-acetic acid solution. The aggregation of the covalently immobilized naphthalimide groups within the hydrogel structure caused intense yellow-green emission, with increasing pH reducing emission intensity (Fig. 7h) [104].

A summary of chemically prepared fluorescent hydrogels with their key features is presented in Table 3.

3.3 Physically and Ionically Prepared Fluorescent Hydrogels

As mentioned before, the mixing of AIEgens with hydrogels, mainly those that physically turn into a 3D network, results in the development of fluorescent hydrogels. It is well known that the preparation of hydrogels by this route is not efficient, with the leaching of AIEgens as the main drawback. However, researchers have employed this approach or prepared AIEgen-hydrogel bioconjugate systems to study hydrogel processing and nanoscopy [105]. Tang et al. used TPE-chitosan bioconjugates with different degrees of labelling to study the gelation process in a chitosan-alkaline urea system and monitored the layer-by-layer self-assembly process of natural hydrogels [106, 107]. The visual characterization of the gelation process in situ, using a fluorescent microscope, was introduced as a powerful research tool to identify the formation of junction points during the gelation process [108]. These researchers identified two individual, but integrated, phases of thermal gelation and rinse (polymer–solvent interaction) that play a role in the gelation process. They reported that an embryonic and crystalline structure was formed due to the temperature change, while intramolecular hydrogen bonds were developed during the rinse phase (Fig. 8a). Tavakoli et al. [50] physically mixed TPE-2BA with polyvinyl alcohol (PVA) hydrogels and reported an increase in the fluorescent intensity of hydrogels that were exposed to a different number of freeze–thaw cycles. Relying on the formation and growth of crystalline regions as a function of freeze–thaw cycles, their results revealed that the increase in fluorescent intensity reflected changes in the crystallinity of PVA hydrogels. These observations were further evidenced by fluorescent microscope images indicating that the increase in the number of freeze–thaw cycles led to the formation of larger and brighter regions (Fig. 8b) [50]. Lin et al. also physically mixed di(4-propoxyphenyl) dibenzofulvene AIEgen with vinylpyrrolidone and acrylamide hydrogel and observed an increase in fluorescent intensity during freeze–thaw cycles. Interestingly, they found different fluorescent colors occurring at different numbers of freeze–thaw cycles as a result of increasing hydrogel viscosity [109]. A similar approach indicating crystallization-induced emission was also employed by Zheng et al. when developing fluorescent hydrogels with switchable colors that occurred at different temperatures [110]. Gao et al. synthesized an AIEgen that was sensitive to calcium ion and, by physically

Table 3 Summary of chemically prepared hydrogels with fluorescent properties

AIEgen	Key features	References
Poly(γ-glutamic acid) and o-carborane	Water-sensitive luminescence property	[92]
TPE-polyacrylamide-poly acrylic acid	Thermo-sensitive fluorescent property	[94]
Acrylamide and TPE-acrylate	Thermo-sensitive fluorescent property	[95]
TPE-[P(DMA-stat-DAA)]$_2$	Thermo-sensitive fluorescent characteristic with self-healing property	[96]
Phenyldiaminotriazine with vinyl and acrylic monomers	Water-sensitive luminescence property	[97]
TPE-peptide	pH-sensitive fluorescent property	[98]
Naphthalimide-acrylamide	Two simultaneous optical properties regulated by just a single stimulus	[99]
TPE-acrylamide	Thermo-sensitive fluorescent property	[100]
Tetra(4-pyridylphenyl)ethylene	Bi-layered hydrogel formed with pH-sensitive fluorescent property	[101]
AIE-grafted-dopamine	Photoacoustic imaging characteristic	[102]
TPE derivatives	Hydrogel composite with different fluorescent colors and a robust self-healing property	[103]
Naphthalimide-acrylamide	pH-sensitive fluorescent property	[104]

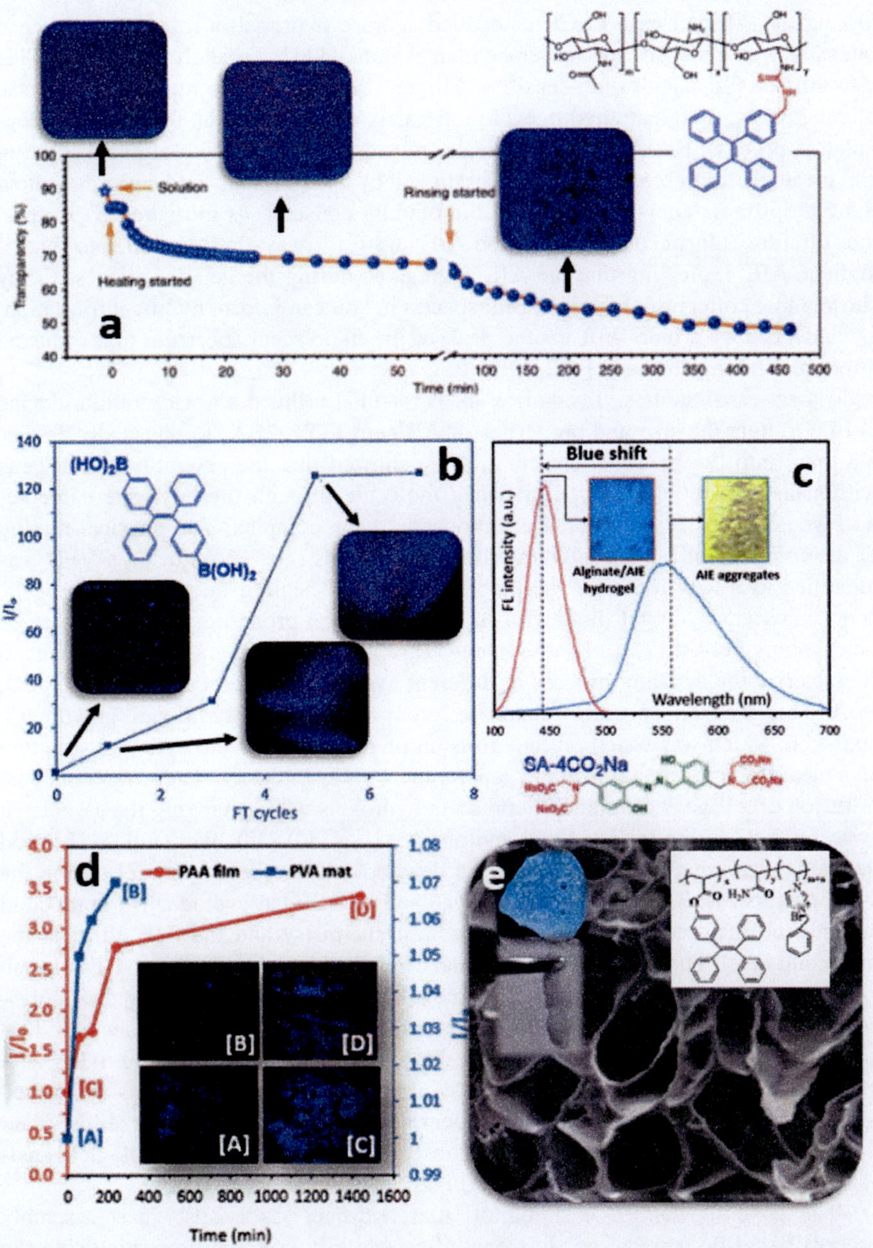

Fig. 8 Application of AIEgens for **a** visualization of the gelation process in chitosan hydrogels (modified from [108] with permission), **b** identification of crystalline regions in polyvinyl alcohol (PVA) hydrogels forming through freeze-thaw cycles (modified from [50] with permission), **c** revealing gel formation in alginate-based hydrogels (modified from [112] with permission), and **d** swelling characterization of different PVA hydrogels through the addition of AIEgen into the swelling media with swelling causes AIEgen diffusion (fluorescent microscopic images included) (modified from [49] with permission). **e** Scanning electron microscopy and photographs (daylight and UV exposure) of an AIE-active polymer based on acrylamide and developed a supramolecular hydrogel through physical mixing with cucurbit[8]uril molecules (modified from [119] with permission)

mixing the AIEgen with $CaCl_2$-embedded-agarose hydrogel, a robust fluorescence intensity was found for the ion-embedded regions [111]. Tavakoli et al. [112, 113] also utilized the calcium ion-sensitive AIEgen (SA-4CO_2Na) to monitor the release of calcium from alginate hydrogels to ultimately characterize the swelling and degradation process. Besides, they physically mixed the AIEgen with alginate solution and created a fluorescent hydrogel, crosslinked by calcium ions, and emitting yellow light at 550 nm. They observed that the binding constant, as measured by isothermal titration calorimetry, between the AIE and Ca^{2+} was almost similar to that of alginate-AIE, indicating that the AIE aggregated during the gelation process. They showed that collecting the fluorescent spectra in situ can inform hydrogel formation, as evidenced by a blue shift for the peak of the fluorescent spectrum that occurred after crosslinking (Fig. 8c) [112, 113].

In a series of studies, Tavakoli et al. [114–117] utilized a vortex fluidic device (VFD) to tune the size and properties of AIEgens (TPE-2BA) to physically assemble them into the hydrogel structure. They showed that the assembly of AIEgens within the structure of a hyperbranched molecule through the host–guest interaction would intensify the fluorescent property of the complex. The physical mixing of the complex with physically crosslinked hydrogels (i.e., PVA) using a VFD was identified as a new route for the preparation of self-healing fluorescent hydrogels, and fluorescent hydrogel films with strong mechanical properties and tuned surface morphology [114–117]. In another study, Tavakoli et al. [49] employed AIEgens to characterize the swelling process of different hydrogels (film, fiber, superabsorbent, etc.). This was an accurate technique, compared to the traditional gravimetric approach, which was based on the diffusion of AIEgens into the hydrogel structure and measurement of the swelling environment's FL intensity. They observed that diffusion of AIEgens developed fluorescent hydrogels while changing the associated concentration within the swelling environment (Fig.I8d) [49]. Wei et al. [118] used the same technique as Tavakoli et al. to develop a fluorescent hydrogel during the swelling process. These researchers developed an AIEgen-cyclodextrin compound and studied the synergic effect of host–guest chemistry and the AIE effect at the molecular level. They immersed a cyclodextrin-based hydrogel into the AIE solution with the hydrogel pore size controlling the host–guest reaction affecting the fluorescent intensity [118]. Liu et al. [119] synthesized an AIE-active polymer based on acrylamide and developed a supramolecular hydrogel through physical mixing with cucurbit[8]uril molecules. Aggregation of the AIE-active polymer was performed through a guest–host interaction with cucurbit[8]uril molecule playing a host function. A reversible sol–gel transition was observed for the hydrogel; rheological analysis indicated its self-healing property (Fig. 8e) [119].

Zhao et al. developed a supramolecular hydrogel via a host–guest assembly process based on tripodal gelators exhibiting white fluorescent property, using the hydrogel as a sensor to detect low concentrations (10^{-9} M) of Fe^{3+} and F^- ions [120]. These researchers also developed another fluorescent hydrogel based on naphthalimide derivative and tri-(pyridine-4-yl)-functionalized trimesic amide that was formed via hydrogen bonding. They reported a self-healing property for the hydrogel, which irradiated an orange color with a high quantum yield (0.578) [121]. Others also developed supramolecular polymeric hydrogels to selectively detect and separate different ions by increasing their concentration inside the hydrogel networks to tune the AIE effect [122–124]. Aschman et al. [125] used guadiniumcarbonylpyrrole molecules in combination with aromatic thioethers as an AIEgen to develop a hydrogelator that can self-assemble into fibrils in aqueous media. Their results revealed a weak fluorescent property for the fibrils with no ability to form a gel. However, when the fibrils were crosslinked using different stimuli, such as heat or pH changes, a highly emissive stable hydrogel was formed [125]. Zhao et al. [126] developed a core–shell pH-responsive hydrogel with negatively charged sulfonated 9,10-distyrylanthracene derivative, as an AIEgen, ionically attached to the surface of nanoparticles [126]. A summary of physically and ionically prepared fluorescent hydrogels with their key features and applications is presented in Table 4.

3.4 Clustering-Triggered and Aggregation-Enhanced Fluorescent Hydrogels

In 2013, Gong et al. observed a bright blue light emission in some natural components, including rice, starch, cellulose, and glucose, illustrating AIE characteristics [127]. Electron cloud overlap and spatial conformation rigidity followed by the clustering of nonconventional chromospheres was suggested as the emission mechanism. Thus, the clustering-triggered emission mechanism was proposed to explain the associated emission behavior [128, 129]. The unexpected emission observed in these natural components prompted a few studies to explore whether specific macromolecules exhibit fluorescent properties in the gel state. Dou et al. studied the emission behavior of sodium alginate and found a distinct concentration-enhanced-emission and AIE characteristics influenced by the ratio of mannuronic to guluronic acids [71]. They found that sodium alginate exhibited a fluorescent property in high concentrated solutions, solid and film states with clustering of oxygen atoms and carboxylates as well as a chain–chain interaction and entanglement held responsible for the observed emission (Fig. 9a, b) [71]. Inspired by the clustering-triggered fluorescent in natural polymers, others have examined some synthetic polymeric gels. Yang and Zhang [124] synthesized polyacrylic acid polymer to examine whether clustered carbonyl acids with potential hydrogen bonds were emissive. They found that polyacrylic molecules exhibited fluorescent properties in dimethylformamide (DMF) with the emission enhanced as the concentration was increased. Both polyacrylic acid powder and hydrogel were also emissive, with the hydrogel exhibiting higher intensity and emitting brighter blue light compared to those in solution and powder. These researchers suggested that clustering of carbonyl acids and electron cloud overlap among C=O and –OH groups were responsible for the associated emission [129]. Zhou et al. also observed the cluster-triggered fluorescence effect in polyacrylic acid, polyacrylamide, and poly(*N*-isopropylacrylamide) (Fig. 9c) [69].

These researchers found a fluorescent property for both polyacrylic acid and polyacrylamide solids at room temperature with poly(*N*-isopropylacrylamide) demonstrating negligible fluorescent property in air due to oxygen quenching. They showed that the fluorescent property of these polymers can be tuned by precisely controlling ionization, pressurization, and oxygen isolation through modulating the intermolecular interactions or isolation of triplet quenchers [69]. Jiang et al. [130] reported a

Table 4 Summary of physically and ionically prepared fluorescent hydrogels with fluorescent properties

	AIEgen	Key features	References
1	TPE-chitosan	Visualization of gelation process	[106]
		Monitoring layer-by-layer self-assembly process of natural hydrogels	[107]
2	TPE-2BA	Visualization of crystalline regions in hydrogels	[50]
		Swelling characterization of hydrogels	[49]
3	Di(4-propoxyphenyl) dibenzofulvene	Different fluorescent colors occurring at different numbers of freeze–thaw cycles	[109]
		with crystallization-induced emission	[110]
4	SA-4CO_2Na	Visualization of crosslinked regions and swelling-degradation characterization in hydrogels crosslinked by calcium ions	[110–113]
5	Hyperbranch-TPE-2BA	Formation of PVA-based fluorescent hydrogels with self-healing property	[114–117]
6	AIEgen-cyclodextrin ssssss8]uril	Host–guest chemistry affecting fluorescent property	[118]
			[119]
7	Naphthalimide derivative and tri-(pyridine-4-yl)-functionalized trimesic amide	Developing self-healing fluorescent hydrogels and sensors to detect low concentration of iron (10^{-9} M)	[120, 121]
8	Guadiniumcarbonylpyrrole and aromatic thioethers	pH- and thermos-sensitive fibrous sensors	[125]
9	Sulfonated 9,10-distyrylanthracene derivative	Developing a core–shell pH-responsive nano hydrogel	[126]

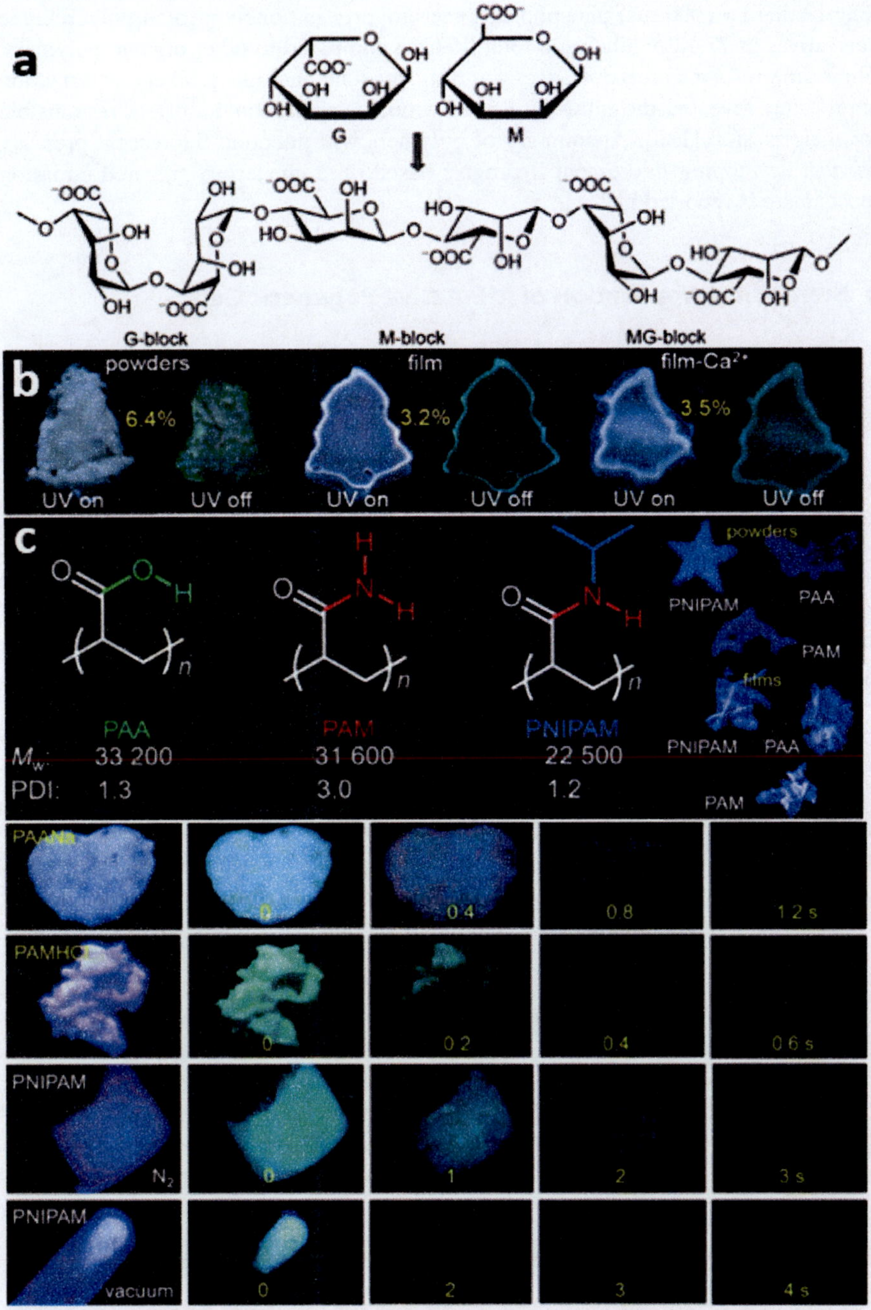

Fig. 9 **a** Chemical structure of alginate and **b** photographs of alginate solid powder, cast film, and cross-linked film (by Ca^{2+}) with M/G = 1.53 taken under 312 nm UV light (modified from [147, 148] with permission). **c** Chemical structures of amorphous nonaromatic polymers with their fluorescent properties in powder and film states and their tunable persistent room temperature phosphorescence (modified from [147, 148] with permission)

long-lasting (> 1000 ms) phosphorescence property in nonconjugated polyurethane derivatives at 77 K, exhibiting a long lifetime compared to other organic polymers. Their time-resolved emission spectra and steady-state photophysical characterization approaches revealed the intra- and/or intermolecular carbonyl clusters responsible for the emission [130]. A summary of polymers with inherent fluorescent property used in developing fluorescent hydrogels based on a clustering-triggered emission mechanism is reported in Table 5.

4 Biomedical Application of AIE-Active Polymeric Gels

The most appealing applications of fluorescent polymeric gels, especially hydrogels with an AIE feature, are in the areas of life sciences and biomedical engineering, useful for in vitro and in vivo imaging of biogenic molecules, cells, tissues, and organisms. With bright emission, high efficiency, high photostability, superb biocompatibility, and large optical nonlinearity, AIE-active hydrogels hold promise for in-depth and long-term imaging and tracing applications. The development of multimodality sensing systems and the combination with other medication processes (i.e., controlled or targeted drug delivery) will further widen the scope of biomedical applications of AIE-active hydrogels in such important areas as clinical diagnosis and treatment [131–136]. Naturally based hydrogels in combination with AIEgens were also proposed for different biomedical applications [137]. The sustained release of drugs can be accurately monitored in vivo using AIE-active thermoresponsive hydrogels, maintaining the effective dosage in the optimal therapeutic window [138]. A recent review by Lai revealed the application of nonconjugated fluorescent polymers in different dosage forms (tablets, gels, hydrogel, and films) for drug delivery [139]. Kim et al. [140] developed silica gel nanoparticles containing AIEgens and photosynthesizing drugs to simultaneously track tumor cells and generate single oxygen to induce a cytotoxicity effect under two-photon irradiation. Ikeda et al. [141] developed a dipeptide-based fluorescent hydrogel demonstrating coupling with glucose oxidase to further prepare a hydrogel with a glucose-responsive insulin-releasing feature.

The high efficacy of emerging multifunctional carrier-free nanomedicine has attracted remarkable attention in cancer treatment. Usually, carrier-free nanomedicines host an AIE complex that can self-assemble into a fluorescent hydrogel with the AIE effect. Exposing the AIE-based hydrogelator to external stimuli (i.e., irradiation), the assembly is cleaved and releases the AIEgen at the tumor site. AIEgens exhibit anti-cancer properties by upregulating the reactive oxygen species levels in cancer tumors [142]. Li et al. developed selenium-AIE molecules through a four-step synthetic route containing TPE groups (hydrophobic) and ethylene glycol groups (hydrophilic) with a self-assembly characteristic. Their results revealed that exposing the hydrogelator to γ-radiation cleaved the diselenide bonds in the assembly, forming seleninic acid with anti-cancer characteristics through the chemotherapy route. Meanwhile, TPE enables the detection of a tumor site and monitoring of drug distribution, which underlines the efficacy of nanomedicine [143]. In another study, Zhao et al. [144] used a TPE-functionalized, diselenide-crosslinked hydrogel

Table 5 Summary of polymers with inherent fluorescent property used in developing fluorescent hydrogels

Natural chemicals	Key features	References
Sodium alginate	Concentration-enhanced emission and AIE characteristics influenced by the ratio of mannuronic to guluronic acids	[71]
Poly acrylic acid	Fluorescent properties in DMF with the emission was enhanced as the concentration was increased	[69, 129]
Polyacrylamide Poly(*N*-isopropylacrylamide)	The fluorescent property of these polymers can be tuned by precisely controlling ionization, pressurization, and oxygen isolation	[69]
Polyurethane derivatives	Exhibited a long lifetime (> 1000 mS) compared to other organic polymers	[130]

DMF dimethylformamide

based on hydroxyethyl methacrylate using diselenide. The hydrogel was degraded upon exposure to hydrogen peroxide or dithiothreitol with TPE-containing polymer chains that were released at the tumor site.

The interaction between the analyte and AIEgen, controlled by the physicochemical properties of the hydrogel and leading to the enhancement of the fluorescent property, is well-known as the mechanism of detection. Detection of hydrogen peroxide released by living cells along with other biomolecules has been reported [145, 146].

AIE-active hydrogels can also be utilized as molecular luminescence thermometers [80]. Heo et al. [147, 148] developed an injectable fluorescent hydrogel for the long-term monitoring of glucose in vivo. They used an AIEgen composed of diboronic acids, anthracene acid, polyethyleneglycol (PEG), and a vinyl group as a glucose-responsive monomer to form the hydrogel (Fig. 10). They also added encapsulated glucose-responsive fluorescent dyes within the gel structure to develop glucose-responsive fluorescent hydrogel beads (Fig. 10) [147, 148].

Recently, Zhang et al. [149] reviewed the design and synthesis methods for nano cellulose-based fluorescent hydrogels creating sensing platforms and being responsive to different stimuli, including pH, temperature, toxic compounds, and metal cations. Xu et al. synthesized a self-assembled gelator based on TPE that exhibited AIE behavior when exposed to melamine, forming fibrous structures. A high concentration of melamine and its accumulation leads to kidney damage [150]. Hsu et al. [151] developed a functional supramolecular hydrogel with high contrast cell-material imaging in 3D tissue-engineered scaffolds. Several studies have reported the application of AIE-active hydrogels for single-cell tracking with high brightness

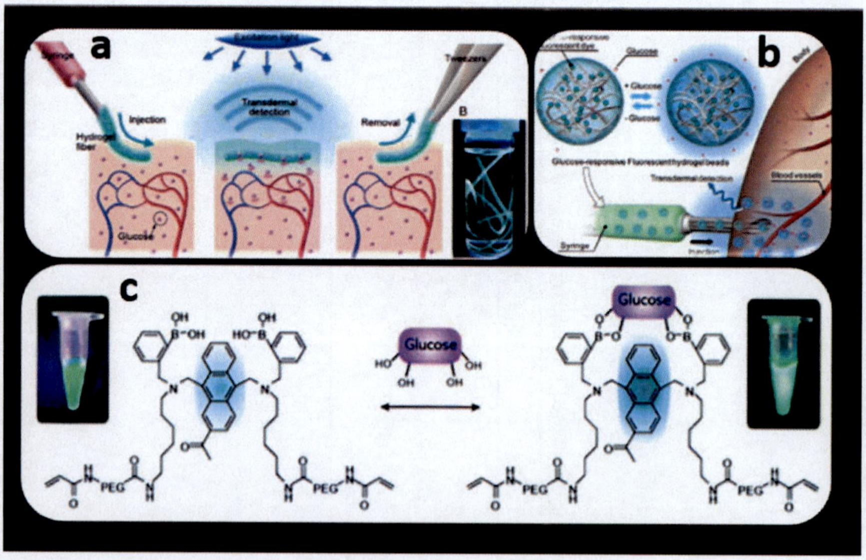

Fig. 10 Schematic drawing of **a** a chemically prepared injectable fluorescent hydrogel and **b** a hydrogel containing fluorescent nanoparticles. **c** The AIE mechanism of a hydrogel through glucose molecules causing aggregation (modified from [147, 148] with permission)

and enhanced quantum yield [152]. The development of a fluorescent hydrogel with amplified far and near infra-red fluorescence for targeted in vivo imaging was reported by Ding et al. [153]. By chemical bonding between TPE and chitosan through a reaction between the amine and isothiocyanate groups, a naturally based fluorescent hydrogel was developed for cell imaging [106]. A self-assembled chitosan-AIE nanoparticle was shown to diffuse into the cell through lipid-raft endocytosis, exhibiting high fluorescent property upon aggregation in the lysosome, demonstrating its application for subcellular imaging [154].

In similar studies, the fabrication of TPE-chitosan nanoparticles was reported for long-term live cell imaging [155, 156]. One study showed that glycol chitosan conjugated a dipolar tricyanostilbene derivative, consisting of highly hydrophilic backbone polymer and hydrophilic pendant, which can self-assemble to a nanohydrogel exhibiting a near-infrared fluorescent property. Endowed with an enhanced fluorescent property and high cellular uptake, the hydrogel has enormous potential for bioimaging [157]. Hydroxyapatite coated with TPE-chitosan hydrogel through electrostatic interaction was also used for cell imaging. The nanoparticles adhered to the 293 T cell membrane or endocytosed through phagocytotic vesicles, showing a strong emission [158]. Endocytosis, as a cellular process by which fluorescent hydrogels are diffused into cells, facilitates labelling cells with a contrast medium with the fluorescent property. The diffusion of hydrogels into the cell forms a fluorescent fibrillar network with a distinct emission compared to cells under fluorescent microscopic observation.

New directions utilizing fluorescent hydrogels, as either tissue culture media or tissue-engineered scaffolds, assist to effectively reveal both cell imaging and cell-scaffold interaction through overlayed cell and hydrogelator imaging [159, 160]. Kong et al. [161] developed a series of non-fluorescent oligopeptides that were aggregated into fluorescent nanoparticles through a self-assembly process with rainbow colors. Their results showed that the sensitivity of the fluorescent oligopeptides to pH, along with the associated biocompatibility and photostability, made them highly appropriate for bioimaging. Mitra et al. [162] reported grafting of fluorescent guar gum on acrylic acid and acrylamide based polymer to create self-assembled hydrogels with AIE effect. They proposed these hydrogels for prospective five-in-one applications, including sensing (Pb), security ink, logic function, ion removal (Pb), and normal and cancer cell imaging [162]. Huang et al. [163] developed a fluorescent hydrogel based on carboxymethyl chitosan with strong fluorescent properties when exposed to Zn^{2+} and Cd^{2+} ions. It seems that the proposed hydrogel can be used for the development of antibacterial wound dressings based on the superior biomedical property of the chitosan; however, without the incorporation of an AIEgen, the main drawback is the low quantum yield (6.8%) of the hydrogel [163]. Functionalized AIEgens were also used for the degradation tracking of the injectable hydrogels important for tissue engineering [102]. Talloj et al. constructed self-assembled fluorescent supramolecular nanobelts based on tetraphenylethylene dipeptides demonstrating cell adhesion selectivity under different ranges of pH, which presented remarkable applications in the fabrication of smart scaffolds for tissue engineering [164]. Chu et al. developed a shape memory polymer with long-term fluorescent properties based on crosslinking 6-arm poly(ethylene glycol)–poly(ε-caprolactone) using a croconate dye. The scaffold exhibited the AIE effect with a promising NIR fluorescent deep tissue imaging function [165]. The degradation of fluorescent polylactic-based scaffolds was monitored during the hydrolytic degradation process of the scaffold [166]. Wang et al. [167] developed a fluorescent PVA-based hydrogel using TPE by the tertiary amine method. The scaffold exhibited low cytotoxicity with enhanced broad-spectrum antimicrobial activity. Several studies have shown the effectiveness of self-assembled hydrogels for bacterial detection or fabricating antibacterial tissue-engineered scaffolds [168–170]. A summary of biomedical applications of fluorescent polymeric gels is presented in Table 6.

5 Current Challenges and Future Strategies

Compared to the traditional fluorescent dyes or luminescent quantum dots, polymeric AIEgens with high sensitivity and excellent stability have demonstrated superior fluorescent properties and so the associated conventional drawbacks are resolved. When it comes to fluorescent gels, the use of traditional fluorescent dyes in low concentrations leads to a weak fluorescent property, with increasing concentration cause quenching. The leakage of fluorescent dyes, photobleaching, and the impact of dyes on gel structural stability are other disadvantages. Besides, while luminescent quantum dots exhibited high quantum yield with brighter fluorescent emission in the solution state, their fluorescent properties are significantly reduced upon incorporation into the gel structure. This observation is known to be related to gel chains shielding the emission of embedded dots, with high gel volume fraction increasing the shielding effect. Due to the unique fluorescent property of AIEgens with tunable emissive colors, their AIE effect leads to enhanced fluorescent properties, even when embedded in the gel structure; therefore, AIE-based gels have attracted attention in recent years (Fig. 11).

Among different methods involved in the fabrication of AIE-based fluorescent gels, AIE polymers with self-assembly properties or those that can be used as crosslinking agents to form fluorescent gels are more efficient. The other advantage of fluorescent gels, especial hydrogels made of AIE polymers, compared to quantum dots, is their higher biocompatibility, making them more attractive for biomedical applications. Reports have indicated the long-term cytotoxicity of fluorescent hydrogels based on quantum dots, which is associated with the release of metal ions over time and the surface property of the nano-sized particles, which are substantial concerns. To tune the biocompatibility of the quantum dots-based fluorescent gel, it is suggested to reduce the associated fraction of quantum dots to gel, or to control the density of the gel network to regulate the release rate, avoiding toxic accumulation. However, such optimization significantly affects the final emission, mechanical and physicochemical properties of the gel. Since the most appealing applications of fluorescent gels, including hydrogels, are in the life sciences and biomedical engineering, the biocompatibility of AIE-based fluorescent hydrogels needs more attention. This influences future strategies for developing more reliable fluorescent hydrogels. Extensive efforts to engineer fluorescent gels using AIEgens have been exerted through different chemical routes but, unfortunately, commercially available

Table 6 Summary of biomedical applications of fluorescent polymeric gels

Fluorescent polymeric gel	Key features	References
DDS		
Poly(PEG/PPG/TPE urethane) with TPE	AIE thermos-gel exhibited advantages over typical DDS wand revealing drug release status and matrix erosion	[138]
Nonconjugated luminescent polymers	Developing dosage forms (tablets, gels, hydrogel, and films) with optimal therapeutic window	[139]
Silica gels	Two-photon photodynamic therapy (tumor cell tracking and treatment)	[140]
Dipeptide-based fluorescent hydrogel	Developing a hydrogel with a glucose-responsive insulin-releasing feature	[141]
Cancer treatment		
Multifunctional carrier-free nanomedicine	Hydrogels hosting an AIE complex releasing the AIEgen at the tumor site	[142]
Selenium-TPE-PEG	Exposure to γ-radiation or hydrogen peroxide cleaved self-assembly and released seleninic acid with anti-cancer characteristics	[143, 144]
Biosensors		
Indium tin oxide and PET	Hydrogen peroxide detection at the cell level	[145, 146]
Silver and PEG	Molecular luminescence thermometer	[80]
AIEgen composed of diboronic acids, anthracene acid, PEG	Glucose-responsive fluorescent hydrogel beads and injectable hydrogels	[147, 148]
Nano cellulose-based fluorescent hydrogels	Sensing platform for pH, temperature, toxic compounds, and metal cations	[149]
TPE-melamine	Detecting melamine for kidney disease	[150]
Oligopeptides	pH sensing and cell imaging	[161]
Carboxymethyl chitosan	Zn^{2+} and Cd^{2+} ions detection	[163]
Py-CN-TFMBE	pH and thermos-sensitive gelator	[84]
Tissue engineering		
Peptide-AIEgens	3D tissue-engineered scaffolds	[151]
Polyester-AIEgens	Cell-scaffold imaging	[159, 160]
Carboxymethyl chitosan	Antibacterial wound dressing	[163]
Peptide-TPE	Smart scaffold with superior cell adhesion	[164]
Poly(ethylene glycol)–poly(ε-caprolactone) with croconate	Fluorescent scaffold with tissue imaging function	[165]

Table 6 (continued)

Fluorescent polymeric gel	Key features	References
Polyvinyl alcohol-TPE	Fluorescent scaffold with antibacterial property	[166]
Bioimaging		
Multicolor conjugated polymer dots	Single-cell tracking	[152]
TPE-TPA-DCM	Targeted in vivo imaging	[153]
TPE-chitosan	Cell imaging	[106]
	Subcellular imaging	[154]
	Live cell imaging	[155]
Glycol chitosan conjugated tricyanostilbene derivative	Bio- and cell imaging	[157]
Hydroxyapatite coated with TPE-chitosan	Bioimaging (cell, 293 T cell membrane, endocytose)	[158]
Guar gum/ PAA and PAAm	Cancer cell imaging	[162]

CN-TFMBE 1-cyano-trans-1,2-bis-(3 0,5 0 -bis-trifluoro- methyl-biphenyl)ethylene, DCM , *DDS* drug delivery system, *PAA* poly acrylic acid, *PEG* polyethylene glycol, *PET* polyethylene terephthalate, *PPG* polypropylene glycol, *TPE-TPA-DCM* (2-(2,6-bis((E)-4-(phenyl(4'-(1,2,2-triphenylvinyl)-[1,1'-biphenyl]-4-yl)amino)styryl)-4H-pyran-4-ylidene)malononitrile

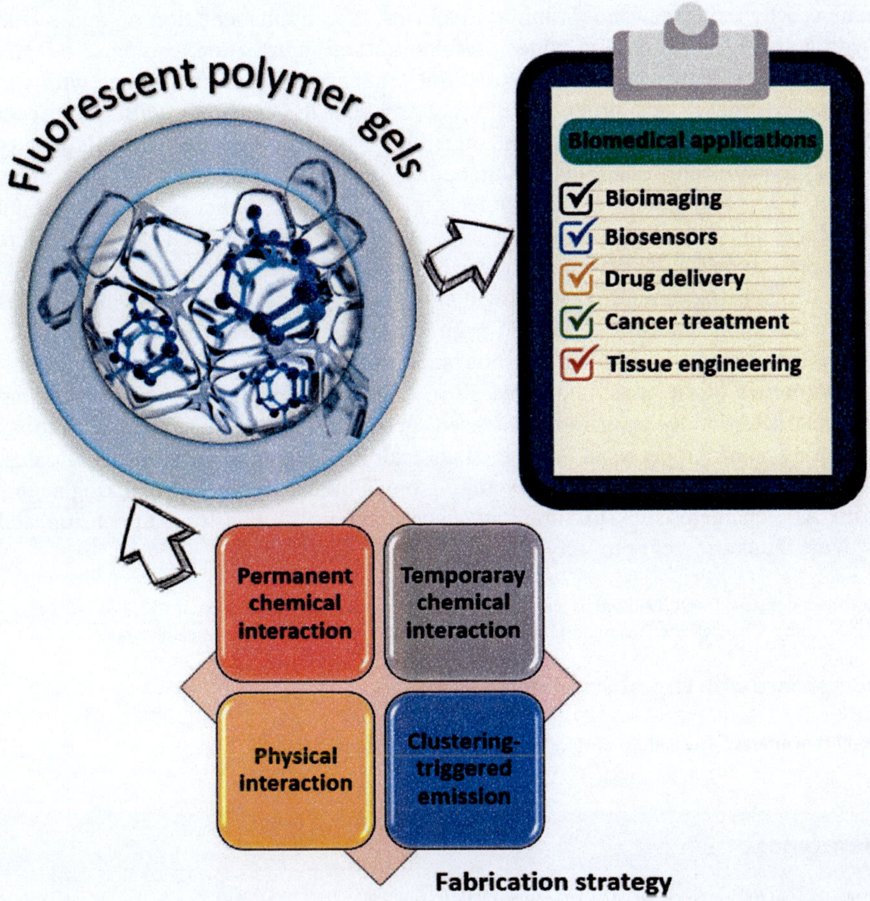

Fig. 11 Fabrication strategies, types and biomedical applications of AIE-based fluorescent hydrogels

fluorescent gels with AIE features are yet to be introduced. The main drawbacks limiting large-scale production have been the complex chemistry and multi-step processes required for the synthesis of AIEgen-based fluorescent gels with enhanced emission properties. On the other hand, the functionalization of AIEgens is a difficult process, making the production of fluorescent gels expensive.

It is important to note that the development of water-soluble AIEgens is critical for biomedical applications. The use of organic solvents to dissolve AIEgens is believed to decrease their biocompatibility and limit their applications. The opportunity of employing AIEgens in numerous biomedical applications, such as drug delivery and tissue engineering, has been underlined by investigating their biocompatibility. Controlling AIEgen particle size is one of the major challenges to be addressed for driving applications in biology and medical areas. To develop fluorescent hydrogels, tuning the size of the AIEgens, leading to smaller and brighter moieties, can be achieved by using new microfluidic platforms. The implementation of fluorescent hydrogels for 3D printing provides useful structural information leading to a better understanding of construct microstructural organization. Current studies with traditional hydrogels lack structural information about the construct, with fluorescent microscopes only employed for cell imaging. Using fluorescent hydrogels to create tissue-engineered scaffolds simultaneously enables cell and scaffold characterization. Current research utilizes fluorescent hydrogels to gather information about the shape and size of the construct [171–173]. However, it is believed that future attempts can focus on the intensity of the emitted light to address construct stiffness, which is important for cell growth and differentiation. Also, observing changes in the emission intensity of the 3D printed fluorescence may help to gather more information about the diffusion of nutrients, as well as the swelling and mechanical properties of the construct as placed in cell culture media. It is still challenging to construct stimuli-responsive fluorescent gels exhibiting simultaneous alterations in fluorescence property and shape, specifically in response to a single stimulus. Another future direction can be developing novel biocompatible crosslinking agents with AIE characteristics to simultaneously stabilize the construct dimension and provide the fluorescent property.

Acknowledgment Javad Tavakoli is grateful for the support of the University of Technology Sydney (UTS) with a Chancellor's Postdoctoral Research Fellowship (CPDRF) for the research work.

Compliance with Ethical Standards

Conflict of interest The authors declare no conflict of interest.

References

1. Liu Z, Ran Y, Xi J, Wang J (2020) Polymeric hybrid aerogels and their biomedical applications. Soft Matter 250:20
2. Cook MT, Brown MB (2018) Polymeric gels for intravaginal drug delivery. J Control Release 270:145–157
3. Cao Z, Dobrynin AV (2016) Nanoparticles as adhesives for soft polymeric materials. Macromolecules 49(9):3586–3592
4. Panja S, Panja A, Ghosh K (2020) Supramolecular gels in cyanide sensing: a review. Mater Chem Front 20:20
5. Appel EA, del Barrio J, Loh XJ, Scherman OA (2012) Supramolecular polymeric hydrogels. Chem Soc Rev 41(18):6195–6214
6. Chivers PR, Smith DK (2019) Shaping and structuring supramolecular gels. Nat Rev Mater 4(7):463–478
7. Lim J, Lin Q, Xue K, Loh X (2019) Recent advances in supramolecular hydrogels for biomedical applications. Mater Today Adv 3:100021
8. Eelkema R, Pich A (2020) Pros and cons: supramolecular or macromolecular: what is best for functional hydrogels with advanced properties? Adv Mater 32(20):1906012
9. Erol O, Pantula A, Liu W, Gracias DH (2019) Transformer hydrogels: a review. Adv Mater Technol 4(4):1900043
10. Distler T, Boccaccini AR (2020) 3D printing of electrically conductive hydrogels for tissue engineering and biosensors—a review. Acta Biomater 101:1–13
11. Merino S, Martín C, Kostarelos K, Prato M, Vázquez E (2015) Nanocomposite hydrogels: 3D polymer-nanoparticle synergies for on-demand drug delivery. ACS Nano 9(5):4686–4697

12. Tavakoli J, Tang Y (2017) Honey/PVA hybrid wound dressings with controlled release of antibiotics: structural, physico-mechanical and in-vitro biomedical studies. Mater Sci Eng C 77:318–325
13. Chuah C, Wang J, Tavakoli J, Tang Y (2018) Novel bacterial cellulose-poly (acrylic acid) hybrid hydrogels with controllable antimicrobial ability as dressings for chronic wounds. Polymers 10(12):1323
14. Park SH, Park JY, Ji YB, Ju HJ, Min BH, Kim MS (2020) An injectable click-crosslinked hyaluronic acid hydrogel modified with a BMP-2 mimetic peptide as a bone tissue engineering scaffold. Acta Biomater 20:20
15. Patel A, Zaky SH, Schoedel K, Li H, Sant V, Beniash E, Sfeir C, Stolz DB, Sant S (2020) Design and evaluation of collagen-inspired mineral-hydrogel nanocomposites for bone regeneration. Acta Biomater 112:262–273
16. Mehrban N, Pineda Molina C, Quijano LM, Bowen J, Johnson SA, Bartolacci J, Chang JT, Scott DA, Woolfson DN, Birchall MA, Badylak SF (2020) Host macrophage response to injectable hydrogels derived from ECM and α-helical peptides. Acta Biomater 111:141–152
17. Tavakoli J (2017) Tissue engineering of the intervertebral disc's annulus fibrosus: a scaffold-based review study. Tissue Eng Regener Med 14(2):81–91
18. Basu S, Pacelli S, Paul A (2020) Self-healing DNA-based injectable hydrogels with reversible covalent linkages for controlled drug delivery. Acta Biomater 105:159–169
19. Liu C, Guo X, Ruan C, Hu H, Jiang B-P, Liang H, Shen X-C (2019) An injectable thermosensitive photothermal-network hydrogel for near-infrared-triggered drug delivery and synergistic photothermal-chemotherapy. Acta Biomater 96:281–294
20. Chen S, Duan J, Tang Y, Zhang Qiao S (2013) Hybrid hydrogels of porous graphene and nickel hydroxide as advanced supercapacitor materials. Chem A Eur J 19(22):7118–7124
21. Zeng Q, Qi X, Zhang M, Tong X, Jiang N, Pan W, Xiong W, Li Y, Xu J, Shen J (2020) Efficient decontamination of heavy metals from aqueous solution using pullulan/polydopamine hydrogels. Int J Biol Macromol 145:1049–1058
22. Wahlström N, Steinhagen S, Toth G, Pavia H, Edlund U (2020) Ulvan dialdehyde-gelatin hydrogels for removal of heavy metals and methylene blue from aqueous solution. Carbohyd Polym 249:116841
23. Keshawy M, Mahmoud A-R, Abdel-Raouf ME-S (2020) Polystyrene-based magnetic hydrogels for elimination of some toxic metal cations from aqueous solutions. Environ Sci Pollut Res Int 20:20
24. Gao Y, Gu S, Jia F, Wang Q, Gao G (2020) "All-in-one" hydrolyzed keratin protein-modified polyacrylamide composite hydrogel transducer. Chem Eng J 30:125555
25. Feng Q, Gao H, Wen H, Huang H, Li Q, Liang M, Liu Y, Dong H, Cao X (2020) Engineering the cellular mechanical microenvironment to regulate stem cell chondrogenesis: insights from a microgel model. Acta Biomater 113:393–406
26. Dehbari N, Tavakoli J, Khatrao SS, Tang Y (2017) In situ polymerized hyperbranched polymer reinforced poly (acrylic acid) hydrogels. Mater Chem Front 1(10):1995–2004
27. Tavakoli J, Tang Y (2017) Hydrogel based sensors for biomedical applications: an updated review. Polymers 9(8):364
28. Lei H, Dong L, Li Y, Zhang J, Chen H, Wu J, Zhang Y, Fan Q, Xue B, Qin M, Chen B, Cao Y, Wang W (2020) Stretchable hydrogels with low hysteresis and anti-fatigue fracture based on polyprotein cross-linkers. Nat Commun 11(1):4032
29. Zhang Y, Le X, Jian Y, Lu W, Zhang J, Chen T (2019) 3D fluorescent hydrogel origami for multistage data security protection. Adv Func Mater 29(46):1905514
30. Pandey PK, Ulla H, Satyanarayan MN, Rawat K, Gaur A, Gawali S, Hassan PA, Bohidar HB (2020) Fluorescent MoS2 quantum dot–DNA nanocomposite hydrogels for organic light-emitting diodes. ACS Appl Nano Mater 3(2):1289–1297
31. Wei S, Li Z, Lu W, Liu H, Zhang J, Chen T, Tang BZ (2020) Multicolor fluorescent polymeric hydrogels: colorfulness is more shining than homochromy. Angew Chem Int Ed (online ahead of print). https://doi.org/10.1002/anie.202007506
32. Jiang H, Qin Z, Zheng Y, Liu L, Wang X (2019) Aggregation-induced electrochemiluminescence by metal-binding protein responsive hydrogel scaffolds. Small 15(18):1901170
33. Cengiz N (2020) Glutathione-responsive multifunctionalizable hydrogels via amine-epoxy "click" chemistry. Eur Polym J 123:109441
34. Khandai S, Siegel RA, Jena SS (2020) Probing the microenvironment of polyacrylamide hydrogel matrix using turbidity and fluorescence recovery after photobleaching: one versus two phases. Colloids Surf A 593:124618

35. Vanheusden M, Vitale R, Camacho R, Janssen KP, Acke A, Rocha S, Hofkens J (2020) Fluorescence photobleaching as an intrinsic tool to quantify the 3D expansion factor of biological samples in expansion microscopy. ACS Omega 5(12):6792–6799
36. Zhang X, Ballem MA, Hu ZJ, Bergman P, Uvdal K (2011) Nanoscale light-harvesting metal-organic frameworks. Angew Chem 123(25):5847–5851
37. Zhang X, Ballem MA, Ahrén M, Suska A, Bergman P, Uvdal K (2010) Nanoscale Ln (III)-carboxylate coordination polymers (Ln= Gd, Eu, Yb): temperature-controlled guest encapsulation and light harvesting. J Am Chem Soc 132(30):10391–10397
38. Shi J, Wang M, Sun Z, Liu Y, Guo J, Mao H, Yan F (2019) Aggregation-induced emission-based ionic liquids for bacterial killing, imaging, cell labeling, and bacterial detection in blood cells. Acta Biomater 97:247–259
39. Chen Y, Lam JWY, Kwok RTK, Liu B, Tang BZ (2019) Aggregation-induced emission: fundamental understanding and future developments. Mater Horizons 6(3):428–433
40. Hong Y, Lam JWY, Tang BZ (2011) Aggregation-induced emission. Chem Soc Rev 40(11):5361–5388
41. Zhu C, Kwok RTK, Lam JWY, Tang BZ (2018) Aggregation-induced emission: a trailblazing journey to the field of biomedicine. ACS Appl Bio Mater 1(6):1768–1786
42. Yang J, Chi Z, Zhu W, Tang BZ, Li Z (2019) Aggregation-induced emission: a coming-of-age ceremony at the age of eighteen. Sci China Chem 62(9):1090–1098
43. Mei J, Hong Y, Lam JWY, Qin A, Tang Y, Tang BZ (2014) Aggregation-induced emission: the whole is more brilliant than the parts. Adv Mater 26(31):5429–5479
44. Jiang N, Shen T, Sun JZ, Tang BZ (2019) Aggregation-induced emission: right there shining. Sci China Mater 62(9):1227–1235
45. Guo F, Gai W-P, Hong Y, Tang BZ, Qin J, Tang Y (2015) Aggregation-induced emission fluorogens as biomarkers to assess the viability of microalgae in aquatic ecosystems. Chem Commun 51(97):17257–17260
46. Qin A, Jim CKW, Tang Y, Lam JWY, Liu J, Mahtab F, Gao P, Tang BZ (2008) Aggregation-enhanced emissions of intramolecular excimers in disubstituted polyacetylenes. J Phys Chem B 112(31):9281–9288
47. Cheng X, Jiang J, Liang G (2020) Covalently conjugated hydrogelators for imaging and therapeutic applications. Bioconjug Chem 31(3):448–461
48. Tang Y, Tang BZ (2018) Principles and applications of aggregation-induced emission. Springer, Berlin
49. Tavakoli J, Zhang H-P, Tang BZ, Tang Y (2019) Aggregation-induced emission lights up the swelling process: a new technique for swelling characterisation of hydrogels. Mater Chem Front 3(4):664–667
50. Tavakoli J, Gascooke J, Xie N, Tang BZ, Tang Y (2019) Enlightening freeze-thaw process of physically cross-linked poly(vinyl alcohol) hydrogels by aggregation-induced emission fluorogens. ACS Appl Polym Mater 1(6):1390–1398
51. Lee CY, Farha OK, Hong BJ, Sarjeant AA, Nguyen ST, Hupp JT (2011) Light-harvesting metal–organic frameworks (MOFs): efficient strut-to-strut energy transfer in bodipy and porphyrin-based MOFs. J Am Chem Soc 133(40):15858–15861
52. Haldar R, Rao KV, George SJ, Maji TK (2012) exciplex formation and energy transfer in a self-assembled metal–organic hybrid system. Chem A Eur J 18(19):5848–5852
53. Zeng X, Liang J, Wang C, Yu Z, Zhao X, Lu H, Wang Q (2020) Gelation process visualized by synchronous fluorescence enhancement of polyhydroxy benzoylhydrazone-based organogel. J Lumin 224:117259
54. Du X, Zhou J, Shi J, Xu B (2015) Supramolecular hydrogelators and hydrogels: from soft matter to molecular biomaterials. Chem Rev 115(24):13165–13307
55. Suzuki M, Hanabusa K (2010) Polymer organogelators that make supramolecular organogels through physical cross-linking and self-assembly. Chem Soc Rev 39(2):455–463
56. Li B, Zhang Y, Yan B, Xiao D, Zhou X, Dong J, Zhou Q (2020) A self-healing supramolecular hydrogel with temperature-responsive fluorescence based on an AIE luminogen. RSC Adv 10(12):7118–7124
57. Zhang C, Zhang T, Jin S, Xue X, Yang X, Gong N, Zhang J, Wang PC, Tian J-H, Xing J (2017) Virus-inspired self-assembled nanofibers with aggregation-induced emission for highly efficient and visible gene delivery. ACS Appl Mater Interfaces 9(5):4425–4432

58. Sukul PK, Singh PK, Maji SK, Malik S (2013) Aggregation induced chirality in a self assembled perylene based hydrogel: application of the intracellular pH measurement. J Mater Chem B 1(2):153–156
59. Wang J, Lin X, Shu T, Su L, Liang F, Zhang X (2019) Self-assembly of metal nanoclusters for aggregation-induced emission. Int J Mol Sci 20(8):1891
60. Externbrink M, Riebe S, Schmuck C, Voskuhl J (2018) A dual pH-responsive supramolecular gelator with aggregation-induced emission properties. Soft Matter 14(30):6166–6170
61. Gu L, Liu X, Dong S, Chen Z, Han R, He C, Wang D, Zheng Y (2020) Natural lignin nanoparticles: a promising nano-crosslinker for constructing fluorescent photoswitchable supramolecular hydrogels. Polym Chem 11(11):1871–1876
62. Feng H-T, Lam JWY, Tang BZ (2020) Self-assembly of AIEgens. Coord Chem Rev 406:213142
63. Tian Y, Wang Q, Wang K, Ke M, Hu Y, Shen L, Geng Q, Cheng J, Zhang J (2020) From biomass resources to functional materials: a fluorescent thermosetting material based on resveratrol via thiol-ene click chemistry. Eur Polymer J 123:109416
64. Wu J-L, Zhang C, Qin W, Quan D-P, Ge M-L, Liang G-D (2019) Thermoresponsive fluorescent semicrystalline polymers decorated with aggregation induced emission luminogens. Chin J Polym Sci 37(4):394–400
65. Li B, He T, Shen X, Tang D, Yin S (2019) Fluorescent supramolecular polymers with aggregation induced emission properties. Polym Chem 10(7):796–818
66. Lou X-Y, Yang Y-W (2018) Manipulating aggregation-induced emission with supramolecular macrocycles. Adv Opt Mater 6(22):1800668
67. Qin A, Lam JWY, Tang BZ (2012) Luminogenic polymers with aggregation-induced emission characteristics. Prog Polym Sci 37(1):182–209
68. Hu R, Qin A, Tang BZ (2020) AIE polymers: synthesis and applications. Prog Polym Sci 100:101176
69. Zhou Q, Wang Z, Dou X, Wang Y, Liu S, Zhang Y, Yuan WZ (2019) Emission mechanism understanding and tunable persistent room temperature phosphorescence of amorphous nonaromatic polymers. Mater Chem Front 3(2):257–264
70. Chen X, Luo W, Ma H, Peng Q, Yuan WZ, Zhang Y (2018) Prevalent intrinsic emission from nonaromatic amino acids and poly(amino acids). Sci China Chem 61(3):351–359
71. Dou X, Zhou Q, Chen X, Tan Y, He X, Lu P, Sui K, Tang BZ, Zhang Y, Yuan WZ (2018) Clustering-triggered emission and persistent room temperature phosphorescence of sodium alginate. Biomacromol 19(6):2014–2022
72. López D, García-Frutos EM (2015) Aggregation-induced emission of organogels based on self-assembled 5-(4-nonylphenyl)-7-azaindoles. Langmuir 31(31):8697–8702
73. Bhattacharya S, Samanta SK (2012) Unusual salt-induced color modulation through aggregation-induced emission switching of a bis-cationic phenylenedivinylene-based π hydrogelator. Chem A Eur J 18(52):16632–16641
74. Zhao Z, Lam JW, Tang BZ (2013) Self-assembly of organic luminophores with gelation-enhanced emission characteristics. Soft Matter 9(18):4564–4579
75. Yang X, Liu Y, Li J, Wang Q, Yang M, Li C (2018) A novel aggregation-induced-emission-active supramolecular organogel for the detection of volatile acid vapors. New J Chem 42(21):17524–17532
76. Liu Y, Lam JW, Mahtab F, Kwok RT, Tang BZ (2010) Sterol-containing tetraphenylethenes: synthesis, aggregation-induced emission, and organogel formation. Front Chem China 5(3):325–330
77. Chen Q, Zhang D, Zhang G, Yang X, Feng Y, Fan Q, Zhu D (2010) Multicolor tunable emission from organogels containing tetraphenylethene, perylenediimide, and spiropyran derivatives. Adv Func Mater 20(19):3244–3251
78. Wang M, Zhang D, Zhang G, Zhu D (2009) Fluorescence enhancement upon gelation and thermally-driven fluorescence switches based on tetraphenylsilole-based organic gelators. Chem Phys Lett 475(1):64–67
79. Wan J-H, Mao L-Y, Li Y-B, Li Z-F, Qiu H-Y, Wang C, Lai G-Q (2010) Self-assembly of novel fluorescent silole derivatives into different supramolecular aggregates: fibre, liquid crystal and monolayer. Soft Matter 6(14):3195–3201
80. Sun P, Wang Z, Bi Y, Sun D, Zhao T, Zhao F, Wang W, Xin X (2020) Self-assembly-driven aggregation-induced emission of silver nanoclusters for light conversion and temperature sensing. ACS Appl Nano Mater 3(2):2038–2046

81. Yoon S-J, Kim JH, Chung JW, Park SY (2011) Exploring the minimal structure of a wholly aromatic organogelator: simply adding two β-cyano groups to distyrylbenzene. J Mater Chem 21(47):18971–18973
82. An B-K, Lee D-S, Lee J-S, Park Y-S, Song H-S, Park SY (2004) Strongly fluorescent organogel system comprising fibrillar self-assembly of a trifluoromethyl-based cyanostilbene derivative. J Am Chem Soc 126(33):10232–10233
83. Seo J, Chung JW, Jo E-H, Park SY (2008) Highly fluorescent supramolecular gels with chirality transcription through hydrogen bonding. Chem Commun 24:2794–2796
84. Chung JW, An B-K, Park SY (2008) A thermoreversible and proton-induced Gel−Sol phase transition with remarkable fluorescence variation. Chem Mater 20(21):6750–6755
85. Zhang C, Liu C, Xue X, Zhang X, Huo S, Jiang Y, Chen W-Q, Zou G, Liang X-J (2014) Salt-responsive self-assembly of luminescent hydrogel with intrinsic gelation-enhanced emission. ACS Appl Mater Interfaces 6(2):757–762
86. Zhang Y-M, Zhu W, Huang X-J, Qu W-J, He J-X, Fang H, Yao H, Wei T-B, Lin Q (2018) Supramolecular aggregation-induced emission gels based on pillar[5]arene for ultrasensitive detection and separation of multianalytes. ACS Sustain Chem Eng 6(12):16597–16606
87. Felip-León C, Galindo F, Miravet JF (2018) Insights into the aggregation-induced emission of 1,8-naphthalimide-based supramolecular hydrogels. Nanoscale 10(36):17060–17069
88. Yao H, Wang J, Fan Y-Q, Zhou Q, Guan X-W, Kan X-T, Zhang Y-M, Lin Q, Wei T-B (2019) Supramolecular hydrogel-based AIEgen: construction and dual-channel recognition of negative charged dyes. Dyes Pigm 167:16–21
89. Wang H, Ji X, Li Y, Li Z, Tang G, Huang F (2018) An ATP/ATPase responsive supramolecular fluorescent hydrogel constructed via electrostatic interactions between poly(sodium p-styrenesulfonate) and a tetraphenylethene derivative. J Mater Chem B 6(18):2728–2733
90. Xing P, Chen H, Xiang H, Zhao Y (2018) Selective coassembly of aromatic amino acids to fabricate hydrogels with light irradiation-induced emission for fluorescent imprint. Adv Mater 30(5):1705633
91. Cheng Q, Cao Z, Hao A, Zhao Y, Xing P (2020) Fluorescent imprintable hydrogels via organic/inorganic supramolecular coassembly. ACS Appl Mater Interfaces 12(13):15491–15499
92. Kokado K, Nagai A, Chujo Y (2010) Poly(γ-glutamic acid) hydrogels with water-sensitive luminescence derived from aggregation-induced emission of o-carborane. Macromolecules 43(15):6463–6468
93. Jiang Y, Yang X, Ma C, Wang C, Chen Y, Dong F, Yang B, Yu K, Lin Q (2014) Interfacing a tetraphenylethene derivative and a smart hydrogel for temperature-dependent photoluminescence with sensitive thermoresponse. ACS Appl Mater Interfaces 6(7):4650–4657
94. Zhou H, Liu F, Wang X, Yan H, Song J, Ye Q, Tang BZ, Xu J (2015) Aggregation induced emission based fluorescence pH and temperature sensors: probing polymer interactions in poly(N-isopropyl acrylamide-co-tetra(phenyl)ethene acrylate)/poly(methacrylic acid) interpenetrating polymer networks. J Mater Chem C 3(21):5490–5498
95. Tang L, Jin JK, Qin A, Yuan WZ, Mao Y, Mei J, Sun JZ, Tang BZ (2009) A fluorescent thermometer operating in aggregation-induced emission mechanism: probing thermal transitions of PNIPAM in water. Chem Commun 33:4974–4976
96. Wang X, Xu K, Yao H, Chang L, Wang Y, Li W, Zhao Y, Qin J (2018) Temperature-regulated aggregation-induced emissive self-healable hydrogels for controlled drug delivery. Polym Chem 9(40):5002–5013
97. Galindo JM, Leganés J, Patiño J, Rodríguez AM, Herrero MA, Díez-Barra E, Merino S, Sánchez-Migallón AM, Vázquez E (2019) Physically cross-linked hydrogel based on phenyl-1,3,5-triazine: soft scaffold with aggregation-induced emission. ACS Macro Lett 8(10):1391–1395
98. Zhang C, Li Y, Xue X, Chu P, Liu C, Yang K, Jiang Y, Chen W-Q, Zou G, Liang X-J (2015) A smart pH-switchable luminescent hydrogel. Chem Commun 51(20):4168–4171
99. Liu H, Wei S, Qiu H, Zhan B, Liu Q, Lu W, Zhang J, Ngai T, Chen T (2020) Naphthalimide-based aggregation-induced emissive polymeric hydrogels for fluorescent pattern switch and biomimetic actuators. Macromol Rapid Commun 20:2000123
100. Hou F, Xi B, Wang X, Yang Y, Zhao H, Li W, Qin J, He Y (2019) Self-healing hydrogel with cross-linking induced thermo-response regulated light emission property. Colloids Surf B 183:110441
101. Li Z, Liu P, Ji X, Gong J, Hu Y, Wu W, Wang X, Peng H-Q, Kwok RTK, Lam JWY, Lu J, Tang BZ (2020) Bioinspired simultaneous changes in fluorescence color, brightness, and shape of hydrogels enabled by AIEgens. Adv Mater 32(11):1906493

102. Zhang M, Wang Z, Huang P, Jiang G, Xu C, Zhang W, Guo R, Li W, Zhang X (2020) Real-time and noninvasive tracking of injectable hydrogel degradation using functionalized AIE nanoparticles. Nanophotonics 9(7):2063
103. Ji X, Li Z, Hu Y, Xie H, Wu W, Song F, Liu H, Wang J, Jiang M, Lam JW (2020) Bioinspired hydrogels with muscle-like structure for AIEgen-guided selective self-healing. CCS Chem 20:1146–1156
104. Li P, Zhang D, Zhang Y, Lu W, Zhang J, Wang W, He Q, Théato P, Chen T (2019) Aggregation-caused quenching-type naphthalimide fluorophores grafted and ionized in a 3D polymeric hydrogel network for highly fluorescent and locally tunable emission. ACS Macro Lett 8(8):937–942
105. Dang D, Zhang H, Xu Y, Xu R, Wang Z, Kwok RTK, Lam JWY, Zhang L, Meng L, Tang BZ (2019) Super-resolution visualization of self-assembling helical fibers using aggregation-induced emission luminogens in stimulated emission depletion nanoscopy. ACS Nano 13(10):11863–11873
106. Wang Z, Chen S, Lam JWY, Qin W, Kwok RTK, Xie N, Hu Q, Tang BZ (2013) Long-term fluorescent cellular tracing by the aggregates of AIE bioconjugates. J Am Chem Soc 135(22):8238–8245
107. Jia J, Wang Z, Lu W, Yang L, Wu Q, Qin W, Hu Q, Tang BZ (2014) Monitoring layer-by-layer self-assembly process of natural polyelectrolytes by fluorescent bioconjugate with aggregation-induced emission characteristic. J Mater Chem B 2(47):8406–8411
108. Wang Z, Nie J, Qin W, Hu Q, Tang BZ (2016) Gelation process visualized by aggregation-induced emission fluorogens. Nat Commun 7(1):12033
109. Lin Y, Li C, Song G, He C, Dong YQ, Wang H (2015) Freezing-induced multi-colour emissions of AIE luminogen di(4-propoxyphenyl) dibenzofulvene. J Mater Chem C 3(11):2677–2685
110. Zheng H, Li C, He C, Dong YQ, Liu Q, Qin P, Zeng C, Wang H (2014) Luminescent hydrogels based on di(4-propoxyphenyl)-dibenzofulvene exhibiting four emission colours and organic solvents/thermal dual-responsive properties. J Mater Chem C 2(29):5829–5835
111. Gao M, Li Y, Chen X, Li S, Ren L, Tang BZ (2018) Aggregation-induced emission probe for light-up and in situ detection of calcium ions at high concentration. ACS Appl Mater Interfaces 10(17):14410–14417
112. Tavakoli J, Raston CL, Tang Y (2020) Tuning surface morphology of fluorescent hydrogels using a vortex fluidic device. Molecules 25(15):3445
113. Tavakoli J, Laisak E, Gao M, Tang Y (2019) AIEgen quantitatively monitoring the release of Ca2+ during swelling and degradation process in alginate hydrogels. Mater Sci Eng C 104:109951
114. Tavakoli J, Raston CL, Ma Y, Tang Y (2020) Vortex fluidic mediated one-step fabrication of polyvinyl alcohol hydrogel films with tunable surface morphologies and enhanced self-healing properties. Sci China Mater 63(7):1310–1317
115. Tavakoli J, Joseph N, Chuah C, Raston CL, Tang Y (2020) Vortex fluidic enabling and significantly boosting light intensity of graphene oxide with aggregation induced emission luminogen. Mater Chem Front 20:20
116. Tavakoli J, Joseph N, Raston CL, Tang Y (2020) A hyper-branched polymer tunes the size and enhances the fluorescent properties of aggregation-induced emission nanoparticles. Nanosc Adv 2(2):633–641
117. Tavakoli J, Pye S, Reza AM, Xie N, Qin J, Raston CL, Tang BZ, Tang Y (2020) Tuning aggregation-induced emission nanoparticle properties under thin film formation. Mater Chem Front 4(2):537–545
118. Wei P, Li Z, Zhang J-X, Zhao Z, Xing H, Tu Y, Gong J, Cheung TS, Hu S, Sung HHY, Williams ID, Kwok RTK, Lam JWY, Tang BZ (2019) Molecular transmission: visible and rate-controllable photoreactivity and synergy of aggregation-induced emission and host-guest assembly. Chem Mater 31(3):1092–1100
119. Liu Y, Zhou M, Liu Y, Han X, Zhang X, Liu S (2020) Host–guest interaction-mediated fabrication of aggregation-induced emission supramolecular hydrogel for use as aqueous light-harvesting systems. Supramol Chem 32(8):445–451
120. Zhao Q, Dai X-Y, Yao H, Zhang Y-M, Qu W-J, Lin Q, Wei T-B (2020) Stimuli-responsive supramolecular hydrogel with white AIE effect for ultrasensitive detection of Fe^{3+} and as rewritable fluorescent materials. Dyes Pigm 20:108875
121. Yang H-L, Zhang Q-P, Zhang Y-M, Gong G-F, Chen Y-Y, Qi Z, Yao H, Wei T-B, Lin Q (2019) A novel strong AIE bi-component hydrogel as a multi-functional supramolecular fluorescent material. Dyes Pigm 171:107745

122. Zhang Y-M, Zhu W, Zhao Q, Qu W-J, Yao H, Wei T-B, Lin Q (2020) Th4+ tuned aggregation-induced emission: a novel strategy for sequential ultrasensitive detection and separation of Th^{4+} and Hg^{2+}. Spectrochim Acta Part A Mol Biomol Spectrosc 229:117926
123. Fan Y-Q, Liu J, Chen Y-Y, Guan X-W, Wang J, Yao H, Zhang Y-M, Wei T-B, Lin Q (2018) An easy-to-make strong white AIE supramolecular polymer as a colour tunable photoluminescence material. J Mater Chem C 6(48):13331–13335
124. Zhu CN, Bai T, Wang H, Bai W, Ling J, Sun JZ, Huang F, Wu ZL, Zheng Q (2018) Single chromophore-based white-light-emitting hydrogel with tunable fluorescence and patternability. ACS Appl Mater Interfaces 10(45):39343–39352
125. Aschmann D, Riebe S, Neumann T, Killa D, Ostwaldt J-E, Wölper C, Schmuck C, Voskuhl J (2019) A stimuli responsive two component supramolecular hydrogelator with aggregation-induced emission properties. Soft Matter 15(36):7117–7121
126. Zhao Y, Pang B, Chen J, Xiao L, Liu H, Lian W, Sun T, Jiang Y, Lin Q (2020) Polystyrene@poly(ar-vinylbenzyl)trimethylammonium-co-acrylic acid core/shell pH-responsive nanoparticles for active targeting and imaging of cancer cell based on aggregation induced emission. Microchim Acta 187(3):166
127. Gong Y, Tan Y, Mei J, Zhang Y, Yuan W, Zhang Y, Sun J, Tang BZ (2013) Room temperature phosphorescence from natural products: crystallization matters. Sci China Chem 56(9):1178–1182
128. Zhou Q, Cao B, Zhu C, Xu S, Gong Y, Yuan WZ, Zhang Y (2016) Clustering-triggered emission of nonconjugated polyacrylonitrile. Small 12(47):6586–6592
129. Zhang Yuan W, Zhang Y (2017) Nonconventional macromolecular luminogens with aggregation-induced emission characteristics. J Polym Sci Part A Polym Chem 55(4):560–574
130. Jiang N, Li G-F, Zhang B-H, Zhu D-X, Su Z-M, Bryce MR (2018) Aggregation-induced long-lived phosphorescence in nonconjugated polyurethane derivatives at 77 K. Macromolecules 51(11):4178–4184
131. Li Y, Young DJ, Loh XJ (2019) Fluorescent gels: a review of synthesis, properties, applications and challenges. Mater Chem Front 3(8):1489–1502
132. Li B, Lin C, Lu C, Zhang J, He T, Qiu H, Yin S (2020) A rapid and reversible thermochromic supramolecular polymer hydrogel and its application in protected quick response codes. Mater Chem Front 4(3):869–874
133. Zhang Z, Du P, Pu G, Wei L, Wu Y, Guo J, Lu X (2019) Utilization and prospects of electrochemiluminescence for characterization, sensing, imaging and devices. Mater Chem Front 3(11):2246–2257
134. Zhang Y, Qi Y, Ulrich S, Barboiu M, Ramström O (2020) Dynamic covalent polymers for biomedical applications. Mater Chem Front 4(2):489–506
135. Chen X, Li J (2020) Bioinspired by cell membranes: functional polymeric materials for biomedical applications. Mater Chem Front 4(3):750–774
136. Wan Q, Huang Q, Liu M, Xu D, Huang H, Zhang X, Wei Y (2017) Aggregation-induced emission active luminescent polymeric nanoparticles: non-covalent fabrication methodologies and biomedical applications. Appl Mater Today 9:145–160
137. Wang Y, Nie J, Fang W, Yang L, Hu Q, Wang Z, Sun JZ, Tang BZ (2020) Sugar-based aggregation-induced emission luminogens: design. Struct Appl Chem Rev 120(10):4534–4577
138. Liow SS, Dou Q, Kai D, Li Z, Sugiarto S, Yu CYY, Kwok RTK, Chen X, Wu Y-L, Ong ST, Kizhakeyil A, Verma NK, Tang BZ, Loh XJ (2017) Long-term real-time in vivo drug release monitoring with AIE thermogelling polymer. Small 13(7):1603404
139. Lai W-F (2020) Non-conjugated polymers with intrinsic luminescence for drug delivery. J Drug Deliv Sci Technol 59:101916
140. Kim S, Ohulchanskyy TY, Pudavar HE, Pandey RK, Prasad PN (2007) Organically modified silica nanoparticles co-encapsulating photosensitizing drug and aggregation-enhanced two-photon absorbing fluorescent dye aggregates for two-photon photodynamic therapy. J Am Chem Soc 129(9):2669–2675
141. Ikeda M, Tanida T, Yoshii T, Hamachi I (2011) Rational molecular design of stimulus-responsive supramolecular hydrogels based on dipeptides. Adv Mater 23(25):2819–2822
142. Wang D, Lee MMS, Xu W, Kwok RTK, Lam JWY, Tang BZ (2018) Theranostics based on AIEgens. Theranostics 8(18):4925–4956
143. Li T, Pan S, Zhuang H, Gao S, Xu H (2020) Selenium-containing carrier-free assemblies with aggregation-induced emission property combine cancer radiotherapy with chemotherapy. ACS Appl Bio Mater 3(2):1283–1292

144. Zhao J, Pan X, Zhu J, Zhu X (2020) Novel AIEgen-functionalized diselenide-crosslinked polymer gels as fluorescent probes and drug release carriers. Polymers 12(3):551
145. Jiang X, Wang H, Yuan R, Chai Y (2018) Functional three-dimensional porous conductive polymer hydrogels for sensitive electrochemiluminescence in situ detection of H(2)O(2) released from live cells. Anal Chem 90(14):8462–8469
146. Gao W, Muzyka K, Ma X, Lou B, Xu G (2018) A single-electrode electrochemical system for multiplex electrochemiluminescence analysis based on a resistance induced potential difference. Chem Sci 9(16):3911–3916
147. Heo YJ, Shibata H, Okitsu T, Kawanishi T, Takeuchi S (2011) Long-term in vivo glucose monitoring using fluorescent hydrogel fibers. Proc Natl Acad Sci USA 108(33):13399–13403
148. Shibata H, Heo YJ, Okitsu T, Matsunaga Y, Kawanishi T, Takeuchi S (2010) Injectable hydrogel microbeads for fluorescence-based in vivo continuous glucose monitoring. Proc Natl Acad Sci USA 107(42):17894–17898
149. Zhang Z, Liu G, Li X, Zhang S, Lü X, Wang Y (2020) Design and synthesis of fluorescent nanocelluloses for sensing and bioimaging applications. ChemPlusChem 85(3):487–502
150. Xu Z, Liu Y, Qian C, Wu L, Wu Z, Zhang J, Cheng J, Miao M, Zhang D (2020) Tuning the morphology of melamine-induced tetraphenylethene self-assemblies for melamine detecting. Org Electron 76:105476
151. Hsu S-M, Wu F-Y, Cheng H, Huang Y-T, Hsieh Y-R, Tseng DT-H, Yeh M-Y, Hung S-C, Lin H-C (2016) Functional supramolecular polymers: a fluorescent microfibrous network in a supramolecular hydrogel for high-contrast live cell-material imaging in 3D environments. Adv Healthc Mater 5(18):2406–2412
152. Wu C, Bull B, Szymanski C, Christensen K, McNeill J (2008) Multicolor conjugated polymer dots for biological fluorescence imaging. ACS Nano 2(11):2415–2423
153. Ding D, Li K, Qin W, Zhan R, Hu Y, Liu J, Tang BZ, Liu B (2013) Conjugated polymer amplified far-red/near-infrared fluorescence from nanoparticles with aggregation-induced emission characteristics for targeted in vivo imaging. Adv Healthc Mater 2(3):500–507
154. Mandal K, Jana D, Ghorai BK, Jana NR (2018) Functionalized chitosan with self-assembly induced and subcellular localization-dependent fluorescence 'switch on' property. New J Chem 42(8):5774–5784
155. Li M, Hong Y, Wang Z, Chen S, Gao M, Kwok RT, Qin W, Lam JW, Zheng Q, Tang BZ (2013) Fabrication of chitosan nanoparticles with aggregation-induced emission characteristics and their applications in long-term live cell imaging. Macromol Rapid Commun 34(9):767–771
156. Zhang X, Zhang X, Yang B, Liu M, Liu W, Chen Y, Wei Y (2013) Facile fabrication and cell imaging applications of aggregation-induced emission dye-based fluorescent organic nanoparticles. Polym Chem 4(16):4317–4321
157. Lim C-K, Kim S, Kwon IC, Ahn C-H, Park SY (2009) Dye-condensed biopolymeric hybrids: chromophoric aggregation and self-assembly toward fluorescent bionanoparticles for near infrared bioimaging. Chem Mater 21(24):5819–5825
158. Wang Z, Liu Y, Jia J, Chen S, Qin W, Hu Q, Tang BZ (2016) Fabrication of hybridized nanoparticles with aggregation-induced emission characteristics and application for cell imaging. J Mater Chem B 4(31):5265–5271
159. Mehwish N, Dou X, Zhao Y, Feng C-L (2019) Supramolecular fluorescent hydrogelators as bio-imaging probes. Mater Horizons 6(1):14–44
160. Grafahrend D, Heffels K-H, Beer MV, Gasteier P, Möller M, Boehm G, Dalton PD, Groll J (2011) Degradable polyester scaffolds with controlled surface chemistry combining minimal protein adsorption with specific bioactivation. Nat Mater 10(1):67–73
161. Kong J, Zhang J, Wang Y, Qi W, Huang M, Su R, He Z (2020) Bioinspired fluorescent peptidyl nanoparticles with rainbow colors. ACS Appl Mater Interfaces 12(28):31830–31841
162. Mitra M, Mahapatra M, Dutta A, Deb M, Dutta S, Chattopadhyay PK, Roy S, Banerjee S, Sil PC, Singha NR (2020) Fluorescent guar gum-g-terpolymer via in situ acrylamido-acid fluorophore-monomer in cell imaging, Pb(II) sensor, and security ink. ACS Appl Bio Mater 3(4):1995–2006
163. Huang J, Wang Y-L, Yu X-D, Zhou Y-N, Chu L-Q (2020) Enhanced fluorescence of carboxymethyl chitosan via metal ion complexation in both solution and hydrogel states. Int J Biol Macromol 152:50–56
164. Talloj SK, Mohammed M, Lin H-C (2020) Construction of self-assembled nanostructure-based tetraphenylethylene dipeptides: supramolecular nanobelts as biomimetic hydrogels for cell adhesion and proliferation. J Mater Chem B 8(33):7483–7493

165. Chu C, Xiang Z, Wang J, Xie H, Xiang T, Zhou S (2020) A near-infrared light-triggered shape-memory polymer for long-time fluorescence imaging in deep tissues. J Mater Chem B 8(35):8061–8070
166. Ma H, Zhang A, Zhang X, Zhao H, Cui Z, Fu P, Liu M, Zhou S, Pang X, Qiao X (2020) Novel platform for visualization monitoring of hydrolytic degradation of bio-degradable polymers based on aggregation-induced emission (AIE) technique. Sens Actuators B Chem 304:127342
167. Wang J, Gao M, Cui Z-K, Jia Y-G, Liu S, Chen K-F, Chen X, Zhang Y, Fang Z, Chen Y, Wang K, Zhang H, Wang L, Ren L (2020) One-pot quaternization of dual-responsive poly(vinyl alcohol) with AIEgens for pH-switchable imaging and killing of bacteria. Mater Chem Front 4(9):2635–2645
168. Liu J, Liu L, Li S, Kang Q, Zhang R, Zhu Z (2021) Self-assembled nanogels of luminescent thiolated silver nanoclusters and chitosan as bactericidal agent and bacterial sensor. Mater Sci Eng C 118:111520
169. Tang Y, Huang H, Xue W, Chang Y, Li Y, Guo X, Zhong C (2020) Rigidifying induced fluorescence enhancement in 2D porous covalent triazine framework nanosheets for the simultaneously luminous detection and adsorption removal of antibiotics. Chem Eng J 384:123382
170. An H, Yang H, Bo H, Ma X, Wang Y, Liu L, Wang H, He Y, Qin J (2021) Fabrication of self-healing hydrogel from quaternized *N*-[3(dimethylamino)propyl]methacrylamide copolymer for antimicrobial and drug release applications. J Biomed Mater Res Part A 109:42–53
171. Jia W, Gungor-Ozkerim PS, Zhang YS, Yue K, Zhu K, Liu W, Pi Q, Byambaa B, Dokmeci MR, Shin SR, Khademhosseini A (2016) Direct 3D bioprinting of perfusable vascular constructs using a blend bioink. Biomaterials 106:58–68
172. Skardal A, Zhang J, Prestwich GD (2010) Bioprinting vessel-like constructs using hyaluronan hydrogels crosslinked with tetrahedral polyethylene glycol tetracrylates. Biomaterials 31(24):6173–6181
173. Unagolla JM, Jayasuriya AC (2020) Hydrogel-based 3D bioprinting: a comprehensive review on cell-laden hydrogels, bioink formulations, and future perspectives. Appl Mater Today 18:100479

REVIEW

Functional Polymer Systems with Aggregation-Induced Emission and Stimuli Responses

Ting Han[1] · Xinnan Wang[2] · Dong Wang[1] · Ben Zhong Tang[1,2]

Received: 28 November 2020 / Accepted: 18 December 2020 / Published online: 11 January 2021

Abstract

Functional polymer systems with stimuli responses have attracted great attention over the years due to their diverse range of applications. Such polymers are capable of altering their chemical and/or physical properties, such as chemical structures, chain conformation, solubility, shape, morphologies, and optical properties, in response to single or multiple stimuli. Among various stimuli-responsive polymers, those with aggregation-induced emission (AIE) properties possess the advantages of high sensitivity, fast response, large contrast, excellent photostability, and low background noise. The changes in fluorescence signal can be conveniently detected and monitored using portable instruments. The integration of AIE and stimuli responses into one polymer system provides a feasible and effective strategy for the development of smart polymers with high sensitivity to environmental variations. Here, we review the recent advances in the design, preparation, performance, and applications of functional synthetic polymer systems with AIE and stimuli responses. Various AIE-based polymer systems with responsiveness toward single physical or chemical stimuli as well as multiple stimuli are summarized with specific examples. The current challenges and perspectives on the future development of this research area will also be discussed at the end of this review.

Keywords Aggregation-induced emission · Stimuli responses · Functional polymers · Fluorescence sensors

Chapter 9 was originally published as Han, T., Wang, X., Wang, D. & Tang, B. Z. Topics in Current Chemistry (2021) 379: 7. https://doi.org/10.1007/s41061-020-00321-7.

✉ Dong Wang
wangd@szu.edu.cn

✉ Ben Zhong Tang
tangbenz@ust.hk

1 Center for AIE Research, Shenzhen Key Laboratory of Polymer Science and Technology, Guangdong Research Center for Interfacial Engineering of Functional Materials, College of Materials Science and Engineering, Shenzhen University, Shenzhen 518060, China

2 Department of Chemistry, Hong Kong Branch of Chinese National Engineering Research Center for Tissue Restoration and Reconstruction, and Institute for Advanced Study, The Hong Kong University of Science and Technology, Clear Water Bay, Kowloon, Hong Kong, China

1 Introduction

Stimuli-responsive materials widely exist in living systems and play an important role for the sustainability of life and maintaining biological function. For instance, human bodies can respond to changes in the environment with the assistance of complicated biological processes of proteins and enzymes. Driven by the desire to mimic nature, scientists have developed diverse stimuli-responsive polymer systems [1–4]. These smart polymers are capable of altering their chemical and/or physical properties in response to single or multiple external stimuli, such as mechanical force, temperature, pH, light irradiation, electric/magnetic fields, and specific analytes (gases, ions, bioactive molecules, etc.). The stimuli-induced changes can be chemical structures, chain conformation, solubility, wettability, shape (i.e., contraction or bending), surface charges, morphologies, optical properties, etc. Taking advantage of the stimuli responsiveness, such polymers have been utilized as controlled drug-delivery systems, smart coatings, artificial muscles and actuators, as well as fluorescent sensors [5–7]. Generally speaking, there are at least two parts for a responsive polymer system. One is the stimuli acceptor (input), and the other is the signaling part (output). The stimuli acceptor generally needs to possess high sensitivity, and the signaling part is better to produce readily detectable or visible signals. For instance, mechano-responsive polymer materials that are sensitive enough to spontaneously detect and report the occurrence and propagation of mini-cracks are highly desirable, so that workers can take early actions before small-scale damage develops into macro-damage that compromises the material's performance. Thermo-responsive polymers also require highly sensitive and visualized responses to temperature change. Particularly for those used for drug delivery applications, the direct monitoring of the release process and location of drugs is important for the understanding of the mechanisms of action. However, the responses of many traditional pure polymer systems are often insensitive or hard to detect. Although some detection methods such as UV–vis spectroscopy, electron microscopy, and mechanical performance tests have been employed to evaluate the response signals, these techniques are generally expensive and often require troublesome operation and sample preparation procedures. Therefore, there is still much room for improvement in terms of the signaling part of responsive polymer systems. It is highly demanded to develop functional polymer systems whose responses can be directly and sensitively visualized and monitored in real time, onsite, and in situ using portable instruments.

Stimuli-responsive fluorescent systems generally have the advantages of high sensitivity, fast response, less invasive nature, and large contrast [8–11]. The changes in fluorescence signal can be conveniently detected using portable instruments. In most cases, the fluorescence responses can be directly visualized by the naked eye in real time and in situ. Attracted by these advantages, scientists and researchers have applied fluorescence probes and labels in polymer materials to detect the conformational changes of polymers in response to microenvironment variations via fluorescence signals. For example, the temperature- and viscosity-dependent fluorescence change of molecular rotors can be utilized for the detection of glass transition temperature and polymerization degree [12–15]. The distance-dependent fluorescence of the Förster resonance energy transfer (FRET) process has been introduced as a "spectroscopic ruler" for the measurement of distances in macromolecules [16]. Besides, fluorescence responses have also been applied for the study of complicated protein and DNA folding/unfolding phenomena and processes [17]. However, conventional chromophores often suffer from poor fluorescence efficiency and a photo-bleaching problem in aggregate or solid states due to the aggregation-caused quenching (ACQ) effect, which greatly hinders their use in solid-state responsive systems. The weak fluorescence signal of ACQ dyes was detrimental to both the sensitivity and selectivity of optical indicators, and the stimuli responses were often in a turn-off mode. In the past two decades, materials with aggregation-induced emission (AIE) properties have attracted increasing attention and found significant applications in chemo-/biosensing and imaging [18, 19]. AIE luminogens (AIEgens) show highly efficient fluorescence at high concentrations or in aggregate states, which increases the sensitivity of optical indicators in thin films and other solid materials. Moreover, AIE-based sensing systems generally possess excellent photostability and low background noise. The well-accepted working mechanism for AIEgens is the restriction of intramolecular motion (RIM) model [20, 21]. When the intramolecular motion of the rotor-rich AIEgens is restricted, such as at low temperatures or in confined spaces, the fluorescence is turned on or markedly enhanced. Therefore, the fluorescence of AIEgens can exhibit an off–on switching property under various external stimuli, such as organic solvent vapor, pressure, heat, and metal ions [22, 23] The turn-on responses and good photostability can benefit the long-term visualization and monitoring of the response processes. These properties enable AIEgens to serve as excellent signaling components for developing sensitive stimuli-responsive systems. Notably, AIE polymers have many advantages over low-mass AIEgens, including good processability, easy functionalization, complicated structural features, and excellent thermal stability [24, 25]. Hence, by integrating AIE and stimuli responses into one polymer system, we can achieve robustly engineered materials with high sensitivity to environmental variations.

The combinatorial impact of AIE and stimuli-responsive polymers can be independent and synergistic. In some cases, the stimuli-responsive functionalities of polymers and AIE units work independently to perform their own functions as stimuli acceptor and signaling part, respectively. Through elaborative and rational design, the performance of responsive polymer materials and the functionalities of AIE materials can be mutually enhanced and enriched. Actually, the synergistic effect has been observed in most of the AIE-based stimuli-responsive polymer systems. For example, the aggregates of AIEgen-containing polymers often exhibit a superamplification effect toward explosives or metal ions when they are used as fluorescence chemosensors [24, 26]. This phenomenon has rarely been observed in small molecular AIEgens, indicating that the polymeric structure can enhance the sensing performance of AIEgens. On the other hand, the inherent stimuli responsiveness of AIE components can further enhance the responsive performance and enrich the functionalities of the parent polymer matrix or backbones [27, 28]. In such circumstances, the AIE unit can serve as both signaling component and responsive acceptor. Moreover, AIEgens can also function as additives to improve the inherent physical properties of polymers, such as solubility and shrinking rates [29, 30]. The combination of stimuli-responsive AIEgens and stimuli-responsive polymers makes the development of multi-responsive materials with high sensitivity much easier to realize.

Attracted by the abovementioned advantages, researchers have devoted much effort to the development of functional polymer systems with AIE and stimuli responses, and very impressive progress has been made in recent years. Polymers range from biomacromolecules, synthetic polymers to supramolecular polymer systems. The applications of AIE in stimuli-responsive biomacromolecules and supramolecular polymer systems have already been summarized in several comprehensive review articles [31–37]. Due to the limited space, this review will mainly introduce the recent progress on synthetic polymer systems with AIE and stimuli responses, including the preparation strategies and their responsive behaviors to different stimuli (i.e. physical stimuli, chemical stimuli, and multiple stimuli). The current challenges and perspectives on the future development of this research area will also be discussed. From this review, we hope readers obtain a general grasp on what has been done in this area and meanwhile gain some insights for the future design and development of AIE-based stimuli-responsive polymer systems.

2 Preparation Strategies

According to the mechanism of AIE, most of the AIE polymer systems have potential for use used as stimuli-responsive materials. Their fluorescence properties are supposed to be sensitive to various environmental changes, such as temperature, pH, light, forces, and solvents. Therefore, the strategies for preparing AIE polymer systems, including both chemical synthesis and physical fabrication, are generally applicable to the fabrication of AIE-based stimuli-responsive polymers.

2.1 Chemical Synthesis

Chemical synthesis strategies toward AIE-based stimuli-responsive polymers can be roughly divided into two types. One is covalently linking polymers with AIEgens, and the other is the formation of clusteroluminescent AIE polymers [38]. By chemically incorporating typical AIEgens, such as tetraphenylethene (TPE) and its derivatives, into polymer structures, stimuli-responsive AIE polymers with various compositions and architecture can be obtained. Till now, diverse AIE polymers with linear, graft, hyperbranched, cross-linked, or dendrimeric structures have been constructed by chemical synthesis. Some AIE-active metal–organic frameworks and covalent organic frameworks have also been synthesized by similar strategies [36, 39, 40]. Taking polymers with linear structures as an example, they can be readily synthesized by the homo-polymerization of AIEgen-containing monomers (Fig. 1a).

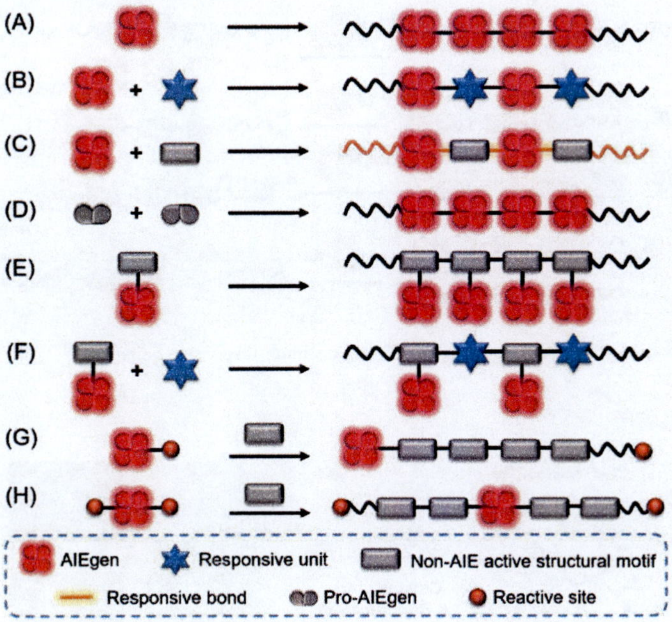

Fig. 1 Polymerization strategies towards AIE-based stimuli-responsive polymers

To improve or tune the polymer properties, co-monomers with stimuli-responsive units are often used to copolymerize with the AIEgen-containing monomers (Fig. 1b). Besides the use of stimuli-responsive monomers, the generation of stimuli-responsive chemical bonds in the backbone of AIE polymers is also an effective strategy toward stimuli-responsive fluorescent polymers (Fig. 1c). In some cases, AIEgen core structures can be generated in situ in the polymer main chains through the polymerization of AIE-inactive precursors (Fig. 1d). By attaching AIEgens on a polymerizable monomer, the AIE unit can also be attached on the polymers as side chains (Fig. 1e, f). Another interesting design is using AIEgen-containing initiators to initiate polymerization to produce stimuli-responsive polymers with AIEgens in the terminal or the middle of polymer chains (Fig. 1g, h).

Besides direct polymerizations, AIE-active stimuli-responsive polymers can also be generated through post-modification strategies of polymeric reactants. As schematically illustrated in Fig. 2a, polymer-containing reactive side groups can react with AIEgens to generate AIE polymers with stimuli-responsive chemical bonds. In addition to AIE initiators, AIEgen-containing terminators can also specifically introduce AIE moieties on the terminal of polymers (Fig. 2b). The combined uses of AIE initiator and AIE terminator strategies can benefit the regulation or enhancement of the stimuli-responsive performances (Fig. 2c). Furthermore, it is also possible to achieve AIE diblock copolymers with an AIE unit in the middle of the polymer chains through coupling the reaction of two different polymer block chains with AIEgens (Fig. 2d). The strategies shown in Fig. 2c, d have been rarely reported but are worthy of future exploration.

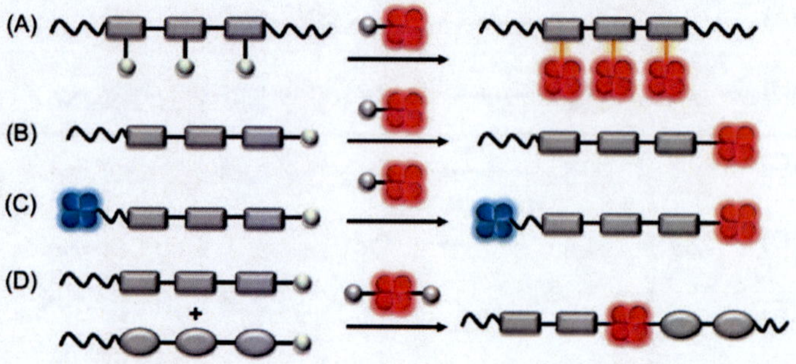

Fig. 2 Post-modification strategies towards AIE-based stimuli-responsive polymers

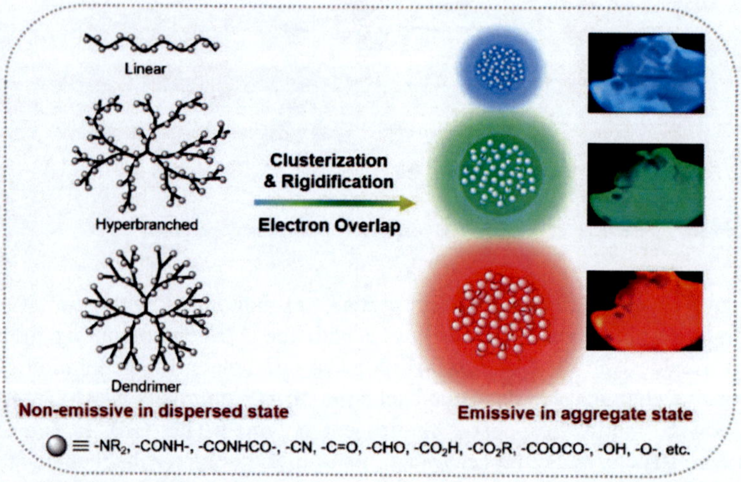

Fig. 3 Schematic illustration of the working mechanism for clusteroluminescent AIE polymers. Insets: films of clusteroluminescent polymers under different excitation wavelengths. Reproduced from [43] with permission from Springer Nature

In addition to AIEgen-containing polymers, some non-conjugated polymers have also been reported to show AIE characteristics and stimuli responsiveness. As depicted in Fig. 3, such polymers typically contain multiple electron-rich subgroups, such as N, O, S, and P with lone-pair electrons and/or unsaturated nitrile, carbonyl, aldehyde, carboxylic acid, ester, anhydride, etc. [41]. The uncommon fluorescence of these non-conjugated polymers is often explained by the clusterization-triggered emission (CTE) mechanism [42]. When the electron-rich polymers are molecularly dispersed, such as in dilute solutions, they are weakly or even non-emissive due to the active molecular motion and the lack of efficient electronic conjugation. However, the flexible conformations are rigidified upon aggregation. In aggregate state, the electron-rich subgroups tend to form clusters with different sizes, where their electron clouds are overlapped and shared to extend the electronic conjugation in a through-space way. Clusteroluminescent polymers generally exhibit excitation-dependent emission due to the heterogeneity of emissive clusters [43]. The larger the cluster, the brighter and redder the luminescence will be. According to the CTE mechanism, external stimuli that can affect the clusterization and rigidification processes of clusteroluminescent polymers can readily change their fluorescence. Therefore, clusteroluminescent polymers are promising functional materials with stimuli responsiveness.

2.2 Physical Fabrication

Chemical strategies normally require the attachment of a specific motif such as AIEgens or stimuli-responsive units to polymer structures via extensive synthetic efforts, which can increase the preparation cost and limit the application scope. By contrast, AIEgens can also be physically combined with polymer materials to construct smart flexible or rigid polymeric optical indicators to external stimuli. The physical blending strategies generally enjoy the advantages of facile fabrication procedures, good

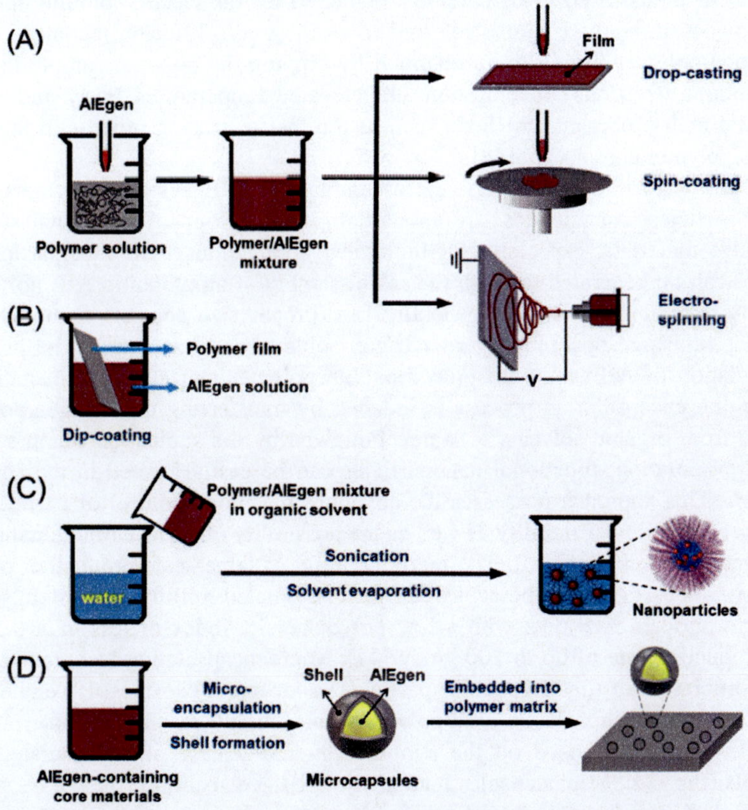

Fig. 4 Fabrication of stimuli-responsive AIEgen-polymer composites by physical blending

economics for raw materials, and wide application range. Taking advantage of the microenvironment-sensitive characteristic of AIEgens and the excellent mechanical processibility and stimuli-responsiveness of the polymer matrix, smart materials with diverse forms can be readily accessed. As shown in Fig. 4, AIEgen-based stimuli-responsive polymer composites can be engineered into films, coatings, fibers, micro/nanoparticles, microcapsules, bulk materials, etc.

The films of AIEgen-embedded stimuli-responsive polymers can be obtained by different processing methods ranging from drop-casting, spin-coating, and electrospinning (Fig. 2a) to dip-coating (Fig. 2b). Among them, drop-casting is one of the most convenient methods. Thin films can be cast into different shapes by simply dropping the blend solution or melt of polymer and AIEgen on a flat substrate or in an open mold, but this method often suffers from the nonuniformity of film products. On the other hand, spin-coating method can produce uniform thin films with the loss of some raw materials during the spinning process. Electro-spinning method is suitable to fabricate stimuli-responsive fibers from the AIEgen/polymer blend solutions or melts. By depositing the nanofibers on a collector, porous polymer films with high surface-to-volume ratio can be obtained. In order to produce high-quality fibers or films, the preparation conditions, such as the concentration and viscosity of solutions or melts, need to be carefully regulated for the electro-spinning method. The dip-coating method is also referred to as the guest-diffusion technique. AIEgen-doped polymer films can be obtained by dipping the polymer sample in AIEgen solutions at various concentrations and elevated temperatures. Early studies also employed melt-processing methods such as the melt-extrusion process to fabricate AIEgen/polymer composites [44].

In addition to films and fibers, the assemblies of AIEgens and polymers, such as nanoparticles and micelles, are another important group of functional stimuli-responsive materials, especially for biological applications. AIE nanoparticles or micelles can be generated through the self-assembly of amphiphilic AIE polymers, but the obtained nanoparticles generally lack proper size control. A more popular and promising approach toward AIEgen/polymer nanoparticles is the physical encapsulation of AIEgens with biocompatible polymer matrices. As illustrated in Fig. 4c, the encapsulation process is induced by transferring the AIEgen/polymer mixture from organic solvents to water. Followed by the sonication treatment and solvent evaporation, functional nanoparticles can be easily formed in the aqueous solutions. This approach possesses the advantages of good control of particle size and excellent colloidal stability as well as the possibility of surface functionalization for targeted delivery [45, 46]. The reproducibility of large-scale production of AIE polymer nanoparticles can be achieved by an automated millifluidic system, which can form uniform AIE dots with a low polydispersity index of less than 0.1 in a high production rate of up to 200 g/day [47]. Microencapsulation technique allows the incorporation of tiny droplets or particles inside small capsules. By embedding the obtained microcapsules in bulk polymer films, stimuli-responsive materials can be facilely achieved based on the stimuli-triggered release of encapsulated core materials (Fig. 2d). Microcapsules loaded with AIEgens can in principle be readily incorporated into any polymer of interest by simple mixing and without the need of chemical modification. This approach is quite simple and versatile in terms of activation modes and responses that can be achieved [48]. In some cases, the chemical synthesis and physical encapsulation methods are combined together to fabricate stimuli-responsive materials. For example, AIEgen-doped hydrogels are often fabricated by mixing AIEgens with monomers and initiators to physically encapsulate AIEgens inside the hydrogels during the polymerization process [29, 49]. Recently, advanced manufacture techniques such as 3D and 4D printing have also been utilized to prepare AIE-based soft actuators [29].

3 Responses to Physical Stimuli

Most examples of stimuli-responsive AIE polymers only show efficient response to a single stimulus. Materials with sensitive response to multiple stimuli are important but are challenging to develop. For convenience, we will discuss functional AIE polymers with responsiveness from a single stimulus first and then to multiple stimuli. The stimuli are divided into two groups, including physical stimuli and chemical stimuli. In this section, the structure design, responsive performance, and potential applications of functional AIE polymers with responses to a single physical stimulus, such as mechanical force, temperature, light, electricity, γ-rays, and morphology, will be introduced with representative examples.

3.1 Force

Mechanochromic luminescent polymer materials whose fluorescence properties can change in response to external mechanical forces can be used as built-in fluorescent force sensors, security inks, damage-reporting coatings, flexible optical devices, etc. The utilization of the AIE effect endows the mechanochromic polymer materials with higher contrast due to their better solid-state luminescence than the conventional ACQ materials. Till now, a variety of AIE-based mechanochromic polymer materials have been reported, including AIE-active linear and cross-linked polymers, clusteroluminescent AIE polymers, blended AIEgen/polymer films, and polymer composites with AIEgen-loaded microcapsules [44].

For example, Chi and coworkers recently reported a linear AIE-active polymer with remarkable mechanochromism in the solid state [28]. As shown in Fig. 5, polymer **P1** contains conjugated fluorene and 9,10-distyrylanthracene moieties in the polymer backbone, and TPE units are employed as bulky pendant groups to hinder the π-π stacking of the rigid backbone. The solid powder of **P1** exhibits strong yellow fluorescence peaked at 541 nm. Upon grinding with a pestle, the fluorescence obviously changed from yellow to red with a large redshift of 61 nm in the emission maximum (Fig. 5b, c). By heating the ground sample at 200 °C or fuming it with dichloromethane vapor for 30 min, the emission maximum can shift back to 580 nm and 574 nm, respectively. The mechanochromism of **P1** was explained by the destruction of its crystalline structure after grinding, which results in the conformational planarization and the increase of molecular conjugation of the polymer backbone. The poor reversibility may result from the slow crystalline rate of the

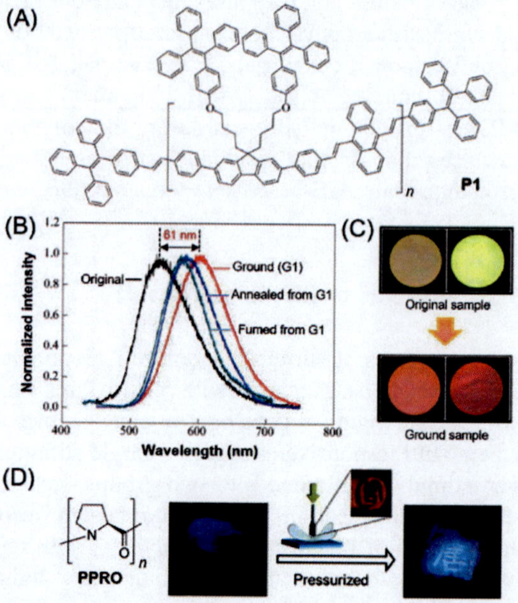

Fig. 5 **a** Structure of P1. **b** PL spectra of P1 in different states. **c** Images of P*1* powder in different states without (left) and with (right) the illumination of UV light. Reproduced from [28] with permission from Springer Nature. **d** Illustration of fluffy poly-L-proline (PPRO) in response to pressure. Reproduced from [50] with permission from the Royal Society of Chemistry

ground sample due to the high molecular weight and the segmental motion of polymers. Besides the AIEgen-bonded polymers, AIE-active polymers based on the CTE mechanism have also been reported to be sensitive to mechanical forces [50]. As depicted in Fig. 5d, the non-conjugated poly-L-proline (PPRO) is weakly emissive under UV illumination, with a fluffy fibrous appearance. However, when PPRO was pressurized and patterned using a traditional Chinese stamp, the wrinkles created by the character on the stamp showed much brighter emission than the surrounding parts. The mechanochromic phenomenon of clusteroluminescent PPRO was attributed to the clustering of the electron-rich amide groups in close distance and the rigidification of molecular conformations after the pressurization process.

In addition to pure polymer systems, AIEgen/polymer composites are another main group of AIE-based mechano-responsive polymer materials. In earlier studies, such polymer composites were commonly achieved by directly blending mechanochromic AIEgens into various polymer matrices [51]. A more recent approach to realize the mechano-response is dispersing the AIEgen-containing microcapsules into polymeric matrices. Since the first work reported by Moore et al. in 2016 [52], the microencapsulation approach has been demonstrated to be a simple yet powerful strategy to fabricate polymer materials with autonomous detection of damage. As illustrated in Fig. 6a, core–shell microcapsules containing a dilute solution of AIEgen are embedded in a polymeric material. The intact polymer composite is non-emissive under UV light illumination. When the matrix material was subjected to mechanical damage, the microcapsules collapsed to release the AIEgen solution by capillary force. The subsequent solvent evaporation led to the deposition of solid AIEgen at the cracked area to show bright fluorescence. Such self-reporting coatings can provide clear visual indication of

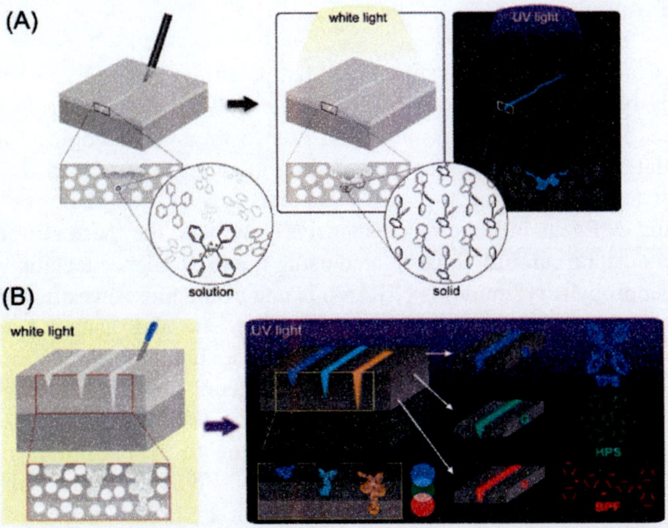

Fig. 6 **a** Schematic of the autonomous damage-reporting polymer composites using AIEgen-loaded microcapsules. **b** Schematic of the damage indication in multilayer polymeric materials with varying crack depth using AIEgens. Reproduced from [52, 53] with permission from the American Chemical Society

surface damage in a single layer, but cannot detect the depth information of crack penetration. In order to autonomously report microcracks and meanwhile differentiate the varying penetration depth, they further developed a self-reporting multilayered polymer system based on a similar strategy [53]. As shown in Fig. 6b, three AIEgens with red, green, and blue emission colors were selected, and their solutions were fabricated into microcapsules. Then, these microcapsules were incorporated into different layers of polymeric coatings. In this way, scratches with various penetration depths can result in the release of different combinations of AIEgens to show a diverse blend fluorescence colors. The microscopic damage depth can thus be visually assessed by correlating the luminescent colors with crack depth. This AIEgen-based detection strategy is applicable to a variety of polymer coatings and can provide high contrast between the damaged and undamaged regions.

In addition to the self-reporting function, the AIE-based microcapsule systems can also be endowed with self-healing properties to achieve dual functions of self-sensing and self-repairing [54, 55]. The damage to the polymer surface or even inside the polymer materials not only can be signaled but can also be instantaneously repaired. This kind of polymeric material can significantly improve the safety and reliability of critical engineering components. Moreover, the life cycle costs resulting from the regular maintenance and inspection of engineering components can be reduced to some extent. Therefore, such materials can have a significant economic and societal impact in terms of structural health monitoring, especially in the aeronautic, automotive, and construction industries.

3.2 Temperature

Among all available stimuli, the stimulus of temperature plays a unique role because it can easily be applied and tuned from outside. Polymers with thermo-responsive fluorescence have attracted great attention due to their technological significance and potential applications as fluorescent thermometers, carriers in drug release, intracellular temperature imaging, soft actuators, etc. By chemically or physically incorporating AIEgens into thermo-responsive polymers, the thermal transitions of the polymers can be carefully investigated using the fluorescence technique.

Poly(*N*-isopropylacrylamide) (PNIPAM) is one of the most investigated synthetic thermo-responsive polymers, exhibiting a characteristic and distinct lower critical solution temperature (LCST) in water. At LCST, a sharp transition from a stable one-phase to an unstable two-phase system will occur. The polymer starts to phase separate from solution to make the solution become turbid at LCST because of the transition from an enthalpy-favored coiled structure to an entropy-favored dense globular structure [56]. By chemically labelling the PNIPAM chain with AIEgens, the temperature-induced microstructure change in PNIPAM chains can directly affect the aggregation behavior of the grafted AIEgens, thus leading to the fluorescence change. In this way, the delicate details of the thermally induced liquid–solid phase transition process can be clearly revealed or monitored according to the change in fluorescence signal. For example, in 2009, Tang et al. prepared a TPE-labelled PNIPAM system through direct polymerizations and utilized its thermo-responsive fluorescence change to probe the thermal transitions of PNIPAM in water [57]. Similarly, Zhu et al. recently reported an AIEgen-grafted copolymer P(NIPAM-*co*-EM) for constructing a fluorescent thermometer [58]. As shown in Fig. 7a, P(NIPAM-*co*-EM) is composed of a quinolinemalononitrile-based AIEgen as the signal reporting unit and NIPAM unit as the thermo-responsive unit. The temperature-dependent absorption experiments revealed that the aqueous solution of the copolymer exhibits a specific LCST at about 30 °C. In the temperature region of 30–45 °C, coil–globule transition occurs along with the volume shrinkage, resulting in the formation of compact aggregates. This microstructure change leads to the activation of the RIM process of the AIEgen label to enhance the fluorescence. As depicted in Fig. 7b, c, the fluorescence intensity remained almost unchanged as temperature increased below the LCST value. When heating to 30–45 °C, a sharp increase in fluorescence intensity for about 3.7-fold was observed. This AIE-active fluorescent thermometer fundamentally solves the serious fluorescence quenching problem of conventional fluorescent thermometers. It is worth mentioning that further heating the polymer mixture to above 45 °C led to fluorescence decrease. Because the phase transition process has already been completed at the LCST, further increasing the temperature may have little effect on the compactness of the polymer aggregates. Instead, the thermally activated molecular motion starts to take effect in the high-temperature region to consume energy of the excited state through the non-radiative pathway and thus result in the continuous decrease in fluorescence intensity. Moreover, the fluorescence signal of P(NIPAM-*co*-EM) can be switched reversibly. These results demonstrated the great potential of AIEgen-labelled PNIPAM as a nondestructive fluorescent thermometer. By carefully adjusting the hydrophilicity of copolymers,

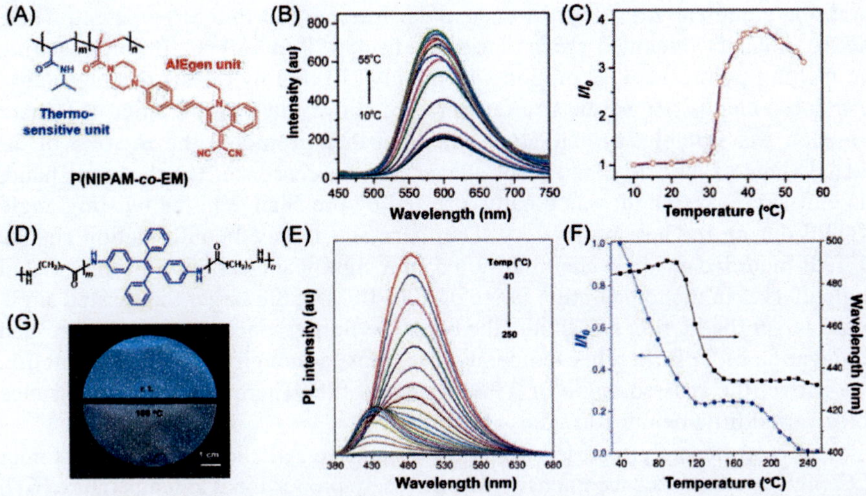

Fig. 7 **a** Chemical structure of the fluorescent thermometer P(NIPAM-*co*-EM) containing the side-chain grafted AIEgen of quinolinemalononitrile. **b** Temperature-dependent emission spectra of P(NIPAM-*co*-EM) (14 mM) in an aqueous solution upon heating. **c** The relative fluorescent intensity at 607 nm as a function of temperature. Reproduced from [58] with permission from the Royal Society of Chemistry. **d** Chemical structure of TPE-labeled nylon-6. **e** Fluorescence spectra of TPE-labeled nylon-6 at different temperatures. **f** Plots of relative PL intensity and emission wavelength of TPE-labeled nylon-6 versus temperature. **g** Fluorescence photographs of the TPE-labeled nylon-6 at room temperature and 160 °C. Reproduced from [60] with permission from Wiley-VCH

the LCST as well as the detection temperature range of AIEgen-labelled PNIPAM systems can be precisely controlled towards the physiological temperature. In this way, the application of water-soluble thermo-responsive AIE copolymers can be further extended to a series of biological studies such as cell imaging [59].

In addition to PNIPAM and other polyacrylamides, nylon-6 has also been modified by AIEgens to endow it with thermally responsive fluorescence (Fig. 7d) [60]. Different from the PNIPAM systems, the resulting TPE-labeled nylon-6 exhibits remarkable change in both fluorescence intensity and emission wavelength. As shown in Fig. 7e–g, at low temperature, the TPE-labelled nylon-6 showed strong blue fluorescence, with the emission maximum at 485 nm. With the increase of temperature, the photoluminescence (PL) intensity gradually decreased, with little change in emission wavelength. However, a blueshift of the emission began to occur above 120 °C, while little change was seen in the PL intensity. This thermo-responsive fluorescence property arises from the synergistic effect of the chemical structure and conformation of the TPE-labeled nylon-6. PL intensity is closely related to molecular motion. When the temperature was increased from 30 to 120 °C, the intensive hydrogen bonds between nylon chains were gradually destroyed, and the volume between polymer chains expanded. Therefore, the rotation of the phenyl rings of TPE was gradually activated, and the motion rate became faster as the temperature increased, which consequently led to the decreased fluorescence. At 120 °C, the hydrogen bonds might be completely destroyed to cause the free motion to a large extent. Thus, the PL intensity remained almost the same from 120 to 180 °C. Before reaching the melting point, the TPE motion was mainly affected by the segmental motion of nylon-6 chains. However, at a temperature above the melting point, the chain diffusion was completely activated, which further promoted the rotation of all phenyl rings of TPE to completely quench the fluorescence. On the other hand, the emission wavelength was greatly affected by the change in the twisting angle of TPE during the heating process. The thermally induced conformation change of TPE-modified nylon-6 chains resulted in a significant variation in the twisted angle of TPE in the temperature range of 100–140 °C: the larger the twisted angle or the larger the motion amplitude, the bluer the fluorescence due to the decreased conjugation of TPE. In other temperature regions, although the molecular motion was active, the twisted angle of TPE changed little. Therefore, the fluorescence color varied little below 100 °C or above 140 °C.

Yang et al. recently reported a thermo-responsive AIE liquid crystal elastomer (LCE), whose fluorescence intensity was inversely proportional to temperature [61]. As illustrated in Fig. 8A, an AIE-active LCE soft actuator was fabricated through the direct polymerization of a liquid crystal monomer and a TPE-containing cross-linker. The obtained LCE sample exhibited both shape change and fluorescence variation in response to thermal stimuli (Fig. 8B). The PL intensity decreased linearly from 35 to 135 °C, while the longitudinal shape deformations (L/L_{iso}) of the LCE sample presented a jump change around the LC-to-isotropic phase transition temperature (110 °C). This result demonstrated that the fluorescence variation of the AIEgen-functionalized LCE was mainly determined by the temperature effect rather than the macroscopic shape change. Increasing temperature can provide activation energy for the intramolecular motion of TPE, thus leading to the decrease in fluorescence intensity. This work demonstrates the potential applications of AIE-based LCE materials as thermomechanically controllable fluorescent soft actuators.

Taking advantage of the thermo-responsive fluorescence property of AIEgens, in 2017, Tang et al. successfully detected the glass transition temperatures of various polymer systems through the fluorescence-based method by physically doping AIEgens into polymer matrix [62]. More recently, Ji et al. further reported an AIEgen-based fluorescence method to accurately measure the topology freezing transition temperature (T_v) of vitrimers [63]. Vitrimers are reprocessable cross-linked polymers, and their T_v refers to the upper limit temperature for service and lower temperature for recycling. Below T_v, the verified cross-linked network of vitrimers will restrict the intramolecular motion of AIEgens. Therefore, strong fluorescence is expected to be observed in this state (Fig. 9a). However, the intramolecular motion will be significantly activated above T_v due to the network rearrangement of vitrimers, leading to the weakened fluorescence of AIEgens. This design principle works well, as evidenced by the experimental results (Fig. 9b, c). After doping or swelling AIEgens into vitrimers, the fluorescence of the AIEgen-vitrimer composites changed dramatically below and over T_v with different decreasing rate. Thus, T_v can be determined by the turning point of the decreasing rate. This AIEgen-based detection method is straightforward, sensitive, universal, and reliable. Although T_v is measured in a static state in the presented experiments, this method

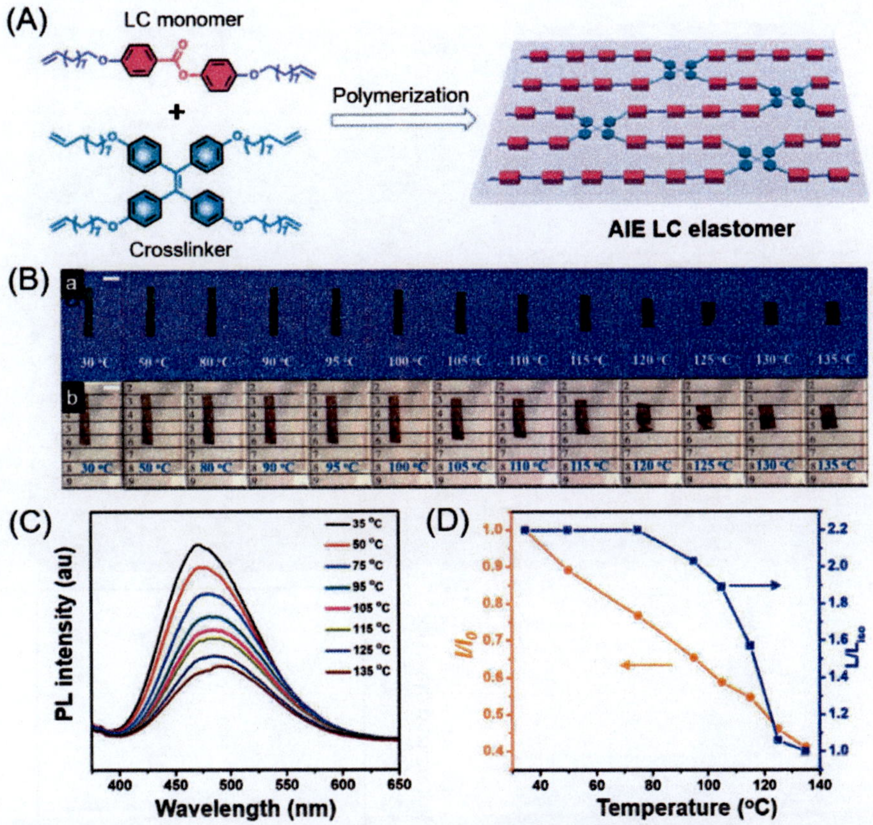

Fig. 8 **A** Chemical composition of an AIE-active LCE film. **B** Photo images of thermally induced actuation behaviors of one LC elastomer ribbon sample exposed to (a) UV 365 nm radiation and (b) ambient light (scale bar = 5 mm). **C** Fluorescence spectra of the LC elastomer sample at different temperatures. **D** Relative fluorescence intensity and the longitudinal shape deformation (L/L_{iso}) of the LC elastomer ribbon versus temperature. Reproduced from [61] with permission from the American Chemical Society

can also be used to monitor the effect of external force on T_v under stress by modifying the sample holder.

Besides the abovementioned examples, a series of other functional polymer systems with thermal responsiveness and AIE characters have also been reported, including AIEgen-modified multiarm star polymers, crystalline polymers, clusteroluminescent AIE polymers, blended AIEgen/polymer films, etc. [64–69] These advancements demonstrate the power of AIE for facilitating the understanding of polymer thermodynamics and advancing the development of functional stimuli-responsive polymers.

3.3 Light

Photo-responsive polymers have attracted great attention due to their significance in smart optical and biological applications. The stimulus of light irradiation

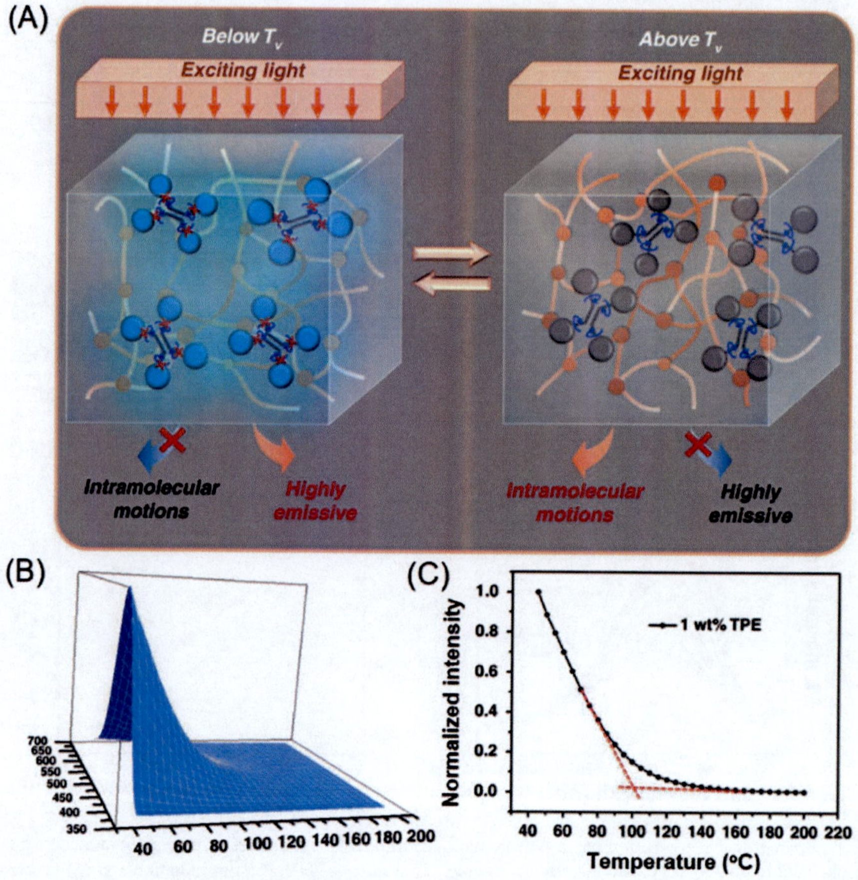

Fig. 9 **a** Schematic mechanism of detecting the topology freezing transition temperature (T_v) of AIEgen-doped vitrimers. **b** PL spectra of 1 wt% TPE-doped epoxy vitrimer at a series of rising temperatures. **c** The variation of PL intensity at 470 nm at different temperatures. All the fluorescence intensities are normalized to 45 °C. Reproduced from [63] with permission from Springer Nature

can be easily and precisely controlled to allow a localized stimulation within a defined area or volume. By changing the irradiation time and light intensity as well as the wavelength range of the light source, the strength of the response can be finely tuned. The most studied photo-responsive molecules include azobenzene, spiropyran, spirooxazine, fulgide, and dithienylethene (DTE) derivatives [70]. Functional photo-responsive polymers are generally constructed by chemically incorporating these photo-responsive units into polymer structures or physically blending the photo-responsive molecules in polymer matrices. The combination of AIE and photo-responsive properties can endow polymers with photo-responsive solid-state fluorescence and promising biomedical applications as photosensitizers.

Taking the DTE-containing photo-responsive polymers as an example, traditional photochromic DTE systems often suffer from the ACQ problem when using fluorescence as the read-out signal, which limits their practical applications in optical systems with high sensitivity and high precision [71]. In 2017, Lin et al. integrated AIEgens into a DTE-based photochromic polymer system to solve the ACQ issue [72]. As shown in Fig. 10a, polymer P-PHT contains DTE, TPE moieties, and a triazole linker in the repeating unit. When the DTE unit was in the open form, the AIE behavior of TPE took effect to enable the solid films of P-PHT to show bright blue fluorescence. Upon UV-irradiation of the P-PHT films, the cyclized DTE completely quenched the solid-state fluorescence of TPE via the non-radiative energy transfer process from TPE (as a fluorophore donor) to the nearby cyclized DTE unit (as an acceptor). In this way, a high fluorescence on/off ratio can be achieved for the photochromic P-PHT solid films. In 2018, Liu et al. reported a physical doping method to generate AIE-DTE-based photo-responsive polymer systems (Fig. 10b) [73]. The intramolecular photochromic property of DTE and the intermolecular AIE behaviors were integrated together using a DTE-bridged bispyridinium salt (DTE-BP). The molecular conformation of the bispyridinium salt located at both sides of the DTE core is sensitive to the restricted environment to endow the open-form DTE-BP with AIE properties. The DTE-BP-embedded polyvinylidene fluoride (PVDF) thin film showed strong green fluorescence. Upon irradiation with 254-nm light for 4 min, the fluorescence of DTE-BP was greatly quenched with a high fluorescence quenching efficiency of up to 91%, which was due to the photocyclization of the

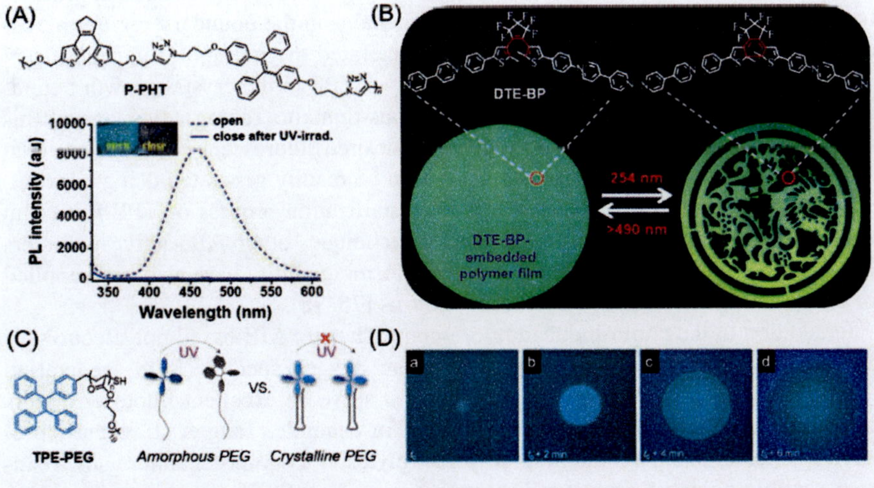

Fig. 10 **a** Chemical structure of P-PHT and its UV-induced PL spectral changes in solid films before and after UV irradiation. Inset: photo images of P-PHT at 0 min and 40 min UV irradiation. Reproduced from [72] with permission from the Royal Society of Chemistry. **b** Fluorescent photographs of DTE-BP-embedded PVDF film upon alternating 254-nm UV and visible light irradiations. Reproduced from [73] with permission from the American Chemical Society. **c** Chemical structure of TPE-PEG and its pattern formation mechanism. **d** Real-time fluorescent micrographs of pattern evolution assisted by intermittent UV irradiation during the isothermal crystallization of TPE-PEG film at 21 °C from a melt. Scale bars = 100 μm. Reproduced from [74] with permission from the American Chemical Society

DTE core from the emissive open form to non-emissive closed form. In comparison to the photochromic behavior of DTE-BP in glycerin (fluorescence quenching efficiency = 76%), the PVDF matrix greatly accelerated the ring-closing reaction and thus significantly improved the quenching efficiency. Moreover, the quenched fluorescence can be almost completely restored after irradiation with visible light (> 490 nm) for 20 s. Taking advantage of the reversible photoswitchable fluorescence behavior, the DTE-BP-embedded PVDF films can serve as excellent photo-writable and photo-erasable materials with controlled fluorescence. The photo-writing-erasing process could be repeated for many times, indicating the good recyclability of the luminescent film.

Besides the incorporation of photochromic molecules, some AIEgens themselves can show remarkable photo-responsive fluorescence change. For example, TPE, one of the most well-known AIEgens, can undergo photocyclization and photo-oxidation reaction to form 9,10-diphenylphenanthrene (DPP) upon UV light irradiation. TPE shows blue fluorescence with maximum emission at 460 nm, while the fluorescence of cyclized-TPE is largely blueshifted to the invisible region (~ 375 nm). Recently, Tang et al. reported a manipulation strategy on the photocyclization activity of TPE using a crystallizing polymer matrix to generate controlled fluorescent patterns [74]. As illustrated in Fig. 10c, TPE was attached to one end of semicrystalline polyethylene glycol (TPE-PEG). The soft amorphous domain of PEG can facilitate the intramolecular motion of TPE, which is favorable for quenching fluorescence and increasing photoreactivity. By contrast, the intramolecular motion of TPE was restricted in the rigid crystalline phase to allow intense blue fluorescence and low photocyclization activity. Moreover, during the melt-crystallization process, the dynamic movement of PEG segments and chains in the boundary regions across crystalline and amorphous phases further accelerated the intramolecular motion of TPE. Therefore, the photocyclization tendency of TPE in the crystal growth boundary layer was higher than that in the amorphous domain. Taking advantage of this motion-dominated photo-responsive process, desired fluorescent patterns with distinct fluorescence contrast (bright vs. dark) can be readily generated using intermittent UV irradiation during the isothermal crystallization process of TPE-PEG film (Fig. 10d). Based on the photolithography technique, other AIE-active polymers with excellent photosensitivity and good film-forming ability have also been applied for the generation of diverse fluorescent patterns [75–78].

In addition to the photo-induced fluorescence change, AIE-based polyelectrolytes also possess extraordinary potency of reactive oxygen species (ROS) generation under light irradiation, which enables them to serve as excellent photosensitizers for various biomedical applications [79, 80]. For example, Tang et al. recently prepared a series of azonia-containing polyelectrolytes in a one-pot manner using readily accessible nonionic reactants in the presence of cheap ionic species (Fig. 11a) [81]. Among these polymers, **P1/2c/3** with a strong donor–acceptor conjugated structure showed significant $^1O_2$ generation ability and excellent photostability. The micelles of **P1/2c/3** can efficiently inhibit the Gram-positive *S. aureus* under white light irradiation with a minimal inhibition concentration of 15 μg/mL. Furthermore, this polyelectrolyte possessed antimicrobial ability against drug-resistant bacteria. As illustrated in Fig. 11b, the high charge density in the polymer main chains

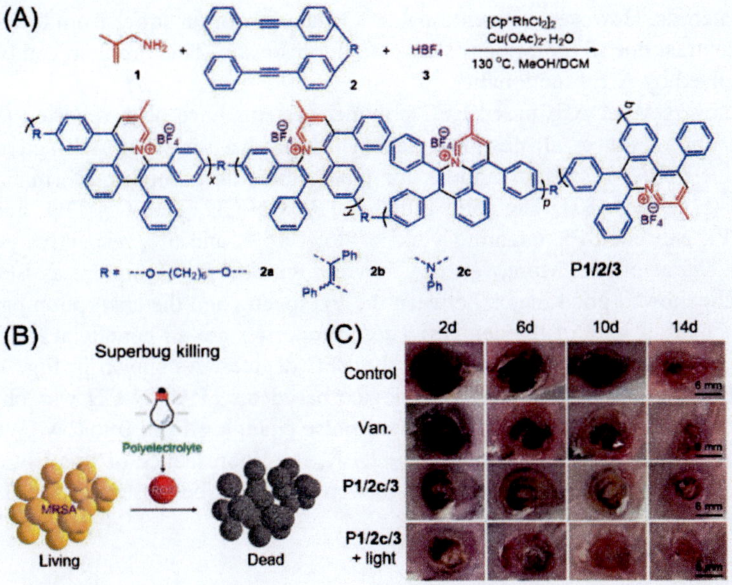

Fig. 11 **a** Synthesis of azonia-containing polyelectrolytes (**P1/2/3**). **b** Schematic illustration for the application of the polyelectrolyte as a photosensitizer in superbug killing. **c** Typical photographs of MRSA-infected burn sites upon different treatments under dark or light irradiation during the therapeutic processes. Reproduced from [81] with permission from the American Chemical Society

endowed **P1/2c/3** micelles with high affinity to the negatively charged bacteria membrane, which could physically damage the bacteria membrane to some extent. Upon white light irradiation, **P1/2c/3** could efficiently generate ROS. The physical killing together with the photodynamic effect of **P1/2c/3** led to strong inhibition against methicillin-resistant Staphylococcus aureus (MRSA), a drug-resistant superbug. The in vivo MRSA-infected mouse burn model further demonstrated that the photodynamic effect of **P1/2c/3** could effectively inhibit the infection of MRSA and accelerate the healing process in the burn sites (Fig. 11c).

Apart from the abovementioned applications, the photo-responsiveness of AIE-based polymer systems can also find applications in anti-counterfeiting materials, super-resolution fluorescent imaging agents, microlasers, integrated photonics, erasable optical memory, logical gates, light-harvesting films, etc. [71, 82, 83]

3.4 Electricity

Electrofluorochromic (EFC) materials with changeable fluorescence in response to the applied potential have attracted great attention due to their potential applications in various optoelectronic devices, including electrofluorochromic devices, organic phototransistor-based memory devices, circularly polarized organic light-emitting diodes (CP-OLED), etc. [84] These practical applications generally require fast response speed, high fluorescence contrast, and stable switching stability for the

EFC materials. However, conventional EFC materials often suffer from low fluorescence contrast due to their poor solid-state fluorescence. This problem can be effectively solved by AIE functionality.

Till now, several AIE-based EFC polymer systems have been reported [84–86]. For example, Liou et al. designed and synthesized a series of AIE-active triphenylamine (TPA)-based polyamides for the application of high-performance EFC devices (Fig. 12a) [87]. The solid films of TPA-CN-CH, TPA-CN-TPE, and TPA-OMe-TPE exhibited PL quantum yield of 46%, 16%, and 5%, respectively, due to the AIE characteristic. Moreover, EFC devices with these polyamides as EFC films (Fig. 12b) showed good match between the PL spectra and the absorption pattern in the switched-on state of the devices. These properties are all beneficial for enhancing the fluorescence contrast ratios of the EFC devices. As shown in Fig. 12c, the bright blue fluorescence of the EFC device based on TPA-CN-CH was obviously quenched to a dark state with the applied pulse changing from 0 to 2 V. When the potential was subsequently applied at −2.1 V, the fluorescence of this EFC device was well recovered. A similar phenomenon can also be observed for the EFC

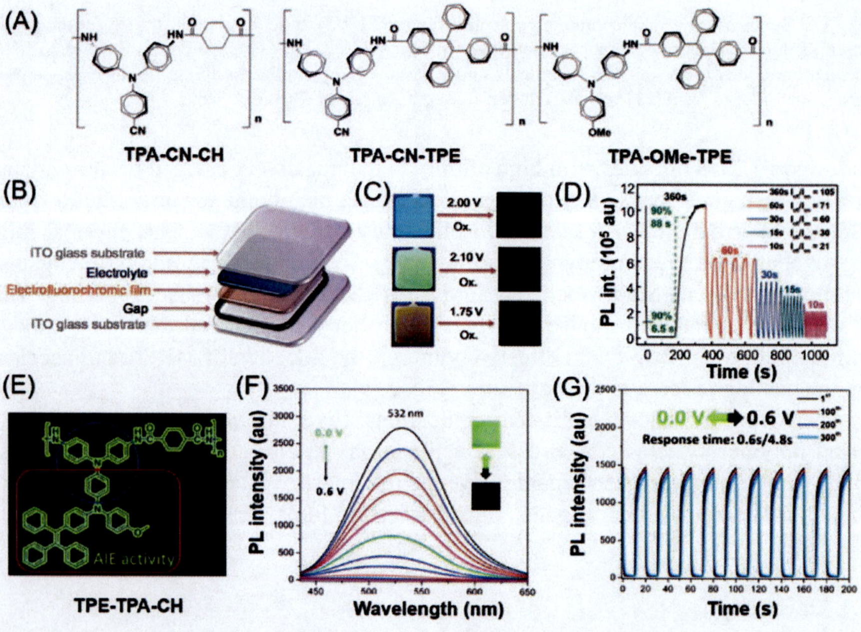

Fig. 12 **a** Chemical structures of three polyamides. **b** Schematic diagram of electrofluorochromic (EFC) devices based on the polyamides. **c** Behaviors of EFC devices based on TPA-CN-CH (top), TPA-CN-TPE (middle), and TPA-OMe-TPE (bottom) with a specific applied voltage for oxidation. **d** Estimation of the fluorescence switching time at different step cycle times of TPA-CN-CH/*n*-heptyl viologen between 1.4 V and −1.5 V. Reproduced from [87] with permission from the Royal Society of Chemistry. **e** Chemical structure of TPE-TPA-CH. **f** Fluorescence spectra of TPE-TPA-CH film at different applied potentials. **g** Fluorescence switching response under applied potentials between 0.0 and 0.6 V at the cycling time of 20 s for 300 cycles. Reproduced from [86] with permission from the American Chemical Society

devices based on TPA-CN-TPE and TPA-OMe-TPE. The fluorescence quenching process could be attributed to the structural planarization and optical absorption shift of TPA units from the neutral to oxidized states during electrochemical switching. With the introduction of *n*-heptyl viologen (HV) into the device system as a counter electrochromic (EC) layer for balancing charges, the resulting EFC device based on TPA-CN-CH showed the highest fluorescence contrast ratio (I_{off}/I_{on}) of up to 105 due to its strongest solid-state fluorescence (Fig. 12d). The shortest response time was achieved in the HV-containing TPA-OMe-TPE-based EFC device, whose response time was less than 4.9 s thanks to its conjugated structures and low oxidation potential. These results suggested that the combination of EC and AIE features in one polymer system is a feasible approach for preparing highly efficient EFC devices. However, the advantages of AIE-TPA unit have not been fully utilized in this work.

Recently, Chen et al. reported an electro- and AIE-active polyamide with a bulky and asymmetric pendant group containing TPE and TPA moieties (Fig. 12e) [86]. The introduction of TPE can improve the fluorescence on/off contrast, while the highly conjugated TPA structure can function as a stable electrochemical modulator to switch the dual emission and color. Moreover, the highly conjugated TPA and bulky pendant substituent can work together to shorten the response time. Such a rational combination of TPE and TPA allowed the resulting polymer (TPE-TPA-CH) to show highly integrated EC and EFC performances, including outstanding emission/color switching behavior (Fig. 12f), high optical contrast of 82 at the duration time of 20 s, and fast response speed (1.8/1.1 s for EC and 0.4/2.9 s for EFC process), as well as excellent long-term switching stability over 300 cycles (Fig. 12g). Therefore, the integration of AIE functionality with a highly conjugated redox unit could be an effective strategy to prepare high-performance emission/color dual-switchable materials in response to the external electric stimulus. In addition, the better device performance of TPE-TPA-CH than those of the polyamides in Fig. 12a also demonstrated that the device structure stability can be improved by attaching electron-donating groups, such as a methoxyl group and TPE, to stabilize the oxidized state of TPA.

Despite these innovations, there is still much room to explore further applications of AIE in EFC materials. With the guidance of the abovementioned strategies, we anticipate that more work could be done to promote the applications of AIE-based EFC materials in sensors, smart windows, and displays.

3.5 Gamma Rays

Gamma (γ)-ray radiation is one type of electromagnetic radiation originating from the atomic nucleus and from particle annihilation. γ-rays play an important role in many areas, including astrophysics research, nuclear power industry, sterilization, materials preparation and processing, biomedical applications, etc. For instance, γ-ray-based radiotherapy is an effective therapy for cancer treatment [88]. However, γ-rays can also be very hazardous to human health [89]. For the sake of security, it is of great significance to develop facile and convenient detection approaches for the detection of γ-ray radiation.

Taking advantage of the AIE characteristics and γ-ray-induced polymer degradation, Zhang et al. developed a simple and efficient fluorescence-based detection method for γ-ray radiation [90]. As illustrated in Fig. 13a, b, the negatively charged polyelectrolyte containing sulfone ($-SO_2-$) groups in the main chain is sensitive to γ-rays. Under γ-ray radiation, the C–S bonds in this polyelectrolyte can be cleaved to degrade the polymer into small fragments. When the cationic AIEgen (silole-NI) was mixed with the aqueous solution of polyelectrolyte, bright blue fluorescence could be observed due to the formation of an aggregation complex with the assistance of electrostatic and hydrophobic interactions. However, when the polyelectrolyte was degraded by γ-ray radiation, the interactions between silole-NI and the polymer residues became weak, consequently reducing the coaggregation tendency and releasing the intramolecular motion of the AIEgen to some extent. Therefore, the fluorescence of the mixture was much weaker after exposure to γ-rays. As shown in Fig. 13c, the fluorescence intensity of the ensemble of silole-NI and polyelectrolyte gradually decreased with the increase in the dose of γ-rays. This detection method can be performed at ambient conditions, and the detection limit of γ-ray radiation was as low as 0.13 kGy. The sensitivity can be further improved by optimizing the chemical structures of the AIE probe and the γ-ray-sensitive polymer.

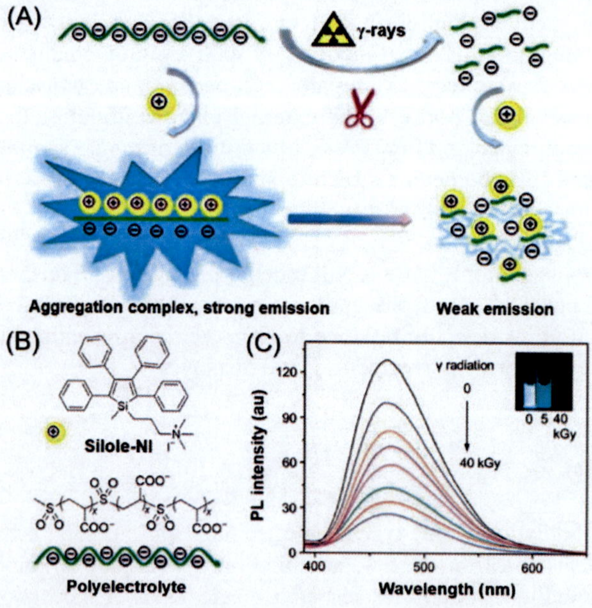

Fig. 13 **a, b** The design rationale and the chemical structures of silole-NI and polyelectrolyte for the fluorescence detection of γ-ray radiation. **b** Fluorescence response of the aqueous solution of silole-NI and the polyelectrolyte before and after exposure to different doses of γ-ray radiation. Insets: the corresponding fluorescent images. Reproduced from [90] with permission from the Royal Society of Chemistry

Besides the sulfone groups, diselenide bonds have also been reported to be sensitive to γ-radiation. Under γ-ray radiation, diselenide bonds can be cleaved and oxidized to generate seleninic acid [91]. Seleninic acid was reported to be a chemotherapeutic drug that can upregulate the ROS levels in cancer cells to cause cell apoptosis [92]. Therefore, the diselenide-containing systems are promising for use in the combination applications of radiotherapy and chemotherapy with improved anticancer activity. The incorporation of an AIE unit into the diselenide systems further endows the drugs with bioimaging capability. For example, Xu et al. synthesized an amphiphilic AIEgen with diselenide bonds in the middle of the structure (Fig. 14) [93]. This AIEgen could self-assemble into spherical assemblies, but the assemblies tended to collapse under γ-ray radiation due to the presence of diselenide bonds. This γ-radiation-induced self-assembly change could be monitored in situ through the fluorescence change. The fluorescence intensity of the assemblies dramatically decreased under γ-ray radiation, suggesting the disassembly of the assemblies. The cleaved selenium-containing assemblies showed excellent anticancer activities because of the anticancer efficacy of the generated seleninic acid. This carrier-free assembly system demonstrated the great potential of AIE-based γ-irradiation-responsive materials as nanomedicines for cancer treatment with combination therapy. Although only two examples of hazardous species detection were showcased in this section, countless species can be sensed by virtue of AIE through appropriate structure designs [19].

3.6 Morphology

The study of polymer morphologies, including the conformations, domain size, shape, composition, texture, as well as phase distribution, is of great significance due to their direct influences on materials properties. Taking advantage of the high

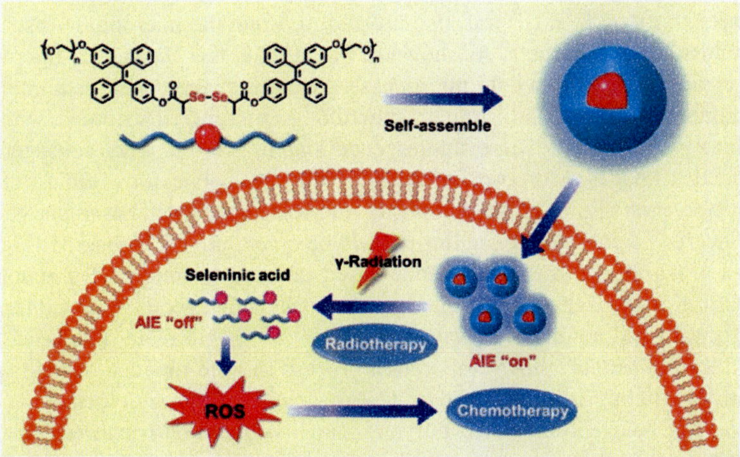

Fig. 14 Schematic illustration of selenium-containing AIE assemblies combining cancer radiotherapy with chemotherapy. Reproduced from [93] with permission from the American Chemical Society

sensitivity of AIEgens to subtle microenvironment change, a few AIE-based polymer systems, such as AIEgen/polymer composites and AIEgen-functionalized polymers, have been exploited for the visualization and monitoring of polymer morphologies. Based on the distinct PL behaviors of AIEgens between molecularly dispersed and aggregated states, the conformational change, phase separation structures, self-healing structures, and swelling process of polymers have been clearly detected via fluorescence signals [27, 94–98].

Very recently, Tang et al. reported an AIE-based polymer system that can visualize polymer microstructures based on the distinct emission colors in different phases [99]. As depicted in Fig. 15a, an amphiphilic AIEgen (TPE-EP) with polymorphic characteristics was prepared and doped within the semicrystalline polymer of poly(*L*-lactide) (PLLA). The fluorescence of TPE-EP aggregates varied from yellow to green depending on the change of polymorphic forms from the metastable crystalline state (Y aggregates) to the thermodynamically stable crystalline state (G aggregates). Through quick and slow evaporation from chloroform solutions, the TPE-EP-loaded amorphous and crystalline PLLA films were produced, respectively. During the film-forming process, the polymorphs of TPE-EP were grown within the polymer network simultaneously. To minimize the influence of TPE-EP on polymer morphology without sacrifice of emission behaviors, 0.1 wt% TPE-EP was used in these films. The amphiphilic TPE-EP molecules were demonstrated to form segregated nanocrystals in the hydrophobic PLLA matrix. In amorphous PLLA, the G aggregates of TPE-EP were stabilized to give a green-emissive film (Fig. 15b) because the loose and mobile network of amorphous PLLA allows the embedded TPE-EP molecules to orient and pack stably. Oppositely, in crystalline PLLA, the Y aggregates were confined between the crystalline lamellae to show yellow emission (Fig. 15c, inset). Under a fluorescence microscope, a number of interlinked spherulites with yellow fluorescence were clearly observed. Therefore, TPE-EP can operate as a good fluorescent marker to distinguish the amorphous and crystalline phases of PLLA by the marked emission color difference. Furthermore, the polymorph selectivity of TPE-EP was also utilized to detect the microphase distribution and composition of polymers. As shown in Fig. 15d, e, the 2D and 3D microscopic images clearly reveal the whole microphase structure of the semicrystalline PLLA films: numerous yellow and circular spherulites randomly distributed within the amorphous green region. These fluorescence imaging results were consistent with those in SEM images. Moreover, a linear relationship between the emission maximum and the crystallinity of the composite film was established, based on which the average degree of polymer crystallinity could be quantitatively measured (Fig. 15f). The high sensitivity of the fluorescence-based imaging technique also enabled the high-contrast visualization of banded spherulites with specific handedness in a straightforward and noninvasive way. As depicted in Fig. 15g and h, the PLLA film prepared from a suitable slow evaporation process showed obvious banded spherulites with alternating bright- and dark-yellow spirals in the anticlockwise (ACW) direction. The AIEgen-embedded polymer films with chiral structured spherulites can efficiently generate circularly polarized light (CPL) upon UV excitation. The film inversion results in the enantiomeric chirality switching of the spiral spherulites to provide an opposite CPL response while the enantiomeric symmetry was

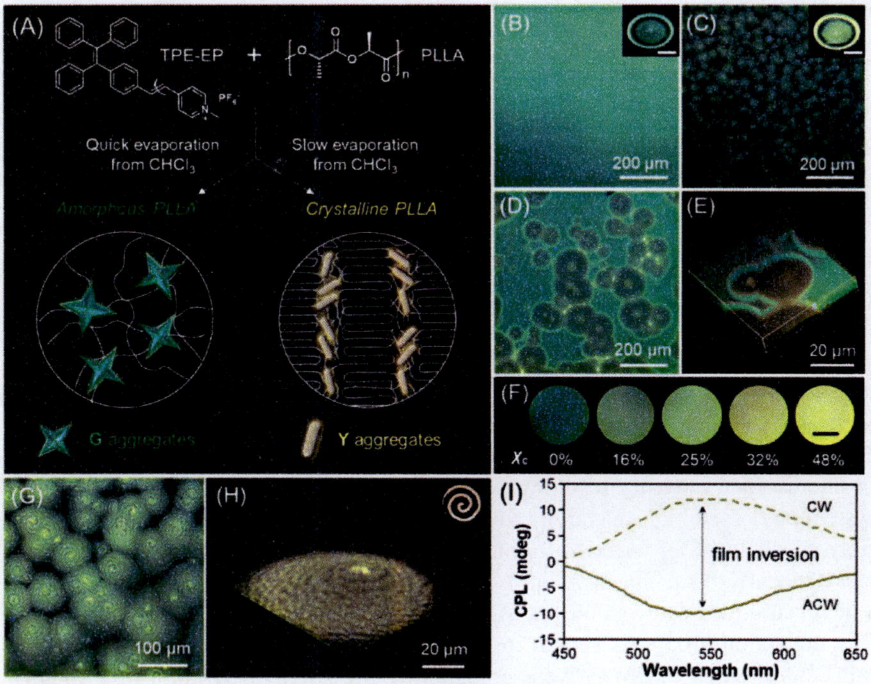

Fig. 15 **a** Schematic representation of the preparation of amorphous and crystalline PLLA embedded with TPE-EP and the polymer phase determination. Fluorescence photos of **b** amorphous and **c** crystalline polymer films and (insets) their corresponding magnified images. Inset scale bar: 1 cm. **d** Overview fluorescence micrograph of the outer surface of the PLLA film with a mixture of crystalline and amorphous regions. **e** Confocal fluorescence 3D images of crystalline spherulites. **f** Fluorescence images of TPE-EP-embedded PLLA at various degrees of crystallinity. Scale bar: 5 mm. **g** Representative microscopic fluorescence (2D) and **h** confocal (3D) representation of the PLLA spherulites with alternative bright and dark yellow spirals in the anticlockwise direction. **i** Enantiomeric CPL switch upon film inversion. Reproduced from [99] with permission from the Royal Society of Chemistry

kept. These results suggested that the morphology-dependent AIEgen/polymer systems can provide direct and detailed information on polymer microstructures to benefit the understanding of the morphology–property relationship of various polymer materials.

4 Responses to Chemical Stimuli

Besides the abovementioned physical stimuli, the detection and measurement of chemical stimuli is also of great significance. In this section, we will discuss the representative examples of functional AIE polymer systems with responses to single chemical stimuli, including gas/vapor, pH, ions, explosives, and other chemicals or chemical reactions.

4.1 Gas/Vapor

The detection of gas like CO_2 and volatile organic chemicals (VOCs) such as NH_3 and acid vapor is fundamentally important for a diverse array of real-world applications. In the past years, various AIE-based polymer systems with gas/vapor responsiveness have been reported [100–103]. They are promising for use as selective CO_2 detection, monitoring of morphology transformations of nanoassemblies, detection of biogenic amines and seafood spoilage, multi-scale humidity sensing, etc. [104]

For example, Chen and coworkers reported a fluorescent turn-on chemosensor for CO_2 gas detection based on AIE (Fig. 16a) [105]. This sensor was constructed by the post-modification of branched polyethyleneimine (PEI) with TPE. Upon purging CO_2 into the ethanol solution of TPE-PEI, the transparent solution immediately turned into the turbid precipitation because of the reaction of CO_2 with alkylamines moieties of PEI to produce the poorly soluble carbamate salts (TPE-PEI-CO_2). This solution-to-precipitation phase transition process induced a turn-on response in fluorescence intensity resulting from the AIE feature of TPE-PEI. This sensing system possessed high selectivity toward CO_2 gas as it is highly tolerant to the possibly coexisting CO, water, sulfurous gases, and many common VOC interferents. More recently, Li et al. developed an AIE-active amphiphilic copolymer system, which can undergo reversible transition between polymersomes and micelles upon CO_2 activation [106]. As shown in Fig. 16b, amphiphilic block copolymers PEG-b-P(DEAEMA-*co*-TPEMA) were prepared by the copolymerization of 2-(diethylamino)ethyl methacrylate (DEAEMA) and TPE-pended methacrylate (TPEMA) in the presence of mPEG-CTA. The obtained copolymers can self-assemble into polymersomes with bright fluorescence by the nanoprecipitation method in THF/water or dioxane/water mixtures. Upon bubbling CO_2 into the polymersome solution, the DEAEMA segments were protonated and experienced hydrophobic-to-hydrophilic transition, which transformed the polymer vesicles into small spherical micelles. The transmittance of the solution increased and finally reached 97% due to the decrease of the nanoparticle size, whereas the fluorescence intensity slightly decreased due to the higher freedom of intramolecular rotation of TPE moieties in micelles (Fig. 16c). Moreover, this polymersome–micelle transition was demonstrated to be reversible upon the subsequent Ar bubbling. Similar morphology transformation strategy was also reported by Xing et al. They utilized a polymerization-induced self-assembly method to construct AIE-active polymeric nano-objects. With the promotion of CO_2, the nano-objects underwent morphology transformation from spheres to a mixture of "jellyfish" and vesicles. This morphology transformation resulted in the variation of AIE behaviors, thus making it a promising CO_2-responsive chemosensor with tunable morphologies and sizes [107].

Taking advantage of the high sensitivity and excellent photostability of AIE polymers, Tang et al. developed an effective fluorescence-based technique to monitor acid–base homeostasis [108]. As shown in Fig. 17a, TPE-containing polyheterocycles (**P1a/2a/3a**) were synthesized through a metal-free multicomponent polymerization of terminal diynes, dialdehydes, and ureas. The resulting polymers exhibited typical AIE behaviors, and their thin films showed intense green fluorescence. Interestingly, the protonation of the thin film of **P1/2a/3a** by HCl vapor caused a marked

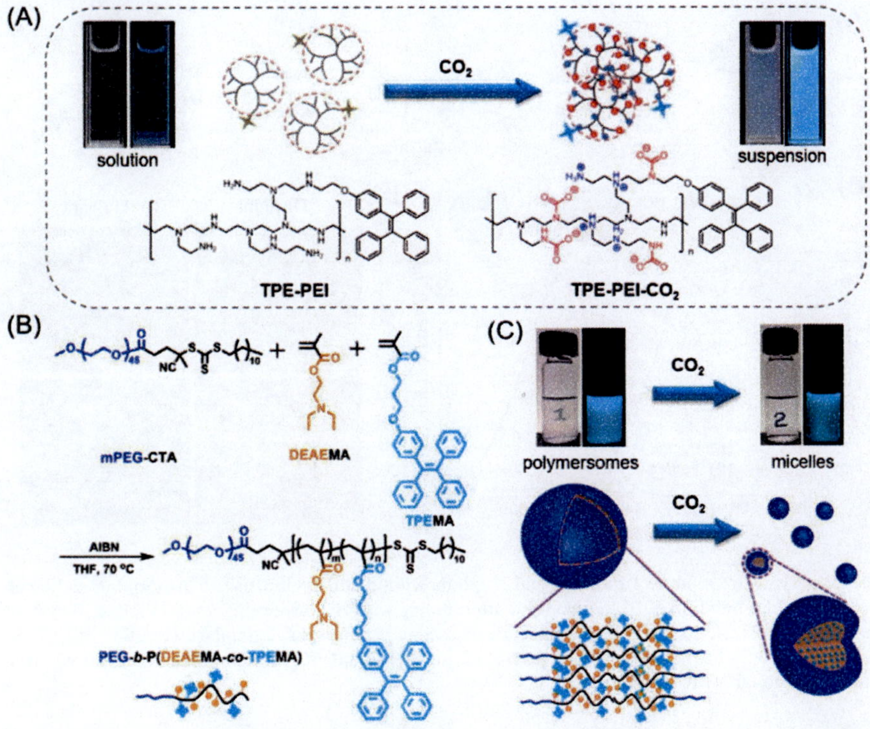

Fig. 16 **a** Structure of the sensory polymer TPE-PEI and its chemical reaction with CO_2 to form poly-ammonium carbamates salts TPE-PEI-CO_2 at ambient conditions and the schematic representation of its sensing process. Insets: photographs of TPE-PEI in ethanol before and after being purged with CO_2. Reproduced from [105] with permission from Elsevier. **b** Synthesis of the CO_2-responsive amphiphilic block copolymer PEG-*b*-P(DEAEMA-*co*-TPEMA). **c** Photograph of the polymer self-assembly solution before and after CO_2 treatment and the reversible polymersome–micelle transition with the schematic molecular organization in the vesicle membrane. Reproduced from [106] with permission from Wiley-VCH

redshift in the absorption and emission. The fluorescence color changed from chartreuse to dark-red accompanied with a large contrast in emission intensity (Fig. 17b). Similar phenomena were observed on quartz optical fiber coated with **P1/2a/3a**. When the acid-treated polymer thin film was exposed to ammonia vapor (NH_3), the absorption and emission properties were completely recovered. Based on this fast and reversible fluorescence response, the protonated H_2**P1/2a/3a**$^{2+}$ was developed into a sensitive fluorescence sensor for ammonia with a detection limit of 960 ppb. The acid-treated thin film of **P1/2a/3a** can serve as a food spoilage sensor based on its sensitive and selective responses toward ammonia and biogenic vapors. As depicted in Fig. 17c, after being sealed in containers with fresh shrimps and scallops for 12 h at room temperature, the H_2**P1/2a/3a**$^{2+}$ obviously changed from non-emissive orange film to emissive orange-yellow film. The remarkable color change in daylight and turn-on fluorescence response change under UV irradiation made this

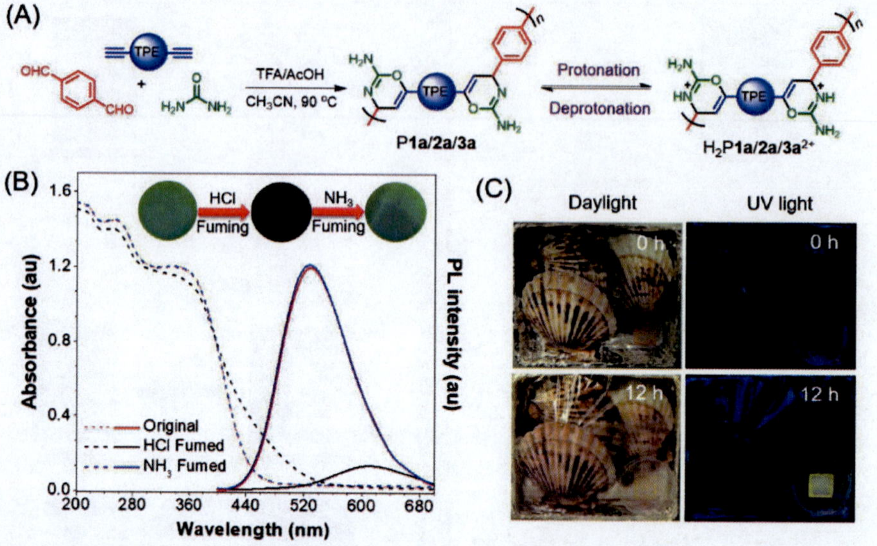

Fig. 17 **a** Synthetic route to **P1a/2a/3a** and its protonation and deprotonation. **b** Absorption and emission spectra of a thin film of **P1a/2a/3a** (red) after fuming with HCl vapor (black, 10 s) followed by NH_3 vapor (blue, 0.08 M, 30 s). Inset: fluorescent microscopy photographs. **c** Spoilage detection of scallops in sealed packages for 12 h at room temperature using H_2**P1a/2a/3a**$^{2+}$ thin film. Reproduced from [108] with permission from Wiley-VCH

polymer system a promising chemosensor for the in situ visualization of the food spoilage process. Apart from the abovementioned examples, AIEgen-doped polymer films and fibers with gas/vapor responses have also been constructed through facile physical preparation strategies [23].

4.2 pH

The stimulus of pH is a critical factor for cellular functions and other life forms. A subtle variation in environmental pH can have devastating effects on plant and animal lives. Moreover, pH value also plays a crucial role for the function and quality of medicines, foods, drinking water, etc. Therefore, the sensitive detection and monitoring of pH level is very important.

One of the most commonly used strategies for developing multifunctional pH-responsive polymers is the integration of AIEgens and pH-sensitive units in one polymer system. For example, Jin and Ji et al. have designed and synthesized a zwitterionic phosphorylcholine-TPE conjugate for pH-responsive drug delivery and AIE imaging (Fig. 18) [109]. The TPE unit was linked with the polymer backbone via acid-cleavable hydrazone bonds. The obtained AIE-active copolymer can self-assemble into spherical micelles and encapsulate doxorubicin (DOX) through hydrophobic interactions to form the theranostic PC-*hyd*-TPE-DOX micelles. Under endo/lysosomal acidic conditions, the cleavage of the hydrophobic TPE unit induced the disassembly of micelles followed by the release of DOX. Moreover, the introduction of AIE function enabled this system to show high-quality fluorescence imaging results. Combined with the efficient pH-responsive drug delivery, the multifunctional PC-*hyd*-TPE-DOX micelles were demonstrated to be a promising nanoplatform for a new generation of cancer theranostics.

The incorporation of pH-sensitive AIEgen into polymers can also achieve pH-responsive functional polymers. As depicted in Fig. 19a, TPE-OX, a pH-sensitive AIEgen, was attached to an amphiphilic copolymer chain with a hydrophilic poly(ethylene glycol) methyl ether methacrylate and hydrophobic polystyrene block [110]. Under acid/base stimuli, TPE-OX switched between the cyan-emissive closed form and the protonated open structure with red emission, thus contributing to the pH-dependent dual-emission of the amphiphilic copolymer. The PL spectra of the self-assembled polymeric micelle of the copolymer showed that its emission intensity at about 489 nm gradually decreased with the decrease of pH value in Tris–HCl buffer solutions (Fig. 19b). A new emission band centered at 640 nm was observed at pH 4.42, and this long-wavelength emission peak became stronger when pH further decreased, resulting in pH-responsive dual emission from cyan to red with a high fluorescence contrast (Fig. 19c). The fluorescence change was attributed to the extended conjugation and the emergence of an intramolecular charge transfer effect when the spiro-ring of the oxazolidine moiety was opened. The cell imaging experiments demonstrated that such polymeric micelles have potential applications in the detection of lysosomal activity and further autophagy in cancer cells based on their pH-dependent reversible AIE fluorescence.

Very recently, Tang et al. utilized the pH-responsive fluorescence of AIEgens and the interactions between AIEgens and hydrogels to fabricate a stimuli-responsive polymer system showing simultaneous changes in fluorescence color, brightness, and shape in response to a single stimulus [29]. As illustrated in Fig. 20a, a simple bilayer hydrogel actuator was designed and prepared with the chemically cross-linked ionomer PAS as the actuator matrix and the pH responsive tetra-(4-pyridylphenyl)ethylene (TPE-4Py) as the dopant molecule of the

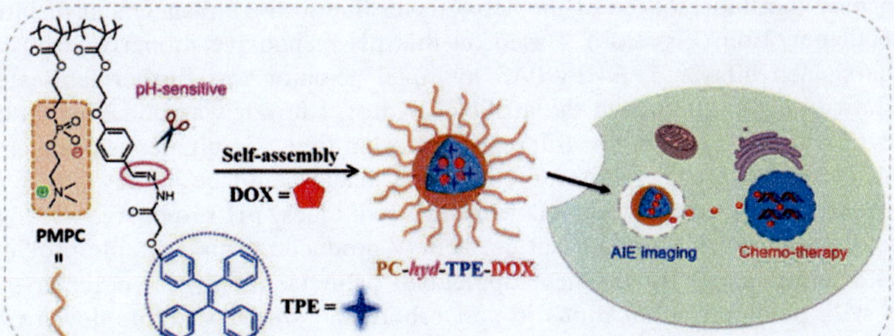

Fig. 18 Schematic illustration of phosphorylcholine-TPE conjugate as a cancer theranostic nano-platform for combined AIE imaging and pH-responsive drug delivery. Reproduced from [109] with permission from the American Chemical Society

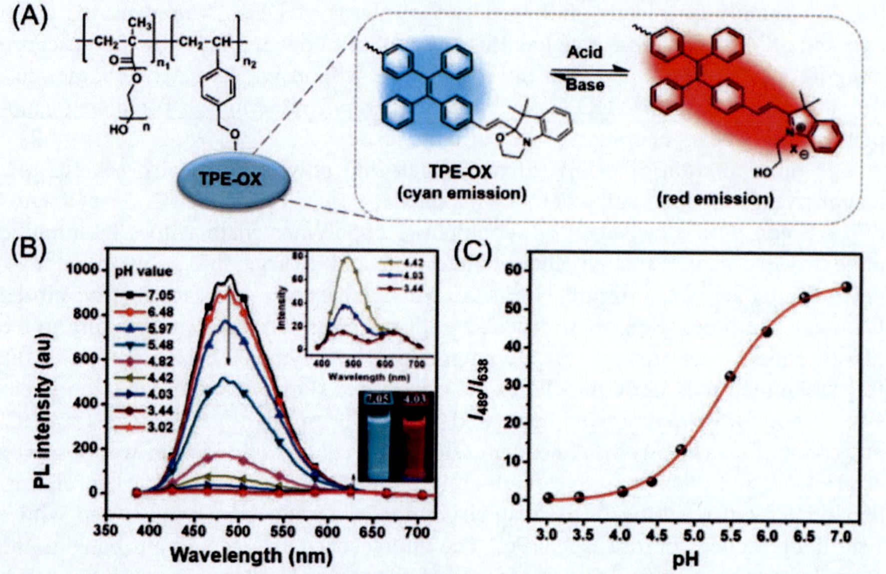

Fig. 19 **a** Chemical structures of the TPE-OX-pended copolymer and its acidichromism mechanism. **b** pH-dependent PL spectra of the polymeric micelles in Tris–HCl buffer solution. **c** Plot of I_{489}/I_{638} versus pH. Reproduced from [110] with permission from the Royal Society of Chemistry

active layer. Under acidic stimuli, the protonation of TPE-4Py led to the redshift in fluorescence due to the intramolecular charge transfer. The protonated TPE-4Py acted as the extra cross-linking point because of its electrostatic interactions with the PAS network, causing the shrinkage and the gradual deformation of the active layer. These interactions restricted the intramolecular motion of TPE-4Py to some extent, thus making the redshift emission visible. Further lowering the pH led to the increased solubility of the protonated TPE-4Py, consequently quenching the fluorescence of the active layer due to the enhancement of intramolecular motion (Fig. 20b). Based on this pH-responsive property, a furling flower-shaped bilayer TPE-4Py/PAS hydrogel actuator was further fabricated. As shown in Fig. 20c, when the artificial hydrogel flower was immersed into a pH 3.12 aqueous solution for different immersion time, simultaneous changes in fluorescence brightness, emission color, and shape could be clearly observed. By using the high-tech 3D or 4D printing techniques, pH-responsive hydrogel actuators with diverse shapes can be facilely produced. This example provided an effective strategy for the development of artificial intelligent polymer systems with potential applications in soft robots and smart wearable devices. In addition to the abovementioned examples, functional polymers with AIE and pH responses have also been applied for ratiometric fluorescence sensing of intestinal pH, monitoring of polymer degradation, biomedical probes, etc. [111–114]

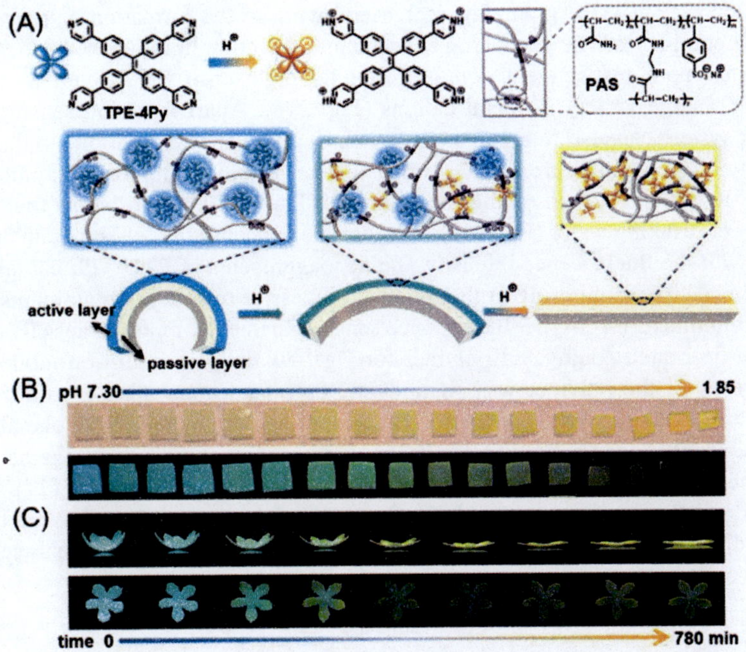

Fig. 20 **a** Chemical structure of the pH-responsive AIEgen TPE-4Py and the chemically cross-linked ionomer PAS and the schematic illustration of design strategy of a TPE-4Py/PAS-based bilayer fluorescent hydrogel actuator. **b** Photographs of monolayer hydrogel samples after immersion in aqueous solutions with different pH values for 17 h taken under daylight (upper line) and 365-nm UV light illumination (lower line). **c** Flat view (upper line) and plane view (lower line) showing the simultaneous emission change and complex shape deformation of the pH-responsive hydrogel actuator. Reproduced from [29] with permission from Wiley-VCH

4.3 Ions

Selective fluorescence sensors for ion detection are of great significance to environmental and health issues. In the past decades, diverse AIE-based polymers with sensitive and selective response to Ca^{2+}, Pd^{2+}, Pb^{2+}, Cu^{2+}, Hg^{2+}, Al^{3+}, Ru^{3+}, CN^{-}, etc. have been developed [23, 115]. The main principle of ion detection is based on molecular recognition. Therefore, the main design strategy for ion-responsive fluorescence sensors is the incorporation of AIE-active moieties and ion recognition groups or units into polymers.

For instance, thiourea compounds are well-known ligands for heavy metal ions with strong binding to mercury ion. Recently, Tang and Hu et al. prepared an AIE-active polythiourea through the catalyst-free multicomponent polymerization of sulfur, aliphatic diamines, and TPE-containing diisocyanides at room temperature (Fig. 21a) [116]. The nanoaggregates of the obtained polymer (PTU-TPE) in 50 vol% DMF/water mixtures showed intense fluorescence due to the AIE effect. With the gradual addition of Hg^{2+} from 0 to 10 μM, the fluorescence of the aqueous suspension of PTU-TPE gradually decreased owing to the formation of polythiourea-Hg^{2+} complexes. The quenching constant and detection limit reached 224,900 L/mol and 0.1 ppm, respectively, demonstrating the high sensitivity of mercury detection over a large variety of metal cations (Fig. 21b). Apart from ion sensing, this polymer system can also function as the mercury removal adsorbent due to the poor solubility of the polythiourea-Hg^{2+} complex. After filtration, the removal efficiency reached up to 99.99%, decreasing the amount of Hg^{2+} to 0.8 μg/L below the standard limit of drinking water (Fig. 21c). Because the concentration of Hg^{2+} can be correlated with the fluorescence intensity, the nanoaggregates of PTU-TPE can also be used for real-time monitoring of the removal process of mercury contaminants. Different from this turn-off-type fluorescence sensor, Ruan et al. reported an AIE-active thioketal-decorated conjugated polymer for Hg^{2+} working in a turn-on mode. The fluorescence of the polymer was significantly enhanced upon mixing with a trace amount of Hg^{2+} due to the Hg^{2+}-promoted deprotection reaction of thioketal. The response of this fluorescence turn-on probe was very fast and efficient together with good selectivity towards Hg^{2+} because of the specific chemical reaction [117].

Based on the ion-induced conformation change of polymers and AIE, Fumitaka and coworkers developed a solid-state fluorescent sensor for extracellular

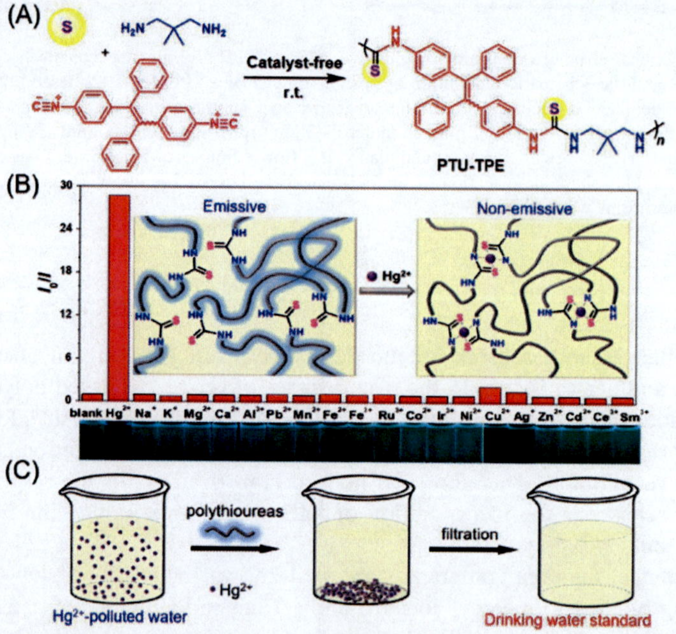

Fig. 21 **a** Synthesis of the AIE-active polythiourea PTU-TPE by catalyst-free multicomponent polymerization. **b** Relative intensity (I_0/I) at 493 nm of PTU-TPE in a DMF/H_2O mixture (v/v, 1/1, 10 μM) in the presence of different metal ions (10 μM) and the corresponding fluorescence photos taken under UV irradiation. I_0 = fluorescence intensity in the absence of metal ions. Inset: the proposed mechanism for the fluorescence detection and removal of mercury ion with PTU-TPE. **c** Schematic diagram of facile mercury removal process. Reproduced from [116] with permission from the American Chemical Society

Ca^{2+} imaging [118]. As shown in Fig. 22a, this solid-state (gel) sensor consists of a chemically cross-linked polyacrylic acid (PAA) block and TPE pendant. When the TPE pendants were attached to PAA, the obtained linear polymers (PAA-TPE) and chemically cross-linked gel (g-PAA-TPE) became highly fluorescent in the presence of Ca^{2+} with excellent selectivity over various interference species such as physiological ions, glucose, and amino acids. The PL intensity was intensified with the increase of Ca^{2+} concentrations. Different from the conventional design strategy of host–guest chemistry using tailored Ca^{2+}-binding sites, the sensing mechanism of this gel sensor relies on polymer chain dynamics triggered by Ca^{2+}. The binding of Ca^{2+} with PAA chains triggered the folding of non-emissive random coils to form aggregates, thereby activating the AIE effect of TPE pendants to turn on the intense fluorescence (Fig. 22b). The g-PAA-TPE gel can be used in various sizes and shapes. For example, a gel sheet fabricated from g-PAA-TPE can visualize the spatial distribution of Ca^{2+} concentrations. As depicted in Fig. 22c, a stamp experiment was conducted by attaching the Ca^{2+}-loaded filter papers on the g-PAA-TPE gel. The filter papers were impregnated in aqueous solutions with different concentrations of Ca^{2+}.

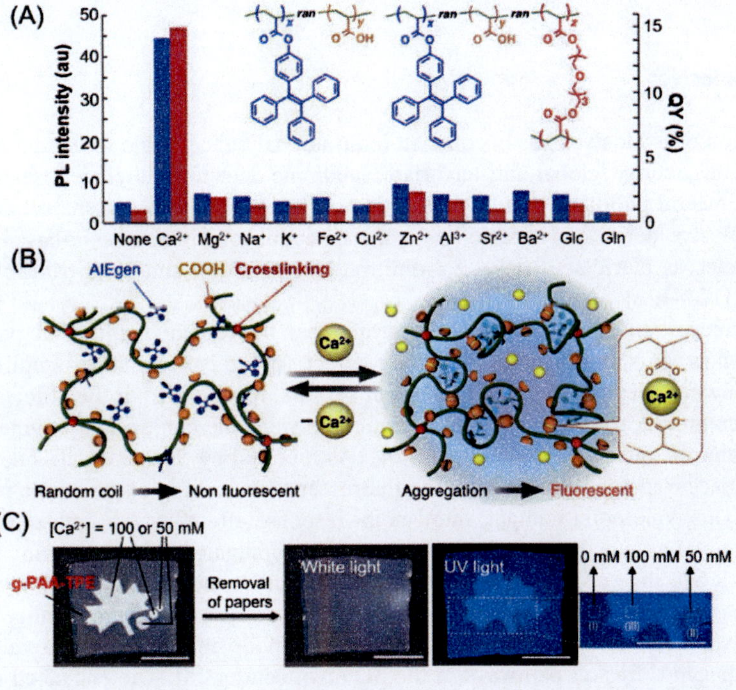

Fig. 22 **a** Fluorescence intensities of PAA-TPE (blue bars) and fluorescence quantum yields of g-PAA-TPE (red bars) in the presence of various metal chlorides, glucose (Glc), and glutamine (Gln). Inset: chemical structures of PAA-TPE and g-PAA-TPE. **b** Schematic illustration of the mechanism of Ca^{2+} sensing with g-PAA-TPE. **c** Stamp experiment using filter papers impregnated with $CaCl_2$ aqueous solution with different concentrations. Scale bars: 1.0 cm. Reproduced from [118] with permission from Springer Nature

After removal of the filter papers, the gel exhibited different fluorescence intensity that can be distinguished by the naked eye under UV irradiation. Considering its tailorable size and shapes as well as high sensitivity and selectivity towards Ca^{2+}, this gel sensor can serve as an excellent Ca^{2+}-imaging fluorescence sensor to benefit the understanding of Ca^{2+}-involved biological and environmental events. Later, Shibayama and coworkers further studied the relationship between the mesoscopic structure of the TPE-appended PAA derivatives (PAA-TPE) and its fluorescence behavior, and elucidated the origin of the Ca^{2+} selectivity on the basis of dynamic and static light-scattering data. By investigating the hydrodynamic radius and molar mass of PAA-TPE in the presence of different metal ions at various concentrations, they concluded that Ca^{2+} can cause polymer chain folding more effectively than other ions to form aggregates with a much higher inner density, thus leading to more pronounced AIE behaviors and better selectivity in Ca^{2+} detection [119].

Apart from the abovementioned examples, clusteroluminescent AIE polymers with sensitive responses to Fe^{3+} have also been reported because the paramagnetic nature of Fe^{3+} can effectively quench the fluorescence of polymer aggregates [120]. However, examples of ion-responsive clusteroluminescent AIE polymers are still very limited. More efforts are needed for the technical investigations of biocompatible AIE polymers for ion detection.

4.4 Explosives

Detection of explosives has become an international concern due to the demands of homeland security, global anti-terrorism, landmine detection, forensic research, and environmental pollution assays. AIE-active polymers have been extensively applied for explosive detection, especially for the detection of nitroaromatic-based explosives, such as picric acid (PA), 2,4-dinitrotoluene (DNT), and 2,4,6-trinitrotoluene (TNT) [121–123].

In contrast to small molecular fluorophores, fluorescent conjugated polymers often show an amplification-quenching effect (or so-called "super-amplification effect") when they are used as fluorescent sensors, resulting from the efficient electronic communication and exciton migration between the conjugated polymer backbones and the quenchers (Fig. 23a) [124]. As reported by Swager et al., one explosive quencher molecule resulted in the fluorescence quenching of the whole polymer chain. This "one-point contact, multi-point response effect" or the "molecular wire effect" could not occur in small molecules but in conjugated polymers owing to their unique long-range conjugated structures [125]. Furthermore, comparing to the linear analogues, hyperbranched polymers often show improved sensing ability toward explosives because their numerous branches provide multiple exciton migration channels and diffusion pathways in the 3D architectures [8]. The enhanced exciton migration as well as the efficient interaction between the explosive analyte and the 3D polymer structure thereby lead to the super-amplification effect of hyperbranched polymers (Fig. 23b) [126].

The super-amplification quenching effect was commonly observed in fluorescence sensors based on hyperbranched polymers, but it is worth noting that

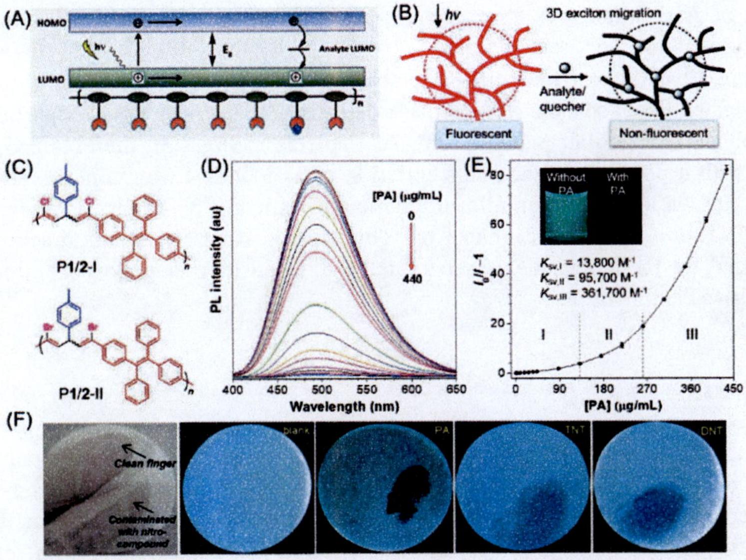

Fig. 23 **a** Schematic diagram illustrating the "the one-point contact, multi-point response effect" of a linear conjugated polymer as an explosive sensor. Reproduced from [124] with permission from the Royal Society of Chemistry. **b** Schematic illustration of the fluorescence-quenching processes of 3D hyperbranched polymers by the analytes. Reproduced from [126] with permission from Springer Nature. **c** Structures of **P1/2**-I and **P1/2**-II. **d** PL spectra of **P1/2**-II in THF/water mixtures with 80% water fraction containing different amounts of picric acid (PA). **e** Stern–Volmer plots of $(I_0/I-1)$ of **P1/2**-II versus the PA concentration, where I_0 = peak intensity without PA. Inset: fluorescent photographs taken under 365-nm UV illumination [127]. Reproduced from [127] with permission from MDPI. **f** Detection of nitro-compound contamination by checking the fluorescent images. Reproduced from [128] with permission from the Royal Society of Chemistry

linear polymers with AIE characteristics have also been reported to show the super-amplification quenching effect in response to explosives. For instance, Tang et al. reported a linear poly(dihalopentadiene) system with AIE activity (**P1/2**-I and **P1/2**-II, Fig. 23c) [127]. The strong fluorescence of their nanoaggregates in aqueous media could be sensitively quenched by PA with a low detection limit (Fig. 23d). The Stern–Volmer plot of the AIE polymer bent upward when a large amount of PA was added (Fig. 23e), demonstrating a remarkable super-amplification quenching effect. This phenomenon can be rationalized by the unique working mechanism of AIE. AIE polymers are used in the form of nanoaggregates, which have more cavities to interact with more quenchers and provide additional inter-chain diffusion pathways for exciton migration. In addition to the electronic effect, physical factors also contribute to the super-amplification quenching phenomenon of AIE systems. For example, the detection of PA was generally conducted in aqueous solutions due to the good water solubility of PA. In this circumstance, water molecules could enter into the polymer networks together with PA molecules to result in the swelling of polymer nanoaggregates. The consequent larger free volume between polymer chains facilitated the free motion of AIE units, thus strengthening the fluorescence quenching of polymer nanoaggregates.

Taking advantage of the solid-state sensing of AIE polymers, Xu and coworkers fabricated paper sensors for explosive detection by spraying the nanoparticles of AIE polymers onto filter papers [128]. As depicted in Fig. 23f, the obtained paper sensor with a polymer surface concentration of as low as 1.0 μg cm^{-2} is effective in detecting various common nitro-aromatic contaminants on fingers, including PA, TNT, DNT, at a scale of less than 1 ng. This type of paper sensor can function as a useful tool for the quick, inexpensive, and highly sensitive detection of nitro-compound-based explosives.

4.5 Others

Beside the different chemical stimuli mentioned above, the detection of organic pollutants, redox reaction, and various biomolecules using AIE-based polymer systems have also been widely reported. These systems are too specific and too numerous to be mentioned individually, so they are discussed in just one category with selected examples in this section.

The sensitive detection of organic pollutants in aqueous media is of great significance for the quality evaluation and safety improvement of water resources. Conventional detection methods for organic pollutants often require time-consuming procedures and the use of bulky and expensive devices. Recently, Tang and Liang et al. described a type of sticky nanopad made of crystallizable fluorescent polymers for the facile, rapid, and sensitive detection of toxic organic pollutants in water [129]. As shown in Fig. 24a, b, TPE was chemically linked to one terminal of polyethylene (PE) chains. The sticky nanopads comprised a single layer of the TPE-modified crystalline PE with TPE units located at the surfaces of the nanopads. Through hydrophobic and π–π interactions, the sticky nanopads were able to absorb organic pollutants in water. The absorbed organic pollutants on the surface of the nanopads subsequently quenched the fluorescence of the AIEgens (Fig. 24c), possibly due to the consumption of the excited-state energy of PE-TPE through collision between the excited chromophore with xylene molecules. The detection limit of PE-TPE nanopads to xylene can reach as low as 7 μg/L, and the response was observed in the order of seconds (Fig. 24d). Later, Liang et al. further developed an amphiphilic fluorescent copolymer micelle system with a hydrophobic TPE core, which could absorb the aromatic pollutants and thereby quench the fluorescence. Based on the capture-report strategy, these types of swellable fluorescent polymer micelles can be used for the rapid detection of aromatic pollutants in water in the order of seconds at a concentration of 1 μg/L [130]. Besides the detection of organic pollutants in aqueous media, functional AIE polymers with fast and sensitive fluorescence responses to organophosphorus pesticide, acids, aliphatic amines, etc. have also been reported [131–133].

Redox-responsive polymer micelles can function as versatile nano-platforms for on-demand drug delivery. The design of redox-responsive polymer materials is often based on the electrochemical processes of redox-sensitive groups or units, such as

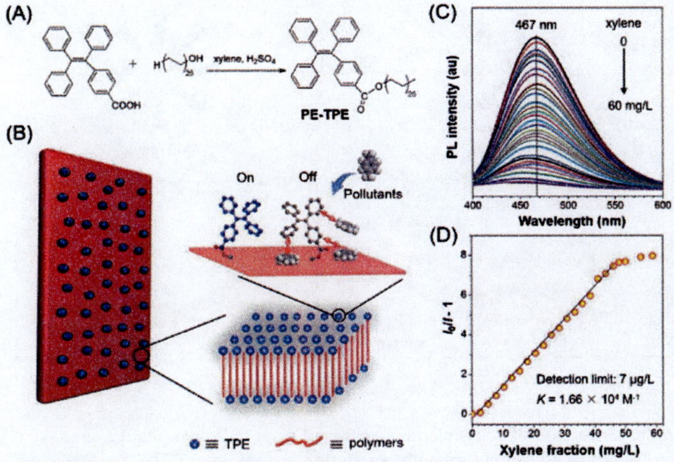

Fig. 24 **a** Synthetic route for polyethylene monoterminated with tetraphenylethene (PE-TPE). **b** Schematic of sticky PE-TPE nanopads for sensing organic pollutants in water. **c** PL spectra of PE-TPE nanopads with various amounts of xylene. **d** Variations in the PL intensity against the xylene concentration. The concentration of PE-TPE was 0.1 mg/mL. Reproduced from [129] with permission from the Royal Society of Chemistry

dithienylethenes, boronate ester, ferrocene, or disulfides, which causes changes in their oxidation states in response to environmental redox variation [134, 135]. However, the in situ evaluation of drug release is challenging. By introducing AIE and FRET effects into the redox-responsive polymer systems, the drug release process from micelles could be well monitored through the variations in fluorescence signal. For example, Zhao and coworkers reported a redox-responsive probe for detection of the intracellular drug release based on the FRET-AIE approach [136]. As illustrated in Fig. 25, poly(ethylene glycol)-polylysine (mPEG-PLys) was selected as the backbone, and the AIE probe (TPE) was chemically linked to the polymer main chain via an amide bond to act as the FRET donor. Meanwhile, curcumin (Cur) was chosen as the model drug and FRET receptor. The ACQ-active Cur was covalently conjugated to the block copolymer backbone via a redox-sensitive disulfide linker. The obtained amphiphilic polymer conjugate could effectively self-assemble into micelles. Upon the supplement of reductants, such as the glutathione (GSH) and tris(2-carboxyethyl)phosphine) (TCEP), the disulfide bond was cleaved to release the Cur drug. This drug release process could turn off the FRET signal and meanwhile turn on the bright AIE fluorescence (Fig. 25). Therefore, both TPE and Cur showed increased fluorescence intensity. This FRET-AIE strategy provided a useful tool for real-time analysis of drug release via stimuli-responsive nano-platforms. However, restricted by the penetration ability of the excitation wavelength, the current system cannot be utilized in situ for in vivo assessment of drug release. Further efforts can be put toward developing redox-responsive polymer systems with NIR absorption and emission properties.

Functional polymer systems with fluorescence response to biomolecules, such as glucose, enzyme, heparin, adenosine triphosphate (ATP), amino acids, proteins,

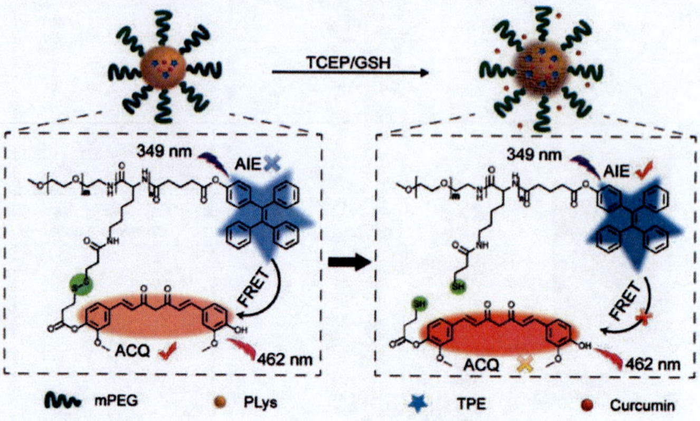

Fig. 25 Schematic illustration of drug release assessment from redox-responsive micelles via integrating FRET with AIE. Reproduced from [136] with permission from Wiley-VCH

and lysosomal HClO, are promising candidates for bioprobes [19, 137]. In this aspect, water-soluble AIE polyelectrolytes have been proven to be powerful biosensors for the sensitive and specific detection of various biogenic small molecules and biomacromolecules. For example, Wang et al. reported an AIE-active cationic poly(diketopyrrolopyrrole-*co*-ethynylfluorene) [138]. This polymer system showed sensitive and selective fluorescence turn-on response to bovine serum albumin (BSA) stimulus based on the BSA-induced deaggregation process of this copolymer. Other biological species bearing negative charge such as alkaline phosphatase, DNA, Con A, lysozyme, ATP, and various anions (Cl^-, Br^-, I^-, NO^{3-}, AcO^-, CO_3^{2-}, NO_3^-, and PO_4^{3-}) have no significant effect on the fluorescence, because they failed to induce the deaggregation of the AIE polymer. Compared with conventional ACQ-type turn-off bioprobes, AIE-based bioprobes generally possess turn-on/light-up responses to biomolecules as well as lower background interference and higher resistance to photobleaching. Therefore, biomolecule-responsive AIE polymer probes are promising for use in continuous monitoring of biological processes with high sensitivities and resolution.

5 Responses to Multiple Stimuli

Intelligent polymer systems that can respond to multiple stimuli are intensively explored in biomimetic design, analytical research, nanotechnology, and biological applications. Compared with the abovementioned single-stimuli-responsive polymer materials, dual- or multi-stimuli-responsive polymers are more intriguing because they can achieve more functions and finer modulations through more parameters. The most commonly used design strategy for developing multi-stimuli-responsive polymers is the combination of two or more stimuli-responsive groups or components into one polymer system [6]. The combinatorial impact of the stimuli-responsive moieties can be independent, serial, or synergistic, depending on the influence of one group's response on the other's.

For example, Kim et al. reported a dual-stimuli-responsive polymer material for chemical reservoir coatings [139]. The AIEgen-loaded microcapsules and commercial pH indicator were physically embedded into the polymer matrix for the detection of cracks and pH variation, respectively. These two stimuli-responsive components work independently in this polymer system. The mechanical response has no influence on the capability of pH response because their detection mechanisms are totally different. Zhang and Hadjichristidis chemically combined the pH-responsive building blocks with the thermo-sensitive AIE unit to construct a pH- and thermo-dual-responsive star copolymer system [140]. As depicted in Fig. 26a, the core cross-linked miktoarm star copolymer $(PE)_n$-$(PMAA)_m$-P(TPE-2St) was synthesized, where the hydrophilic poly(methacrylic acid) (PMAA) arm functions as the pH-responsive block, and the hydrophobic polyethylene (PE) arm was introduced to enable the formation of an AIEgen-containing core–shell structure. The PL intensity of the obtained star copolymers showed a negative correlation with temperature (Fig. 26b, c). The decreased fluorescence at high temperatures was attributed to the swelling of the shell, which activates the motion of TPE units in the core to release the excited-state energy in a non-radiative pathway. The pH-responsiveness of the star copolymers was also related to the core–shell structure. As shown in Fig. 26d, e, the PL intensity of $(PE)_n$-$(PMAA)_m$-P(TPE-2St) aggregates gradually increased when pH increased from 1.1 to 5.0, because the PMAA arms were protonated at $pH < 4.2$, which decreased the solubility of the copolymer to form contracted micelles. In the pH range of 5.0–10.0, the PL intensity changed little. When pH further increased from 10.0 to 13.3, a dramatically decreased PL intensity was observed due to the higher charge density and better hydrophilicity of PMAA in these conditions. The swelled and extended polymer conformations at high pH provide increased space between the TPE moieties by the separation of solvent molecules, thus leading to the decreased fluorescence. In this example, the microstructure of polymers plays a key role in both the temperature- and pH-responsive properties. Therefore, the stimuli-responsive moieties work synergistically in this dual-stimuli-responsive system.

Similar strategies have also been applied for the development of redox- and pH-dual-responsive polymer systems. For example, Wang et al. developed an AIE-active polymeric micelle system based on AIEgen-pended copolymers with both pH- and redox-sensitive units. This copolymer system can be utilized for ultrasensitive pH and redox triggered drug release and bioimaging [141]. Later, Tang et al. physically combined AIE photosensitizers with stimuli-responsive polymers to form multi-stimuli-responsive nanoparticles for photodynamic therapy [142]. As illustrated in Fig. 27, a far red-emissive AIEgen named MeTTMN with high ROS generation efficiency was used as a photosensitizer. A pH-responsive polymer P-Hyd and a redox-responsive polymer P-SS were designed and synthesized with a hydrazine bond and disulfide bond as the stimuli-responsive sites, respectively. Meanwhile, a nonresponsive polymer P-control with hydrophilic poly(ethylene glycol) and hydrophobic polycaprolactone moieties was also synthesized as a control. All the prepared polymers can spontaneously self-assemble into core–shell-structured

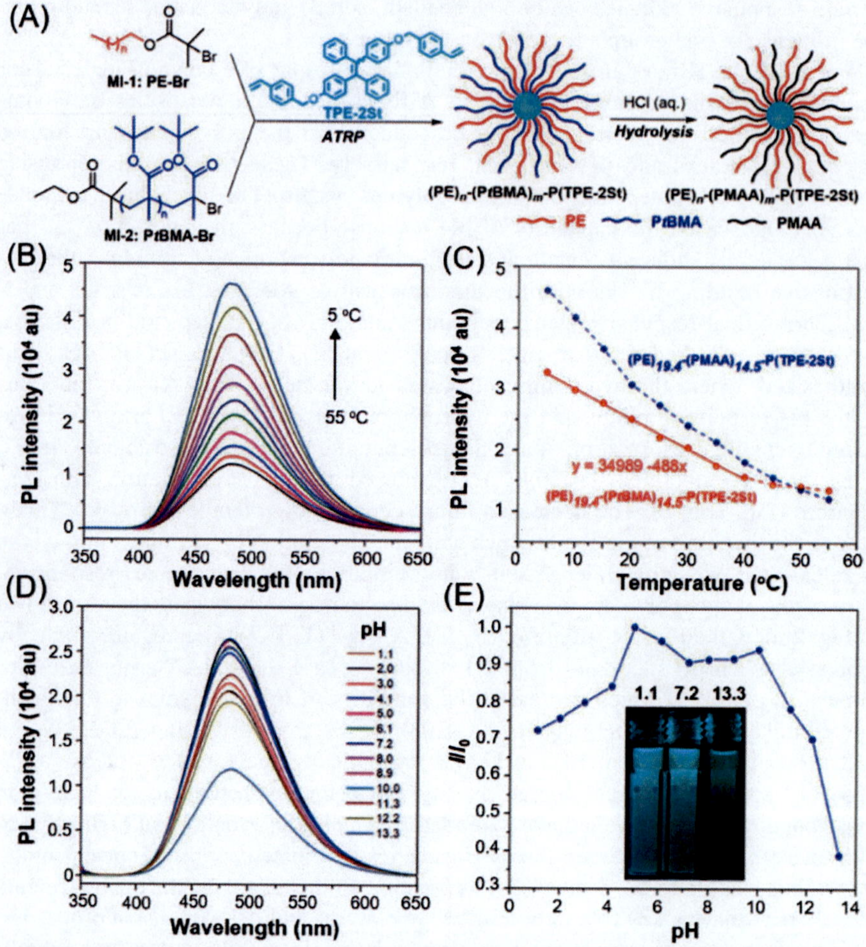

Fig. 26 **a** Synthetic route for the AIE-active core cross-linked multi-miktoarm star copolymers. **b** PL spectra of $(PE)_{19.4}$-$(PMAA)_{14.5}$-P(TPE-2St) in 50/50 vol % THF/*n*-hexane solution at different temperature. **c** Plot of PL intensity vs. temperature. **d** PL spectra of $(PE)_{19.4}$-$(PMAA)_{14.5}$-P(TPE-2St) in THF–buffer mixture (10/90 vol %). **e** Plot of I/I_0 vs. pH value; I_0 is PL intensity at pH 5.0. Reproduced from [140] with permission from the American Chemical Society

nanomicelles in aqueous solution. These nanomicelles with good biocompatibility, excellent stability, and appropriate nanoparticle size can serve as extraordinary carriers for the AIE photosensitizer in high loading efficiency. After efficient uptake of the MeTTMN-loaded nanomicelles by cancer cells, the pH-responsive nanomicelles can release the photosensitizer in a lysosome acid environment due to the cleavage of the hydrazone bond, while the redox-responsive nanomicelles tend to disassemble in cytoplasm of cancer cells to release MeTTMN because of the breakage of the disulfide bond at high concentrations of GSH. The stimuli-responsiveness towards tumor microenvironments can achieve controlled release behaviors. The released

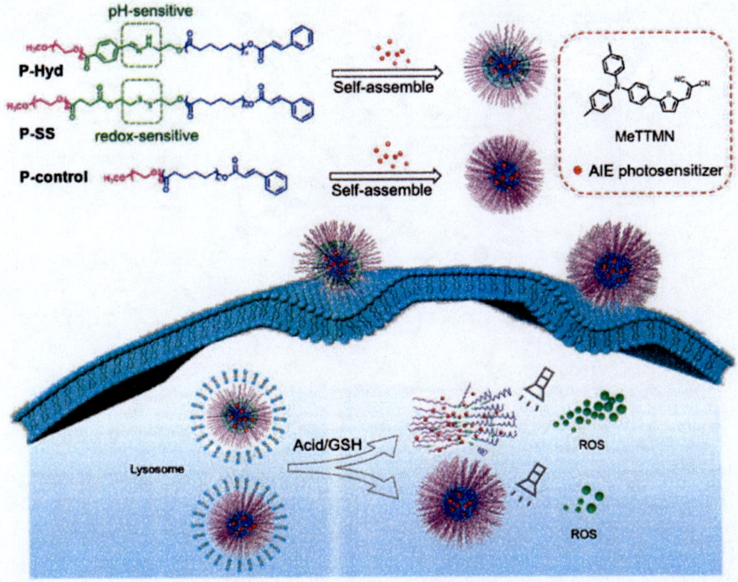

Fig. 27 Illustration of stimuli-responsive AIE polymer nanoparticles with high ROS generation efficiency and boosted PDT effect. Reproduced from [142] with permission from Elsevier

MeTTMN exposed more oxygen than in the core of nanomicelles. Therefore, the MeTTMN-loaded stimuli-responsive assemblies exhibited much higher ROS generation efficiency and significantly promoted PDT performance than those of the P-control nanoassemblies. This example revealed that the addition of another stimuli-responsive assembly can greatly improve or enhance the original response performance to a single stimulus. It can be anticipated that when the stimuli-responsive polymers (P-Hyd and P-SS) were used together to form self-assemblies with the AIE photosensitizers, higher delivery efficiency of the photosensitizer could be achieved, leading to more efficient ROS generation and better photodynamic therapeutic effect.

Some other design strategies have also been reported occasionally. For instance, Hu, et al. took advantage of the high environmental sensitivity of AIEgen and the inherent stimuli-responsiveness of shape-memory polyurethane (PU) to develop a multi-stimuli-responsive chromic polymer [143]. The TPE units (0.1 wt%) were covalently incorporated into the poly(ε-caprolactone)diol soft segments through the polymerization of TPE-diol with other co-monomers. As shown in Fig. 28, the obtained TPE-containing PU films showed remarkable fluorescence change in response to the stimuli of mechanical force, temperature, and organic solvent. The fluorescence of the shape-memory PU obviously decreased when stretched, which was attributed to the deaggregation of TPE units and the change in the film thickness. Upon heating or solvent treatment, the fluorescence of the PU film was restored together with the almost 100% shape recovery of the film. Actually, the mechano-responsiveness of AIE-based polymer systems was often susceptible to

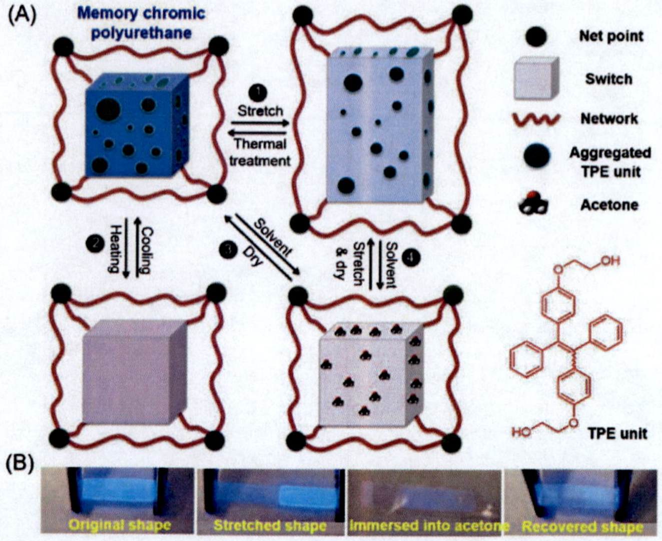

Fig. 28 **a** Schematic illustration of the molecular mechanism during the stretch-recovery process (1), heating–cooling process (2), solvent-dry process (3), and solvent-induced shape recovery process (4) of the TPE-functionalized polyurethane copolymers. **b** The fluorescent images of the polymer film which was first stretched and then immersed into acetone and then removed from solvent and dried at room temperature. Reproduced from [143] with permission from Wiley-VCH

other stimuli, such as solvent vapor and temperature, because the other stimuli can also influence the polymer conformations and the intramolecular motion of AIE-gens, thus affecting the fluorescence signals [44].

6 Summary and Outlook

Thanks to the enthusiastic endeavors of scientists and researchers, a large variety of functional polymer systems with AIE and stimuli responses have been developed in the past decades. In this review, we summarized the recent advances in this field with selected examples. The combination of AIE property and stimuli-responsive ability in one polymeric system can revitalize the long-history study on stimuli-responsive polymers and advance their applications in different areas. Encouraged by the remarkable progress and fruitful achievements, more explorations and break-throughs with numerous possibilities can be made in this field. Future directions and challenges might include the following aspects, among others.

Firstly, the fabrication methods can be further advanced. For example, the development of facile and efficient synthetic strategies is fundamentally important to provide powerful tools for the creation of new stimuli-responsive AIE polymers. By choosing suitable chemical methods, different stimuli receptors and signaling units can be readily integrated in one system to enhance materials performance and enrich the functionalities. For physical blending methods, more efforts can be made on the utilization of melt-processing methods such as the melt-extrusion process as well as high-tech fabrication methods such as 3D printing to advance the industrialization of AIE-based stimuli-responsive polymers. In particular, the application of 3D printing in this field might benefit the study of molecular machines. For instance, our group has developed a light-responsive solid-state molecular actuator based on small molecular AIEgen crystals. Remarkable fluorescence change was observed in the moving parts of the crystal actuators due to the ACQ-AIE-transformation under UV irradiation [144]. This proof-of-concept example indicated that the AIE effect may help to visualize the moving positions and mini-motion or monitor the detailed movement processes of molecular machines. By incorporating the AIEgen-based molecular actuators into soft hydrogel matrix, and with the assistance of 3D or even 4D printing techniques, macroscopic actuators with various sizes and shapes as well as amplified responses might be facilely prepared. Secondly, persistent endeavors are demanded for the development of intelligent AIE polymer systems that can show multiple responses to single or multiple stimuli. In other words, the output signals can also be diverse. The appearance, fluorescence color and/or intensity, macroscopic size and/or shape, and other properties and functions of the desired polymers can change simultaneously upon exposure to external stimuli. Such responsive systems are more in line with the responsive behaviors of living systems in nature. The combination of several recognition moieties, stimuli-responsive groups, and functional agents into one AIE polymeric system will surely broaden their applications. Regarding the choice of AIE polymer systems, clusteroluminescent polymers are worthy of more attention and investigations. These kinds of polymers are inherently responsive to external stimuli such as pressure and temperature based on the CTE mechanism. Further efforts are needed to enhance their AIE performance, including the improvement of solid-state fluorescence intensity and the redshift of emission wavelength. Last but not least, the application forms of stimuli-responsive AIE polymers can be further diversified. For instance, the moisture-sensitive and antibacterial AIE polymer fibers can be further made into smart textiles to benefit our daily life.

It is not possible to list all the examples and address all the possibilities in this review due to the limit of space. We hope this review inspires more scientists to investigate AIE-based stimuli-responsive polymer materials and realize their full potential over the conventional ACQ luminescent systems and conventional stimuli-responsive polymers.

Acknowledgements We acknowledge the funding from the National Natural Science Foundation of China (21905176, 21801169), the Natural Science Foundation for Distinguished Young Scholars of Guangdong Province (2020B1515020011), the Research Grants Council of Hong Kong (16304819, 16305618, and C6009-17G), the Innovation and Technology Commission (ITC-CNERC14SC01), the National Key Research and Development Program of China (2018YFE0190200), and the Science and Technology Plan of Shenzhen (JCYJ20190808142403590, JCYJ20190808153415062).

Compliance with ethical standards

Conflict of interest On behalf of all authors, the corresponding author states that there is no conflict of interest.

References

1. Wei ML, Gao YF, Li X, Serpe MJ (2017) Polym Chem 8:127–143
2. Bauri K, Nandi M, De P (2018) Polym Chem 9:1257–1287
3. Cabane E, Zhang XY, Langowska K, Palivan CG, Meier W (2012) Biointerphases 7
4. Vazquez-Gonzalez M, Willner I (2020) Angew Chem Int Ed 59:15342–15377
5. Stuart MAC, Huck WTS, Genzer J, Muller M, Ober C, Stamm M, Sukhorukov GB, Szleifer I, Tsukruk VV, Urban M, Winnik F, Zauscher S, Luzinov I, Minko S (2010) Nat Mater 9:101–113
6. Schattling P, Jochum FD, Theato P (2014) Polym Chem 5:25–36
7. Smith AE, Xu XW, McCormick CL (2010) Prog Polym Sci 35:45–93
8. Wang JA, Mei J, Qin AJ, Sun JZ, Tang BZ (2010) Sci China Chem 53:2409–2428
9. Li CH, Liu SY (2012) Chem Commun 48:3262–3278
10. Uchiyama S, Gota C, Tsuji T, Inada N (2017) Chem Commun 53:10976–10992
11. Hu JM, Zhang GY, Ge ZS, Liu SY (2014) Prog Polym Sci 39:1096–1143
12. Ellison CJ, Kim SD, Hall DB, Torkelson JM (2002) Eur Phys J E 8:155–166
13. Kim S, Torkelson JM (2011) Macromolecules 44:4546–4553
14. Ellison CJ, Mundra MK, Torkelson JM (2005) Macromolecules 38:1767–1778
15. Ellison CJ, Torkelson JM (2002) J Polym Sci Pol Phys 40:2745–2758
16. Sobakinskaya E, Busch MSA, Renger T (2018) J Phys Chem B 122:54–67
17. Ciruela F (2008) Curr Opin Biotechnol 19:338–343
18. Luo JD, Xie ZL, Lam JWY, Cheng L, Chen HY, Qiu CF, Kwok HS, Zhan XW, Liu YQ, Zhu DB, Tang BZ (2001) Chem Commun 1740–1741.
19. Mei J, Leung NLC, Kwok RTK, Lam JWY, Tang BZ (2015) Chem Rev 115:11718–11940
20. Mei J, Hong YN, Lam JWY, Qin AJ, Tang YH, Tang BZ (2014) Adv Mater 26:5429–5479
21. Leung NLC, Xie N, Yuan WZ, Liu Y, Wu QY, Peng Q, Miao Q, Lam JWY, Tang BZ (2014) Chem Eur J 20:15349–15353
22. Cai XL, Liu B (2020) Angew Chem Int Ed 59:9868–9886
23. Hu R, Qin AJ, Tang BZ (2020) Prog Polym Sci 100:101176
24. Hu R, Leung NLC, Tang BZ (2014) Chem Soc Rev 43:4494–4562
25. Zhan RY, Pan YT, Manghnani PN, Liu B (2017) Macromol Biosci 17:1600433
26. Liu JZ, Zhong YC, Lam JWY, Lu P, Hong YN, Yu Y, Yue YN, Faisal M, Sung HHY, Williams ID, Wong KS, Tang BZ (2010) Macromolecules 43:4921–4936
27. Han T, Gui C, Lam JWY, Jiang MJ, Xie N, Kwok RTK, Tang BZ (2017) Macromolecules 50:5807–5815
28. Chen JR, Zhao J, Xu BJ, Yang ZY, Liu SW, Xu JR, Zhang Y, Wu YC, Lv PY, Chi ZG (2017) Chin J Polym Sci 35:282–292
29. Li Z, Liu PC, Ji XF, Gong JY, Hu YB, Wu WJ, Wang XN, Peng HQ, Kwok RTK, Lam JWY, Lu J, Tang BZ (2020) Adv Mater 32:1906493
30. Taniguchi R, Yamada T, Sada K, Kokado K (2014) Macromolecules 47:6382–6388
31. Wang YJ, Nie JY, Fang W, Yang L, Hu QL, Wang ZK, Sun JZ, Tang B (2020) Chem Rev 120:4534–4577
32. Li B, He T, Shen X, Tang DT, Yin SC (2019) Polym Chem 10:796–818
33. Li J, Wang JX, Li HX, Song N, Wang D, Tang BZ (2020) Chem Soc Rev 49:1144–1172
34. Wang H, Ji XF, Li ZT, Huang FH (2017) Adv Mater 29:1606117
35. Li YY, Liu SJ, Han T, Zhang HK, Chuah C, Kwok RTK, Lam JWY, Tang BZ (2019) Mater Chem Front 3:2207–2220
36. Ma L, Feng X, Wang S, Wang B (2017) Mater Chem Front 1:2474–2486
37. Chen H, Li MH (2019) Chin J Polym Sci 37:352–371
38. Du LL, Jiang BL, Chen XH, Wang YZ, Zou LM, Liu YL, Gong YY, Wei C, Yuan WZ (2019) Chin J Polym Sci 37:409–415
39. Dalapati S, Gu C, Jiang DL (2016) Small 12:6513–6527
40. Feng HT, Yuan YX, Xiong JB, Zheng YS, Tang BZ (2018) Chem Soc Rev 47:7452–7476
41. Yuan WZ, Zhang YM (2017) J Polym Sci Part A: Polym Chem 55:560–574
42. Zhang HK, Zhao Z, McGonigal PR, Ye RQ, Liu SJ, Lam JWY, Kwok RTK, Yuan WZ, Xie JP, Rogach AL, Tang BZ (2020) Mater Today 32:275–292
43. Wang RB, Yuan WZ, Zhu XY (2015) Chin J Polym Sci 33:680–687
44. Han T, Liu LJ, Wang D, Yang JL, Tang BZ (2020) Macromol Rapid Commun 22:2000311

45. Chen SJ, Wang H, Hong YN, Tang BZ (2016) Mater Horizons 3:283–293
46. Battistelli G, Cantelli A, Guidetti G, Manzi J, Montalti M (2016) Wiley Interdiscip Rev-Nanomed Nanobiotechnol 8:139–150
47. Feng GX, Liu B (2018) Acc Chem Res 51:1404–1414
48. Calvino C, Weder C (2018) Small 14:1802489
49. Zheng HY, Li CY, He CC, Dong YQ, Liu QS, Qin PF, Zeng C, Wang HL (2014) J Mater Chem C 2:5829–5835
50. Ye RQ, Liu YY, Zhang HK, Su HF, Zhang YL, Xu LG, Hu RR, Kwok RTK, Wong KS, Lam JWY, Goddard WA, Tang BZ (2017) Polym Chem 8:1722–1727
51. Yoon SJ, Chung JW, Gierschner J, Kim KS, Choi MG, Kim D, Park SY (2010) J Am Chem Soc 132:13675–13683
52. Robb MJ, Li W, Gergely RCR, Matthews CC, White SR, Sottos NR, Moore JS (2016) ACS Cent Sci 2:598–603
53. Lu XC, Li WL, Sottos NR, Moore JS (2018) ACS Appl Mater Interfaces 10:40361–40365
54. Chen SS, Han T, Zhao Y, Luo WJ, Zhang Z, Su HB, Tang BZ, Yang JL (2020) ACS Appl Mater Interfaces 12:4870–4877
55. Song YK, Lee TH, Lee KC, Choi MH, Kim JC, Lee SH, Noh SM, Park YI (2020) Appl Surf Sci 511:145556
56. Doberenz F, Zeng K, Willems C, Zhang K, Groth T (2020) J Mater Chem B 8:607–628
57. Tang L, Jin JK, Qin AJ, Yuan WZ, Mao Y, Mei J, Sun JZ, Tang BZ (2009) Chem Commun 4974–4976.
58. Yang JF, Gu KZ, Shi CX, Li M, Zhao P, Zhu WH (2019) Mater Chem Front 3:1503–1509
59. Li TZ, He SC, Qu JN, Wu H, Wu SZ, Zhao ZJ, Qin AJ, Hu RR, Tang B (2016) J Mater Chem C 4:2964–2970
60. Xu JR, Ji WX, Li C, Lv YF, Qiu ZJ, Gao LC, Chen EQ, Lam JWY, Tang BZ, Jiang L (2018) Adv Opt Mater 6:1701149
61. Liu L, Wang M, Guo LX, Sun Y, Zhang XQ, Lin BP, Yang H (2018) Macromolecules 51:4516–4524
62. Qiu ZJ, Chu EKK, Jiang MJ, Gui C, Xie N, Qin W, Alam P, Kwok RTK, Lam JWY, Tang BZ (2017) Macromolecules 50:7620–7627
63. Yang Y, Zhang S, Zhang XQ, Gao LC, Wei Y, Ji Y (2019) Nat Commun 10:3165
64. Song ZH, Lv XL, Gao LC, Jiang L (2018) J Mater Chem C 6:171–177
65. Bao SP, Wu QH, Qin W, Yu QL, Wang J, Liang GD, Tang BZ (2015) Polym Chem 6:3537–3542
66. Echeverri M, Ruiz C, Gamez-Valenzuela S, Alonso-Navarro M, Gutierrez-Puebla E, Serrano JL, Delgado MCR, Gomez-Lor B (2020) ACS Appl Mater Interfaces 12:10929–10937
67. Lai CT, Chien RH, Kuo SW, Hong JL (2011) Macromolecules 44:6546–6556
68. Saha B, Ruidas B, Mete S, Das Mukhopadhyay C, Bauri K, De P (2020) Chem Sci 11:141–147
69. Zhang Z, Bilalis P, Zhang HF, Gnanou Y, Hadjichristidis N (2017) Macromolecules 50:4217–4226
70. Luo WH, Wang GJ (2020) Adv Opt Mater 2001362.
71. Xie NH, Chen Y, Ye H, Li C, Zhu MQ (2018) Front Optoelectron 11:317–332
72. Singh R, Wu HY, Dwivedi AK, Singh A, Lin CM, Raghunath P, Lin MC, Wu TK, Wei KH, Lin HC (2017) J Mater Chem C 5:9952–9962
73. Liu GX, Zhang YM, Zhang L, Wang CH, Liu Y (2018) ACS Appl Mater Interfaces 10:12135–12140
74. Liu SJ, Cheng YH, Li YY, Chen M, Lam JWY, Tang BZ (2020) ACS Nano 14:2090–2098
75. Han T, Yao ZS, Qiu ZJ, Zhao Z, Wu KY, Wang JG, Poon AW, Lam JWY, Tang BZ (2019) Nat Commun 10:5483
76. Zhu JC, Han T, Guo Y, Wang P, Xie HL, Meng ZG, Yu ZQ, Tang BZ (2019) Macromolecules 52:3668–3679
77. Han T, Deng HQ, Qiu ZJ, Zhao Z, Zhang HK, Zou H, Leung NLC, Shan GG, Elsegood MRJ, Lam JWY, Tang B (2018) J Am Chem Soc 140:5588–5598
78. Chen S, Ma TJ, Bai J, Ma XD, Yin J, Jiang XS (2020) Adv Sci 7:2002372
79. Wu WB (2018) Chem 4:1762–1764
80. Zhao Z, Zhang HK, Lam JWY, Tang BZ (2020) Angew Chem Int Ed 59:9888–9907
81. Liu XL, Li MG, Han T, Cao B, Qiu ZJ, Li YY, Li QY, Hu YB, Liu ZY, Lam JWY, Hu XL, Tang BZ (2019) J Am Chem Soc 141:11259–11268
82. Liu WW, Yu HK, Hu RR, Xu T, Lun YP, Gan JL, Xu SH, Yang ZM, Tang BZ (2020) Small 16:1907074

83. Nakasha K, Fukuhara G (2020) ACS Appl Polym Mater 2:2303–2310
84. Sun NW, Su KX, Zhou ZW, Tian XZ, Wang DM, Vilbrandt N, Fery A, Lissel F, Zhao XG, Chen CH (2019) J Mater Chem C 7:9308–9315
85. Sun NW, Su KX, Zhou ZW, Yu Y, Tian XZ, Wang DM, Zhao XG, Zhou HW, Chen CH (2018) ACS Appl Mater Interfaces 10:16105–16112
86. Sun NW, Su KX, Zhou ZW, Tian XZ, Zhao JH, Chao DM, Wang DM, Lissel F, Zhao XG, Chen CH (2019) Macromolecules 52:5131–5139
87. Cheng SW, Han T, Huang TY, Tang BZ, Liou GS (2018) Polym Chem 9:4364–4373
88. Zhang XD, Chen XK, Jiang YW, Ma NN, Xia LY, Cheng XT, Jia HR, Liu PD, Gu N, Chen Z, Wu FG (2018) ACS Appl Mater Interfaces 10:10601–10606
89. Yang CH, Ni X, Mao D, Ren CH, Liu JJ, Gao Y, Ding D, Liu JF (2019) Biomaterials 188:107–117
90. Liu ZT, Xue WX, Cai ZX, Zhang GX, Zhang DQ (2011) J Mater Chem 21:14487–14491
91. Cao W, Zhang XL, Miao XM, Yang ZM, Xu HP (2013) Angew Chem Int Ed 52:6233–6237
92. Ma N, Xu HP, An LP, Li J, Sun ZW, Zhang X (2011) Langmuir 27:5874–5878
93. Li TY, Pan SJ, Zhuang H, Gao SQ, Xu HP (2020) ACS Appl Bio Mater 3:1283–1292
94. Iasilli G, Battisti A, Tantussi F, Fuso F, Allegrini M, Ruggeri G, Pucci A (2014) Macromol Chem Phys 215:499–506
95. Cheng YH, Liu SJ, Song FY, Khorloo M, Zhang HK, Kwok RTK, Lam JWY, He ZK, Tang BZ (2019) Mater Horiz 6:405–411
96. Sun JM, Wang JG, Chen M, Pu X, Wang G, Li L, Chen GY, Cai YJ, Gu XG, Tang BZ (2019) Chem Mater 31:5683–5690
97. Wang GJ, Zhang RC, Xu C, Zhou RY, Dong J, Bai HT, Zhan XW (2014) ACS Appl Mater Interfaces 6:11136–11141
98. Tavakoli J, Gascooke J, Xie N, Tang B, Tang YH (2019) ACS Appl Polym Mater 1:1390–1398
99. Khorloo M, Cheng YH, Zhang HK, Chen M, Sung HHY, Williams ID, Lam JWY, Tang BZ (2020) Chem Sci 11:997–1005
100. Yao W, Tebyetekerwa M, Bian XH, Li WL, Yang SY, Zhu MF, Hu R, Wang ZM, Qin AJ, Tang BZ (2018) J Mater Chem C 6:12849–12857
101. Jiang XY, Gao HF, Zhang XQ, Pang JH, Li YQ, Li K, Wu YC, Li SZ, Zhu J, Wei Y, Jiang L (2018) Nat Commun 9:3799
102. Tao L, Li ML, Yang KP, Guan Y, Wang P, Shen ZH, Xie HL (2019) ACS Appl Mater Interfaces 11:15051–15059
103. Huang HC, Qiu ZJ, Han T, Kwok RTK, Lam JWY, Tang BZ (2017) ACS Macro Lett 6:1352–1356
104. Cheng YH, Wang JG, Qiu ZJ, Zheng XY, Leung NLC, Lam JWY, Tang BZ (2017) Adv Mater 29:1703900
105. Lu W, Xiao P, Gu JC, Zhang JW, Huang YJ, Huang Q, Chen T (2016) Sens Actuators B Chem 228:551–556
106. Zhang DP, Fan YJ, Chen H, Trepout S, Li MH (2019) Angew Chem Int Ed 58:10260–10265
107. Qiu L, Zhang HR, Wang B, Zhan Y, Xing CF, Pan CY (2020) ACS Appl Mater Interfaces 12:1348–1358
108. Hu YB, Han T, Yan N, Liu JK, Liu XL, Wang WX, Lam JWY, Tang BZ (2019) Adv Funct Mater 29:1902240
109. Chen YJ, Han HJ, Tong HX, Chen TT, Wang HB, Ji J, Jin Q (2016) ACS Appl Mater Interfaces 8:21185–21192
110. Qi QK, Li Y, Yan XY, Zhang FL, Jiang S, Su J, Xu B, Fu XQ, Sun LK, Tian WJ (2016) Polym Chem 7:5273–5280
111. Zhao YM, Zhu W, Ren LX, Zhang K (2016) Polym Chem 7:5386–5395
112. He N, Chen ZJ, Yuan J, Zhao L, Chen MH, Wang T, Li XH (2017) ACS Appl Mater Interfaces 9:32534–32544
113. Ma HH, Zhang AD, Zhang XM, Zhao HT, Cui Z, Fu P, Liu MY, Zhou SZ, Pang XC, Qiao XG (2020) Sensor Actuators B Chem 304
114. Huang W, Bender M, Seehafer K, Wacker I, Schroder RR, Bunz UHF (2019) Macromol Rapid Commun 40:1800774
115. Dong WH, Wu HQ, Chen M, Shi Y, Sun JZ, Qin AJ, Tang BZ (2016) Polym Chem 7:5835–5839
116. Tian T, Hu RR, Tang BZ (201a8) J Am Chem Soc 140:6156–6163
117. Shan YR, Yao WJ, Liang ZQ, Zhu LH, Yang SB, Ruan ZJ (2018) Dyes Pigm 156:1–7
118. Ishiwari F, Hasebe H, Matsumura S, Hajjaj F, Horii-Hayashi N, Nishi M, Someya T, Fukushima T (2016) Sci Rep 6:24275

119. Morishima K, Ishiwari F, Matsumura S, Fukushima T, Shibayama M (2017) Macromolecules 50:5940–5945
120. Wang YZ, Bin X, Chen XH, Zheng SY, Zhang YM, Yuan WZ (2018) Macromol Rapid Commun 39:1800528
121. Gao MX, Wu Y, Chen B, He BR, Nie H, Li TY, Wu FP, Zhou WJ, Zhou J, Zhao ZJ (2015) Polym Chem 6:7641–7645
122. He BZ, Su HF, Bai TW, Wu YW, Li SW, Gao M, Hu RR, Zhao ZJ, Qin AJ, Ling J, Tang BZ (2017) J Am Chem Soc 139:5437–5443
123. Chen T, Yin H, Chen ZQ, Zhang GF, Xie NH, Li C, Gong WL, Tang BZ, Zhu MQ (2016) Small 12:6547–6552
124. Zhou H, Chua MH, Tang BZ, Xu JW (2019) Polym Chem 10:3822–3840
125. Ryu SH, Lee DH, Ko YJ, Lee SM, Kim HJ, Ko KC, Son SU (2019) Macromol Chem Phys 220:1900157
126. Wu YW, Qin AJ, Tang BZ (2017) Chin J Polym Sci 35:141–154
127. Han T, Zhang Y, He BZ, Lam JWY, Tang BZ (2018) Polymers 10:821
128. Zhou H, Wang XB, Lin TT, Song J, Tang BZ, Xu JW (2016) Polym Chem 7:6309–6317
129. Liang GD, Ren F, Gao HY, Zhu FM, Wu Q, Tang BZ (2017) J Mater Chem A 5:2115–2122
130. Zhou YS, Zhang L, Gao HY, Zhu FM, Ge ML, Liang GD (2019) Sens Actuators B Chem 283:415–425
131. Qi CX, Zheng C, Hu RR, Tang BZ (2019) ACS Macro Lett 8:569–575
132. Zhou YS, Gao HY, Zhu FM, Ge ML, Liang GD (2019) J Hazard Mater 368:630–637
133. Chen JL, Chen XJ, Huang QY, Li WL, Yu QX, Zhu LJ, Zhu TW, Liu SW, Chi ZG (2019) ACS Appl Mater Interfaces 11:32689–32696
134. Hirose A, Tanaka K, Yoshii R, Chujo Y (2015) Polym Chem 6:5590–5595
135. Zhang GY, Liao QB, Liu YF, Wang L, Gou HL, Ke C, Huang X, Xi K, Jia XD (2018) Nanoscale 10:5503–5514
136. Wang XL, Li JJ, Yan Q, Chen YR, Fan AP, Wang Z, Zhao YJ (2018) Macromol Biosci 18:1700339
137. Chen C, Ni X, Jia SR, Liang Y, Wu XL, Kong DL, Ding D (2019) Adv Mater 31:1904914
138. Wang LY, Yang LL, Zhu LH, Cao DR, Li L (2016) Sens Actuators B Chem 231:733–743
139. Lee TH, Song YK, Park SH, Park YI, Noh SM, Kim JC (2018) Appl Surf Sci 434:1327–1335
140. Zhang Z, Hadjichristidis N (2018) ACS Macro Lett 7:886–891
141. Zhuang WH, Xu YY, Li GC, Hu J, Ma BX, Yu T, Su X, Wang YB (2018) ACS Appl Mater Interfaces 10:18489–18498
142. Li YM, Wu Q, Kang MM, Song N, Wang D, Tang BZ (2020) Biomaterials 232:119749
143. Wu Y, Hu JL, Huang HH, Li J, Zhu Y, Tang BZ, Han JP, Li LB (2014) J Polym Sci Pol Phys 52:104–110
144. Wang HR, Xing H, Gong JY, Zhang HK, Zhang J, Wei PF, Yang GJ, Lam JWY, Lu R, Tang BZ (2020) Mater Horiz 7:1566–1572

Enzyme-Responsive Peptide-Based AIE Bioprobes

Juliang Yang[1] · Jiaming Wei[1] · Fan Luo[2] · Jun Dai[3] · Jing-Jing Hu[1] · Xiaoding Lou[1] · Fan Xia[1]

Received: 24 June 2020 / Accepted: 18 September 2020 / Published online: 7 October 2020

Abstract
Enzyme, which exists widely in organisms, has high specificity and high catalytic efficiency for its substrates. The absence, the reduced activity, or the overexpression of enzyme are closely related to the occurrence and development of diseases. Therefore, enzyme is often used as markers for disease detection and treatment. To detect enzyme activity and track drug release, aggregation-induced emission (AIE) bioprobes have been developed because of their excellent photostability and high signal-to-noise ratio (SNR). Among them, peptide-based AIE bioprobes with great biocompatibility and specificity are favored by an increasing number of researchers. Enzymatic hydrolysis of peptide can cause aggregation of AIE molecules and drug release. In this review, enzyme-responsive peptide-based AIE bioprobes used for biomedical application are summarized according to the three aggregation strategies triggered by various reaction between peptide and enzyme, including enzyme-triggered precipitate, enzyme-catalyzed coupling, and enzyme-instructed self-assembly. By giving some representative examples, we discuss how each aggregation strategy detects enzyme activity and treats the diseases under imaging guidance. Finally, we comment on the current problems and future prospects of enzyme-responsive peptide-based AIE bioprobes.

Chapter 10 was originally published as Yang, J., Wei, J., Luo, F., Dai, J., Hu, J-J, Lou, X. & Xia, F. Topics in Current Chemistry (2020) 378: 47. https://doi.org/10.1007/s41061-020-00311-9.

✉ Jing-Jing Hu
hujingjing@cug.edu.cn

✉ Xiaoding Lou
louxiaoding@cug.edu.cn

[1] Engineering Research Center of Nano-Geomaterials of Ministry of Education, Faculty of Materials Science and Chemistry, China University of Geosciences, Wuhan 430074, China

[2] China National Tobacco Corporation Hunan Branch Logistics Management Office, Changsha 410004, China

[3] Department of Obstetrics and Gynecology, Tongji Hospital, Tongji Medical College, Huazhong University of Science and Technology, Wuhan 430074, China

Graphic Abstract

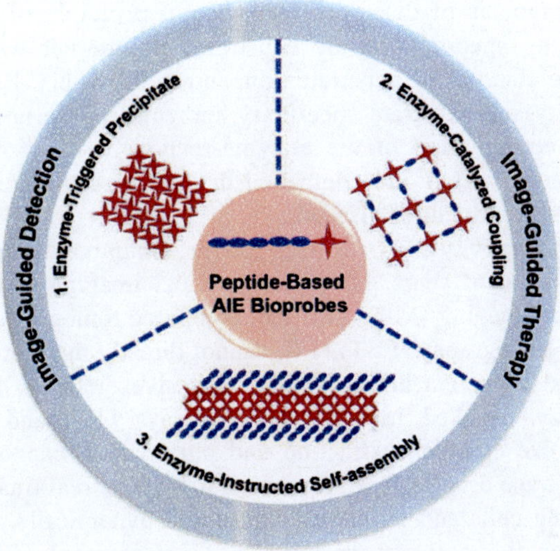

Keywords Peptide · AIE bioprobes · Enzyme · Precipitate · Coupling · Self-assembly

1 Introduction

The enzyme is a class of biocatalysts with protein structure produced from bacteria, molds, fungi, yeasts, and plants that has a high specificity and catalytic efficiency for its substrates [1]. Although there are about 25,000 enzymes in nature, only approximately 4000 are recognized by The International Union of Biochemistry and Molecular Biology. They dominate many catalytic processes, such as the metabolism, nutrition, and energy conversion of organisms. Owing to the wide existence of enzymes, chemical reactions in organisms could proceed efficiently and specifically under mild conditions, as well as in an orderly manner. In turn, the absence, reduced activity, or overexpression of enzyme would induce an abnormal catalyzed reaction, thus resulting in metabolic disorder and even the occurrence and development of the diseases. According to the characteristic of different reactions catalyzed by the enzyme, they are classified into six groups: hydrolases, oxidoreductases, isomerases, transferases, lyases, and ligases [2]. For example, matrix metalloproteinases (MMPs) play an important role in the development of diseases, such as tissue repair and reconstruction, arthritis, and cancer [3, 4]. The overexpression of urokinase-type plasminogen activator (uPA) is closely related to angiogenesis, tumor invasion, and metastasis [5]. Caspases are intracellular cysteine proteases that regulate programmed cell death and apoptosis [6, 7]. Because of the dynamic change, the expression and composition of different enzymes are different in different cells and tissues. Therefore, the precise detection of enzyme activity and enzyme localization is crucial for the diagnosis of diseases. The general principle of enzyme activity analysis is utilizing specific enzyme substrates to interact with enzyme, and then detecting the changes of substrate concentration [8–11]. Besides, owing to the high selectivity and substrate specificity, enzyme can be used as an internal stimulus for the construction of the enzyme-responsive drug delivery systems [12, 13]. Achieving targeted and controlled drug release via enzyme is of great significance for the treatment of diseases.

To detect enzyme activity and track drug release, a number of imaging techniques have been used in recent years. However, classical imaging techniques, such as magnetic resonance imaging (MRI) [14, 15], computed tomography (CT) [16], and positron emission tomography (PET) [17], cannot directly monitor many biological phenomena in real time. Furthermore, other invasive methods, like colorimetric method [18], enzyme-linked immunosorbent assays [19], and electrochemical method [20], involve complex extraction and purification, cannot detect enzyme activity in situ, or trace drug release. Therefore, to achieve real-time in-situ imaging of enzyme in living cells and monitor drug release dynamically, researchers have paid more attention to fluorescent probes [21–23]. Commonly, fluorescent probes emit bright fluorescence at low concentrations, but at high concentrations or in solid states, the fluorescence signal is quenched. The aggregation-caused quenching (ACQ) effect would reduce the detection sensitivity of fluorescent probes greatly. Fortunately, in 2001, Tang and co-workers first came up with the concept of aggregation-induced emission (AIE) [24]. The AIE molecules have little emission in solution, but in a solid state or aggregation state the AIE molecules could emit bright fluorescence due to the restriction of intramolecular motion [25–27]. Over the past two decades, the novel AIE molecules have gained wide application space and show great vitality [28–34]. Inspired by the unique luminescent properties of AIE molecules, researchers have developed a series of AIE bioprobes for enzyme detection and treatment [35–41]. Among them, peptide-modified AIE bioprobes have attracted numerous interests owing to obvious advantages of peptide, such as inherent biological activity, excellent biodegradability, good biocompatibility, and abundant diversity [41–49]. Specifically, the introduction of peptide could improve the hydrophilicity of AIE molecules, which is beneficial to the application of bioprobes in the field of biomedicine [50–55]. Moreover, after the peptide is specifically cleaved by enzyme, the fluorescence signal change caused by the aggregation of AIE bioprobes can be used to detect the activity of the enzyme, and the released drug can be used to treat diseases.

In this review, recent research advancements in enzyme-responsive peptide-based AIE bioprobes for detection and treatment were summarized. According to the different aggregation states after reaction with enzyme, the AIE bioprobes based on peptide were classified into three categories: enzyme-triggered precipitate, enzyme-catalyzed coupling, and enzyme-instructed self-assembly (Fig. 1). In each section, the aggregation processes and representative examples for detecting enzyme activity or treating the diseases under imaging guidance were introduced. The current problems and future prospects of enzyme-responsive peptide-based AIE bioprobes

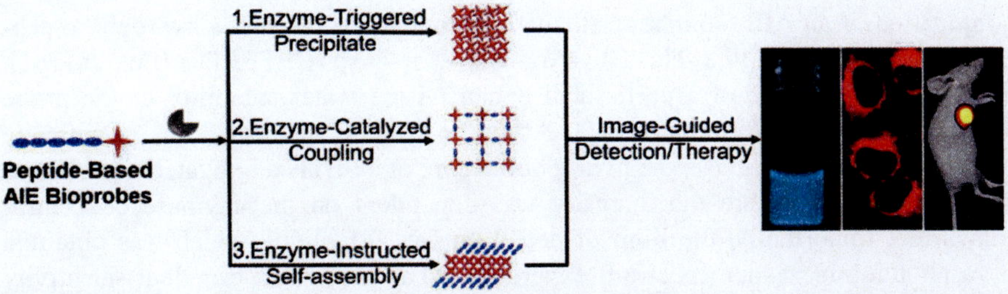

Fig. 1 Three aggregation strategies and applications of enzyme-responsive peptide-based AIE bioprobes

were also discussed. We hope this review can inspire creative ideas and encourage continuing developments in this promising area.

2 Enzyme-Triggered Precipitate

Enzymatic cleavage of AIE-peptide conjugates to induce AIE molecules precipitate is the most common strategy for peptide-based AIE bioprobes. Briefly, the covalent binding of hydrophilic peptide and the AIE molecules make the bioprobes in a dispersed state with non-fluorescence in aqueous solution. When the hydrophilic peptide substrate is hydrolyzed by the enzyme, the residual hydrophobic AIE part precipitates and aggregates rapidly, resulting in the emission of bright fluorescence. Thus, the fluorescence signal change could closely relate to the activity and location of the enzyme. This strategy could not only promote the biocompatibility and specificity of the bioprobes but also improve the in situ and real-time detection capabilities of the bioprobes.

What's more, by further introducing therapeutic agents, this strategy can also be used for imaging-guided therapy. When the enzyme activates the drug delivery systems, the release of the drug is accompanied by the output of the fluorescent signal of the bioprobes. As a result, the enzyme can control the drug release and the AIE molecules can track drug delivery in real time. In this section, we will summarize the recent application of this strategy in imaging-guided detection and imaging-guided therapy, as well as analyze its potential application value.

2.1 Image-Guided Detection

The occurrence of diseases often leads to the abnormality of biological indicators of liquid samples. Compared to the direct detection of relatively inaccessible organs through complicated operations, detecting the relevant liquid samples is easier and more convenient to understand in the occurrence of diseases [56–59]. For example, human chymase is abundant in mast cell granules. The increase of chymase and some serine proteases is very much related to inflammation and immune regulatory functions [60]. For the quantitative detection of chymase, Liu's group reported a specific light-up probe TPETH-2(CFTERD$_3$), which was developed by covalent conjugation of an AIE tetraphenylethenethiophene (TPETH) and a hydrophilic peptide sequence of Cys–Phe–Thr–Glu–Arg–Asp–Asp–Asp ($CFTERD_3$) (Fig. 2a) [61]. The peptide sequence of aspartic acid enhanced the water solubility of the probe and reduced the background signal. After the CFTER sequence was specifically hydrolyzed by the chymase, the hydrophobic core of TPETH aggregated and yielded red fluorescence, where the intensity was dependent on the chymase concentration. More importantly, the limit of detection was 0.1 ng/ml, which was obtained by a photoluminescence spectrum experiment. The probe had excellent selectivity for chymase by incubating with chymase or other proteins such as β-glucuronidase, lysozyme, cathepsin B, BSA, and pepsin. As a result, this probe showed excellent sensitivity and selectivity and could achieve real-time and direct readout. Lipase in serum is an enzyme secreted by the pancreas and has been adopted as a biomarker

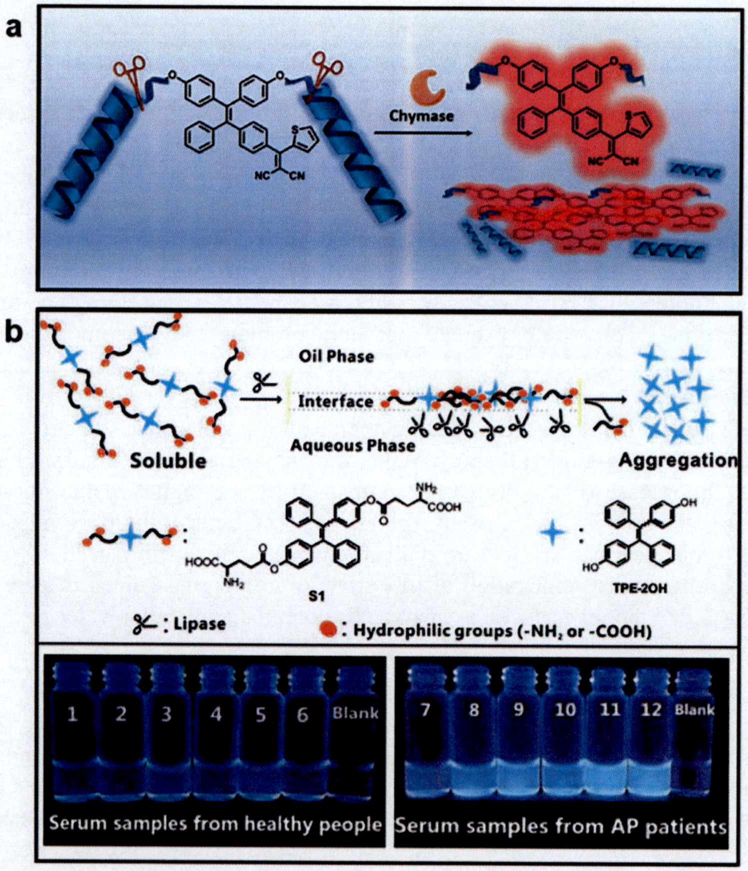

Fig. 2 **a** Sensing mechanism of the probe TPETH-2($CFTERD_3$) in chymase detection (adapted with permission from [61]). **b** Sensing mechanism and chemical structure of probe S1 in lipase detection (*top*); and visual fluorescence of S1 in real human serum samples under ultraviolet light illumination at 365 nm (*bottom*) (adapted with permission from [64])

of acute pancreatitis as well as pancreatic cancer [62, 63]. Its significant elevation always indicates pancreatic injury. To develop robust, convenient, and sensitive fluorescent probes for lipase detection, Tang and coworkers combined the heterogeneous catalytic property of lipase and synthesized a bioprobe S1 [64], which was composed of a tetraphenylethylene (TPE) core and two glutamate units (Fig. 2b). The hydrophobic TPE core of S1 could insert into the oil layer and the hydrophilic glutamine unit of S1 could insert into the water layer, which ensured that S1 was distributed at the oil–water interface and the glutamine unit was in full contact with the lipase in the water phase. When S1 was hydrolyzed by lipase to remove the ester group, the TPE residues were precipitated and emitted bright blue fluorescence. Experimental results demonstrated that due to the presence of interfacial control and the lipase specific digestion of glutamate, S1 could exhibit high sensitivity, low detection limit (0.13 U/l), as well as short response time (7 min), and be utilized to detect the lipase levels in real human serum samples.

The cell is the basic structure and function unit of living organisms. The occurrence of most diseases is related to the change of cell function, so the real-time imaging of enzyme activity in living cells is of great significance [65, 66]. As an example, furin is one of the members of the proprotein convertase family and is mainly located in *trans*-Golgi. Its overexpression is highly related to tumor growth and metastasis [67]. Thus, designing techniques to accomplish in situ detection and imaging of furin is very significant. The peptide sequences Arg–X–Lys/Arg–Arg↓X (X can be any amino acid residue and ↓ shows the cleavage site) were reported as a preferred substrate for furin cleavage [68, 69]. Based on these, Zhang et al. developed a probe HPQF (Fig. 3a) [70]. HPQF consisted of an insoluble fluorescent molecule 2-(2′-hydroxyphenyl)-4(3*H*)-quinazolinone (HPQ) and a peptide sequence RVRR that could be cleaved by furin. HPQF was water soluble and non-fluorescent, but when it was hydrolyzed by the furin, the released Cl-HPQ immediately precipitated and fluoresced brightly. Therefore, HPQF could monitor the activity and localization of furin in living cells in real time and in situ. Moreover, the application of HPQF to monitor the level change of furin in MDA-MB-468 tumor tissues was also proved. In addition to detecting the activity of enzyme in the cell membrane and cytoplasm, tracing enzyme in the nucleus is also important for studying the relationship between the nucleus and diseases. For example, sirtuin type 1 (SIRT1) is a nicotinamide adenine dinucleotide $(NAD)^+$-dependent deacetylase [71]. It is mainly located in the nucleus and participates in a series of physiological and biochemical processes, such as cell proliferation, apoptosis, tumor growth, and metabolism [72]. Wang et al. designed a probe to detect SIRT1 activity in living cells [73]. As shown in Fig. 3b, the probe was comprised of a hydrophobic TPE and a hydrophilic Gly-acetyl-Lys–Tyr–Asp–Asp (G-Ac-KYDD) peptide sequence. When the acetyl group of lysine was deacetylated by SIRT1, KYDD was cut off under the action of lysyl endopeptidase. Finally, the remaining TPE residue was precipitated to induce the light-up fluorescence. Compared with precedent immunofluorescence ways of cell imaging of SIRT1 that required laborious fixing and PBS washing steps, the probe was successfully used to detect and in situ image SIRT1 in the nucleus.

Apoptosis is a strictly controlled active process, which controls cell proliferation and regulates collective homeostasis. In contrast, cancer occurs when uncontrolled

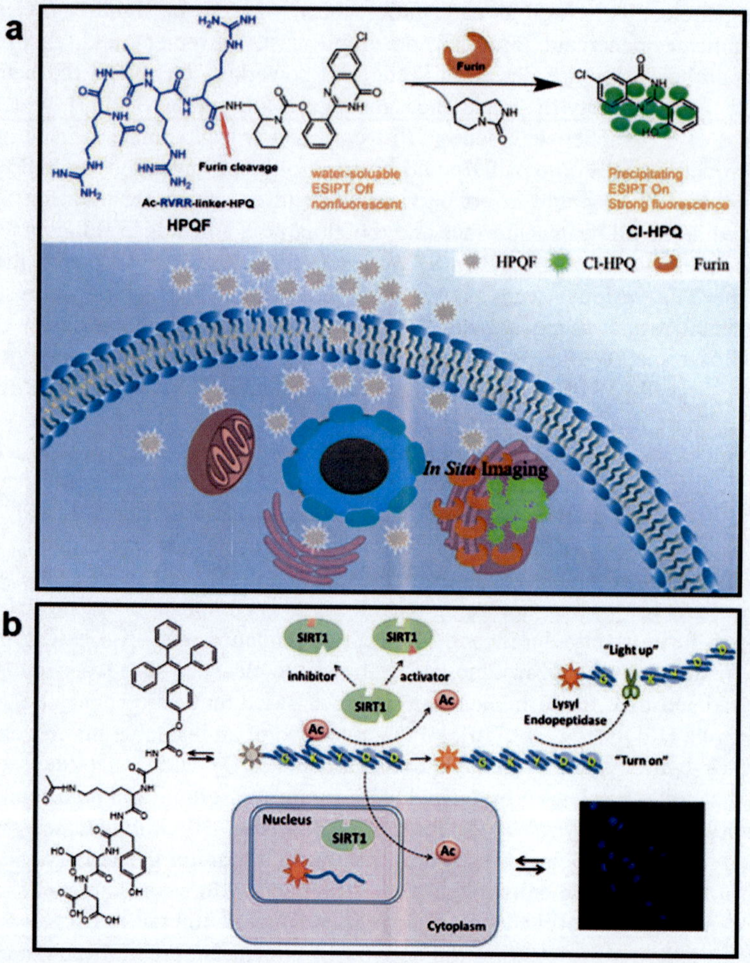

Fig. 3 **a** Molecular design strategy of probe HPQF and its imaging mechanism with furin in living cells (adapted with permission from [70]). **b** Chemical structure of TPE-GK(Ac)YDD and its mechanism for screening both SIRT1 inhibitors and activators in living cells (adapted with permission from [73])

apoptosis leads to excessive cell proliferation [74]. Therefore, cancer treatment is largely aimed at inducing apoptosis of cancer cells [75]. Furthermore, deregulation of the cell apoptosis can also cause myocardial infarction, atherosclerosis, autoimmune diseases, and some other diseases [76, 77]. As a result, apoptosis imaging would contribute to monitoring therapeutic effects and developing new therapeutic methods [78–83]. Despite that the detailed mechanism of the apoptotic process is not fully understood, caspase plays an essential role in the apoptotic process [84–86]. Liu's group synthesized a probe for detecting the activity of caspase-3 in living cells (Fig. 4a) [87]. The probe was synthesized via Cu(I)-catalyzed "click" reaction between a TPE and a DEVD peptide sequence. When the peptide was cleaved by

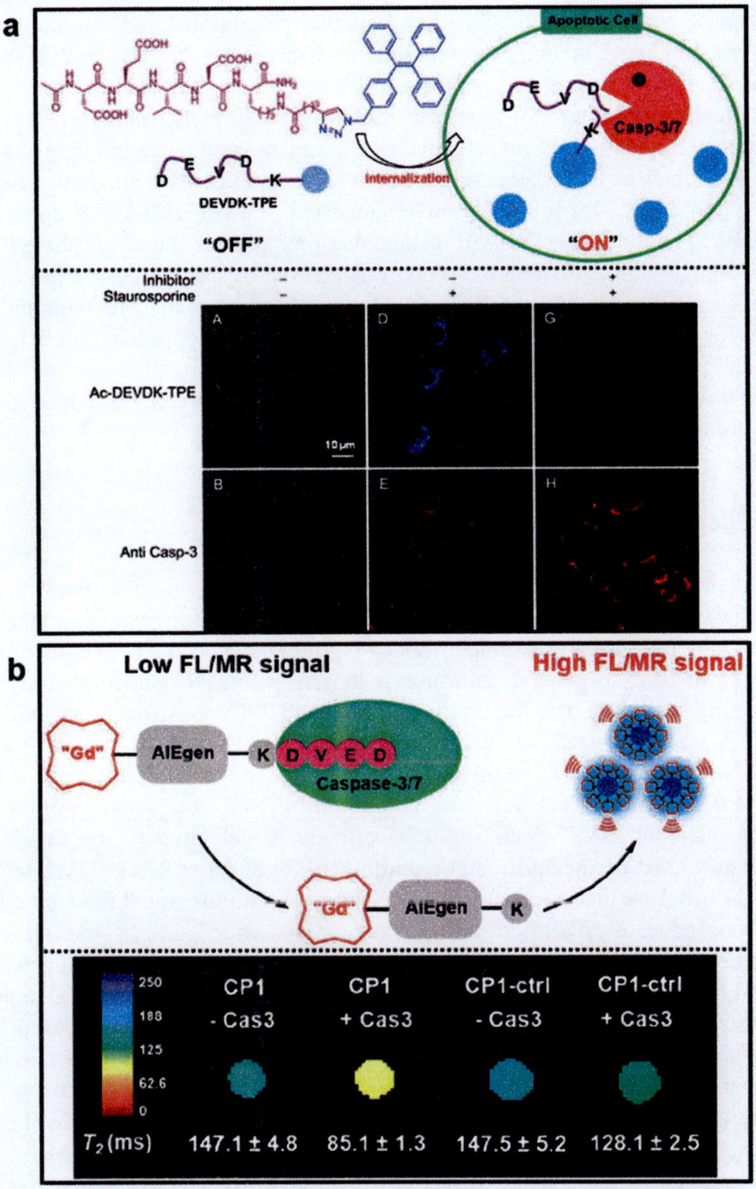

Fig. 4 **a** Chemical structure of the probe DEVDK-TPE and its application for caspase activities study (*top*), and fluorescence microscope images of MCF-7 cells incubated with Ac-DEVDK-TPE and caspase-3 antibody (*bottom*). Staurosporine was utilized to induce cell apoptosis (adapted with permission from [87]). **b** Chemical structures of the probe and caspase-3/7 detective mechanism of multimodal molecular imaging probe (*top*), and T_2-weighted MR images of CP1 and CP1-ctrl solution treating with or without caspase-3 overnight (*bottom*) (adapted with permission from [88])

caspase-3/-7, residual hydrophobic AIE part was precipitated and the fluorescence signal was enhanced. A range of experiments had proved that this probe could be used for real-time sensing of apoptosis in living cells and screening of apoptosis-related medicines. Later, to solve the problems of limited tissue penetration of fluorescence imaging and low sensitivity of magnetic resonance imaging, Meade's group developed a fluorescence-magnetic resonance (FL-MR) bimodal molecular imaging probe (Fig. 4b) [88]. The probe consisted of a Gd-DOTA MR agent, TPE, and a DEVD peptide. Gd (III)-AIE residue was precipitated and displayed enhanced FL-MR signal outputted when the probe was cleaved by caspase-3/-7. Fluorescence imaging could provide accurate MR signal quantification, while MR imaging could be used for imaging deep tissue in vivo. Cellular uptake experiment and MR imaging of cell pellet phantom showed that the probe could effectively enter the cell and produce a significant MR signal. This dual-mode imaging strategy provided the possibility for quantitative monitoring of apoptosis in vivo.

2.2 Image-Guided Therapy

As mentioned, the enzyme-triggered precipitate can not only monitor enzyme activity and locate enzyme position but also realize on-demand drug release and real-time drug tracking [89, 90]. For example, Zhang's group developed a matrix metalloproteinase-2 (MMP-2) responsive colorimetric biosensor for AIE-guided photodynamic therapy (Fig. 5a) [91]. The biosensor (defined as TPPP) consisted of an AIE molecule TPE, a photosensitizer protoporphyrin IX (PpIX), Pro–Leu–Gly–Val–Arg (PLGVR) peptide, and a polyethyleneglycol (PEG) segment. In phosphate-buffered saline (PBS) solution (pH 7.4), the amphiphilic TPPP could self-assemble into spherical nanoparticles, which could be effectively delivered to the tumor cells through enhanced permeability and retention (EPR) effect. After PLGVR in TPPP was hydrolyzed by the overexpressed MMP-2 at the tumor site, TPE emitted blue fluorescence due to AIE effect and the red fluorescence of PpIX had almost no change before and after enzymatic hydrolysis. The colorimetric imaging between TPE and PpIX could not only assess the expression level of MMP-2 but also guided accurate photodynamic therapy of PpIX. TPPP had been successfully used in the treatment of cancer in vivo with minimal side effects. Moreover, to track the release process and improve internalization efficiency of drug, our group reported a prodrug DFP containing cell-penetrating peptide and MMP-2 responsive peptide (Fig. 5b) [92]. The anticancer drug doxorubicin (DOX) was attached to one end of cell-penetrating peptide of DFP, and AIE-active TPE derivative PyTPE was at the other end of MMP-2 responding peptide of DFP. When DFP was cut off by MMP-2, the PyTPE part was precipitated due to AIE effect and turned on fluorescence was used to track drug release, while the DOX residue connected to the cell-penetrating peptide could quickly enter the cell and showed good anticancer effect. The real-time confocal laser scanning microscopy (CLSM) images demonstrated that the yellow fluorescence of PyTPE could effectively monitor the release of DOX. The CLSM imaging of different cell lines verified the MMP-2 selectivity of DFP. This strategy provided a great example for drug release monitoring and efficient internalization.

Apart from utilizing AIE-based fluorescence to monitoring drug release, evaluating the therapeutic effect of drugs and killing bacteria are also important applications by integrating apoptosis imaging as we mentioned above. Liu's group reported a caspase-3 activated AIE molecule precipitation system to monitor the therapeutic effect of Pt (IV) (Fig. 6a) [93]. The system consisted of an AIE molecule tetraphenylsilole (TPS), a chemotherapeutic Pt (IV) prodrug, a targeting cyclic peptide, and DEVD peptide. When the system entered the cell, Pt (IV) was released under reduction conditions and subsequently caused cell apoptosis. After that, the activated caspase-3 cleaved DEVD peptide, and the TPS residue formed a precipitate and emitted bright fluorescence. Experimental results showed that the system could evaluate its own therapeutic effect in situ. In addition to being activated in apoptosis, caspase can also be activated when macrophages are attacked by bacteria. Liu's team developed a bioprobe to monitor the cells for bacterial infection and to kill bacteria (Fig. 6b) [94].

The bioprobe was composed of an AIE fluorophore PyTPE, which could produce reactive oxygen species and had a strong emittance in aggregated state, and a hydrophilic peptide (NEAYVHDAP), which could be cleaved by caspase-1. When macrophages were infected with bacteria, activated caspase-1 cutting bioprobe led to the deposition of PyTPE residues. PyTPE residues emitted fluorescence to guide the photodynamic therapy to kill bacteria. The results showed that the bioprobe was effective in killing bacteria while detecting cell infections. This can greatly simplify the process of handling infected cells. Furthermore, some dual-enzyme-responsive drug delivery systems could play an increasingly important role in personalized medicine because of higher specificity [95, 96].

3 Enzyme-Catalyzed Coupling

In addition to catalyzing hydrolysis reaction of substrate, enzyme can also promote REDOX reaction of substrate, and cause a coupling reaction of substrate. Enzyme-catalyzed coupling uses enzyme to catalyze peptide-based AIE bioprobes from monomers to crosslinked aggregates. The polymerization of AIE bioprobes activated the AIE process and switched on fluorescence emission. Although the enzyme-triggered precipitate has a unique advantage in in situ imaging, it is very limited in improving the signal-to-noise ratio. The enzyme-catalyzed coupling continuously catalyzes the polymerization of the peptide-based AIE bioprobes, which greatly improves imaging sensitivity. Different from common coupling reactions in organic chemistry, enzyme-catalyzed coupling requires milder reaction conditions such as temperature and pressure [97, 98]. Furthermore, with high sensitivity and selectivity, AIE bioprobes based on enzyme-catalyzed coupling are being gradually applied in imaging-guided detection and imaging-guided therapy.

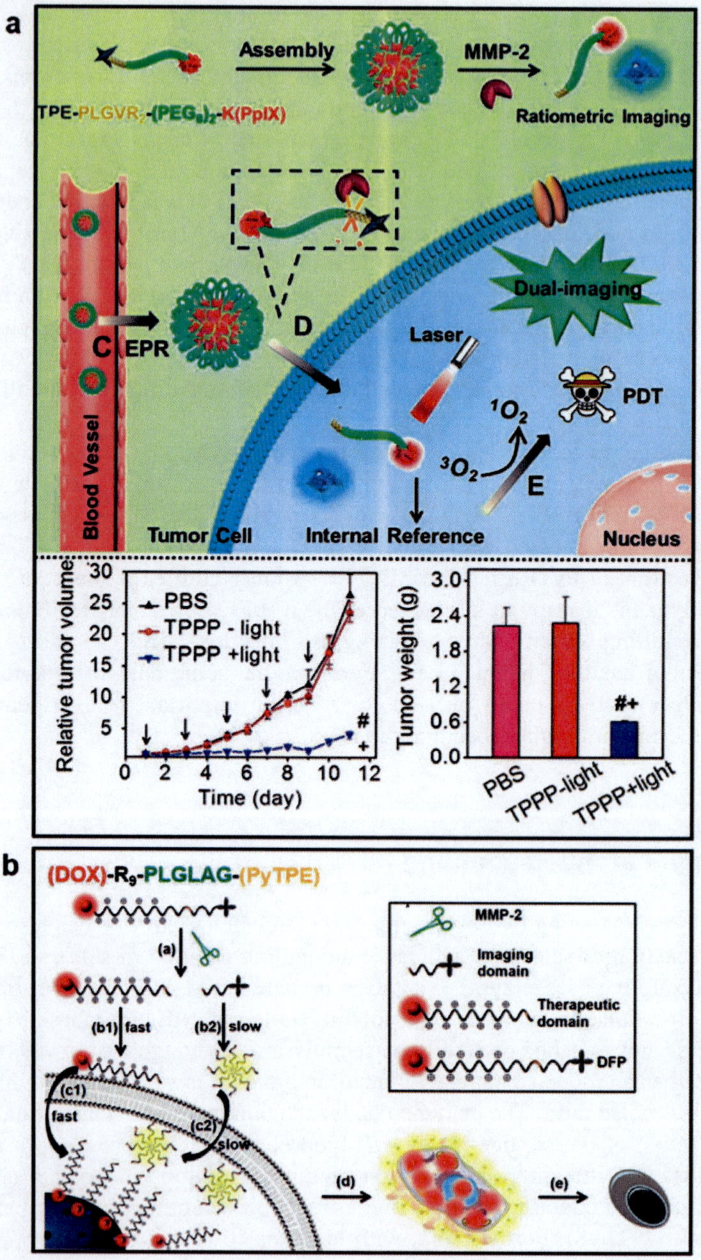

Fig. 5 **a** Self-assembly of TPPP into nanoparticles and TPPP for MMP-2 responsive ratiometric fluorescence imaging and AIE-guided photodynamic therapy (*top*), and volume changes after intravenous treatment and average tumor weight on the 11th day (*bottom*) (adapted with permission from [91]). **b** MMP-2 responsive DFP for fast drug delivery and drug release monitoring in living cells (adapted with permission from [92])

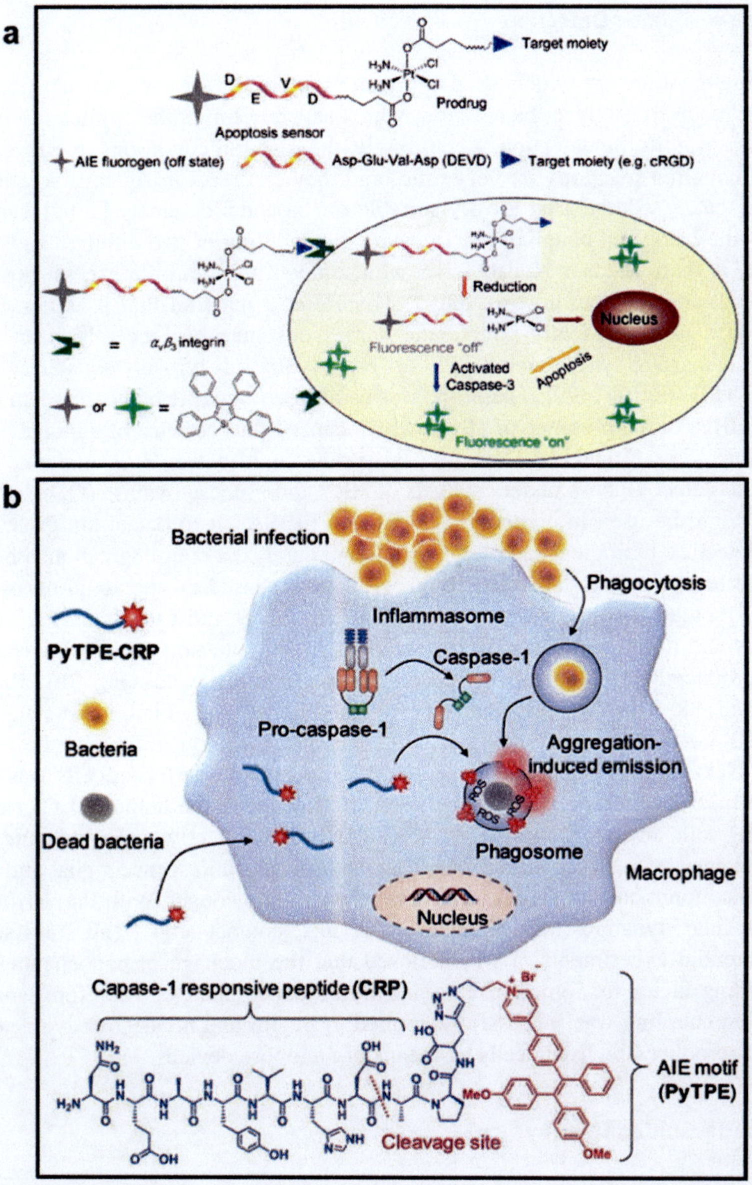

Fig. 6 **a** Simple structure of the sensor and its application in evaluating its own therapeutic effect (adapted with permission from [93]). **b** Diagnosis and clearance of macrophage-mediated intracellular bacterial infection and molecular structure of the probe (adapted with permission from [94])

3.1 Image-Guided Detection

Coupling reactions are widely used in organic synthesis. However, it usually requires organic metal to catalyze the reaction, which severely limits the application of coupling reaction in life detection. Accordingly, the reaction conditions of enzyme-catalyzed coupling reactions are very mild, and they can react in aqueous solution. So enzyme-catalyzed coupling is very suitable for biochemical analysis, but a number of enzyme-catalyzed coupling reactions are the coupling of two different substances and one of them needs to be excessive, which makes it difficult for excess substances to get out and have an adverse effect. Therefore, a reaction that uses horseradish peroxidase to catalytically couple the same substance has recently been developed. Liu's group synthesized a TPE-Tyr probe with four tyrosine covalently bound around TPE (Fig. 7a) [99]. The probe was catalyzed to couple by horseradish peroxidase (HRP) in the presence of H_2O_2, which caused fluorescence of aggregated AIE molecules to be turned on. Therefore, this probe could detect H_2O_2 under action of HRP and detect glucose under catalysis of HRP and glucose oxidase (GOx). Moreover, it could be used in combination with the ELISA kit to detect antigens. Apart from detecting biological samples, the enzyme-catalyzed coupling can also be used to detect enzyme in living cells. In our laboratory, based on the coupling of H_2O_2 and HRP-catalyzed AIE bioprobes to block nanopores and cause changes in ionic currents and fluorescence signal, we achieved in situ and non-invasive detection of H_2O_2 produced by living cells by means of nanochannel devices (Fig. 7b) [100]. TT contained one TPE derivative and two tyrosine derivatives. First, solid-state nanochannels were chemically modified with phenolic hydroxyl groups as the capture probe. Next, polyethylene terephthalate (PET) incubated by living cells was rolled up and inserted into the nanochannel device. Then, with the help of H_2O_2 released by living cells and additional HRP added, TT itself covalently conjugated and covalently reacted with the capture probe. The changes of fluorescence signal and transmembrane ionic current signal were observed simultaneously. With the verification of molecular dynamic (MD) simulations results, contact angle, and solid surface zeta potential experiments, it was believed that the blockage of nanochannels was the leading factor for ionic currents and fluorescence change. Therefore, enzyme-catalyzed coupling was successfully applied to in situ and nondestructive detection of H_2O_2 produced by living cells by means of nanopore devices.

3.2 Image-Guided Therapy

The molecular aggregation caused by enzyme-catalyzed coupling can also cause damage of cellular suborganelles for treatment. Inflammation plays a key role in various stages of tumor occurrence, invasion, and metastasis [101–103]. The enzyme-catalyzed dityrosine coupling in inflammatory cells is associated with oxidative damage to proteins under the catalysis of H_2O_2 and peroxidase. To achieve selective imaging and growth inhibition of inflammatory cells, our laboratory assembled an enzyme-catalyzed coupled probe TT (Fig. 8) [104]. TT comprises a TPE as an AIE emissive core in the middle and two tyrosines as a coupling agent at both ends. Due

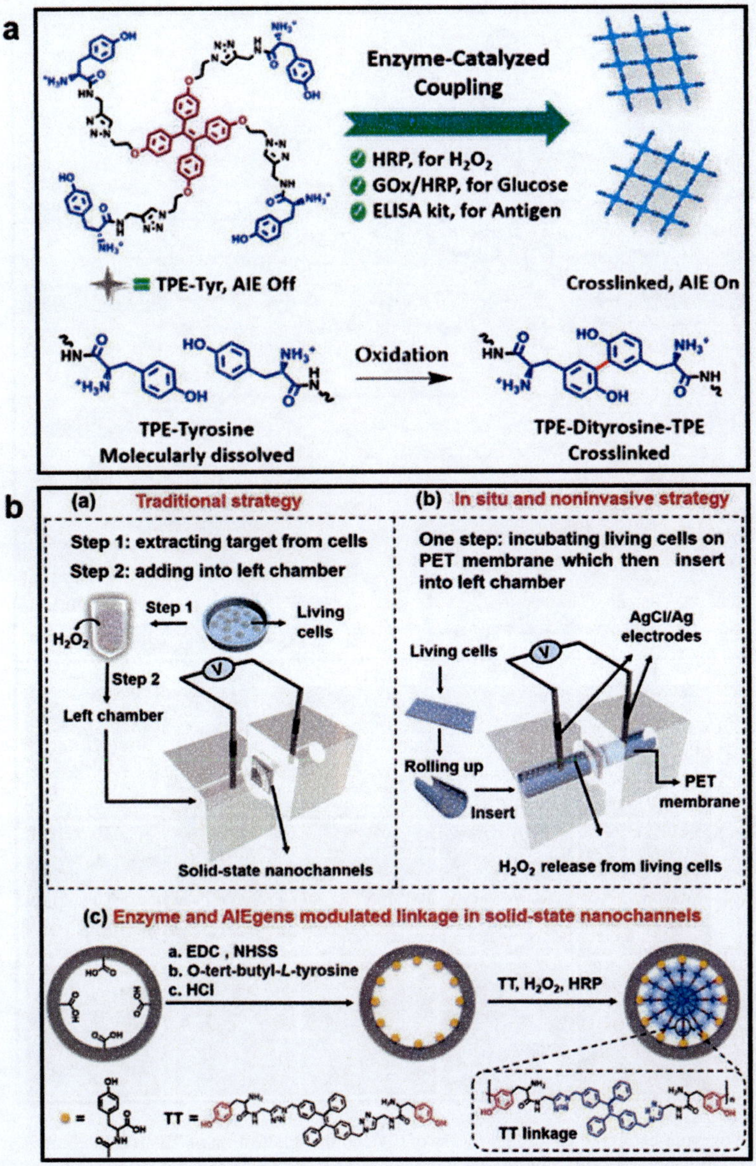

Fig. 7 **a** The application of TPE-Tyr via enzyme-catalyzed coupling and crosslinking mechanism of TPE-Tyr (adapted with permission from [99]). **b** Traditional strategy and new strategy for monitoring H_2O_2 released from living cells based on solid-state nanochannels and mechanism of enzyme-catalyzed coupling using HRP and TT in solid-state nanochannels (adapted with permission from [100])

to the introduction of hydrophilic tyrosine, the hydrophobic TPE becomes hydrophilic TT and TT has weaker fluorescence in aqueous solution. With the help of overexpressed H_2O_2 and myeloperoxidase (MPO) in inflammatory cells, TT initiated

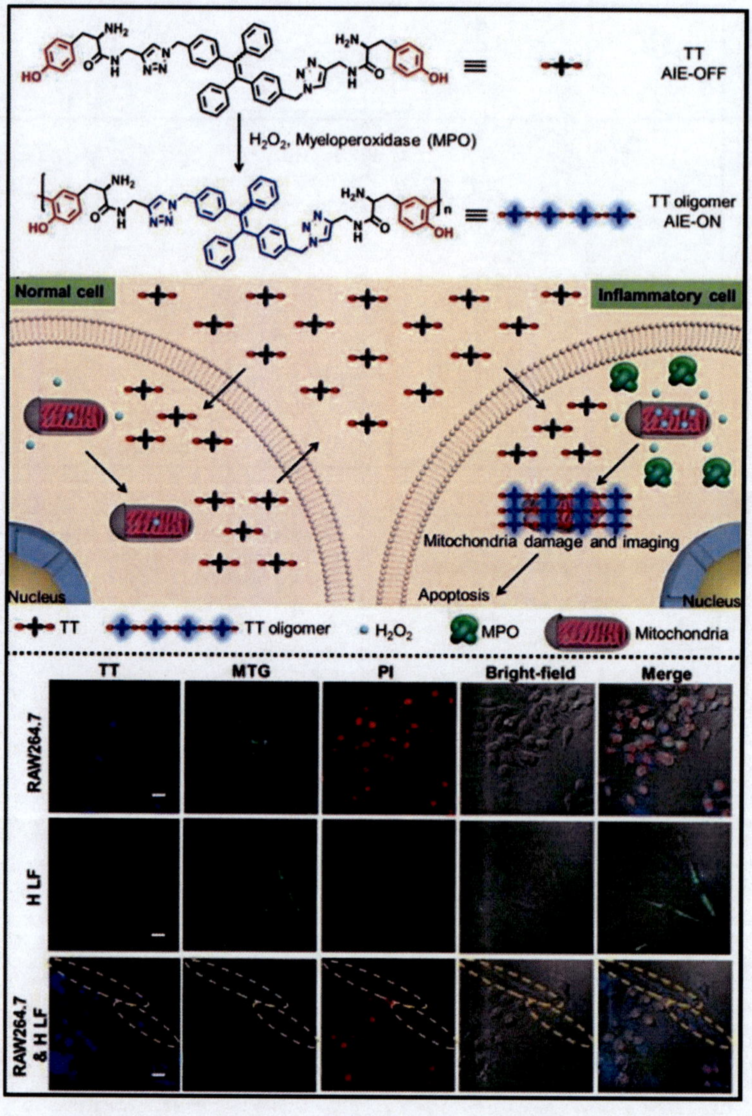

Fig. 8 The enzyme-catalyzed coupling process of TT and the application of TT in cell-selective imaging and promoting apoptosis (*top*), and the CLSM images of RAW264.7 cells, HLF cells and RAW 264.7 and HLF co-cultured cells treated with TT, MTG and PI (*bottom*) (adapted with permission from [104])

coupling reaction and its AIE properties were activated. The fluorescence changes caused by aggregation of TT could be utilized to identify inflammatory cells and normal cells. The CLSM experimental results of Fig. 8 showed that toxicity of TT in RAW264.7 cells (inflammatory cells) was significantly higher than that in HLF cells (normal human lung fibroblast). At the same time, Western-blot results and bio-TEM analysis showed the aggregation of TT caused damage of mitochondria of inflammatory cells, which led to inhibition of growth of inflammatory cells. As far as we know, this is the first report that AIE bioprobes based on the catalytic coupling of H_2O_2 and MPO are used for selective imaging and growth inhibition of inflammatory cells.

4 Enzyme-Instructed Self-assembly

As enzyme-catalyzed coupling exhibited improved sensitivity and enzyme-triggered precipitate possessed in situ imaging ability, enzyme-instructed self-assembly of peptide-based AIE bioprobes could study the dynamic process. This strategy transforms AIE molecules from a dispersed non-fluorescent state to an aggregated luminescent state. Enzyme-instructed self-assembly of peptide-based AIE bioprobes not only has the specificity of enzyme response and the optical advantages of AIE molecules, but can also study dynamic life processes [105–107]. The bioprobes are cleaved by enzyme and they remain bound to the targeted enzyme, allowing dynamic studies of enzyme activation and localization. These advantages provide a promising prospect for the enzyme-instructed self-assembly of peptide-based AIE bioprobes in imaging-guided detection and imaging-guided therapy.

4.1 Image-Guided Detection

Enzyme-instructed self-assembly involves two processes: enzymatic hydrolysis and dynamic self-assembly. Enzymatic hydrolysis allows the system to show specificity and opens up to self-assembly. Self-assembly provides a visual signal amplification and keeps the system from spreading. Therefore, enzyme-instructed self-assembly can be used to detect enzyme specificity and study the dynamic process [108–111]. Ding and coworkers synthesized an enzyme-instructed self-assembled bioprobe for sensitive and selective imaging of caspase-3 (Fig. 9a) [112]. The bioprobe contained an AIE molecule TPE, a peptide sequence DVEDEE-Ac that was digested by caspase-3, a peptide sequence GFFY that induced self-assembly, and an amino acid K that acted as a linker. When the bioprobe reacted with caspase-3, the remaining part of TPE was self-assembled under induction of GFFY, which restricted movement of molecules and caused fluorescence to turn on. Based on the fluorescent data in vitro, the limit of detection for the bioprobe was calculated to be 0.54 pM by using a 3-sigma method. The CLSM experiments of healthy and apoptotic HeLa cells indicated that the bioprobe could specifically respond to caspase-3. Therefore, this bioprobe had a high signal-to-noise ratio and a low detection limit for detection and imaging of caspase-3 in solution and living cells. ALP was a hydrolase that can remove phosphate groups from various substrates [113–115]. As a biomarker of tumor, the activity level of ALP was closely related to the prognosis assessment and diagnosis of malignant tumors [116]. Liang's group developed a probe, TPE-YpYY, to monitor activity and track the status of alkaline phosphatase (ALP) in individual living cells (Fig. 9b) [117]. The peptide sequence YpYY made the probe water soluble, reduced background fluorescence of the probe, and could be cut off by ALP. TPE-YpYY rapidly diffused into whole cells, and

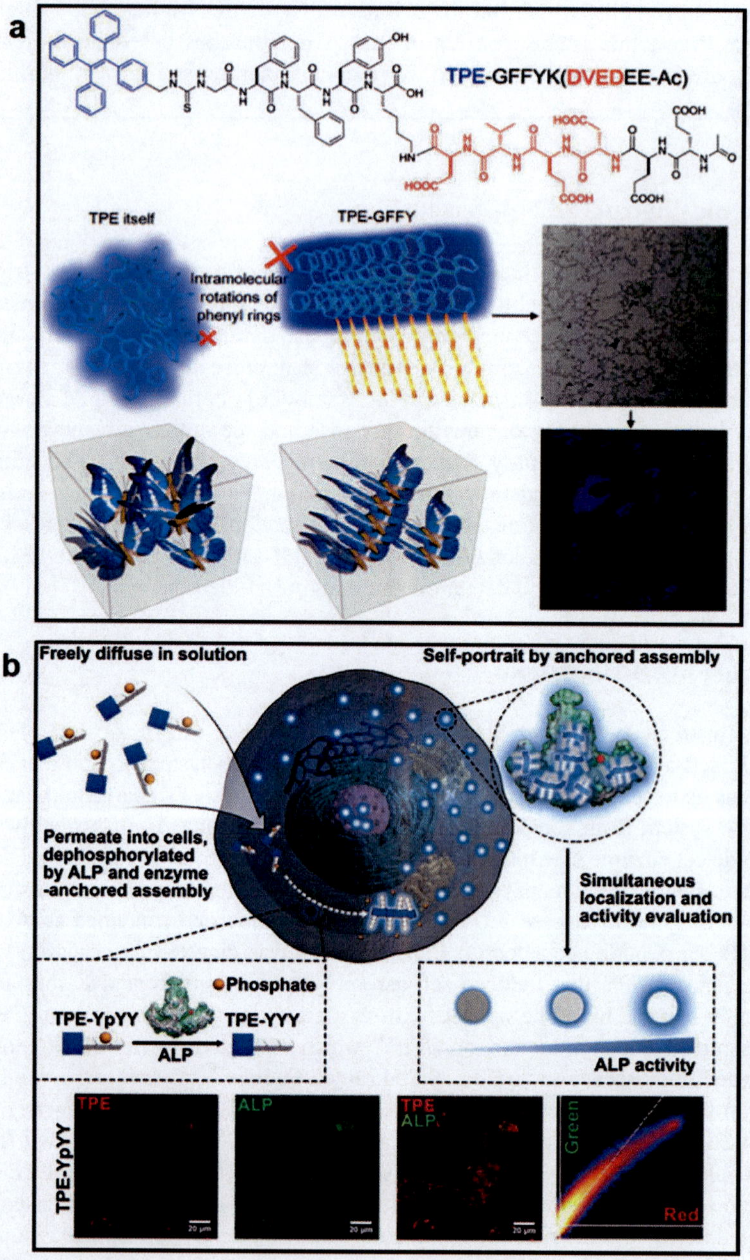

Fig. 9 **a** Chemical structure of the probe and its enzyme-instructed self-assembly for detecting caspase-3 in living cells (adapted with permission from [112]). **b** Enzyme-instructed self-assembly process of TPE-YpYY for in situ detection of ALP in living cells (*top*). Immunofluorescence imaging and 2D plot of colocalization of OS732 cells treated with TPE-YpYY (*red*) and Alexa 488-labeled anti-ALP antibody (*green*) (adapted with permission from [117])

then enzyme-anchored TPE-YYY was self-assembled to aggregate by phosphorylation of ALP in living cells. Thus, the probe was successfully used for imaging ALP in living cells and for dynamic self-assembly in single cells.

4.2 Image-Guided Therapy

Enzyme-instructed self-assembly has also been successfully used in treatment of diseases due to the formation of self-assembled structures that block cell material exchange and affect cell metabolism. Zhang's group designed a transformable chimeric peptide encapsulated with DOX for overcoming tumor multidrug resistance (MDR) (Fig. 10a) [118]. The chimeric peptide CTGP was made up of hydrophilic polyethylene glycol (PEG), 16-carbon alkyl chain embedded in cell membrane, cathepsin B-responsive GFLG sequence, an AIE probe TPE, and GGGH sequence for offering hydrogen bonding interactions. CTGP forms nanomicelles to encapsulate DOX in aqueous solution and form a drug-transport system denoted as CTGP@DOX.

CTGP@DOX eventually accumulated at tumor site through EPR effect as blood circulates. Under hydrolysis of cathepsin B, CTGP@DOX released encapsulated DOX and formed nanofibers by self-assembly on the cell membrane. The formation of nanofibers could effectively prevent efflux of DOX and inhibit MDR. At the same time, the self-assembly of the nanofibers caused the TPE to aggregate and a bright blue fluorescence was followed. In turn, the enhanced blue fluorescence was used to judge fiber formation. The in vivo experiments showed that CTGP@DOX had a significant antitumor efficacy. This strategy has been successfully used in the treatment of cancer with MDR and to determine the process of enzyme-instructed self-assembly through the fluorescence of AIE molecules in the aggregate state. Furthermore, Ding and coworkers used enzyme-instructed self-assembly for both fluorescence imaging of enzyme activity and treatment of cancer (Fig. 10b) [119]. The probe mainly contained an AIE molecule TPE-Py and three phosphorylated tyrosine. TPE-Py not only had bright fluorescence but also produced reactive oxygen species. Phosphorylated tyrosine endowed the probe with the ability to be water soluble and self-assembling. When the probe was dephosphorylated by ALP produced by living cell, fluorescence generated by residual probe self-assembly could be used to detect ALP. When cells were exposed to light, reactive oxygen species were produced that could kill them for treatment of cancer.

5 Conclusions and Outlook

Unlike traditional fluorescent molecules with ACQ effect, AIE molecules are characterized by bright fluorescence in the state of aggregation. Three aggregation strategies of AIE bioprobes based on peptide after reaction with enzyme were summarized in this paper: enzyme-triggered precipitate, enzyme-catalyzed coupling, and enzyme-instructed self-assembly. These three strategies not only give full play to advantages of AIE bioprobes but also endue AIE bioprobes with excellent specificity and significant biocompatibility in detection and treatment of diseases.

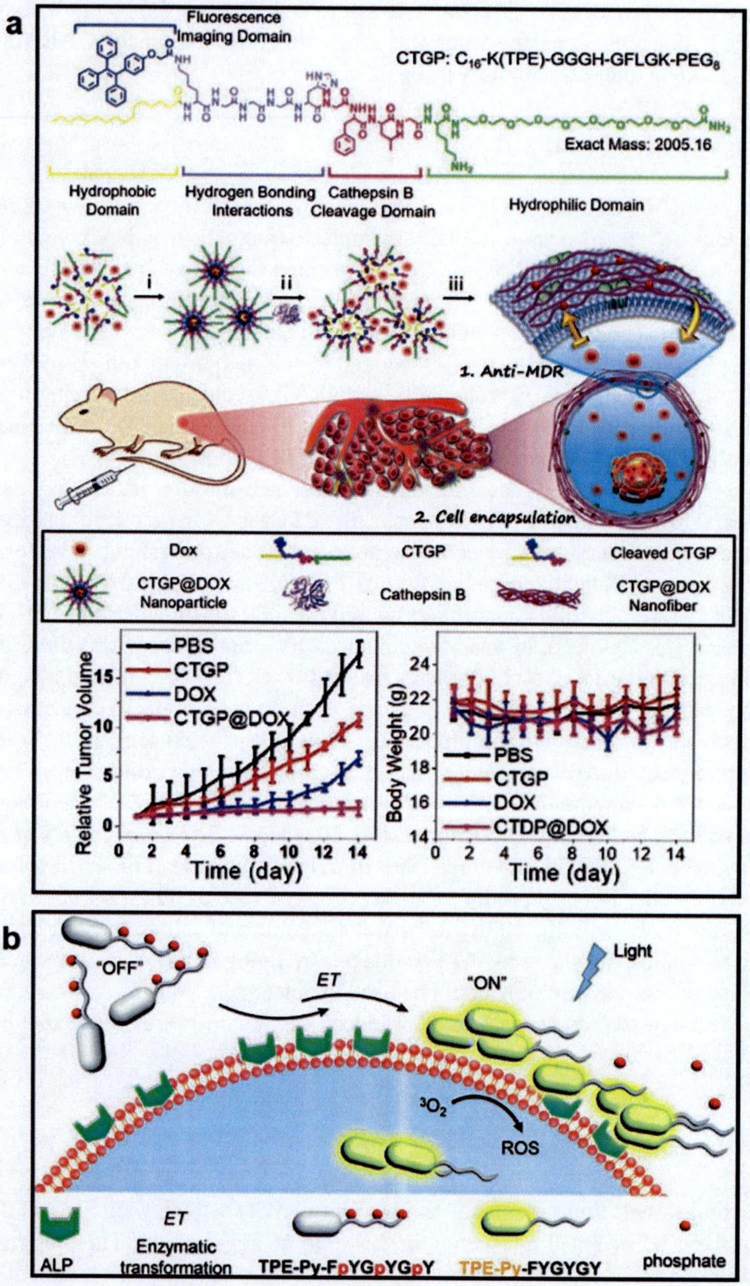

Fig. 10 **a** The structure of CTGP and CTGP@DOX with morphology change from nanomicelles to cathepsin B-responsive self-assembled nanofibers to release DOX (*top*). Relative volume changes and body weight changes of MCF-7R tumor-bearing mice during the 14-day treatment period after intravenous injection with PBS, CTGP, free DOX, and CTGP@DOX (*bottom*) (adapted with permission from [118]). **b** Enzyme-instructed self-assembly of probe for detection of ALP in living cells and treatment of cancer (adapted with permission from [119])

Although there have been many outstanding achievements, a lot of potential values of enzyme-responsive peptide-based AIE bioprobes worth digging, including but not limited to the following issues: (1) in this review, AIE bioprobes used in most systems are green or blue, which greatly limits their application in in vivo imaging. It is necessary to design AIE bioprobes with infrared or near-infrared emission and more diverse functions. (2) According to different reaction properties catalyzed by enzyme, enzymes were divided into six categories. Obviously, three strategies summarized in this paper mainly focus on hydrolase and ligase. Therefore, different aggregation forms can be designed according to different characteristics of other enzymes. (3) Enzyme activation is a dynamic process. How to use enzyme-responsive peptide-based AIE bioprobes for real-time dynamic detection of enzymes is a challenge. (4) It is not clear whether peptide-based AIE probes reacting with enzyme would cause side effects in diseases treatment. Enzyme-responsive peptide-based AIE bioprobes will evolve to be more stable, more versatile, and more suitable for complex systems. Based on previous works, we believe that it will shine in the clinical diagnosis and treatment of diseases in the future.

Acknowledgements The authors acknowledge financial support from the National Natural Science Foundation of China (21722507, 21525523, 21974128, 21874121) and the Natural Science Foundation of Hubei Province (2019CFA043).

Compliance with Ethical Standard

Conflict of interest The author has no conflicts of interest with any known source.

References

1. Eslahi H, Ghaffari-Moghaddam M, Khajehi M, Omay D, Zakipour-Rahimabadi E, Motalleb G (2014) General biography, structure and classification of enzymes. Moghaddam 3:1–83
2. Schomburg D, Schomburg I (2010) Enzyme databases. Methods Mol Biol 609:113–128
3. Hadler-Olsen E, Winberg JO, Uhlin-Hansen L (2013) Matrix metalloproteinases in cancer: their value as diagnostic and prognostic markers and therapeutic targets. Tumor Biol 34:2041–2051
4. Nagase H, Visse R, Murphy G (2006) Structure and function of matrix metalloproteinases and TIMPs. Cardiovasc Res 69:562–573
5. Braun GB, Sugahara KN, Yu OM, Kotamraju VR, Molder T, Lowy AM, Ruoslahti E, Teesalu T (2016) Urokinase-controlled tumor penetrating peptide. J Control Release 232:188–195
6. Boya P, Kroemer G (2008) Lysosomal membrane permeabilization in cell death. Oncogene 27:6434–6451
7. Hengartner MO (2000) The biochemistry of apoptosis. Nature 407:770–776
8. Zhang J, Chai X, He X, Kim H, Yoon J, Tian H (2019) Fluorogenic probes for disease-relevant enzymes. Chem Soc Rev 48:683–722
9. Wang D, Tang BZ (2019) Aggregation-induced emission luminogens for activity-based sensing. Acc Chem Res 52:2559–2570
10. Liu H, Chen L, Xu C, Li Z, Zhang H, Zhang X, Tan W (2018) Recent progresses in small-molecule enzymatic fluorescent probes for cancer imaging. Chem Soc Rev 47:7140–7180
11. Wu X, Shi W, Li X, Ma H (2019) Recognition moieties of small molecular fluorescent probes for bioimaging of enzymes. Acc Chem Res 52:1892–1904
12. Hu J, Zhang G, Liu S (2012) Enzyme-responsive polymeric assemblies, nanoparticles and hydrogels. Chem Soc Rev 41:5933–5949

13. Choi KY, Swierczewska M, Lee S, Chen X (2012) Protease-activated drug development theranostics. Theranostics 2:156–179
14. Decharms RC (2008) Applications of real-time fMRI. Nat Rev Neurosci 9:720–729
15. Werner EJ, Datta A, Jocher CJ, Raymond KN (2008) High-relaxivity MRI contrast agents: where coordination chemistry meets medical imaging. Angew Chem Int Ed 47:8568–8580
16. Elmenoufy AH, Tang Y, Hu J, Xu H, Yang X (2015) A novel deep photodynamic therapy modality combined with CT imaging established via X-ray stimulated silica-modified lanthanide scintillating nanoparticles. Chem Comm 51:12247–12250
17. Ametamey SM, Honer M, Schubiger PA (2008) Molecular imaging with PET. Chem Rev 108:1501–1516
18. Song Y, Wei W, Qu X (2011) Colorimetric biosensing using smart materials. Adv Mater 23:4215–4236
19. Alberti D, Erve MV, Stefania R, Ruggiero MR, Tapparo M, Crich SG, Aime S (2014) A quantitative relaxometric version of the ELISA test for the measurement of cell surface biomarkers. Angew Chem Int Ed 53:3488–3491
20. Chikkaveeraiah BV, Bhirde AA, Morgan NY, Eden HS, Chen X (2012) Electrochemical immunosensors for detection of cancer protein biomarkers. ACS Nano 6:6546–6561
21. Gao P, Pan W, Li N, Tang B (2019) Fluorescent probes for organelle-targeted bioactive species imaging. Chem Sci 10:6035–6071
22. Wang Q, Wang X, Xu M, Lou X, Xia F (2019) One-dimensional and two-dimensional nanomaterials for the detection of multiple biomolecules. Chin Chem Lett 30:1557–1564
23. Tian M, Ma Y, Lin W (2019) Fluorescent probes for the visualization of cell viability. Acc Chem Res 52:2147–2157
24. Luo J, Xie Z, Lam JWY, Cheng L, Chen H, Qiu C, Kwok HS, Zhan X, Liu Y, Zhu D, Tang BZ (2001) Aggregation-induced emission of 1-methyl-1,2,3,4,5-pentaphenylsilole. Chem Commun 18:1740–1741
25. Hong Y, Lam JWY, Tang BZ (2011) Aggregation-induced emission. Chem Soc Rev 40:5361–5388
26. Mei J, Leung NLC, Kwok RTK, Lam JWY, Tang BZ (2015) Aggregation-induced emission: together we shine, united we soar! Chem Rev 115:11718–11940
27. Zhao Z, Zhang H, Lam JWY, Tang BZ (2020) Aggregation-induced emission: new vistas at aggregate level. Angew Chem Int Ed. https://doi.org/10.1002/anie.201916729
28. Ding D, Li K, Liu B, Tang BZ (2013) Bioprobes based on AIE fluorogens. Acc Chem Res 46:2441–2453
29. Wang D, Lee MMS, Xu W, Kwok RTK, Lam JWY, Tang BZ (2018) Theranostics based on AIEgens. Theranostics 8:4925–4956
30. Mei J, Huang Y, Tian H (2017) Progress and trends in AIE-based bioprobes: a brief overview. ACS Appl Mater Interfaces 10:12217
31. Yan L, Zhang Y, Xu B, Tian W (2016) Fluorescent nanoparticles based on AIE fluorogens for bioimaging. Nanoscale 8:2471–2487
32. Lu H, Zhao X, Tian W, Wang Q, Shi J (2014) Pluronic F127–folic acid encapsulated nanoparticles with aggregation-induced emission characteristics for targeted cellular imaging. RSC Adv 4:18460–18466
33. Zhu C, Kwok RTK, Lam JWY, Tang BZ (2018) Aggregation-induced emission: a trailblazing journey to the field of biomedicine. ACS Appl Bio Mater 1:1768–1786
34. Wang YF, Zhang T, Liang XJ (2016) Aggregation-induced emission: lighting up cells, revealing life. Small 12:6451–6477
35. Shi J, Li Y, Li Q, Li Z (2017) Enzyme-responsive bioprobes based on the mechanism of aggregation-induced emission. ACS Appl Mater Interfaces 10:12278
36. Kenry CKC, Liu B (2019) Reactivity-based organic theranostic bioprobes. Acc Chem Res 52:3051–3063
37. Cai X, Liu B (2020) Aggregation-induced emission: recent advances in materials and biomedical applications. Angew Chem Int Ed. https://doi.org/10.1002/anie.202000845
38. Xu S, Liu HW, Hu XX, Huan SY, Zhang J, Liu YC, Yuan L, Qu FL, Zhang XB, Tan W (2017) Visualization of endoplasmic reticulum aminopeptidase 1 under different redox conditions with a two-photon fluorescent probe. Anal Chem 89:7641–7648
39. Huang Y, Zhu L, Ji J, Li Y, Liu T, Lei J (2020) Cleancap-regulated aggregation-induced emission strategy for highly specific analysis of enzyme. Anal Chem 92:4726–4730

40. Li H, Yao Q, Xu F, Li Y, Kim D, Chung J, Baek G, Wu X, Hillman PF, Lee EY, Ge H, Fan J, Wang J, Nam SJ, Peng X, Yoon J (2020) An activatable AIEgen probe for high-fidelity monitoring of overexpressed tumor enzyme activity and its application to surgical tumor excision. Angew Chem Int Ed 59:10186–10195
41. Gu K, Qiu W, Guo Z, Yan C, Zhu S, Yao D, Shi P, Tian H, Zhu W (2019) An enzyme-activatable probe liberating AIEgens: on-site sensing and long-term tracking of β-galactosidase in ovarian cancer cells. Chem Sci 10:398
42. Wu F, Wu X, Duan Z, Huang Y, Lou X, Xia F (2019) Biomacromolecule-functionalized AIEgens for advanced biomedical studies. Small 15:1804839
43. Xia F, Wu J, Wu X, Hu Q, Dai J, Lou X (2019) Modular design of peptide- or DNA-modified AIEgen probes for biosensing applications. Acc Chem Res 52:3064–3074
44. Yuan Q, Cheng Y, Lou X, Xia F (2019) Rational fabrication and biomedical application of biomolecule-conjugated AIEgens through click reaction. Chin J Chem 37:1072–1082
45. Zhang P, Cui Y, Anderson CF, Zhang C, Li Y, Wang R, Cui H (2018) Peptide-based nanoprobes for molecular imaging and disease diagnostics. Chem Soc Rev 47:3490–3529
46. Liang J, Tang BZ, Liu B (2015) Specific light-up bioprobes based on AIEgen conjugates. Chem Soc Rev 44:2798–2811
47. Xu M, Wang X, Wang Q, Hu QY, Huang K, Lou X, Xia F (2019) Analyte-responsive fluorescent probes with AIE characteristic based on the change of covalent bond. Sci China Mater 62:1236–1250
48. Cheng Y, Sun C, Ou X, Liu B, Lou X, Xia F (2017) Dual-targeted peptide-conjugated multifunctional fluorescent probe with AIEgen for efficient nucleus-specific imaging and long-term tracing of cancer cells. Chem Sci 8:4571–4578
49. Cheng Y, Sun C, Liu R, Yang J, Dai J, Zhai T, Lou X, Xia F (2019) A multifunctional peptide-conjugated AIEgen for efficient and sequential targeted gene delivery into the nucleus. Angew Chem Int Ed 58:5049–5053
50. Li L, Qiao ZY, Wang L, Wang H (2018) Programmable construction of peptide-based materials in living subjects: from modular design and morphological control to theranostics. Adv Mater 31:1804971
51. Zhan F, Liu J, Cheng B, Liu Y, Lai T, Lin H, Yeh M (2019) Tumor targeting with DGEA peptide ligands: a new aromatic peptide amphiphile for imaging cancers. Chem Commun 55:1060–1063
52. Zhang X, Ren C, Hu F, Gao Y, Wang Z, Li H, Liu J, Liu B, Yang C (2020) Detection of bacterial alkaline phosphatase activity by enzymatic in situ self-assembly of the AIEgen-peptide conjugate. Anal Chem 92:5185–5190
53. Yang J, Dai J, Wang Q, Cheng Y, Guo J, Zhao Z, Hong Y, Lou X, Xia F (2020) Tumor triggered disassembly of multiple-agent-therapy probe for efficient cellular internalization. Angew Chem Int Ed. https://doi.org/10.1002/ange.202009196
54. Liu X, Sun X, Liang G (2020) Peptide-based supramolecular hydrogels for bioimaging applications. Biomater Sci. https://doi.org/10.1039/D0BM01020K
55. Qi G, Gao Y, Wang L, Wang H (2018) Self-assembled peptide-based nanomaterials for biomedical imaging and therapy. Adv Mater 30:1703444
56. Lin H, Yang H, Huang S, Wang F, Wang DM, Liu B, Tang YD, Zhang CJ (2018) Caspase-1 specific light-up probe with aggregation-induced emission characteristics for inhibitor screening of coumarin-originated natural products. ACS Appl Mater Interfaces 10:12173–12180
57. Hou X, Zeng F, Wu S (2016) A fluorescent assay for γ-glutamyltranspeptidase via aggregation induced emission and its applications in real samples. Biosens Bioelectron 85:317–323
58. Zhang XB, Waibel M, Hasserodt J (2010) An autoimmolative spacer allows first-time incorporation of a unique solid-state fluorophore into a detection probe for acyl hydrolases. Chem Eur J 16:792–795
59. Dhara K, Hori Y, Baba R, Kikuchi K (2012) A fluorescent probe for detection of histone deacetylase activity based on aggregation-induced emission. Chem Comm 48:11534–11536
60. Madjene LC, Pons M, Danelli L, Claver J, Ali L, MaderaSalcedo IK, Kassas A, Pellefigues C, Marquet F, Dadah A, Attout T, El-Ghoneimi A, Gautier G, Benhamou M, Charles N, Daugas E, Launay P, Blank U (2015) Mast cells in renal inflammation and fibrosis: lessons learnt from animal studies. Mol Immunol 63:86–93
61. Zhang R, Zhang CJ, Feng G, Hu F, Wang J, Liu B (2016) Specific light-up probe with aggregation-induced emission for facile detection of chymase. Anal Chem 88:9111–9117

62. Dhiraj Y, Agarwal N, Pitchumoni CS (2002) A critical evaluation of laboratory tests in acute pancreatitis. Am J Gastroenterol 97:1309–1318
63. Catanzaro R, Cuffari B, Italia A, Marotta F (2016) Exploring the metabolic syndrome: nonalcoholic fatty pancreas disease. World J Gastroenterol 22:7660–7675
64. Shi J, Deng Q, Wan C, Zheng M, Huang F, Tang BZ (2017) Fluorometric probing lipase levels as acute pancreatitis biomarker based on interfacial controlled aggregation-induced emission. Chem Sci 8:6188–6195
65. Aw J, Widjaja F, Ding Y, Mu J, Liang Y, Xing B (2017) Enzyme-responsive reporter molecules for selective localization and fluorescence imaging of pathogenic biofilms. Chem Comm 53:3330–3333
66. Huang S, Wu Y, Zeng F, Chen J, Wu S (2018) A turn-on fluorescence probe based on aggregation-induced emission for leucine aminopeptidase in living cells and tumor tissue. Anal Chim Acta 1031:169–177
67. Thomas G (2002) Furin at the cutting edge: from protein traffic to embryogenesis and disease. Nat Rev Mol Cell Biol 3:753–766
68. Hosaka M, Nagahama M, Kim WS, Watanabe T, Hatsuzawa K, Ikemizu J, Murakami K, Nakayama K (1991) Arg-X-Lys/Arg–Arg motif as a signal for precursor cleavage catalyzed by furin within the constitutive secretory pathway. J Biol Chem 266:12127–12130
69. Liu X, Liang G (2017) Dual aggregation-induced emission for enhanced fluorescence sensing of furin activity in vitro and in living cells. Chem Commun 53:1037–1040
70. Li K, Hu X, Liu H, Xu S, Huan S, Li J, Deng T, Zhang X (2018) In situ imaging of furin activity with a highly stable probe by releasing of precipitating fluorochrome. Anal Chem 90:11680–11687
71. Guarani V, Deflorian G, Franco CA, Kruger M, Phng LK, Bentley K, Toussaint L, Dequiedt F, Mostoslavsky R, Schmidt MH, Zimmermann B, Brandes RP, Mione M, Westphal CH, Braun T, Zeiher AM, Gerhardt H, Dimmeler S, Potente M (2011) Acetylation-dependent regulation of endothelial Notch signalling by the SIRT1 deacetylase. Nature 473:234–238
72. Chauhan D, Bandi M, Singh AV, Ray A, Raje N, Richardson P, Anderson KC (2011) Preclinical evaluation of a novel SIRT1 modulator SRT1720 in multiple myeloma cells. Br J Hamaetol 155:588–598
73. Wang Y, Chen Y, Wang H, Cheng Y, Zhao X (2015) Specific turn-on fluorescent probe with aggregation-induced emission characteristics for SIRT1 modulator screening and living cell imaging. Anal Chem 87:5046–5049
74. Wong RS (2011) Apoptosis in cancer: from pathogenesis to treatment. J Exp Clin Cancer Res 30:87
75. Ghobrial IM, Witzig TE, Adjei AA (2005) Targeting apoptosis pathways in cancer therapy. Ca Cancer J Clin 55:178–194
76. Okada H, Mak TW (2004) Pathways of apoptotic and non-apoptotic death in tumour cells. Nat Rev Cancer 4:592–603
77. Riedl SJ, Shi Y (2004) Molecular mechanisms of caspase regulation during apoptosis. Nat Rev Mol Cell Biol 5:897–907
78. Kim T, Jin H, Bae J, Kim Y (2017) An excimer emission-based fluorescent probe targeting caspase-3. Anal Chem 89:10565–10569
79. Hu Y, Shi L, Su Y, Zhang C, Jin X, Zhu X (2017) A fluorescent light-up aggregation-induced emission probe for screening gefitinib-sensitive non-small cell lung carcinoma. Biomater Sci 5:792–799
80. Ding D, Liang J, Shi H, Kwok RTK, Gao M, Feng G, Yuan Y, Tang BZ, Liu B (2014) Light-up bioprobe with aggregation-induced emission characteristics for real-time apoptosis imaging in target cancer cells. J Mater Chem B 2:231–238
81. Liang J, Shi H, Kwok RTK, Gao M, Yuan Y, Zhang W, Tang BZ, Liu B (2014) Distinct optical and kinetic responses from E/Z isomers of caspase probes with aggregation-induced emission characteristics. J Mater Chem B 2:4363–4370
82. Yuan Y, Zhang CJ, Kwok RTK, Mao D, Tang BZ, Liu B (2017) Light-up probe based on AIEgens: dual signal turn-on for caspase cascade activation monitoring. Chem Sci 8:2723–2728
83. Yuan Y, Zhang R, Cheng X, Xu S, Liu B (2016) A FRET probe with AIEgen as the energy quencher: dual signal turn-on for self-validated caspase detection. Chem Sci 7:4245–4250
84. Li J, Yuan J (2008) Caspases in apoptosis and beyond. Oncogene 27:6194–6206
85. Degterev A, Boyce M, Yuan J (2003) A decade of caspases. Oncogene 22:8543–8567
86. Kumar S (2007) Caspase function in programmed cell death. Cell Death Differ 14:32–43

87. Shi H, Kwok RTK, Liu J, Xing B, Tang BZ, Liu B (2012) Real-time monitoring of cell apoptosis and drug screening using fluorescent light-up probe with aggregation-induced emission characteristics. J Am Chem Soc 134:17972–17981
88. Li H, Parigi G, Luchinat C, Meade TJ (2019) Bimodal fluorescence-magnetic resonance contrast agent for apoptosis imaging. J Am Chem Soc 141:6224–6233
89. Yuan Y, Zhang CJ, Gao M, Zhang R, Tang BZ, Liu B (2015) Specific light-up bioprobe with aggregation-induced emission and activatable photoactivity for the targeted and image-guided photodynamic ablation of cancer cells. Angew Chem Int Ed 54:1780–1786
90. Xia B, Yan X, Fang WW, Chen S, Jiang Z, Wang J, Sun TC, Li Q, Li Z, Lu Y, He T, Cao B, Yang CT (2020) Activatable cell-penetrating peptide conjugated polymeric nanoparticles with Gd-chelation and aggregation-induced emission for bimodal MR and fluorescence imaging of tumors. ACS Appl Bio Mater 3:1394–1405
91. Han K, Wang SB, Lei Q, Zhu J, Zhang X (2015) Ratiometric biosensor for aggregation-induced emission-guided precise photodynamic therapy. ACS Nano 9:10268–10277
92. Cheng Y, Huang F, Min X, Gao P, Zhang T, Li X, Liu B, Hong Y, Lou X, Xia F (2016) Protease-responsive prodrug with aggregation-induced emission probe for controlled drug delivery and drug release tracking in living cells. Anal Chem 88:8913–8919
93. Yuan Y, Kwok RTK, Tang BZ, Liu B (2014) Targeted theranostic platinum (IV) prodrug with a built-in aggregation-induced emission light-up apoptosis sensor for noninvasive early evaluation of its therapeutic responses in situ. J Am Chem Soc 136:2546–2554
94. Qi G, Hu F, Kenry SL, Wu M, Liu B (2019) An AIEgen-peptide conjugate as a phototheranostic agent for phagosome-entrapped bacteria. Angew Chem Int Ed 131:16375–16381
95. Yuan Y, Zhang C, Kwok RTK, Xu S, Zhang R, Wu J, Tang BZ, Liu B (2015) Light-up probe for targeted and activatable photodynamic therapy with real-time in situ reporting of sensitizer activation and therapeutic responses. Adv Funct Mater 25:6586–6595
96. Han H, Teng W, Chen T, Zhao J, Jin Q, Qin Z, Ji J (2017) A cascade enzymatic reaction activatable gemcitabine prodrug with an AIE-based intracellular light-up apoptotic probe for in situ self-therapeutic monitoring. Chem Comm 53:9214–9217
97. Abu R, Woodley JM (2015) Application of enzyme coupling reactions to shift thermodynamically limited biocatalytic reactions. Chemcatchem 7:3094–3105
98. Li J, Wang J, Gavalas VG, Atwood DA, Bachas LG (2003) Alumina–pepsin hybrid nanoparticles with orientation-specific enzyme coupling. Nano Lett 3:55–58
99. Wang X, Hu J, Zhang G, Liu S (2014) Highly selective fluorogenic multianalyte biosensors constructed via enzyme-catalyzed coupling and aggregation-induced emission. J Am Chem Soc 136:9890–9893
100. Lou X, Song Y, Liu R, Cheng Y, Dai J, Chen Q, Gao P, Zhao Z, Xia F (2020) Enzyme and AIEgens modulated solid-state nanochannels: in situ and noninvasive monitoring of H_2O_2 released from living cells. Small 4:1900432
101. Shadel GS, Horvath TL (2015) Mitochondrial ROS signaling in organismal homeostasis. Cell 163:560–569
102. Cairns RA, Harris IS, Mak TW (2011) Regulation of cancer cell metabolism. Nat Rev Cancer 11:85–95
103. Barnham KJ, Masters CL, Bush AI (2004) Neurodegenerative diseases and oxidative stress. Nat Rev Drug Discov 3:205–214
104. Cheng Y, Dai J, Sun C, Liu R, Zhai T, Lou X, Xia F (2018) Intracellular H_2O_2-responsive AIEgen with peroxidase-mediated catalysis for inflammatory cell selective imaging and inhibition. Angew Chem Int Ed 57:3123–3127
105. Qian LH, Li L, Yao SQ (2016) Two-photon small molecule enzymatic probes. Acc Chem Res 49:626–634
106. Wang H, Feng Z, Xu B (2019) Assemblies of peptides in complex environment and their applications. Angew Chem Int Ed 58:10423–10432
107. Du X, Zhou J, Shi J, Xu B (2015) Supramolecular hydrogelators and hydrogels: from soft matter to molecular biomaterials. Chem Rev 115:13165–13307
108. Shi J, Xu B (2015) Nanoscale assemblies of small molecules control the fate of cells. Nano Today 10:615–630
109. Lin Y, Qiao S, Wang Y, Zhang R, An H, Ma Y, Rajapaksha RPYJ, Qiao Z, Wang L, Wang H (2017) An in situ intracellular self-assembly strategy for quantitatively and temporally monitoring autophagy. ACS Nano 11:1826–1839

110. Gao Z, Gao H, Zheng D, Xu T, Chen Y, Liang C, Wang L, Ding D, Yang Z (2020) β-galactosidase responsive AIE fluorogene for identification and removal of senescent cancer cells. Sci China Chem 63:398–403
111. Wang X, Liu H, Li J, Ding K, Lv Z, Yang Y, Chen H, Li X (2014) A fluorogenic probe with aggregation-induced emission characteristics for carboxylesterase assay through formation of supramolecular microfibers. Chem Asian J 9:784–789
112. Han A, Wang H, Kwok RTK, Ji S, Li J, Kong D, Tang BZ, Liu B, Yang Z, Ding D (2016) Peptide-induced AIEgen self-assembly: a new strategy to realize highly sensitive fluorescent light-up probes. Anal Chem 88:3872–3878
113. Coleman JE (1992) Structure and mechanism of alkaline phosphatase. Annu Rev Biophys Biomol Struct 21:441–483
114. Shi Y (2009) Serine/threonine phosphatases: mechanism through structure. Cell 139:468–484
115. Liu H, Li K, Hu X, Zhu L, Rong Q, Liu Y, Zhang X, Hasserodt J, Qu F, Tan W (2017) In situ localization of enzyme activity in live cells by a molecular probe releasing a precipitating fluorochrome. Angew Chem Int Ed 129:11788–11792
116. Tong H, Hong YN, Dong YQ, Haussler M, Lam JWY, Li Z, Guo ZF, Guo ZH, Tang BZ (2006) Fluorescent "light-up" bioprobes based on tetraphenylethylene derivatives with aggregation-induced emission characteristics. Chem Commun 35:3705–3707
117. Zhao Y, Zhang X, Li Z, Huo S, Zhang K, Gao J, Wang H, Liang XJ (2017) Spatiotemporally controllable peptide-based nanoassembly in single living cells for a biological self-portrait. Adv Mater 29:1601128
118. Zhang C, Liu L, Qiu W, Zhang Y, Song W, Zhang L, Wang S, Zhang X (2018) A transformable chimeric peptide for cell encapsulation to overcome multidrug resistance. Small 14:1703321
119. Ji S, Gao H, Mu W, Ni X, Yi X, Shen J, Liu Q, Bao P, Ding D (2018) Enzyme-instructed self-assembly leads to the activation of optical properties for selective fluorescence detection and photodynamic ablation of cancer cells. J Mater Chem B 6:2566–2573

Publisher's Note Springer Nature remains neutral with regard to jurisdictional claims in published maps and institutional affiliations.

REVIEW

Imaging, Identification and Inhibition of Microorganisms Using AIEgens

Harini A. Perera[1] · Mingdi Yan[1]

Received: 31 December 2020 / Accepted: 22 March 2021 / Published online: 9 April 2021

Abstract
Microorganisms, including bacteria, viruses and fungi, are ubiquitous in nature. Some are extremely beneficial to life on Earth, whereas some cause diseases and disrupt normal human physiology. Pathogenic microorganisms can also undergo mutations and develop resistance to antimicrobial agents, which complicates diagnostic and therapeutic regimens. This calls for continuing efforts to develop new strategies and tools that can provide fast, sensitive and accurate diagnosis, as well as effective treatment of ever-evolving infectious diseases. Aggregation-induced emission luminogens (AIEgens) have shown promise in imaging, identification and inhibition of various microbial species. Compared to conventional organic fluorophores, AIEgens can offer improved photostability, and have found utilities in imaging microorganisms. AIEgens have been shown to detect microbial viability and differentiate among different microbial strains. Theranostic AIEgens that integrate imaging and killing of microbes have also been developed. This review highlights examples in the literature where AIEgens have been employed as molecular probes in the imaging, discrimination and killing of bacteria, viruses and fungi.

Keywords AIEgen · Imaging · Identification · Inhibition · Bacteria · Viruses · Fungi

1 Introduction

Microorganisms such as bacteria, viruses and fungi are ubiquitous in nature. They benefit humans, plants, animals and the natural environment by producing oxygen, providing nutrients, and decomposing organic matters. For example, the gut microbiome contains a wide variety of microorganisms in the gastrointestinal tract, and is

Chapter 11 was originally published as Perera, H. A. & Yan, M. Topics in Current Chemistry (2021) 379: 21. https://doi.org/10.1007/s41061-021-00333-x.

✉ Mingdi Yan
mingdi_yan@uml.edu

1 Department of Chemistry, University of Massachusetts Lowell, Lowell, MA 01854, USA

now known to play important roles in human health, and is involved in critical processes like metabolism and immune responses [1]. The gut microbiota is responsible for the absorption of nutrients and minerals, production of short-chain fatty acids, and the synthesis of enzymes, amino acids and vitamins [2].

Microorganisms can also be pathogenic, causing diseases in humans, animals and plants. Certain deadly infectious microbial strains can cause widespread disease, leading to high mortality as well as economic burden. In such cases, rapid diagnosis of pathogenic microbes is a critical first step for both infectious disease control and effective therapies. In addition to clinical settings, pathogens are also prevalent in other areas like the food industry, water and the environment.

Aggregation-induced emission luminogens (AIEgens) are a class of solid-state fluorophores that offer high emission stability. These molecules are weakly emissive in solution, behaving quite the opposite to conventional fluorogens, where aggregation-caused quenching (ACQ) often occurs, due mostly to coplanarity in aromatic structures, leading to the formation of aggregates via π–π interactions. AIEgen designs having propeller-shaped, non-planar rotor structures that restrict stacking through π–π interactions and give high intensity stable fluorescence via restriction of intramolecular vibrational and rotational motions in aqueous solutions [3–5]. The high photostability was attributed to the slower motion of AIEgens in the aggregated state compared to the molecular state, thus producing less heat through radiative decay channels [6]. Additionally, unlike free organic fluorophores that are fully exposed to the environment in solution, only the fluorophore molecules on the surface of the AIE aggregates are exposed and those in the interior are protected from the environment. For example, the percentage of the AIEgen on the particle surface is ~ 27%, and ~ 6% for a particle diameter of 10 nm and 100 nm, respectively, assuming a size of 10 Å for a typical AIEgen like tetraphenylethene (TPE, Scheme 1) 7, [8].

Since 2001, a plethora of AIEgens have been developed with diverse tunability and high quantum yields for applications in medical, electronic and energy fields. In addition to the common AIE core structures (Scheme 1), new building blocks such as quinoline-malonitrile (QM) have been developed [9, 10]. New advancements in AIE research include bioAIEgens obtained from natural products or their derivatives with biocompatible properties, X-ray excited AIEgens with better tissue penetration depths and minimal autofluorescence, and click-chemistry based AIEgens for bioconjugation applications [11–17].

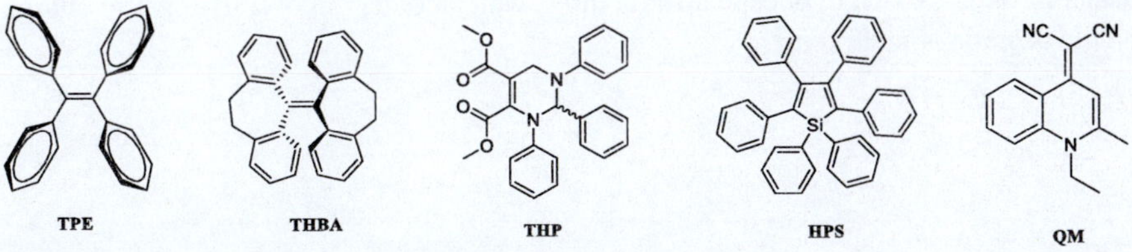

Scheme 1 Common aggregation-induced emission (AIE) core structures

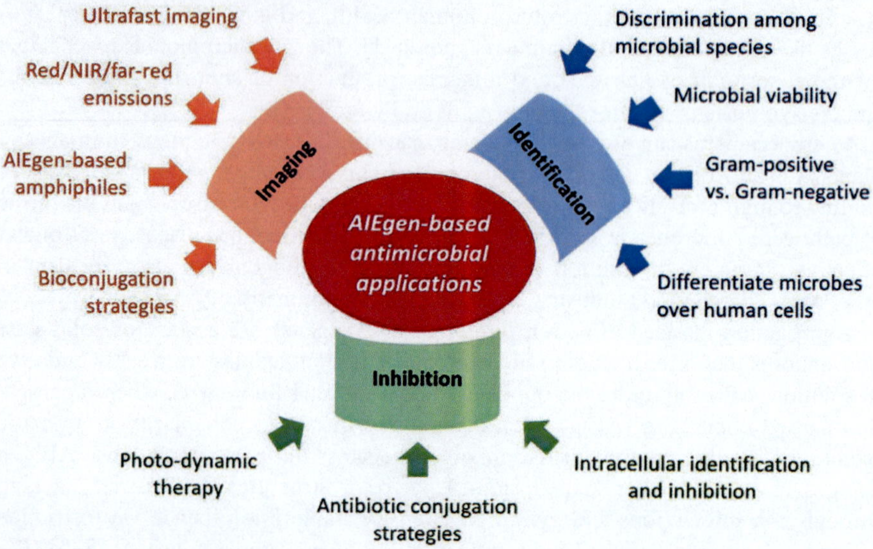

Fig. 1 Aggregation-induced emission luminogens (AIEgens) in antimicrobial applications

This review summarizes applications of AIEgens in imaging, identification and inhibition of bacteria, viruses and fungi (Fig. 1). We highlight examples in the literature where various AIEgen designs have been used for these purposes. As most of the examples are related to bacteria and have targeted bacterial surfaces, it is necessary to clarify some of the terminologies used to describe bacterial surface structures [18]. In bacteria, the plasma membrane, i.e., the inner membrane (IM), encloses the cytoplasm where biological processes take place. The IM and other structures outside the IM are called the bacterial cell envelope. In Gram-positive bacteria, the IM is surrounded by a thick peptidoglycan cell wall that contains teichoic acids. Gram-negative bacteria also have a peptidoglycan layer; however, this is much thinner than that of the Gram-positive bacteria. Additionally, outside the peptidoglycan layer in Gram-negative bacteria is an outer membrane (OM) that contains lipopolysaccharide (LPS). An OM is present in Gram-negative bacteria and is regarded as the major barrier responsible for the ineffectiveness of many antibiotics to Gram-negative infections. Many examples using AIEgens for imaging and killing bacteria have taken advantage of the negative charge of the bacterial surface. The main contributors to the negative surface charge are the teichoic acids in the cell wall of Gram-positive bacteria and LPS in the OM of Gram-negative bacteria. Both are rich in phosphate structures, which confer an overall negative charge on the bacteria at neutral pH.

2 AIEgens for Imaging Microorganisms

2.1 Fluorescence in Bioimaging

Fluorescence-based imaging in biomedical applications offers the advantage of direct visualization of biological molecules and processes. Fluorescence can be used in several different ways in analysis: direct imaging, fluorescence quenching, fluorescence turn-on, and fluorescence energy transfer where the excitation of the fluorophore (acceptor) relies on the energy emitted by a donor molecule [19, 20]. Among the fluorescent tags, organic dyes are the most widely used. They are available in a variety of structures, solubility, emission colors, and functional groups for ease of conjugation. Small molecule-based fluorophores are widely used in biomedical applications, and as contrast agents in positron emission tomography/single photon emission computed tomography (PET/SPECT) and magnetic resonance imaging (MRI) imaging, which are emerging new techniques critical for disease diagnosis [21, 22]. A major drawback of organic small molecule-based fluorophores is photobleaching. High-intensity or prolonged illumination can cause the fluorophore to change its structure so that it can no longer fluoresce (so-called photobleaching), which can reduce and eventually diminish the fluorescence signal [23]. One way to reduce photobleaching is to encapsulate organic dyes in nanoparticles, which serves to protect the dyes from oxygen and the medium, which are the key inducers of photobleaching [24, 25].

New forms of fluorophores that can overcome photobleaching have thus been developed. Examples of such new fluorophores are quantum dots (QDs), <10 nm nanoparticles possessing unique optical and electronic properties as a result of quantum confinement [26]. QDs are bright and are resistant to photobleaching owing to their high extinction coefficients and photochemical stability. These properties, together with their versatility, multiple colors, ease of surface modification, and tunable emissions in near-infrared (NIR) region (900–1400 nm) have made QDs popular fluorophores for various bioimaging applications [27–29]. QDs, however, suffer from ACQ—where the emission decreases drastically upon molecular aggregation or nanoparticle aggregation.

2.2 Imaging Microorganisms with AIEgens

AIE materials have been used for in vitro cell imaging. The modes include one-photon imaging, fluorescence imaging, near-infrared (NIR) imaging and photoacoustic (PA) imaging [30–34]. AIEgens generally have a non-planar structure with rotors. The AIE cores most commonly used are TPE, THBA (10,10,11,11-tetrahydro-5,5-bidibenzo[a,d][7]-annulenylidene), THP (tetrahydropyran), HPS (hexaphenylsilole) and QM (Scheme 1), which are building blocks for many other new AIEgens [35]. Fluorophores have also been transformed into AIEgens by attaching AIE cores to the periphery. Introduction of donor/acceptor groups into the π conjugate system of AIEgen allows fine-tuning absorption/emission wavelengths to cover the visible and NIR regions [36].

AIEgens have been used as probes to detect microorganisms. One mode of such use is aggregation in aqueous solutions through hydrophobic interactions. AIEgens in the molecular form show minimal fluorescence whereas aggregates are highly fluorescent; this helps to achieve high fluorescence signals and minimal background noise after binding to microbes [35]. The initial design of AIE imaging probes included cationic charges to facilitate binding with negatively charged bacteria [37]. However, electrostatic and hydrophobic interactions are relatively weak, which resulted in non-specific fluorescence being turned on. Improved designs were developed later to enhance detection specificity in a complex environment [38]. One such design involved the conjugation to the AIEgen of a targeting ligand, such as a peptide, an aptamer or a protein, that can specifically recognize and bind to the receptor on the microbe [39].

Another design relies on the conjugation of reactive functional groups such as cleavable quenchers or bio-orthogonal clickable groups to AIEgens. In this case, fluorescence turn-on is initiated by a specific chemical or biological reaction, and, as such, false positives resulting from non-specific binding are minimized [30, 32]. These strategies achieve fluorescence turn-on with enhanced specificity while keeping the background noise to a minimum without additional washing steps. Antimicrobial peptides (AMPs) are antimicrobial agents popularized due to their broad-spectrum activities against bacteria [40]. Most cationic AMPs can interact with the negatively charged bacteria cell wall or permeabilize to attack negatively charged internal organelles such as RNA. They disrupt cell wall integrity by pore formation or membrane destabilization [41]. An AIEgen-AMP probe, HHC36-2HBT (Fig. 2a), synthesized by conjugating an antimicrobial peptide HHC36 to an AIEgen 2-(2-hydroxyphenyl)benzothiazole (HBT), was developed to investigate the bactericidal mechanism of AMP [42]. Super-resolution fluorescence images in Fig. 2b showed that probe aggregated on the cell surface of *Escherichia coli*, causing a fluorescence turn-on. *E. coli* incubated with the AIEgen HBT-COOH itself did not show any fluorescence, which indicated that AMP was the cause of the binding of AMP-2HBT. Antimicrobial activity data showed that HHC36-2HBT probe was equally efficient as HHC36 at killing bacteria.

The ability of D-mannose (Man) to bind the FimH lectin on the fimbriae of *E. coli* was exploited to design an AIE probe to image *E. coli* (Fig. 3a) [43]. A

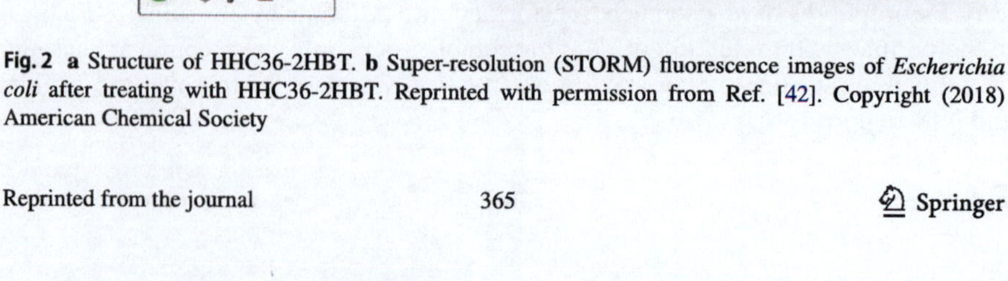

Fig. 2 **a** Structure of HHC36-2HBT. **b** Super-resolution (STORM) fluorescence images of *Escherichia coli* after treating with HHC36-2HBT. Reprinted with permission from Ref. [42]. Copyright (2018) American Chemical Society

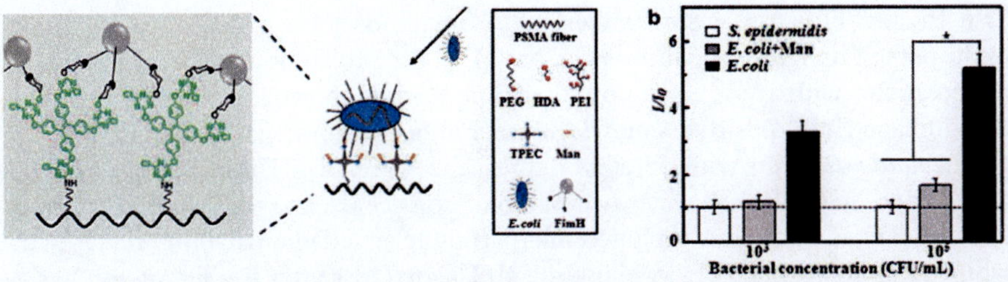

Fig. 3 **a** D-Mannose (Man)- and tetraphenylethene (TPE)-grafted polystyrene-co-maleic anhydride (PSMA) fiber to detect *E. coli*. **b** Fluorescence intensity before (I_0) and after (I) incubation of the fiber with bacteria. Reprinted with permission from Ref. [43]. Copyright (2015) American Chemical Society

mannose- and TPE-conjugated polystyrene-co-maleic anhydride (PSMA) fiber was designed to trap the bacteria and to trigger fluorescence turn-on through specific binding to FimH on *E. coli* (Fig. 3b). The fluorescence intensity was found to be correlated linearly to the *E. coli* concentration. The fibrous mat could be made into a strip for the ease of detection.

Vancomycin, a glycopeptide antibiotic that binds to the peptidoglycan layer in the cell wall of Gram-positive bacteria, was conjugated to a red AIEgen (Fig. 4a) to image the Gram-positive bacterium *Bacillus subtilis*. In this example, incorporating antibiotics along with AIE photosensitizing properties enhanced the bacteria-killing efficiency [44]. Upon binding to the bacteria, the probe AIE-2Van exhibited fluorescence turn-on (Fig. 4b). Both AIE-2Van and RBA-BTPEBT [45] probes were used to detect *B. subtilis* and ricin simultaneously in the same sample. Intrinsic fluorescence of both probes was eliminated by the addition of graphene oxide (GO).

Multiphoton excitation is where two or more photons arrive at the sample at once to excite a fluorophore from the ground to the excited state. This is advantageous for deep tissue imaging and reducing substrate autofluorescence while causing minimal photochemical damage to live tissues [46]. Three TPE-based probes (TPE-Quino,

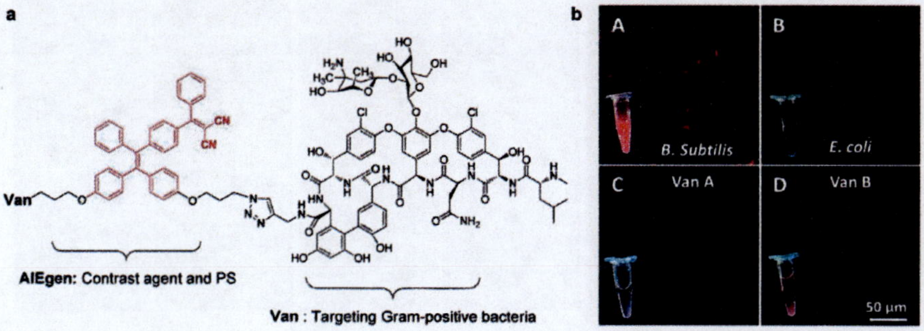

Fig. 4 **a** AIE-2Van structure. **b–d** Confocal laser scanning microscope (CLSM) images of *Bacillus subtilis* (**a**), *E. coli* (**b**), Van A (**c**), and Van B (**d**), incubated with AIE-2Van. *Insets* Images in the presence of graphene oxide (GO). Reprinted with permission from Ref. [44]. Copyright (2015) Royal Society of Chemistry

TPE-Py and TPE-Acr) were synthesized to image bacteria under single-photon or multi-photon imaging modalities (Fig. 5a) [47]. The probe molecules assembled into nanoparticles with average sizes of 20–40 nm. Results showed that the nanoparticles could image Gram-positive *Staphylococcus aureus*, Gram-negative *E. coli*, and fungus *Candida albicans* with different fluorescence (Fig. 5b). These TPE probes were also able to image *E. coli* under two-photon fluorescence imaging modality. Continuous laser irradiation in two-photon microscopy revealed photostability of TPE-Arc, while TPE-Quino and TPE-Py showed a 40% signal loss after irradiation for 30 s.

An increasing number of studies has used red/NIR/far-red emitting AIEgens in microbial imaging due to the advantages of minimal autofluorescence and photodamage to tissues [48–50]. Recently, a NIR emissive AIEgen was reported for ultrafast imaging and discrimination of Gram-positive from Gram-negative bacteria in a mixed sample by controlling the incubation time [48]. Another example of NIR emissive AIEgens was based on the diphenyl isoquinolinium (IQ) motifs [51]. The positive charge acts as an electron acceptor, which makes it feasible to tune emission properties by changing electron donors. Figure 6a shows examples of IQ AIEgens

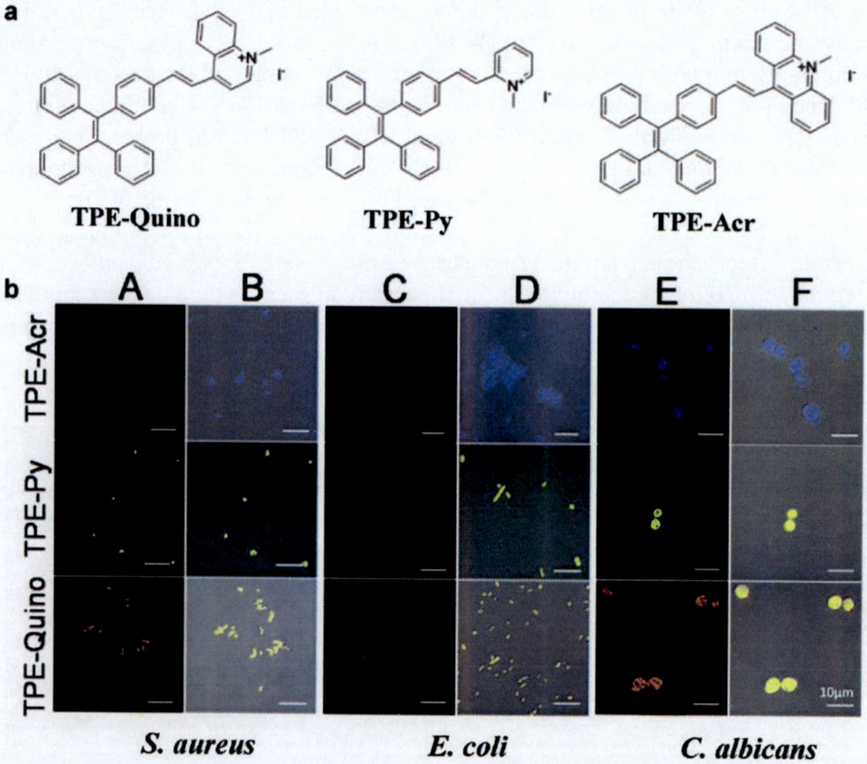

Fig. 5 **a** Structures of TPE-Quino, TPE-Py and TPE-Acr. **b** One-photon images of microbes using TPE nanoparticles. *Staphylococcus aureus* (*A, B*), *E. coli* (*C, D*), *Candida albicans* (*E, F*). Reprinted with permission from Ref. [47]. Copyright (2017) Royal Society of Chemistry

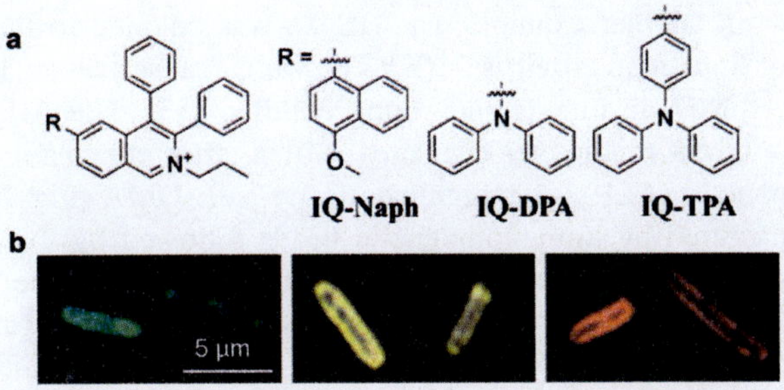

Fig. 6 **a** Structures of IQ-Naph, IQ-DPA, and IQ-TPA. **b** Confocal fluorescence images of *E. coli* stained by IQ-Naph, IQ-DPA, and IQ-TPA (from *left* to *right*), respectively. Reprinted with permission from Ref. [51]. Copyright (2017) John Wiley & Sons Inc

having different donor groups to accomplish longer excitation and emission wavelengths: IQ-Naph (naphthalene), IQ-DPA (diphenylamine), and IQ-TPA (triphenylamine). Experiments were performed to evaluate the wash-free bacterial staining capabilities of these probes. Upon incubation with *E. coli* for 10 min at a probe concentration of 2 μM, the bacteria became fluorescent and the colors were consistent with the emission of the corresponding probes (Fig. 6b). The fluorescence turn-on was postulated as the result of interruption to molecular motions upon binding to the bacteria through electrostatic and hydrophobic interactions. All three probes imaged both live and dead bacteria without differentiation.

AIE probes have also been used for the detection of viruses. One such example is a TPE-based oligosaccharide probe to detect influenza virus [52]. The probe was synthesized by conjugating lactosyl and 6′-sialyllactosyl moieties to a TPE core (Fig. 7a). Upon binding to hemagglutinin (HA) on the cell envelope of the virus, the probe aggregated and showed strong fluorescence (Fig. 7b). In the presence of the virus, emission at 460 nm increased, giving a fluorescence turn-on at virus concentrations higher than 10^5 plaque-forming unit (PFU)/100 μl.

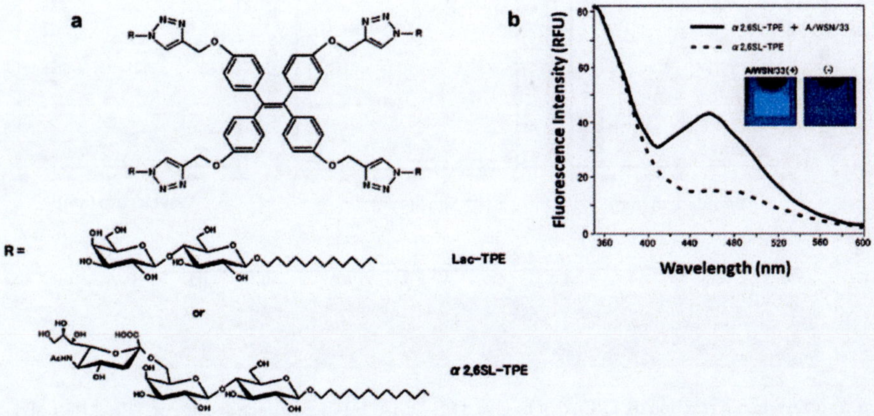

Fig. 7 **a** AIEgen α2,6SL-TPE for the detection of the influenza virus. **b** Fluorescence spectra and photographs of α2,6SL-TPE in the presence or absence of the influenza virus A/WSN/33. Reprinted with permission from Ref. [52]. Copyright (2010) Elsevier

In another example, an AIEgen was coupled to immunoassay for the detection of human enterovirus 71(EV71), which causes severe hand, foot and mouth disease (HFMD) in infants and young children [53]. The AIEgen TPE-APP was designed with enzymatic cleavage sites, which, after enzymatic hydrolysis by alkaline phosphatase (ALP), gave highly emissive TPE-DMA (Fig. 8a–c). The target virions were captured by immunomagnetic beads followed by the binding of rabbit polyclonal antibodies (P-Abs), biotinylated Abs (BAbs), and the signal tags streptavidin-ALP (SA-ALP). The produced TPE-DMA formed aggregates to trigger strong fluorescence with a 380-fold enhancement in intensity. Meanwhile, the hydrolysis product of TPE-APP simultaneously reduced Ag^{+} to form a silver nanoshell around gold nanoparticles. This resulted in a plasmonic color change allowing the direct visual detection of virus (Fig. 8d, e). Thus, this AIEgen immunoassay integrated fluorescence and plasmonic colorimetry in a single detection tool.

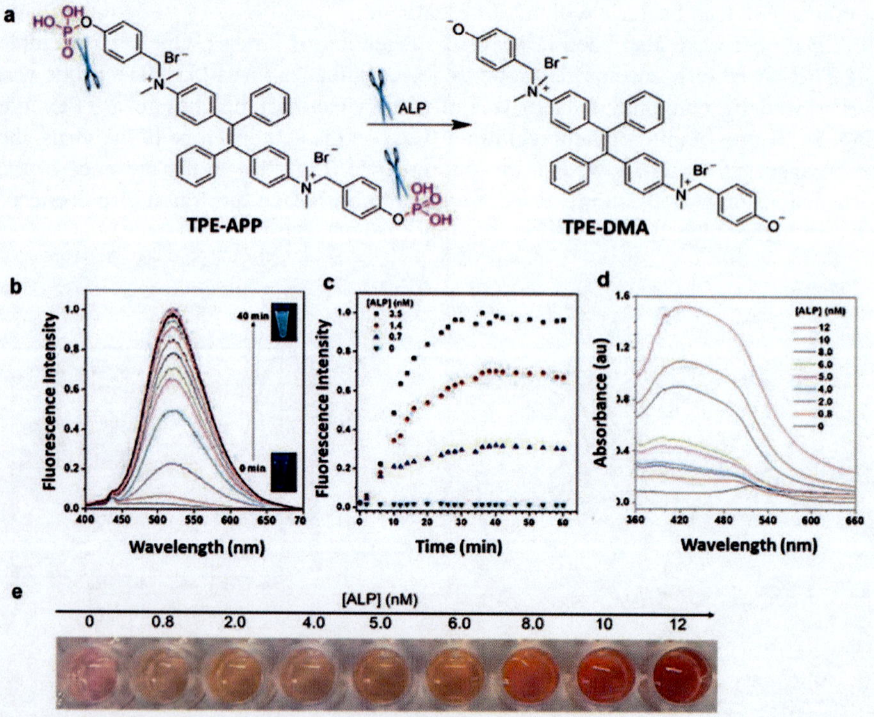

Fig. 8 **a** Enzymatic reaction of TPE-APP to give TPE-DMA. **b** Fluorescence spectra of TPE-APP + alkaline phosphatase (ALP) versus incubation time. **c** Fluorescence intensity versus incubation time in the presence of ALP. **d** Absorption spectra of the mixture of AuNP, Ag^{+}, and TPE-APP with various concentrations of ALP after 30 min incubation. **e** Photograph of the color change in (**d**). Reprinted with permission from Ref. [53]. Copyright (2018) American Chemical Society

3 Discrimination of Microorganisms Using AIEgens

3.1 Methods for Discrimination of Microbial Species

Identification of microbial species is important in applications like food processing, clinical diagnosis, pharmaceutical development, and environment monitoring [54]. Discrimination among Gram-positive vs. Gram-negative bacteria, strain, viability, and total counts are important in surveilling microbial contamination to ensure food safety and to eliminate risks to human health. Therefore, the desired detection tools need to give rapid and accurate results using simple procedures [55–57].

The gold standard for the identification of different microbial species is plating and culturing, where the microorganism is plated on growth medium and then identified based on its characteristic microbiological features. The drawback is that these protocols can be time consuming, which is inconvenient, particularly when a rapid decision is required [58]. Nucleic acid probe-based methods, like polymerase chain reaction (PCR), have also been popular in microbial detection as they give conclusive and unambiguous results, and allow the detection of trace amounts of target molecules [59]. However, PCR cannot differentiate viable and non-viable cells [60]. Immunological methods such as enzyme-linked immunosorbent assay (ELISA) consist of pre-treatment steps to capture the target pathogen from the cocktail using antibodies [61] and coupling with a detection method such as optical or magnetic force microscopy or magnetoresistance [62, 63]. The drawbacks include false positives with non-specific materials in the sample, and the difficulty of differentiating viable from non-viable pathogens [64].

Optical techniques have been the most attractive method because of their non-destructive nature and speed. Light scattering, surface plasmon resonance, and fluorescence-based methods are used widely for this purpose [65]. Light scattering techniques differentiate samples based on refractive index, size, shape, and composition [66]. However, stage of growth, growth medium, growth temperature, aeration, and the final dilution of the suspended medium can affect the reproducibility of this method [67]. Surface plasmon resonance (SPR) and localized surface plasmon resonance (LSPR) are refractive index-based sensing techniques, and are rapid and label-free [68]. Either is inherently selective and, therefore, detection specificity comes from surface-confined biorecognition elements. In common with all surface adsorption-based approaches, the possibility of non-specific adsorption giving false positive signals is a pervasive issue [69]. Fluorescence methods based on exogenous fluorescence for detection involves chemically tagging appropriate sensing element with a fluorophore. Synthetic fluorophores have been engineered to have tunable excitation and emission, high quantum efficiency, good photostability, excellent specificity when conjugated to a recognition element, and the possibility to detect single molecules and microorganisms [69, 70]. Fluorescence also offers the possibility of spectral multiplexing. For example, Wang et al. [13] employed two different antibodies to simultaneously detect three pathogens from artificially contaminated food with a limit of detection (LOD) of 10^3 CFU/mL. More recently, a droplet-based fluorescence immunoassay was developed to selectively detect *E. coli* and

Salmonella typhimurium in a mixture with a LOD of 10^2 CFU /ml for both bacteria [71].

3.2 AIEgens in Discrimination of Microbial Species

AIEgen have been developed for microbial identification and differentiation. For example, an AIE probe, 2-[(diphenylmethylene)hydrazono]methyl naphthalene (M1-DPAN) (Fig. 9), was able to selectively identify Gram-positive bacteria [72]. While minimal fluorescence was observed on Gram-negative (*Pseudomonas aeruginosa, E. coli*) and fungi (*Saccharomyces cerevisiae, C. albicans*) (Fig. 9a–h), M1-DPAN probe showed enhanced fluorescence on Gram-positive bacteria *S. aureus*, *B. subtilis* and *Enterococcus faecalis* (Fig. 9i–o). For *S. aureus*, the emission lasted for 24 h. The probe could also differentiate Gram-positive bacterium *B. subtilis* from its mixture with *P. aeruginosa* and *S. cerevisiae* (Fig. 9g, p).

Positive charges have also been introduced to AIEgens to aid bacterial discrimination [73, 74]. In one case, a TPE probe bearing a pyridinyl group was able to detect *E. coli* [75], whereas in another case, a TPIM probe bearing an imidazolium

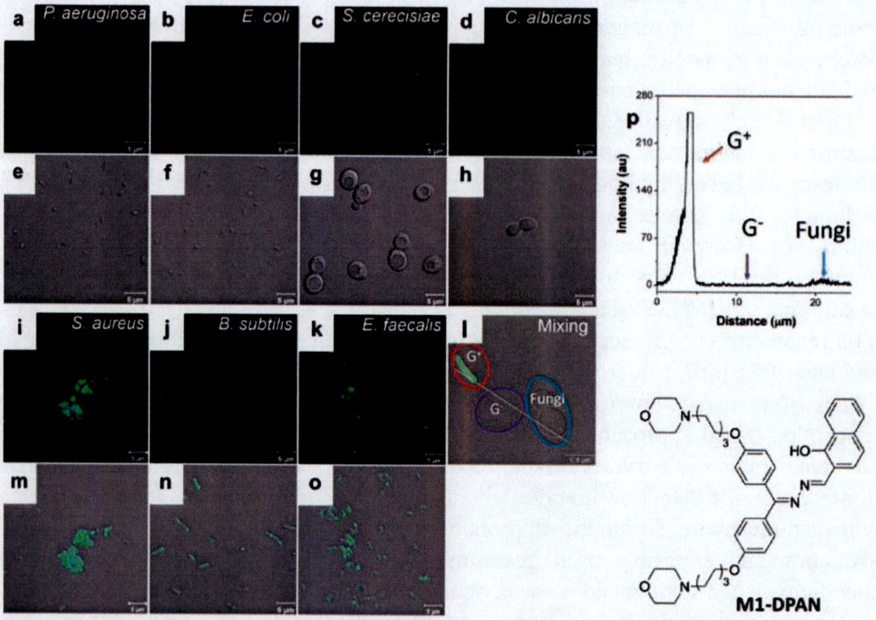

Fig. 9 Differentiation of Gram-positive from Gram-negative bacteria and fungi by M1-DPAN (structure in *bottom right panel*). **a–o** CLSM images of the adsorption of M1-DPAN on the surfaces of *Pseudomonas aeruginosa* (**a, e**), *E. coli* (**b, f**), *Saccharomyces cerevisiae* (**c, g**), *C. albicans* (**d, h**), *S. aureus* (**i, m**), *B. subtilis* (**j, n**), and *Enterococcus faecalis* (**k, o**). **p** Differentiation of Gram-positive (*G*+) and Gram negative (*G*–) bacteria from a mixture. Reprinted with permission from Ref. [72]. Copyright (2018) Elsevier

group selectively recognized Gram-positive *S. aureus* [76]. A positively charged AIEgen was reported with potential to selectively image both Gram-positive and Gram-negative bacteria over mammalian cells, which is useful when studying bacterial infections [77].

Detection selectivity can be improved significantly using sensor arrays and in conjunction with multivariant analysis. A fluorescence sensor that combined an AIEgen and GO was able to identify six microbes, including two fungi (*C. albicans* and *S. cerevisiae*), two Gram-positive bacteria (*S. aureus* and *B. subtilis*), and two Gram-negative bacteria (*P. aeruginosa* and *E. coli*) [78]. Seven AIEgens were used in the sensor array: negatively charged (A1), positively charged (A2, A3, A4, A6, and A7), and neutral (A5) (Fig. 10a). Two strategies were evaluated: direct incubation of seven AIEgens with bacterial lysates (Fig. 10b) and mixing bacteria lysates with AIEgen/GO complex (Fig. 10c). In the AIEgen/GO system, GO served as the fluorescence quencher and for the adhesion of biomolecules. Interaction with microbial lysates would reduce the amount of biomolecules and form a competitive interaction among GO, AIEgen and biomolecules. Using principle component analysis (PCA), six microbes were identified with 100% accuracy (Fig. 10d).

Another series of sensor arrays based on AIE-active TPE derivatives (TPE Ar; Fig. 11a) was developed for the detection and discrimination of microbes [79]. Interactions between the bacterium and the AIEgens hindered the rotation of TPE, enhancing fluorescence emission. All seven TPE-ARs contained quaternary ammonium group, and different hydrophobic substitutions provided tunability of interactions with microbes. TPE-ARs showed diverse fluorescence response

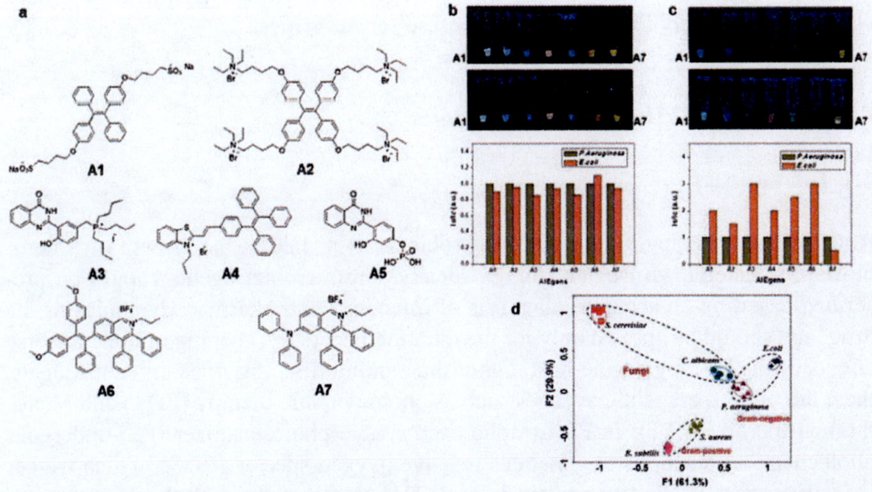

Fig. 10 AIEgen arrays for identification of bacteria. **a** Structures of selected AIEgens. **b** Direct incubation of seven AIEgens with two bacteria lysates (*top P. aeruginosa, bottom E. coli*). **c** Fluorescence of two bacteria lysates incubated with AIEgen/GO complex. **d** Principle component analysis (PCA) plot generated from six microbial lysates. Reprinted with permission from Ref. [78]. Copyright (2018) American Chemical Society

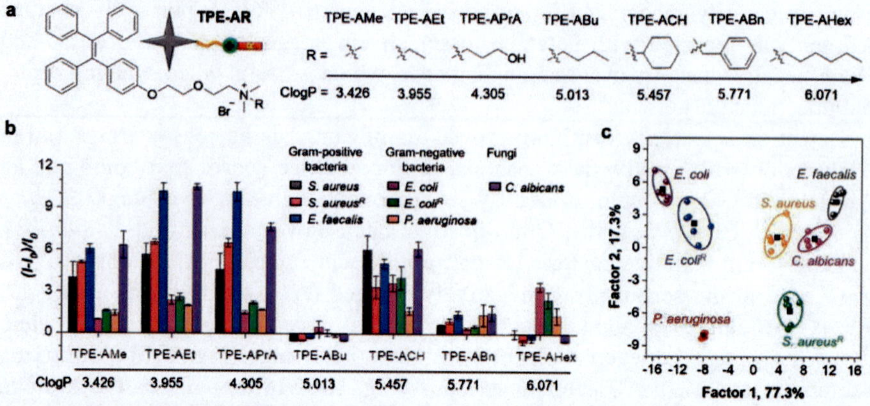

Fig. 11 **a** Structures of AIEgen TPE-ARs. **b** Fluorescence responses of TPE-ARs after the addition of different microbes. **c** Two-dimensional (2D) canonical score plot by linear discriminant analysis (LDA) for TPE-APrA, TPE-ACH, and TPE-AHex. Reprinted with permission from Ref. [79]. Copyright (2018) John Wiley & Sons Inc

signals against the seven microbes due to different microbe–probe interactions, as shown in Fig. 11b. In total, 14 sensor arrays were constructed, and each consisted of three TPE-ARs to balance fluorescence responses. The fluorescence response patterns produced by the sensor array were analyzed using linear discriminant analysis (LDA). Figure 11c is a two-dimensional (2D) canonical score plot by LDA for TPE-APrA, TPE-ACH, and TPE-AHex. The seven pathogens were well separated from each other, falling into seven different groups. Blends of microbes were also separately identifiable using these sensor arrays.

4 Antimicrobial Theranostic AIEgens

4.1 Theranostics

Antimicrobial therapy aims to kill (microbiocidal) or inhibit the growth of (microbiostatic) bacteria, viruses and fungi. Ideally, antimicrobial agents should be prescribed based on an accurate diagnosis of infections and pharmacodynamics of the drug, and should be applied only for the duration necessary, keeping in mind adverse effects of the therapy on the host. Other than antibiotics, antivirals and antifungals, there are other therapeutic avenues such as photodynamic therapy (PDT) and photothermal therapy (PTT). In PDT, a photoactivatable photosensitizer (PS) undergoes photochemical reactions to generates reactive oxygen species (ROS), which trigger oxidative stress leading to protein injury, DNA damage, intracellular lipid peroxidation and, eventually, cell death [80, 81]. In PTT, photon energy is employed to induce heat at a specific localized site to destroy target cells [82].

Theranostics is an emerging approach that integrates diagnostics and therapeutics in a single system [83]. Advantages of theranostics include improved therapy selection on the basis of specific molecular features of disease, greater predictive power for adverse effects, and new ways to objectively monitor therapeutic response [83–85]. These properties are fundamental elements of personalized medicine. For example, in breast cancer, an immunohistochemical test to determine the status of the human epidermal growth factor receptor 2 (*HER2*) expression is a prerequisite for a treatment regimen using the monoclonal antibody trastuzumab (Herceptin) [86]. A theranostic system that couples the therapy with diagnostic information specific for the intended target would be ideal [87].

Nanomaterials and supramolecular structures are well-suited to theranostics because of their modular construction [84, 88]. These nanostructures offer several advantages in structural configurations that are not readily possible with traditional small organic molecules, including the ability to carry targeting agents, imaging moieties, and drugs. The nanostructure itself can constitute the biologically active entity or serve as a carrier to accommodate a wide variety of drug compositions, potentially enabling the use of drug candidates that have been shelved due to poor solubility or pharmacokinetic profiles [88–90]. Numerous nanomaterial-based theranostic systems have been developed especially in cancer therapies. They can be grouped into two general types: untargeted or targeted [91, 92]. The currently United States Food and Drug Administration (FDA)-approved nanoparticle-based therapies fall into the first category, which basically relies on the enhanced permeability and retention (EPR) effect, which is the passive accumulation of nanoparticles at tumor sites through variably leaky neovasculature [93]. Targeting can be achieved by a number of methods, including the use of specific chemical or biologic ligands to target cell surface receptors. Targeted nanomedicines remain the subject of highly active research, particularly for cancer-related diseases, and are likely to enter clinical trials in the near future [89, 91, 92].

4.2 AIEgens in Theranostic Applications

AIEgens have been configured into theranostic platforms through different strategies, for example, by conjugating an antimicrobial agent to the AIEgen, and to enable PDT and PTT capabilities of AIEgens. Bacterial invasion into host cells provides protection of the bacteria from antibiotics and the host immune system. Many AIEgens have been developed successfully to simultaneously detect and eliminate intracellular bacteria [94–96].

Several methods have been used to fabricate AIE nanoparticles. One involves the chemical modification of AIEgens with a hydrophilic polymer or biomolecule to yield amphiphilic AIEgen-polymer conjugates, which self-assemble into micelles. In addition to the need for chemical modification of each AIEgen, the size distribution of the resulting AIE nanoparticles can be difficult to control and predict. An alternative method is to physically encapsulate AIEgens into a matrix to produce AIE nanoparticles with good water dispersibility. The matrix also reduces photobleaching to give brighter emission. One of the most commonly used matrices is amphiphilic polymers, such as 1,2-distearoyl-sn-glycero-3-phosphoethanolamine-*N*-[methoxy(poly(ethylene glycol))] (DSPE-PEG) or derivatives. Nanoparticles can be prepared by nanoprecipitation by adding water to the solution containing the AIEgen and amphiphilic polymer in a water-miscible organic solvent [36, 97].

A bacterial detection and inhibition AIE motif was developed, consisting of a TPE core and bacteria-binding and amphiphilic/cationic poly(DMA-co-BMA-co-Gd) arms prepared from DMA (2-(dimethylamino)ethyl methacrylate) an BMA (butyl methacrylate) copolymer grafted with a quaternary ammonium moiety and the T1-type magnetic resonance (MR) imaging contrast agent Gd (Fig. 12a) [98]. TPE and Gd together led to a synergistic fluorescence turn-on and an enhancement in MR longitudinal relaxation rate. TPE/copolymer/Gd was antibacterial against both Gram-negative (*E. coli* and *P. aeruginosa*) and positive (*S. aureus*) strains (Fig. 12b). The fluorometric and MR imaging modes exhibited LOD of ~8.5×10^5 CFU ml^{-1} and ~5×10^3 CFU ml^{-1} against *E. coli*, respectively. The MR mode was more sensitive than the fluorescence mode because of the magnetic amplification of MR imaging agents as they affect the relaxation properties of all surrounding water protons, whereas fluorescence detects the imaging probe only. Untreated *E. coli* preserved its smooth rod-like shape (Fig. 12c) while treated bacteria exhibited aggregation and deformation (Fig. 12d). No *E. coli* colonies could be seen after 4 h incubation and re-inoculation, whereas untreated bacteria showed ~ 200 colonies under the same experimental conditions.

Antibiotics have also been used to construct theranostic AIEgens. A fluoroquinolone-based AIEgen has been developed that showed antibacterial activity as well as AIE properties [99, 100]. The AIEgens were synthesized by conjugating a

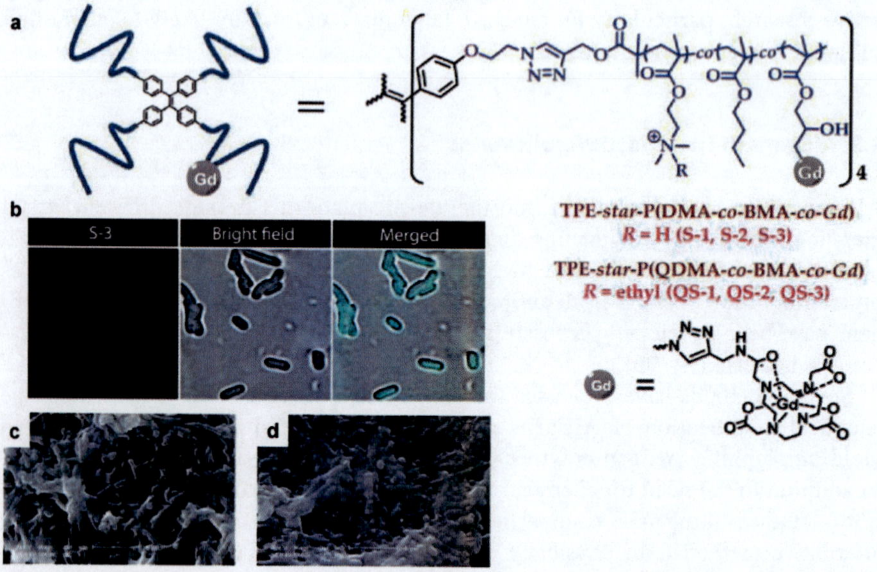

Fig. 12 **a** TPE/copolymer/Gd AIE probe. **b** CLSM images recorded for *E. coli* upon S-3 labeling. **c, d** Scanning electron microscopy (SEM) images of *E. coli* after incubating for 2 h at 37 °C in the absence (**c**) and presence (**d**) of S-3. Reprinted with permission from Ref. [98]. Copyright (2014) John Wiley & Sons Inc

fluoroquinolone antibiotic, like ciprofloxacin or norfloxacin to perfluoroaryl, and a phenyl ring through an azide-mediated amidination reaction to give a product that possesses a propeller-shaped, twisted donor (piperazinyl ring)–acceptor (fluoroquinolone ring) structure (Fig. 13a) [101, 102]. Nanoprecipitation produced spherical nanoaggregates with high stability and strong AIE enhancement with up to 11% quantum yield (Fig. 13b). Minimal inhibitory concentrations (MICs) of the soluble form and the nanoaggregates were compared, where the compounds in the nanoaggregate form resulted in lower MICs than the soluble form (Fig. 13c). For example, compound 5 ®=F, R′=cyclopropyl: ciprofloxacin) had an MIC of 233 μM in the soluble form against *E. coli* ORN178, whereas the nanoaggregates (39–74 nm) gave an MIC of 1.25–8.7 μM, up to two logs lower (Fig. 13c). Upon incubation with *E. coli*, colocalization of nanoaggregates could be seen on the bacteria (Fig. 13d), suggesting that the enhanced antimicrobial activity might be due to higher local concentration of the nanoaggregates around the bacterial cells.

Triphenylethylene-naphthalimide triazole (TriPE-NT) was developed to introduce PDT, by conjugating the inherently fluorescent TriPE and the antibacterial agent NT (Fig. 14a) [103]. The positively charged TriPE-NT probe was capable of imaging and inhibiting both Gram-negative and Gram-positive bacteria at low concentrations. Under light irradiation, the TriPE-NT probe generated ROS, resulting in antibacterial activity of the probe by PDT. Under irradiation with white light, all *E. coli* and more than 96% multi-drug resistant (MDR) *E. coli*, 90% *K. pneumoniae*, 61% MDR *K. pneumoniae*, 98% *S. aureus*, and 97% MDR *S. aureus* were killed

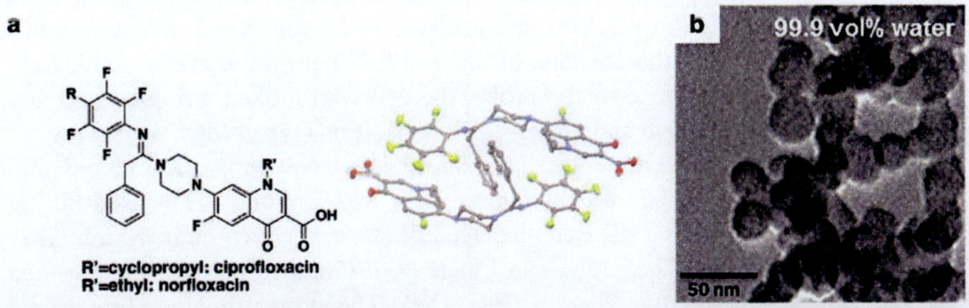

c

	Free drug		Nanoaggregates			
	MIC$_{Free}$ [μM]	n	MIC$_{Aggregate}$, μM	n	Size†, nm	MIC$_{free}$/MIC$_{aggregate}$*
			E. coli ATCC 25922			
5	116–232	3	1.25–8.6	6‡,§	39–74	36
6	29–116	3	5.0	3	49–127	12
7	19–78	3	2.5	3¶	38–51	23
8	75–150	3	5.0–8.9	5‡	39–90	22
			E. coli ORN 178			
5	233	2	1.25–8.7	8#	39–80	44
7	78	2	2.5–5.3	5‡	37–90	25

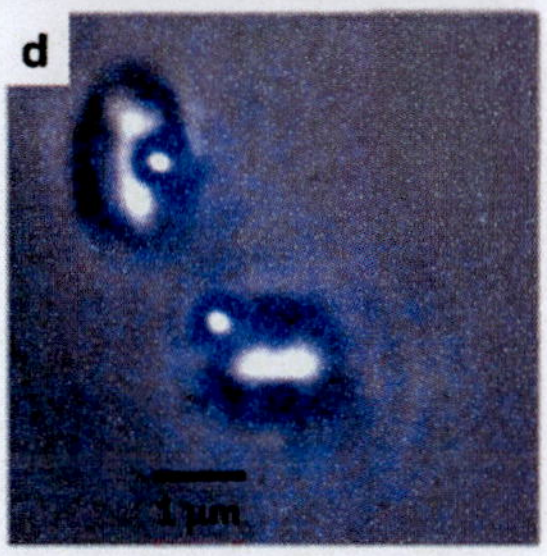

Fig. 13 **a** Fluoroquinolone-AIEgens. **b** SEM images of nanoaggregates prepared by nanoprecipitation. **c** MICs against *E. coli* strains. **d** Confocal fluorescence microscopy images of *E. coli* ORN178 incubated with nanoaggregates. Reprinted with permission from Ref. [99]. Copyright (2017) National Academy of Sciences USA

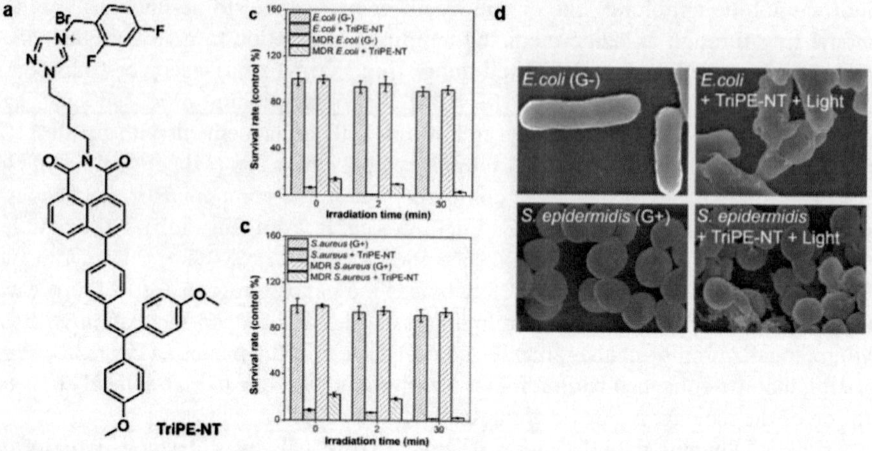

Fig. 14 **a** Triphenylethylene-naphthalimide triazole (TriPE-NT). **b**, **c** Percent survival of *E. coli* (**b**) and *S. aureus* (**c**) treated with TriPE-NT under light irradiation. **d** SEM images of TriPE-NT-treated *E. coli* and *Staphylococcus epidermidis*. Reprinted with permission from Ref. [103]. Copyright (2018) John Wiley & Sons Inc

(Fig. 14b, c). Confocal fluorescence images of TriPE-NT (orange) co-incubated with FM4-64FX (a red lipid membrane dye) revealed that TriPE-NT stained only the cell wall of *E. coli* but stained the whole cell of *Staphylococcus epidermidis*. TriPE-NT stained both live and dead bacteria without differentiation. Field-emission high-resolution transmission electron microscopy (TEM) and SEM images showed no change in the shape in the absence of the TriPE-NT probe; however, with light irradiation and in the presence of the probe, the cell wall looked wrinkled and disrupted in both Gram-negative and Gram-positive bacteria (Fig. 14d).

An amphiphilic fluorescent probe, TPE-Bac, having two long alkyl chains and two positively charged amines was synthesized for imaging and PDT-based killing of bacteria (Fig. 15a) [104]. TPE-Bac displayed fluorescence turn-on in 10 min after incubation with bacteria. The probe can image both Gram-positive *S. epidermidis* (Fig. 15c) and Gram-negative *E. coli* (Fig. 15e). The probe shows antimicrobial activity in dark conditions. In the dark, toxicity towards *S. epidermidis* was higher than towards *E. coli*. The pyridinium cation is electron-withdrawing whereas ether is an electron donating group. The donor–acceptor pair across TPE facilitated intramolecular charge transfer, lowering the energy gap. As a result, the absorption spectrum displayed bathochromic shift. Its broad absorption in the ranges of UV, blue and green regions enables it for photodynamic therapy. Upon light irradiation, the TPE-Bac AIEgen acted as a photosensitizer to generate ROS, which increased the efficiency of the inherent bactericidal effect of the probe. Percent survival for *E. coli* and *S. epidermidis* after incubation for 10 min under light irradiation was 45% and 40%, respectively. SEM images showed membrane disruptions (Fig. 15f–i).

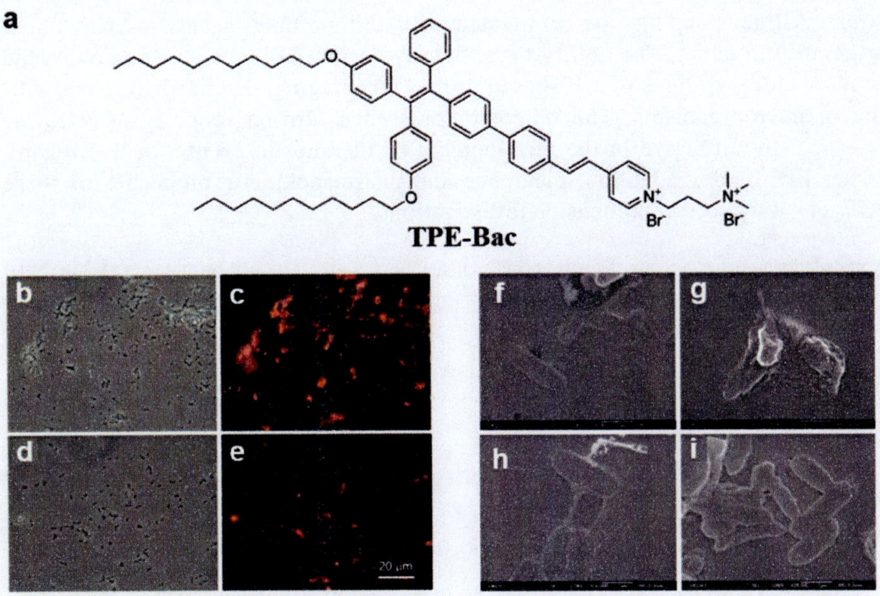

Fig. 15 **a** TPE-Bac. **b–e** Bright field (**b, d**) and fluorescence images (**c, e**) of *S. epidermidis* (**b, c**) and *E. coli* (**d, e**) incubated with 10 μM TPE-Bac for 10 min. **f–i** SEM images of *S. epidermidis* (**f, g**) and *E. coli* (**h, i**) incubated without (**f, h**) and with (**g, i**) 10 μM TPE-Bac for 10 min. Reprinted with permission from Ref. [104]. Copyright (2015) American Chemical Society

5 Conclusions

This review highlighted some recent examples of AIE-based luminogens in imaging, discrimination and growth inhibition of microorganisms. Imaging was achieved mainly by conjugating AIEgens with charged moieties, functional groups or aptamers, or by loading AIEgens into nanoparticles functionalized with such groups targeting virulence factors to form intermolecular interactions with microbes selectively. Such binding would restrict the intramolecular motion of AIEgens to trigger fluorescence turn-on. In most examples, emission properties were modulated by using a suitable donor–acceptor pair of fluorophores across the AIEgens to facilitate intramolecular charge transfer by lowering the energy gap. This would lead to longer emission wavelengths, expanding the absorption range for the AIEgen, as near-IR emissions are more desirable for non-invasive deep-tissue imaging. In differentiating microbial species, differences in the cell wall structures, for example, the peptidoglycan layer of Gram-positive and the OM of Gram-negative bacteria, were exploited. Detection specificity can be enhanced greatly using sensor arrays, where multiple AIEgens showed diverse fluorescence responses against different microbes. Multivariant data analysis allows the identification of each species with high accuracy. AIEgens are excellent candidates for theranostics as they are inherently emissive, and some AIEgens can function as photosensitizers to kill microbes via photo-induced oxidative stress. AIEgens can also be conjugated with antimicrobial agents to bring about a therapeutic effect. The ability to modularly design AIEgens will no doubt lead to more clever designs of AIE-based probes for imaging, discriminating and killing of microorganisms. The inherent florescence turn-on property of AIEgens is especially attractive in the development of theranostic agents. In this regard, issues like biodegradability, clearance and pharmacokinetic properties of these AIEgens warrant comprehensive investigations.

Acknowledgements We thank the partial financial support from the National Institutes of Health (NIH) (GM128164).

References

1. Guinane CM, Cotter PD (2013) Role of the gut microbiota in health and chronic gastrointestinal disease: understanding a hidden metabolic organ. Therap Adv Gastroenterol 6(4):295–308. https://doi.org/10.1177/1756283X13482996
2. Rowland I, Gibson G, Heinken A, Scott K, Swann J, Thiele I, Tuohy K (2018) Gut microbiota functions: metabolism of nutrients and other food components. Eur J Nutr 57(1):1–24. https://doi.org/10.1007/s00394-017-1445-8
3. Ren Y, Xie S, Svensson Grape E, Inge AK, Ramström O (2018) Multistimuli-responsive enaminitrile molecular switches displaying H+-induced aggregate emission, metal ion-induced turn-on fluorescence, and organogelation properties. J Am Chem Soc 140(42):13640–13643. https://doi.org/10.1021/jacs.8b09843
4. Mei J, Hong Y, Lam JWY, Qin A, Tang Y, Tang BZ (2014) Aggregation-induced emission: the whole is more brilliant than the parts. Adv Mater 26(31):5429–5479. https://doi.org/10.1002/adma.201401356
5. Mei J, Leung NL, Kwok RT, Lam JW, Tang BZ (2015) Aggregation-induced emission: together we shine, united we soar! Chem Rev 115(21):11718–11940. https://doi.org/10.1021/acs.chemrev.5b00263
6. Yu J, Sun X, Cai F, Zhu Z, Qin A, Qian J, Tang B, He S (2015) Low photobleaching and high emission depletion efficiency: the potential of AIE luminogen as fluorescent probe for STED microscopy. Opt Lett 40(10):2313–2316. https://doi.org/10.1364/OL.40.002313
7. Jayawardena HSN, Liyanage SH, Rathnayake K, Patel U, Yan M (2021) Analytical methods for characterization of nanomaterial surfaces. Anal Chem 93(4):1889–1911. https://doi.org/10.1021/acs.analchem.0c05208
8. Liyanage SH, Yan M (2020) Quantification of binding affinity of glyconanomaterials with lectins. Chem Comm 56(88):13491–13505. https://doi.org/10.1039/D0CC05899H
9. He Z, Ke C, Tang BZ (2018) Journey of aggregation-induced emission research. ACS Omega 3(3):3267–3277. https://doi.org/10.1021/acsomega.8b00062
10. Zhao Z, Zhang H, Lam JWY, Tang BZ (2020) Aggregation-induced emission: new vistas at the aggregate level. Angew Chem Int Ed 59(25):9888–9907. https://doi.org/10.1002/anie.201916729
11. Gu Y, Zhao Z, Su H, Zhang P, Liu J, Niu G, Li S, Wang Z, Kwok RTK, Ni X-L, Sun J, Qin A, Lam Jacky WY, Tang BZ (2018) Exploration of biocompatible AIEgens from natural resources. Chem Sci 9(31):6497–6502. https://doi.org/10.1039/C8SC01635F
12. Lee MMS, Zheng L, Yu B, Xu W, Kwok RTK, Lam JWY, Xu F, Wang D, Tang BZ (2019) A highly efficient and AIE-active theranostic agent from natural herbs. Mater Chem Front 3(7):1454–1461. https://doi.org/10.1039/C9QM00242A
13. Zhao Z, Wang Z, Tavakoli J, Peng C, Xiong Y, Tang Y, Jacky W, Tang B (2019) A X-ray excitable vibrational AIE system based on platinum (II) salts. https://doi.org/10.26434/chemrxiv.8796098
14. Chen X, Hu R, Qi C, Fu X, Wang J, He B, Huang D, Qin A, Tang BZ (2019) Ethynylsulfone-based spontaneous amino-yne click polymerization: a facile tool toward regio- and stereoregular dynamic polymers. Macromolecules 52(12):4526–4533. https://doi.org/10.1021/acs.macromol.9b00670

15. Huang J, Jiang X (2018) Injectable and degradable pH-responsive hydrogels via spontaneous amino-Yne click reaction. ACS Appl Mater Interfaces 10(1):361–370. https://doi.org/10.1021/acsami.7b18141
16. Shi Y, Bai T, Bai W, Wang Z, Chen M, Yao B, Sun JZ, Qin A, Ling J, Tang BZ (2017) Phenol-yne click polymerization: an efficient technique to facilely access regio- and stereoregular poly(vinylene ether ketone)s. Chemistry 23(45):10725–10731. https://doi.org/10.1002/chem.201702966
17. Guo Z, Yan C, Zhu W-H (2020) High-performance quinoline-malononitrile core as a building block for the diversity-oriented synthesis of AIEgens. Angew Chem Int Ed 59(25):9812–9825. https://doi.org/10.1002/anie.201913249
18. Silhavy TJ, Kahne D, Walker S (2010) The bacterial cell envelope. Cold Spring Harb Perspect Biol 2(5):a000414. https://doi.org/10.1101/cshperspect.a000414
19. Li B, Yu Q, Duan Y (2015) Fluorescent labels in biosensors for pathogen detection. Crit Rev Biotechnol 35(1):82–93. https://doi.org/10.3109/07388551.2013.804487
20. Combs CA (2017) Fluorescence microscopy: a concise guide to current imaging methods. Curr Protoc Neurosci 79:2.1.1–2.1.25.
21. Shen Q, Wang S, Yang N-D, Zhang C, Wu Q, Yu C (2020) Recent development of small-molecule organic fluorophores for multifunctional bioimaging in the second near-infrared window. J Lumin. https://doi.org/10.1016/j.jlumin.2020.117338
22. Nune SK, Gunda P, Thallapally PK, Lin YY, Forrest ML, Berkland CJ (2009) Nanoparticles for biomedical imaging. Expert Opin Drug Deliv 6(11):1175–1194. https://doi.org/10.1517/17425240903229031
23. Lee S, Sun Y, Cao Y, Kang SH (2019) Plasmonic nanostructure-based bioimaging and detection techniques at the single-cell level. TrAC Trends Anal Chem 117:58–68. https://doi.org/10.1016/j.trac.2019.05.006
24. Wang X, Ramström O, Yan M (2011) Dye-doped silica nanoparticles as efficient labels for glycans. Chem Comm 47(14):4261–4263. https://doi.org/10.1039/C0CC05299J
25. Wang X, Matei E, Deng L, Koharudin L, Gronenborn AM, Ramström O, Yan M (2013) Sensing lectin-glycan interactions using lectin super-microarrays and glycans labeled with dye-doped silica nanoparticles. Biosens Bioelectron 47:258–264. https://doi.org/10.1016/j.bios.2013.03.014
26. Smith AM, Nie S (2004) Chemical analysis and cellular imaging with quantum dots. Analyst 129(8):672–677. https://doi.org/10.1039/B404498N
27. Kairdolf BA, Smith AM, Stokes TH, Wang MD, Young AN, Nie S (2013) Semiconductor quantum dots for bioimaging and biodiagnostic applications. Annu Rev Anal Chem 6(1):143–162. https://doi.org/10.1146/annurev-anchem-060908-155136
28. Song Y, Zhu S, Yang B (2014) Bioimaging based on fluorescent carbon dots. RSC Adv 4(52):27184–27200. https://doi.org/10.1039/C3RA47994C
29. Zrazhevskiy P, Sena M, Gao X (2010) Designing multifunctional quantum dots for bioimaging, detection, and drug delivery. Chem Soc Rev 39(11):4326–4354. https://doi.org/10.1039/B915139G
30. He X, Xiong L-H, Zhao Z, Wang Z, Luo L, Lam JWY, Kwok RTK, Tang BZ (2019) AIE-based theranostic systems for detection and killing of pathogens. Theranostics 9(11):3223–3248. https://doi.org/10.7150/thno.31844
31. Mei J, Huang Y, Tian H (2018) Progress and trends in AIE-based bioprobes: a brief overview. ACS Appl Mater Interfaces 10(15):12217–12261. https://doi.org/10.1021/acsami.7b14343
32. Wang H, Liu G (2018) Advances in luminescent materials with aggregation-induced emission (AIE) properties for biomedical applications. J Mater Chem B 6(24):4029–4042. https://doi.org/10.1039/C8TB00674A
33. Yan L, Zhang Y, Xu B, Tian W (2016) Fluorescent nanoparticles based on AIE fluorogens for bioimaging. Nanoscale 8(5):2471–2487. https://doi.org/10.1039/C5NR05051K
34. Liu B, Tang BZ (2019) Themed issue on aggregation-induced emission (AIE). Chem Asian J 14(6):672–673. https://doi.org/10.1002/asia.201900185
35. Wang H, Zhao E, Lam JWY, Tang BZ (2015) AIE luminogens: emission brightened by aggregation. Mater Today 18(7):365–377. https://doi.org/10.1016/j.mattod.2015.03.004
36. Cai X, Liu B (2020) Aggregation-induced emission: recent advances in materials and biomedical applications. Angew Chem Int Ed Engl 59(25):9868–9886. https://doi.org/10.1002/anie.202000845

37. Kwok RTK, Leung CWT, Lam JWY, Tang BZ (2015) Biosensing by luminogens with aggregation-induced emission characteristics. Chem Soc Rev 44(13):4228–4238. https://doi.org/10.1039/C4CS00325J
38. Panigrahi A, Are VN, Jain S, Nayak D, Giri S, Sarma TK (2020) Cationic organic nanoaggregates as AIE luminogens for wash-free imaging of bacteria and broad-spectrum antimicrobial application. ACS Appl Mater Interfaces 12(5):5389–5402. https://doi.org/10.1021/acsami.9b15629
39. Liang J, Tang BZ, Liu B (2015) Specific light-up bioprobes based on AIEgen conjugates. Chem Soc Rev 44(10):2798–2811. https://doi.org/10.1039/C4CS00444B
40. Stach M, Siriwardena TN, Kohler T, van Delden C, Darbre T, Reymond JL (2014) Combining topology and sequence design for the discovery of potent antimicrobial peptide dendrimers against multidrug-resistant *Pseudomonas aeruginosa*. Angew Chem Int Ed Engl 53(47):12827–12831. https://doi.org/10.1002/anie.201409270
41. Kumar P, Kizhakkedathu JN, Straus SK (2018) Antimicrobial peptides: diversity, mechanism of action and strategies to improve the activity and biocompatibility in vivo. Biomolecules 8(1):4. https://doi.org/10.3390/biom8010004
42. Chen J, Gao M, Wang L, Li S, He J, Qin A, Ren L, Wang Y, Tang BZ (2018) Aggregation-induced emission probe for study of the bactericidal mechanism of antimicrobial peptides. ACS Appl Mater Interfaces 10(14):11436–11442. https://doi.org/10.1021/acsami.7b18221
43. Zhao L, Chen Y, Yuan J, Chen M, Zhang H, Li X (2015) Electrospun fibrous mats with conjugated tetraphenylethylene and mannose for sensitive turn-on fluorescent sensing of *Escherichia coli*. ACS Appl Mater Interfaces 7(9):5177–5186. https://doi.org/10.1021/am507593p
44. Feng G, Yuan Y, Fang H, Zhang R, Xing B, Zhang G, Zhang D, Liu B (2015) A light-up probe with aggregation-induced emission characteristics (AIE) for selective imaging, naked-eye detection and photodynamic killing of Gram-positive bacteria. Chem Commun (Camb) 51(62):12490–12493. https://doi.org/10.1039/c5cc03807c
45. Zhang R, Cai X, Feng G, Liu B (2017) Real-time naked-eye multiplex detection of toxins and bacteria using AIEgens with the assistance of graphene oxide. Faraday Discuss 196:363–375. https://doi.org/10.1039/c6fd00169f
46. Larson AM (2011) Multiphoton microscopy. Nat Photonics 5(1):1–1. https://doi.org/10.1038/nphoton.an.2010.2
47. Yang X, Wang N, Zhang L, Dai L, Shao H, Jiang X (2017) Organic nanostructure-based probes for two-photon imaging of mitochondria and microbes with emission between 430 nm and 640 nm. Nanoscale 9(14):4770–4776. https://doi.org/10.1039/c7nr00342k
48. Lee MMS, Xu W, Zheng L, Yu B, Leung ACS, Kwok RTK, Lam JWY, Xu FJ, Wang D, Tang BZ (2020) Ultrafast discrimination of Gram-positive bacteria and highly efficient photodynamic antibacterial therapy using near-infrared photosensitizer with aggregation-induced emission characteristics. Biomaterials 230:119582. https://doi.org/10.1016/j.biomaterials.2019.119582
49. Liu J, Chen C, Ji S, Liu Q, Ding D, Zhao D, Liu B (2017) Long wavelength excitable near-infrared fluorescent nanoparticles with aggregation-induced emission characteristics for image-guided tumor resection. Chem Sci 8(4):2782–2789. https://doi.org/10.1039/C6SC04384D
50. Qi J, Sun C, Li D, Zhang H, Yu W, Zebibula A, Lam JWY, Xi W, Zhu L, Cai F, Wei P, Zhu C, Kwok RTK, Streich LL, Prevedel R, Qian J, Tang BZ (2018) Aggregation-induced emission luminogen with near-infrared-II excitation and near-infrared-I emission for ultradeep intravital two-photon microscopy. ACS Nano 12(8):7936–7945. https://doi.org/10.1021/acsnano.8b02452
51. Jiang M, Gu X, Kwok RTK, Li Y, Sung HHY, Zheng X, Zhang Y, Lam JWY, Williams ID, Huang X, Wong KS, Tang BZ (2018) Multifunctional AIEgens: ready synthesis, tunable emission, mechanochromism, mitochondrial, and bacterial imaging. Adv Funct Mater 28:1704589.
52. Kato T, Kawaguchi A, Nagata K, Hatanaka K (2010) Development of tetraphenylethylene-based fluorescent oligosaccharide probes for detection of influenza virus. Biochem Biophys Res Commun 394(1):200–204. https://doi.org/10.1016/j.bbrc.2010.02.155
53. Xiong LH, He X, Zhao Z, Kwok RTK, Xiong Y, Gao PF, Yang F, Huang Y, Sung HH, Williams ID, Lam JWY, Cheng J, Zhang R, Tang BZ (2018) Ultrasensitive virion immunoassay platform with dual-modality based on a multifunctional aggregation-induced emission luminogen. ACS Nano 12(9):9549–9557. https://doi.org/10.1021/acsnano.8b05270
54. Sauer S, Kliem M (2010) Mass spectrometry tools for the classification and identification of bacteria. Nat Rev Microbiol 8(1):74–82. https://doi.org/10.1038/nrmicro2243
55. Chatterjee A, Abraham J (2018) Chapter 2—Microbial contamination, prevention, and early detection in food industry. In: Holban AM, Grumezescu AM (eds) microbial contamination and

food degradation. Academic, New York, pp 21–47. https://doi.org/10.1016/B978-0-12-811515-2.00002-0
56. Zhao E, Hong Y, Chen S, Leung CW, Chan CY, Kwok RT, Lam JW, Tang BZ (2014) Highly fluorescent and photostable probe for long-term bacterial viability assay based on aggregation-induced emission. Adv Healthc Mater 3(1):88–96. https://doi.org/10.1002/adhm.201200475
57. Kong TT, Zhao Z, Li Y, Wu F, Jin T, Tang BZ (2018) Detecting live bacteria instantly utilizing AIE strategies. J Mater Chem B 6(37):5986–5991. https://doi.org/10.1039/c8tb01390j
58. Lagier J-C, Edouard S, Pagnier I, Mediannikov O, Drancourt M, Raoult D (2015) Current and past strategies for bacterial culture in clinical microbiology. Clin Microbiol Rev 28(1):208–236. https://doi.org/10.1128/CMR.00110-14
59. Järvinen A-K, Laakso S, Piiparinen P, Aittakorpi A, Lindfors M, Huopaniemi L, Piiparinen H, Mäki M (2009) Rapid identification of bacterial pathogens using a PCR- and microarray-based assay. BMC Microbiol 9(1):161. https://doi.org/10.1186/1471-2180-9-161
60. Yamamoto Y (2002) PCR in diagnosis of infection: detection of bacteria in cerebrospinal fluids. Clin Diagn Lab Immunol 9(3):508–514. https://doi.org/10.1128/cdli.9.3.508-514.2002
61. Salazar G, Zhang N, Fu T-M, An Z (2017) Antibody therapies for the prevention and treatment of viral infections. NPJ Vaccines 2:19–19. https://doi.org/10.1038/s41541-017-0019-3
62. Mani V, Chikkaveeraiah BV, Rusling JF (2011) Magnetic particles in ultrasensitive biomarker protein measurements for cancer detection and monitoring. Expert Opin Med Diagn 5(5):381–391. https://doi.org/10.1517/17530059.2011.607161
63. Amemiya Y, Tanaka T, Yoza B, Matsunaga T (2005) Novel detection system for biomolecules using nano-sized bacterial magnetic particles and magnetic force microscopy. J Biotechnol 120(3):308–314. https://doi.org/10.1016/j.jbiotec.2005.06.028
64. Peruski AH, Peruski LF Jr (2003) Immunological methods for detection and identification of infectious disease and biological warfare agents. Clin Diagn Lab Immunol 10(4):506–513. https://doi.org/10.1128/cdli.10.4.506-513.2003
65. Fischer M, Triggs GJ, Krauss TF (2016) Optical sensing of microbial life on surfaces. Appl Environ Microbiol 82(5):1362. https://doi.org/10.1128/AEM.03001-15
66. Brar SK, Verma M (2011) Measurement of nanoparticles by light-scattering techniques. TrAC Trends Analyt Chem 30(1):4–17. https://doi.org/10.1016/j.trac.2010.08.008
67. Haavig DL, Hollen KR, Debruin AE, Lisecki SR, Shoup NC, Stimac MJH, Treloar AM, Jodoin ZN, Bohm MS, Sharp JS (2017) Cellular light scattering for the identification of bacteria and its application to the identification of staphylococcus. J AOAC Int 100(6):1836–1847. https://doi.org/10.5740/jaoacint.17-0097
68. Haes AJ, Van Duyne RP (2004) A unified view of propagating and localized surface plasmon resonance biosensors. Anal Bioanal Chem 379(7):920–930. https://doi.org/10.1007/s00216-004-2708-9
69. Hu J, Bohn PW (2017) Optical biosensing of bacteria and bacterial communities. J Anal Test 1(1):4. https://doi.org/10.1007/s41664-017-0002-z
70. Shashkova S, Leake MC (2017) Single-molecule fluorescence microscopy review: shedding new light on old problems. Biosci Rep 37(4):BSR20170031. https://doi.org/10.1042/BSR20170031
71. Nicolini AM, Fronczek CF, Yoon JY (2015) Droplet-based immunoassay on a “sticky” nanofibrous surface for multiplexed and dual detection of bacteria using smartphones. Biosens Bioelectron 67:560–569. https://doi.org/10.1016/j.bios.2014.09.040
72. Hu R, Zhou F, Zhou T, Shen J, Wang Z, Zhao Z, Qin A, Tang BZ (2018) Specific discrimination of gram-positive bacteria and direct visualization of its infection towards mammalian cells by a DPAN-based AIEgen. Biomaterials 187:47–54. https://doi.org/10.1016/j.biomaterials.2018.09.019
73. Kang M, Zhou C, Wu S, Yu B, Zhang Z, Song N, Lee MMS, Xu W, Xu F-J, Wang D, Wang L, Tang BZ (2019) Evaluation of structure-function relationships of aggregation-induced emission luminogens for simultaneous dual applications of specific discrimination and efficient photodynamic killing of gram-positive bacteria. J Am Chem Soc 141(42):16781–16789. https://doi.org/10.1021/jacs.9b07162
74. Liu GJ, Tian SN, Li CY, Xing GW, Zhou L (2017) Aggregation-induced-emission materials with different electric charges as an artificial tongue: design, construction, and assembly with various pathogenic bacteria for effective bacterial imaging and discrimination. ACS Appl Mater Interfaces 9(34):28331–28338. https://doi.org/10.1021/acsami.7b09848

75. Jiang G, Wang J, Yang Y, Zhang G, Liu Y, Lin H, Zhang G, Li Y, Fan X (2016) Fluorescent turn-on sensing of bacterial lipopolysaccharide in artificial urine sample with sensitivity down to nanomolar by tetraphenylethylene based aggregation induced emission molecule. Biosens Bioelectron 85:62–67. https://doi.org/10.1016/j.bios.2016.04.071
76. Gao T, Cao X, Dong J, Liu Y, Lv W, Li C, Feng X, Zeng W (2017) A novel water soluble multifunctional fluorescent probe for highly sensitive and ultrafast detection of anionic surfactants and wash free imaging of Gram-positive bacteria strains. Dyes Pigm 143:436–443. https://doi.org/10.1016/j.dyepig.2017.05.006
77. Li Q, Li Y, Min T, Gong J, Du L, Phillips DL, Liu J, Lam JWY, Sung HHY, Williams ID, Kwok RTK, Ho CL, Li K, Wang J, Tang BZ (2020) Time-dependent photodynamic therapy for multiple targets: a highly efficient AIE-active photosensitizer for selective bacterial elimination and cancer cell ablation. Angew Chem Int Ed 59(24):9470–9477. https://doi.org/10.1002/anie.201909706
78. Shen J, Hu R, Zhou T, Wang Z, Zhang Y, Li S, Gui C, Jiang M, Qin A, Tang BZ (2018) Fluorescent sensor array for highly efficient microbial lysate identification through competitive interactions. ACS Sens 3(11):2218–2222. https://doi.org/10.1021/acssensors.8b00650
79. Zhou C, Xu W, Zhang P, Jiang M, Chen Y, Kwok RTK, Lee MMS, Shan G, Qi R, Zhou X, Lam JWY, Wang S, Tang BZ (2019) Engineering sensor arrays using aggregation-induced emission luminogens for pathogen identification. Adv Funct Mater. https://doi.org/10.1002/adfm.201805986
80. Yang Z, Sun Z, Ren Y, Chen X, Zhang W, Zhu X, Mao Z, Shen J, Nie S (2019) Advances in nanomaterials for use in photothermal and photodynamic therapeutics (Review). Mol Med Rep 20(1):5–15. https://doi.org/10.3892/mmr.2019.10218
81. Ni J-S, Min T, Li Y, Zha M, Zhang P, Ho CL, Li K (2020) Planar AIEgens with enhanced solid-state luminescence and ROS generation for multidrug-resistant bacteria treatment. Angew Chem Int Ed 59(25):10179–10185. https://doi.org/10.1002/anie.202001103
82. Gellci K, Mehrmohammadi M (2014) Photothermal therapy. In: Schwab M (ed) Encyclopedia of cancer. Springer, Berlin, pp 1–5. https://doi.org/10.1007/978-3-642-27841-9_7097-1
83. Jeelani S, Reddy RCJ, Maheswaran T, Asokan GS, Dany A, Anand B (2014) Theranostics: a treasured tailor for tomorrow. J Pharm Bioallied Sci 6(Suppl 1):S6–S8. https://doi.org/10.4103/0975-7406.137249
84. Kim TH, Lee S, Chen X (2013) Nanotheranostics for personalized medicine. Expert Rev Mol Diagn 13(3):257–269. https://doi.org/10.1586/erm.13.15
85. Jo SD, Ku SH, Won Y-Y, Kim SH, Kwon IC (2016) Targeted nanotheranostics for future personalized medicine: recent progress in cancer therapy. Theranostics 6(9):1362–1377. https://doi.org/10.7150/thno.15335
86. Dendukuri N, Khetani K, McIsaac M, Brophy J (2007) Testing for HER2-positive breast cancer: a systematic review and cost-effectiveness analysis. CMAJ 176(10):1429–1434. https://doi.org/10.1503/cmaj.061011
87. Pysz MA, Gambhir SS, Willmann JK (2010) Molecular imaging: current status and emerging strategies. Clin Radiol 65(7):500–516. https://doi.org/10.1016/j.crad.2010.03.011
88. Prasad M, Lambe UP, Brar B, Shah I, Manimegalai J, Ranjan K, Rao R, Kumar S, Mahant S, Khurana SK, Iqbal HMN, Dhama K, Misri J, Prasad G (2018) Nanotherapeutics: an insight into healthcare and multi-dimensional applications in medical sector of the modern world. Biomed Pharmacother 97:1521–1537. https://doi.org/10.1016/j.biopha.2017.11.026
89. Zhou L, Qiu T, Lv F, Liu L, Ying J, Wang S (2018) Self-assembled nanomedicines for anticancer and antibacterial applications. Adv Healthc Mater 7(20):1800670. https://doi.org/10.1002/adhm.201800670
90. Ventola CL (2012) The nanomedicine revolution: part 1: emerging concepts. P T 37(9):512–525
91. Iqbal J, Abbasi BA, Ahmad R, Mahmood T, Ali B, Khalil AT, Kanwal S, Shah SA, Alam MM, Badshah H, Munir A (2018) Nanomedicines for developing cancer nanotherapeutics: from bench-top to bedside and beyond. Appl Microbiol Biotechnol 102(22):9449–9470. https://doi.org/10.1007/s00253-018-9352-3
92. Din FU, Aman W, Ullah I, Qureshi OS, Mustapha O, Shafique S, Zeb A (2017) Effective use of nanocarriers as drug delivery systems for the treatment of selected tumors. Int J Nanomedicine 12:7291–7309. https://doi.org/10.2147/IJN.S146315
93. Wadajkar AS, Dancy JG, Hersh DS, Anastasiadis P, Tran NL, Woodworth GF, Winkles JA, Kim AJ (2017) Tumor-targeted nanotherapeutics: overcoming treatment barriers for glioblastoma. Wiley Interdiscip Rev Nanomed Nanobiotechnol. https://doi.org/10.1002/wnan.1439

94. Lee MMS, Yan D, Chau JHC, Park H, Ma CCH, Kwok RTK, Lam JWY, Wang D, Tang BZ (2020) Highly efficient phototheranostics of macrophage-engulfed Gram-positive bacteria using a NIR luminogen with aggregation-induced emission characteristics. Biomaterials 261:120340. https://doi.org/10.1016/j.biomaterials.2020.120340
95. Hu F, Qi G, Kenry MD, Zhou S, Wu M, Wu W, Liu B (2020) Visualization and in situ ablation of intracellular bacterial pathogens through metabolic labeling. Angew Chem Int Ed 59(24):9288–9292. https://doi.org/10.1002/anie.201910187
96. Qi G, Hu F, Kenry SL, Wu M, Liu B (2019) An AIEgen-peptide conjugate as a phototheranostic agent for phagosome-entrapped bacteria. Angew Chem Int Ed Eng 58(45):16229–16235. https://doi.org/10.1002/anie.201906099
97. Wu W, Li Z (2021) Nanoprobes with aggregation-induced emission for theranostics. Mater Chem Front. https://doi.org/10.1039/D0QM00617C
98. Li Y, Yu H, Qian Y, Hu J, Liu S (2014) Amphiphilic star copolymer-based bimodal fluorogenic/magnetic resonance probes for concomitant bacteria detection and inhibition. Adv Mater 26(39):6734–6741. https://doi.org/10.1002/adma.201402797
99. Xie S, Manuguri S, Proietti G, Romson J, Fu Y, Inge AK, Wu B, Zhang Y, Hall D, Ramstrom O, Yan M (2017) Design and synthesis of theranostic antibiotic nanodrugs that display enhanced antibacterial activity and luminescence. Proc Natl Acad Sci USA 114(32):8464–8469. https://doi.org/10.1073/pnas.1708556114
100. Xie S, Manuguri S, Ramström O, Yan M (2019) Impact of hydrogen bonding on the fluorescence of *N*-amidinated fluoroquinolones. Chemistry 14(6):910–916. https://doi.org/10.1002/asia.201801916
101. Xie S, Lopez SA, Ramström O, Yan M, Houk KN (2015) 1,3-Dipolar cycloaddition reactivities of Perfluorinated Aryl Azides with Enamines and Strained Dipolarophiles. J Am Chem Soc 137(8):2958–2966. https://doi.org/10.1021/ja511457g
102. Xie S, Zhou J, Chen X, Kong N, Fan Y, Zhang Y, Hammer G, Castner DG, Ramström O, Yan M (2019) A versatile catalyst-free perfluoroaryl azide–aldehyde–amine conjugation reaction. Mater Chem Front 3(2):251–256. https://doi.org/10.1039/C8QM00516H
103. Li Y, Zhao Z, Zhang J, Kwok RTK, Xie S, Tang R, Jia Y, Yang J, Wang L, Lam JWY, Zheng W, Jiang X, Tang BZ (2018) A bifunctional aggregation-induced emission luminogen for monitoring and killing of multidrug-resistant bacteria. Adv Funct Mater 28:1804632.
104. Zhao E, Chen Y, Wang H, Chen S, Lam JWY, Leung CWT, Hong Y, Tang BZ (2015) Light-enhanced bacterial killing and wash-free imaging based on AIE fluorogen. ACS Appl Mater Interfaces 7(13):7180–7188. https://doi.org/10.1021/am509142k